ABBREVIATIONS OF UNITS

cm = centimeter

kg = kilograms

lb = pound

m = metre

mN = mega-Newton

N = Newton

Pa = Pascal

psi = pounds per square inch

sec = second

Manufacturing Processes for Technology

Second Edition

William O. Fellers
American River College, Sacramento

William W. Hunt
American River College, Sacramento

Prentice Hall

Upper Saddle River, New Jersey
Columbus, Ohio

Library of Congress Cataloging-in-Publication Data

Fellers, William O.
Manufacturing processes for technology / William O. Fellers, William W. Hunt.—2nd ed.
p. cm.
ISBN 0-13-017791-1
1. Manufacturing processes. I. Hunt, William W. II. Title.

TS183 .F46 2001
670.42—dc21 00-029852

Vice President and Publisher: Dave Garza
Editor in Chief: Stephen Helba
Associate Editor: Michelle Churma
Production Editor: Louise N. Sette
Production Supervision: Carlisle Publishers Services
Design Coordinator: Robin G. Chukes
Cover Designer: Alan Bumpus
Cover art: © Alan Bumpus
Production Manager: Brian Fox
Marketing Manager: Chris Bracken

This book was set in Dutch 801 by Carlisle Communications, Ltd. It was printed and bound by R. R. Donnelley & Sons Company. The cover was printed by Phoenix Color Corp.

10 9 8 7 6 5 4 3 2 1
ISBN 0-13-017791-1

To the Student

Have you ever wondered: "How did they make that? Why did they make it in such a poor way?" Choosing a manufacturing method is part of the design process for nearly everything that is made. Today, that is a complicated choice, but an essential one.

Humans have been making useful objects ever since they discovered that they could hold tools. Until about two hundred years ago, most objects were manufactured by hand—in fact, the word "manufacturing" literally translated from Latin means "to make by hand." By today's standards, these objects were crudely made. Parts were not interchangeable. Each part was individually fitted to the others so that the total mechanism worked, at least until something broke. Then, the replacement part had to be made specially to fit.

Today's society could not exist without modern manufacturing and production techniques. In fact, one of the indicators used to measure a country's economic strength is its manufacturing and production capability. Mass production with interchangeable parts is vital to our standard of living. One cannot even imagine the difficulty involved in making an automobile, a modern tractor used on a farm, or a commercial aircraft if all the parts had to be made by hand. Without modern manufacturing methods, we could not even feed the present population.

The Industrial Revolution brought about two radical changes: the standardization of measurements and the development of manufacturing techniques that allowed interchangeable parts to be mass produced. Before then, workers measured lengths by cubits, hands, or other imprecise gauges. When measurements became standardized, an inch was an inch or a millimetre was the same no matter whose scale was used.

A result of standardization of measurements was the development of interchangeable parts. Using standardized measurements, workers could make parts that would fit similar mechanisms.

With the advent of the steam engine and the resulting portable power source, workers had power available that could run machine tools, make machine tools handle large capacities not possible with human or animal power, and increase production. Workers could make dozens of the same parts at a time and produce them more cheaply than ever before.

The immediate result was that manufactured goods became cheaper, more people could own those goods, and the standard of living rose.

Modern manufacturing is changing continuously. Standards that were adequate a decade ago are not good enough to make the state-of-the-art machines operate properly. Methods of manufacturing will always improve and new materials will change the way that we make things.

Look around. How many products can you list that have become obsolete in your lifetime? We have gone from the vacuum tube to the transistor to the integrated circuit even before some of the old vacuum tubes have burned out. In cameras, we have progressed from black-and-white film to color film to instant prints to filmless electronic-imaging cameras in less than half a lifetime. Thirty years ago, classroom exams were mimeographed, or simply written on the chalkboard. Now, photocopier machines are the standard. As late as 1975, engineers used slide rules, which now are as obsolete as the abacus. Electric typewriters hailed as state of the art are antiques compared to the modern computer and word processing systems. This latter invention is one for which the authors of this text are truly grateful.

This text is as up-to-date as the authors could make it, yet even the newest manufacturing techniques described in this text might be replaced before you finish this course. Please consider this text only a starting point. *It is the responsibility of each engineer and technician to keep current with new technologies in his/her field!* Professionals should read technical journals and trade magazines in their field and join technical societies. These societies publish magazines and books and offer workshops to keep their members current. The era when a person could learn a skill and be set for life ended long ago. One must keep up-to-date!

Societies, as well as individuals, must keep up with change. The United States is the only major industrial country that does not use the metric system exclusively. Eventually, the United States and American industry will have to join the rest of the world in this standardization of measurement. Therefore, at the request of many industries, we have included some metric system examples and problems in this book. This is a good time to master the metric system if you have not already done so. You are going to need it.

Now, the authors would like to take the liberty of passing on to the students a few words of advice, which they have learned from being students themselves for many years. Learning is not a "passive" experience. You cannot learn anything simply by watching someone else do it. You cannot learn to play the piano by watching a concert pianist perform nor learn to play baseball by simply being in the bleachers watching the game. You have to do it yourself to learn it. Neither can you learn to solve a problem by watching your instructor work the problem on the board. You have learned a new concept when you, by yourself, can use it properly. Even after watching your instructor solve a problem, copy it, take it home, and rework the problem yourself. When you can make it look easy and can even explain it, then you will be a "pro" at it. Remember, your instructors have been working these problems for several years to get to the point where they can teach it and make it look easy.

Learn to take notes in class. You would be surprised by how much you forget between classes. It is also helpful to review the notes immediately each day. Further, the notes provide a good review in preparation for tests.

The format for this text is as follows: A concept is introduced, discussed, and followed immediately by questions and exercises. Get in the habit of answering the questions and working the exercises as you come to them. They have been designed to help you master a block of material before building on it with new concepts. Don't wait until test time to do the work, or the course will get ahead of you.

Above all, stick with it! It takes hard work to master these new concepts and organize all of this material, but you need them in order to be ready to take your place in today's industry.

Good luck in the course.

■ ACKNOWLEDGMENTS

We would like to express our appreciation to the reviewers of this book: Ergun A. Oguz, Gwinnett Technical Institute, GA; Thomas Reyman, Mesa Community College, AZ; Dr. Morteza Sadat-Hossieny, Marshall University, WV; Musasa E. Ssemakula, Wayne State University, MI; and Neil K. Thomas, Ivy Tech State College, IN.

William O. Fellers
William W. Hunt

Contents

SECTION I FUNDAMENTALS OF MANUFACTURING 1

APPENDICES

SECTION

I

Fundamentals of Manufacturing

INTRODUCTION

Look around you. How many things do you see that you wear, sit on, or use that were made by someone? We rarely use "found objects." Therefore, *someone had to make* nearly everything that we use, and they had to make it in some manner. How those things were made, the **manufacturing processes,** is what this book is all about.

Specifically, this book examines the processes by which things can be made. We know that they can be made rapidly, with acceptably high quality, at a cost low enough to be affordable, and with an acceptably small environmental impact if we are wise enough in our choice of manufacturing processes. We can make wise choices only if we clearly understand all of the available options in manufacturing and their consequences.

The design of any object limits the choice of processes by which the object can be manufactured. The designer must bear this fact in mind. Sometimes, the converse is also true: The design may be limited by the small number of manufacturing processes available to the engineer or designer. This is especially true in small industries with limited facilities. So the design choices and the manufacturing process choices are integral parts of a single design process. In this book, the primary focus is on the manufacturing process considerations, viewed as one part of the design process. The two central themes in this book are as follows:

1. For a given *object,* what manufacturing processes could be used to produce it? How many ways could it be made?
2. For a given *manufacturing process,* what are the conditions necessary to use that process? Are those conditions met for this application? Will that process produce the desired results? Does that manufacturing process meet the standards and criteria specified in the design? What are the advantages and disadvantages of each process used in the making of the object?

Answering these two questions may seem to be a tall order, but let us start by developing a conceptual framework so that the information will be easier to organize and understand.

Chapter

1

An Approach to Manufacturing Processes

■ INTRODUCTION

Until the late 1800s people were transported by train, bicycle, horseback, carriages, wagons, or other animal-drawn vehicles. Then came the automobile. Unfortunately, manufacturing methods were slow and inadequate for large-scale production. As a result, automobiles were scarce and expensive. It was even said, at that time, that "the automobile would become a rich man's plaything and a poor man's necessity." Henry Ford sought to make the automobile available to all, at prices they could afford. To accomplish his dream, he had to invent new methods of manufacturing.

One result of his efforts was the assembly line. His mottos were "Keep it off the floor" and "Keep it moving." Instead of one person, or a small group of people, making an entire automobile from start to finish, the vehicle was moved along a line of workers, each one performing a specific task (or small number of tasks) on each car as it moved past them. The worker would then move on to perform the same task on the next unit. The idea of the assembly line was revolutionary and is still in use today, in an improved form that often uses robotics. It has succeeded in increasing production and making products affordable.

To accomplish this innovation, Henry Ford had to study, organize, and classify the tasks involved in the production of the automobile. Similarly, all manufacturing can be broken into several individual tasks. These tasks are applicable to the production of everything from artworks, such as pottery and ceramic tiles; to industrial machinery, such as lathes and milling machines; to jet engines; to musical instruments, from cornets to zithers; to washing machines; to . . . oh, you get the idea. Manufacturing applies to everything that is made.

■ CLASSIFICATION OF PRODUCTION TASKS

What tools would you expect to find in a mechanic's toolbox? Perhaps there would be a hammer, some wrenches, several screwdrivers, tin snips, propane torch, micrometers, and maybe even a flashlight. Why does a mechanic or machinist need so many different tools? They're needed because there are so many different jobs to be done. The same is true in manufacturing. The manufacture of any single item, whether it is a simple screw or a jet aircraft, requires many different operations or jobs during its production. To study all of the possible methods of manufacturing may seem to be an overwhelming task. Fortunately, nearly all **manufacturing processes** can be divided into just six categories:

1. Material removal
2. Material addition
3. Change of form
4. Change of condition
5. Material joining
6. Finishing

A full section in this book is devoted to each of these operations, and most of these sections are further divided into separate chapters. For instance, material joining is divided into chapters on adhesives, welding, and mechanical or other forms of joining. Entire courses are often offered in each of these manufacturing operations. The purpose of this text is to provide a brief introduction to each of the concepts. Although more detailed explanations of these categories appear later in this book, a brief description of them is in order here.

Material Removal

Material removal includes any process by which a part or piece of a material is severed or separated from another section of the same material. This includes the use of such hand tools as saws, chisels, and snips, as well as mechanically driven tools such as lathes, drills, planes, shapers, and grinders. Mechanical, chemical, electrical, thermal, optical, hydraulic, and other methods can also be used to remove material. These techniques are discussed in Section II.

Material Addition

Material addition involves all methods by which a piece of stock can be increased in volume or weight. (**Stock** is any material still in the form and shape in which it comes from the supplier.) These methods include electroplating, dipping, metallizing, electroforming, spraying, and vacuum deposition. Many of these processes are quite sophisticated technologically and require highly skilled personnel to perform them. Section III covers these methods.

Change of Form

Change of form includes the methods by which the shape of a piece of material is altered. Such processes as rolling, forging, bending, and many others are discussed in Section IV.

Change of Condition

Often, the internal structure of metal and other parts can be altered to provide the qualities required in the final product. Steels in particular can be hardened or softened significantly by heat treatments. The mechanical properties of other metals can be altered somewhat by forging and cold rolling. The characteristics of glass can be changed by chemical or thermal means. Any alteration of the properties of a material is considered to be a **change of condition.** Although it is necessary to study the properties and mechanics of materials in separate courses to understand completely the processes that result in a change of condition, the more important concepts of this subject are covered in Section V.

Material Joining

The method by which two or more parts are held together is called **material joining,** which includes, but is not limited to, such methods as welding, riveting, gluing, bolting, and pinning. The number of these methods that are available may be surprising. Section VI discusses this vast array of methods.

Finishing

When a machined product comes right out of the mold or off the machine, the surface it has at that point is not usually suitable for further use. At that point, it is unattractive, probably would not be salable, and might not even perform its intended function very well. It is therefore necessary that a "finish" be applied to the part. **Finishing** can involve anything from painting the part to plating it. Finishes are covered in Section VII.

■ TOOL OR PROCESS SELECTION

Once a decision has been made as to which operation classification is to be used, several other questions must be asked. Suppose that a piece of stock must be cut to start making a part. Which tool should be used? Would a pocket knife do the job, or would a saw, torch, or perhaps a set of shears be required? Can steel, wood, rubber, and ceramic be cut with the same tool? To determine the proper tool with which to remove the material, the following questions must be answered:

1. What are the physical properties of the material being cut, formed, or shaped, and what are the properties of the tools being used?
2. Does the tool or process produce an object or part that meets all of the design specifications given in the plans?
3. Does the tool or process selected have the precision required for the product?
4. Does the tool or process selected meet the required production rate of the job?
5. Is the tool or process economical? In other words, is the per-unit cost of the process low enough to do the job profitably?
6. Does the selected tool or process meet the social or environmental requirements, and are the resulting environmental costs small enough to justify using the process?
7. Will the tool or process be available when it is needed?
8. Is a trained operator required for the process? If so, will one be available when needed?

Questions 7 and 8 are not covered in this text, but they must be considered in actual industrial situations. With regard to Question 7, many small industrial plants do not have and cannot afford many of the tools, production facilities, or high-technology methods currently available. Further, small job runs may preclude the investment of large sums of money to obtain a high-production-rate machine. Therefore, the manufacturing engineer must often "work with the tools at hand." Simply put, to complete a contract, a small company may find it cheaper to hire a machinist to produce the few parts needed using the lathes and milling machines currently in the shop than to buy the latest computer-aided, robotics-controlled equipment.

As for Question 8, many manufacturing processes require specially trained, often licensed personnel to perform them. It would be futile for a manufacturing engineer to specify, for instance, "heliarc welding" if there was no one available who knew how to do it. In large companies, or if the contract is sufficiently large, a management decision might be made to hire a qualified heliarc welder to do the job. On the other hand, management might tell the engineer to figure out some other way to do the job.

What if a wise choice cannot be found among the processes and tools that are available? Then the choice may be between obtaining the part from a subcontractor or vendor or devising a new manufacturing process to do the job. Why not risk the latter? Often the best engineering is innovative and cannot be found in a "cookbook."

■ EXAMPLE OF PROCESS SELECTION

To illustrate a method of selecting a manufacturing process, let us look at a specific example and examine the options available. Suppose that we need to produce a 1-inch-wide wood chisel for a specific job. Further let us assume, for purposes of argument, that we are on an isolated island and the only tools and materials available are hand tools in a toolbox. We examine the questions posed earlier with respect to this task. The shape of the chisel is approximately as illustrated in Figure 1–1.

1. What are the physical properties of the material and the properties of the tools? Here the choice is very limited. Perhaps the only material from which the chisel can be made is an old file. But files are very brittle and cannot be cut with other ordinary hand tools. In this case the file would have to be annealed by heating it to

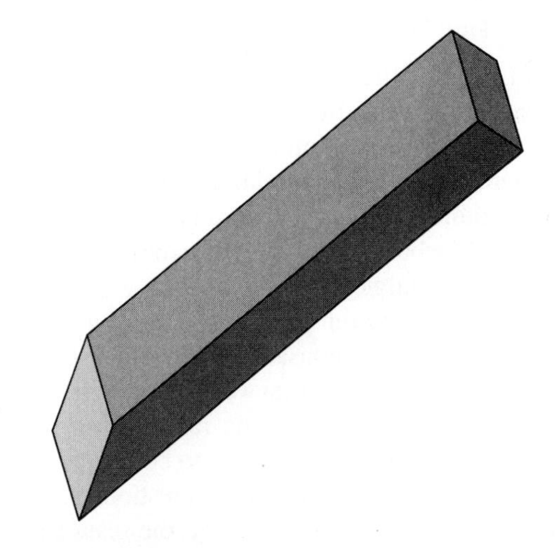

Figure 1-1. Wood chisel.

a red heat in a fire and allowing it to cool slowly in air. The file could then be cut with a hacksaw, drilled with a hand drill, or filed with another file. The edge could be put on the tool with an abrasive stone.

2. Does the tool or process produce an object or part that meets all of the design specifications given in the plans? Since, in this case, the chisel is to be used only by the maker of the tool, it can safely be assumed that the maker will ensure that the tool will do the job for which it was intended.

3. Does the tool or process selected have the precision required for the product? The precision here would be limited by the skill of the worker. Because this job does not require any mating parts, the requirement for precision is not very high.

4. Does the tool or process selected meet the required rate of the job? Production of a tool by hand would be a very slow process. But since only one is needed, the production requirement could be met.

5. Is the tool or process economical? In a word, *no!* However, considering other alternatives, such as importing a chisel from another part of the world and paying shipping charges, it might be cheaper to make the tool than to order it.

6. Does the selected tool or process meet the social or environmental requirements? The production of a tool, such as a chisel, by hand produces few environmental consequences. The smoke and fumes from the annealing fire would seem to be the major pollutants. The shavings and filings from cutting and shaping the tool would not be excessive.

7 and **8.** Will the tools and operators be available? This is a case in which the process is determined by the tools available, not the other way around. If the required whetstone is not available for sharpening the chisel, a suitable piece of sandstone might be found on the beach. Maybe the tool must be sharpened by rubbing it against a large rock.

Now let's change the problem a bit. This time the task is to produce ten thousand 1-inch-wide wood chisels. Further, let us assume that we are working in a large tool-making factory with a wide range of equipment and supplies at hand. The chisel is to be made of high-quality steel. We review the same questions discussed earlier, but this time with respect to the mass production of this chisel.

1. What are the physical properties of the material being cut, formed, or shaped, and what are the properties of the tools being used? A good chisel requires a medium-carbon steel or stainless steel. In both cases, these materials, after hardening, require abrasive cutting tools. In the unhardened state, medium-carbon steel can be cut with a high-speed steel (HSS) or carbide blade. Most stainless steels require carbide-tipped cutting tools. Flame cutting of the stock would soften the steel, thus requiring a further heat treatment process. The material must be cut either with an abrasive tool or carbide-tipped cutting tools. Of course, the chisel could be redesigned to use a low-carbon steel, which could be cut with shears, band saw blades, or other standard cutting tools. The chisel would have to be case hardened after the cutting is done. The edge of the chisel would be ground to the final shape.

2. Does the tool or process produce an object or part that meets all of the design specifications given in the plans? The plans, in this project, call for a 1-inch-wide chisel with no other information given. We therefore assume that other criteria such as finish, coating, and sharpness are not critical issues that need be considered here. However, if these or any other items are specified in the plans, quality control stations must be set up to make sure the design criteria are met.

3. Does the tool or process selected have the precision required for the product? Since a chisel does not require fine tolerances, any of the cutting methods mentioned in paragraph 1 would meet the requirements.

4. Does the tool or process selected meet the required production rate of the job? Cutting a hard steel requires a slow rate of travel of the cutting tool over the stock. Running the cutting tools at too high a rate of speed simply cuts down on tool life, requiring more frequent sharpening, higher cost, and eventually a slower production rate. Cutting speed can be increased by using a fluid-cooled cutting tool. Abrasive cutting may require a further heat treatment process. For mass production in this case, perhaps the best choice of tool would be the carbide-tipped cutting tools.

5. Is the tool or process economical? Although carbide-tipped cutting tools are more expensive than standard HSS blades, they can stand higher temperatures and can be operated at higher speeds. The resulting increase in production rate would offset the higher cost of tools and be justified in this case.

6. Does the selected tool or process meet the social or environmental requirements? For the production of a steel chisel the environmental concerns are minimal. Even the scraps and shavings from the steel could be recycled. However, if the decision were made to electroplate the chisel to give it a better appearance and to prevent corrosion, the disposal of the electroplating chemicals could pose an environmental problem. Perhaps a high chrome or stainless steel should be selected for the chisel so that the electroplating process could be avoided.

7 and **8.** Will the tool, process, and operators be available when needed? Since this example states that this is a large tool-making plant, we assume that the tools and operators would be available at any time during the manufacturing process. Note, however, that the company may be producing other items at the same time so that scheduling of the machinery and operators may be required. Do not assume they will always be available without checking with the production department.

There you have it. A complete manufacturing design for the production of a wood chisel in two completely different situations. You can see that the problems are never solved perfectly. The job of the manufacturing technician or engineer is to *optimize* the selection to get the best possible solution, not the perfect one.

A WARNING ABOUT LISTS

Each of us thinks in a unique way. The classification of production jobs into six categories and the list of eight questions to ask in the selection of a manufacturing process are only the result of the way that the authors think about this subject. If some other scheme for organizing the material makes better sense to you, by all means, use it! We ask only that you try our scheme first. If it works for you, fine! But if it doesn't, then make one that does.

Problem Set 1-1

1. You need to remove some material from a piece of small-diameter, low-carbon steel rod. How many ways can you think of to do it? Can you do it with no tools at all? List the tools required by each method.
2. Visit a hardware or building supply store and list five different tools that can be used for each of the six categories of production jobs. (You might not be able to find five for category 2.) Alternatively, look through a tool catalog and create your list from it.
3. Choose a design for a belt that you would wear. How might you make it? What other way might you make it if you lacked the tools that you prefer to use?
4. How did the original inhabitants of your country produce cutting blades? Hammers? Cooking containers? Clothing? Shelter? Did these tasks involve some lost skills that merit rediscovery?
5. Choose some simple item such as a knife, chisel, or screwdriver. Outline which of the six manufacturing processes would be involved in its manufacture.
6. Look about you. Find three items that were (or could have been) produced by each of the six categories of manufacturing. Some of the items may involve more than one of the categories.
7. From articles in the newspapers or other media, compile a list of items being produced by industry for which the environmental or social cost could be improved by redesign.
8. Consider the manufacture of a silver-plated teaspoon. List the manufacturing processes needed to make the spoon. List your starting materials and tools needed to make the spoon.
9. Classify each of the following tools as to whether they are used for material removal, material addition, change of form, change of condition, or material joining. Some may fit into more than one category.

   a. End wrench
   b. Crosscut saw
   c. Hammer
   d. Drill press
   e. Screwdriver
   f. Torch
   g. Pliers
   h. Soldering iron
   i. Grinder
   j. Vise grip pliers

10. You have been asked to make a footstool for use in the living room. Make a sketch of a footstool, then analyze the project using the criteria listed in this chapter as to which processes could be used with the tools you (or your family or friends) have available at home. Outline a process by which the footstool could be made using the facilities—and abilities—available to you.

11. Many items, products and tools, have become obsolete in the past 10 years or so (in your lifetime). Make a list of as many of these items as possible.

12. Many new products are on the market that have been developed in the past few years (in your lifetime). Make a list of as many of these as you can.

The design of a part and the choice of a manufacturing process to make that part are interrelated issues; each affects and limits the other. Two themes run through this book:

1. To make a given part, you must select a process by which to make it.
2. For a given process, you must decide if it would be the wisest choice for making the part.

This book is organized around the classification of "manufacturing processes" into six categories:

1. Material removal
2. Material addition
3. Change of form
4. Change of condition
5. Material joining
6. Finishing

Once a category is selected, the procedure for choosing the optimum ("wisest") manufacturing process consists of answering eight questions:

1. What are the physical properties of the material?
2. Does the tool or process produce an object or part that meets all of the design specifications given in the plans?
3. Does the tool or process selected have the precision required for the product?
4. Does the tool or process selected meet the required production rate of the job?
5. Is the tool or process economical?
6. Does the tool or process meet the social or environmental requirements, and are these costs small enough to justify using the process?
7. Will the tool or process be available when you need it?
8. Will a trained operator, if needed, be available to operate it?

Chapter 2

Properties of Materials

Any product that is made must be made from some material. That is obvious. Sometimes, however, the material chosen is not the optimum choice. That is, the product is made from a material whose properties are not compatible with the product's intended use. People may become dissatisfied with an item if it breaks, does not do the job properly, or wears out quickly. The problem is that the material used to make the defective merchandise often lacks the properties required for the product to do its job properly. For example, the cast iron wrenches sometimes found at department stores are brittle and are not manufactured to the tolerances required to fit bolt heads properly. As such, the wrenches break easily and leave the user very frustrated. Therefore, to design and manufacture a good, usable, and dependable product, the engineer or technician must understand the various properties of the materials and be able to select the *right material for the right application.*

Good engineering requires the proper selection of materials in order for the product to accomplish its task. To say that all tools must be of the highest quality and standards is not always true. There may be a use for tools made of inexpensive materials. For instance, if a tool is to be used once then discarded, the tool may be made merely to "do the job." The decision to make high-quality tools or less-expensive, one-shot tools is usually made by management.

Although it is not the purpose of this book to present a complete discussion of properties of materials (we will leave that to a separate course in materials science), a quick review of some essential terms associated with materials is in order. This chapter, therefore, covers those physical properties with which a person must deal in the design and manufacture of any product. Since such topics as *solidification in casting* and *heat treatments of metals* are deeply rooted in the theory of materials, a short discussion of the structure of matter is included here. The properties of materials presented in this chapter include stress, strain, tensile strength, compressive strength, safety factor, torsion, flexure, shear, peel, impact, creep, fatigue, modulus of elasticity, hardness, and others.

■ STRUCTURE OF MATTER

A quick review of the structure of matter will help with your understanding of the properties of materials. All properties of materials are a function of their structure. If the **atomic structure, bonding structure, crystal structure,** and the **imperfections** in the material are known, the properties of the material can be determined.

Matter is composed of *atoms.* Atoms are the smallest units of individual *elements.* There are 92 naturally occurring elements on earth. Iron, tin, aluminum, magnesium, sulfur, and chlorine are examples of elements.* Everything on earth is made either by using these elements or by combining the atoms of these elements into molecules to form compounds. All plastics, for example, are made of compounds.

Atoms themselves are composed of *protons, neutrons,* and *electrons.* Protons carry a positive electric charge, neutrons have no charge, and electrons are negatively charged particles. The protons and neutrons form the nucleus of the atom, while the electrons form the outer shells (see Figure 2–1). Most of the mass of the atom is in the protons and neutrons. An electron weighs roughly 1/1836 that of a proton. An analogy would be to say that if an electron weighs one pound, a proton would weigh about a ton.

The type of element is determined by the number of protons contained in the nucleus. Thus, if an atom has only one proton, it is a hydrogen atom. Magnesium atoms have 12 protons, aluminum has 13, iron has 26, and uranium, the largest naturally occurring element, has 92 protons. In atoms with no net electrical charge, the number of protons equals the number of electrons.

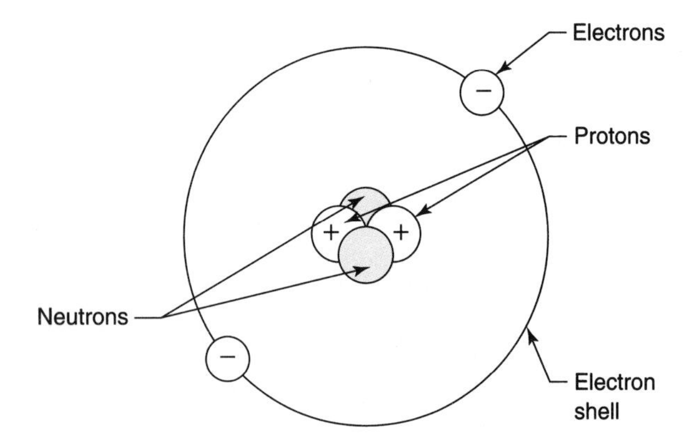

Figure 2-1. Atom.

* Scientists have succeeded in making over a dozen more elements artificially. Neptunium and plutonium are examples of these. However, aside from the atomic bomb, nuclear power plants, and radioisotope applications, these "transuranic" elements are not used in manufacturing and are not discussed here.

Atoms of the same element always have the same number of protons, but the number of neutrons in the nucleus may vary. Atoms with the same number of protons but varying numbers of neutrons are known as *isotopes.* Most hydrogen atoms have one proton and no neutrons. However, some hydrogen atoms have one proton and one neutron and are known as deuterium atoms. Still fewer hydrogen atoms contain one proton and two neutrons and are known as tritium. Hydrogen, deuterium, and tritium are isotopes of each other.

Atoms can combine to form *molecules.* Molecules are the smallest units of chemical compounds. The atoms are held together by chemical "bonds." Present theory categorizes all chemical bonds into four types: *ionic, covalent, metallic,* and *van der Waal.* Ionic bonds are created between charged particles. Elements such as sodium, potassium, aluminum, and other metals contain more electrons than their most stable configuration. Other, nonmetallic elements such as sulfur, chlorine, and oxygen become more stable with the addition of electrons to the atom. In such cases, the atoms with excess electrons will donate electrons to those atoms needing more electrons to create a stable configuration. When this occurs, the "donating" atoms have more protons in the nucleus than electrons and have a net positive charge, whereas those atoms accepting electrons acquire a net negative charge. These charged particles are called *ions.* The resulting charged particles are attracted to each other through what is called an *ionic bond.* Figure 2–2 shows this concept.

Covalent bonds are formed between elements that have too many electrons to be given off or require too many electrons to be added for ionic bonds to form. Carbon and silicon are examples of these types of elements. These types of elements often "share" their electrons to form more stable compounds. These shared electrons form *covalent bonds.* See Figure 2–3 for an illustration of the covalent bond. All organic matter and most bonded plastics involve the covalent bond.

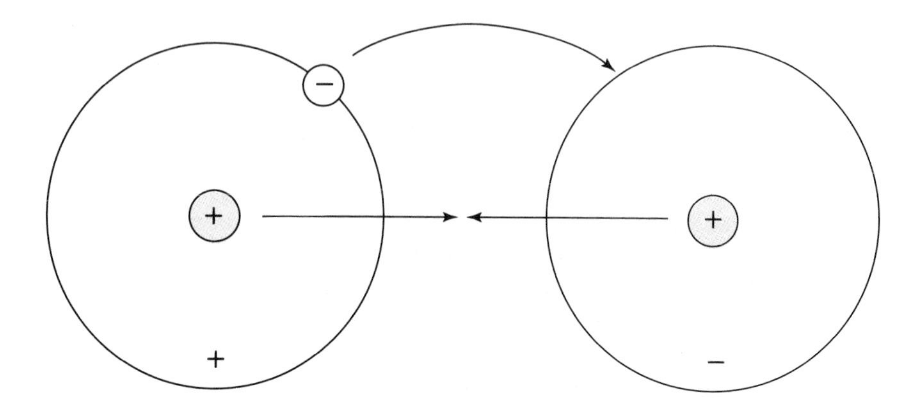

Figure 2–2. Ionic bond.

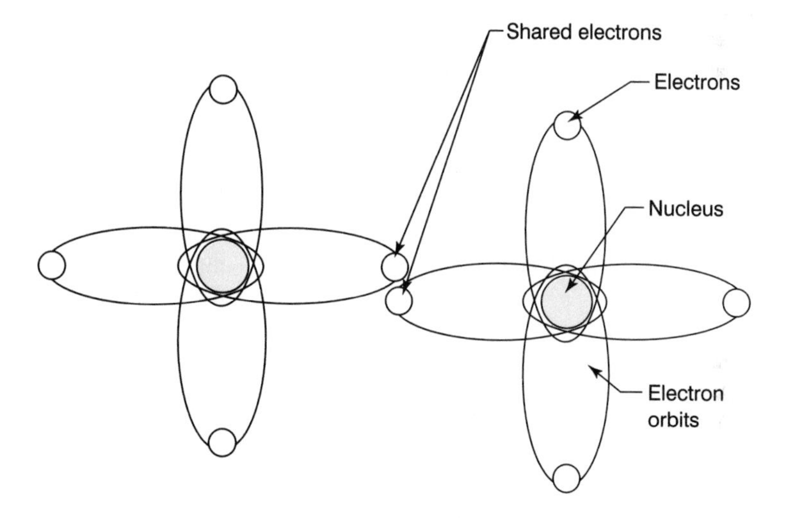

Figure 2–3. Covalent bond.

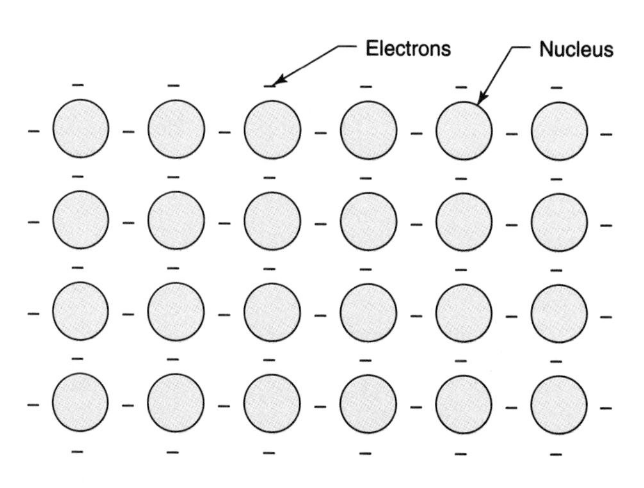

Figure 2–4. Metallic bond.

As implied by their title, *metallic bonds* are always formed between metals. All metals have an excess of electrons over their most stable configuration. The metallic bond theory is that these atoms form a "cloud" of electrons, to which the nuclei are attracted. Figure 2–4 depicts the metallic bond.

The *van der Waal bonds* (sometimes called "London forces" or "polar bonds") do not exist by themselves; they are modifications of other types of bonds. The van der Waal bond is formed when an atom or molecule is asymmetric, creating a net polar moment in the charges. Many compounds exhibit this polarity, which, in turn, modifies the properties of the compound. For example, helium, with its two protons, two neutrons, and two electrons, normally has evenly distributed electrons and is a very stable atom. As such, it does not even want to bond with itself. The atoms remain as a gas. However, at extremely low temperatures (around −452°F), the electronic action is slowed to a point at which the atoms can become asymmetrical, forming this polar bond. The positive side of one atom is slightly attracted to the negative side of the next atom. At that temperature, the atoms of helium can come together to form a solid due to the van der Waal bond. Figure 2–5 illustrates this type of bond.

Note that the formation of compounds through the various types of bonds produces properties in the compounds that are completely different from those of the elements from which the compounds are made. Common table salt is made from sodium and chlorine. Elemental sodium reacts so violently with water that it ignites the hydrogen that is produced in the reaction. Chlorine is a very toxic gas. Yet, ironically, when reacted together, they bond to form sodium chloride, which we need (in small amounts) to survive. Pure carbon is found in charcoal as a black, brittle material. Yet it reacts through covalent bonds to produce all living matter on earth as well as plastics and many other compounds.

Note that few materials have only one type of bond in them. For instance, a material may be primarily ionically bonded, but have some covalent bonds in its structure, or the converse may be true.

The properties that materials exhibit, such as melting point, boiling point, optical transparency or opacity, electrical conductivity, thermal conductivity, crystal structure, and others, often depend on the type of bonds in the material. The ionic bond produces compounds that are optically transparent, electronic insulators (however ions, when in a liquid, can move to form electrolytic conductors), and mechanically brittle materials. They usually have cubic or hexagonal close-packed crystal structures. In contrast, covalently bonded materials are optically transparent, electrical insulators, and usually form crystal structures that are not cubic. Metallic bonds produce electrical conductors, thermal conductors, and optically opaque materials.

■ STATES OF MATTER

All matter exists in one of four "states": *gas, liquid, solid,* and *plasma.* What is the difference between these states of matter? Why is water a solid below 32°F and a liquid above that temperature? Similarly, why is water a liquid below 212°F and a gas above? In manufacturing it is important to know how materials transition from liquids to solids. Many of the principles of material removal, forging, and heat treatments involve these transitions.

Plasma is an ionized gas. It is estimated that well over 90% of all matter in the universe is plasma. Our sun and the stars are all plasma. However, there is not that much plasma on earth and we seldom use it in manufacturing, so it is not discussed here.

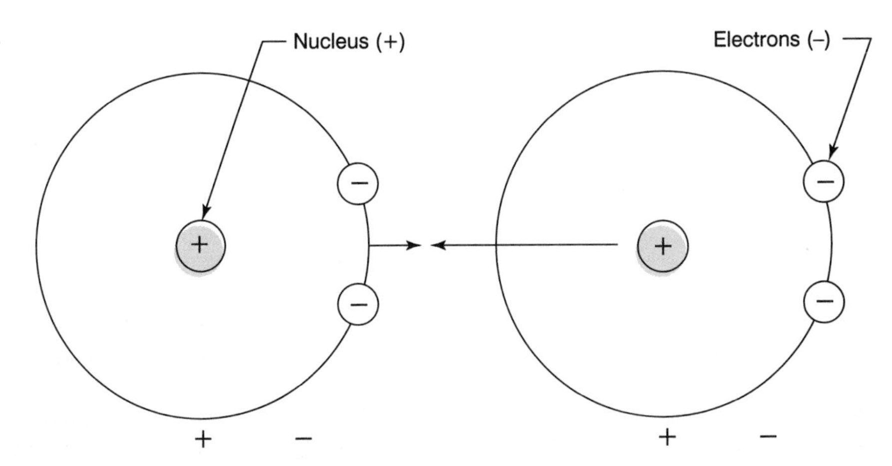

Figure 2-5. Van der Waal bond.

Gaseous State

In the gaseous state of matter, the individual atoms or molecules have little or no attraction to each other. The atoms or molecules of a gas are in constant motion. The particles are continuously "bouncing" from one particle to another or to the container walls. This atomic or molecular motion is due to the energy in the material and is a function of the heat stored in the atoms. The temperature of the gas is a function of the number of collisions per unit of time between the particles or the container. The more collisions, the higher the temperature. Compressing these particles into a smaller volume shoves the particles closer together, which results in more collisions and therefore a higher temperature. Lowering the pressure on a gas or lowering the temperature reduces the number of collisions.

Gases may be comprised of individual atoms or molecules. Many gases including oxygen, hydrogen, nitrogen, and others exist as *diatomic* molecules, that is, two atoms bonded to form an individual gas molecule. Although very few items are manufactured from gases, they are used in welding and other manufacturing operations.

Boiling Point

As energy is removed from gaseous particles as they are cooled, their motion is slowed to the point at which they can begin to bond to each other. The temperature at which this bonding can begin to occur is the *boiling point* of the material. Once the gas has been cooled to its boiling point, the *heat of vaporization* must be removed to convert the gas to a liquid. Similarly, in changing a liquid to a gas, the same energy must be added. For water the heat of vaporization is 540 *calories* per gram or *970* BTUs per pound. The reason a person feels cold when leaving a swimming pool is that the liquid water on the skin is being vaporized and the heat of vaporization for the water is taken from the person's skin, thus cooling the skin.

Liquid State

Liquids lack a definite long-range crystal structure. Instead, their particles are arranged randomly. This random arrangement produces bonds of varying lengths. The longer these bonds, the weaker they are. As heat is applied, the longer bonds break, reducing the *viscosity* of the material. At higher temperatures, shorter bonds break. A rigid material is not necessarily a solid. It may just be a very viscous liquid. To the materials scientist, a solid must have a definite structure.

Solid State

A *solid* is defined as that state which has a definite, long-range crystal structure. A crystal is an orderly array of particles such as atoms, molecules, or ions. The types of crystal structures are discussed later in this chapter. Figure 2–6 illustrates the differences between gases, liquids, and solids.

Melting Point

When heating a solid, a temperature will be reached at which the bonds between the particles can no longer hold the particles together. If there is enough energy to break one bond of a crystal, then there is enough energy to break them all, and a definite melting temperature is established. The temperature at which this crystal structure is formed or broken apart is the *melting point* of the material. All true solids have a definite melting point. This is a test of a solid.

Glass, for instance, does not melt at a single temperature; it just gets softer and less viscous as the temperature rises. The reason for this is that glass is *already* a superviscous liquid at room temperature. Glass fails the test of a solid because it has no definite crystal structure. Given enough time, glass, at room temperature, will "flow" like a liquid. It takes several centuries for glass to flow a millimetre, but it does. Some plastics and other amorphous (amorphous means "without body" and noncrystalline) substances behave in the same way as glass.

■ NUCLEATION OF GRAINS

When the temperature of a molten material is lowered to the melting point, little crystals or *nuclei* are formed at many points in the liquid. This phenomenon is called *nucleation.* These nuclei form individual crystals and start growing by adding more and more atoms from the liquid to the solid (see Figure 2–7).

Figure 2-6. Gas versus liquid versus solid.

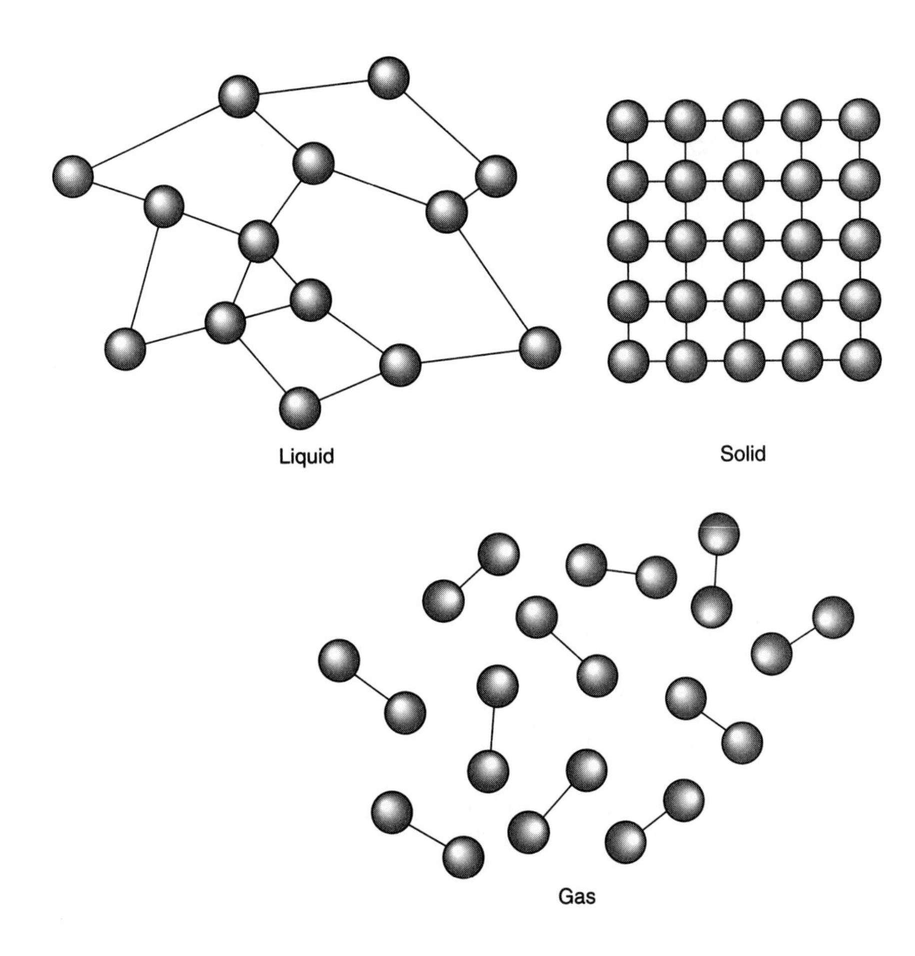

Figure 2-7. Nucleation.

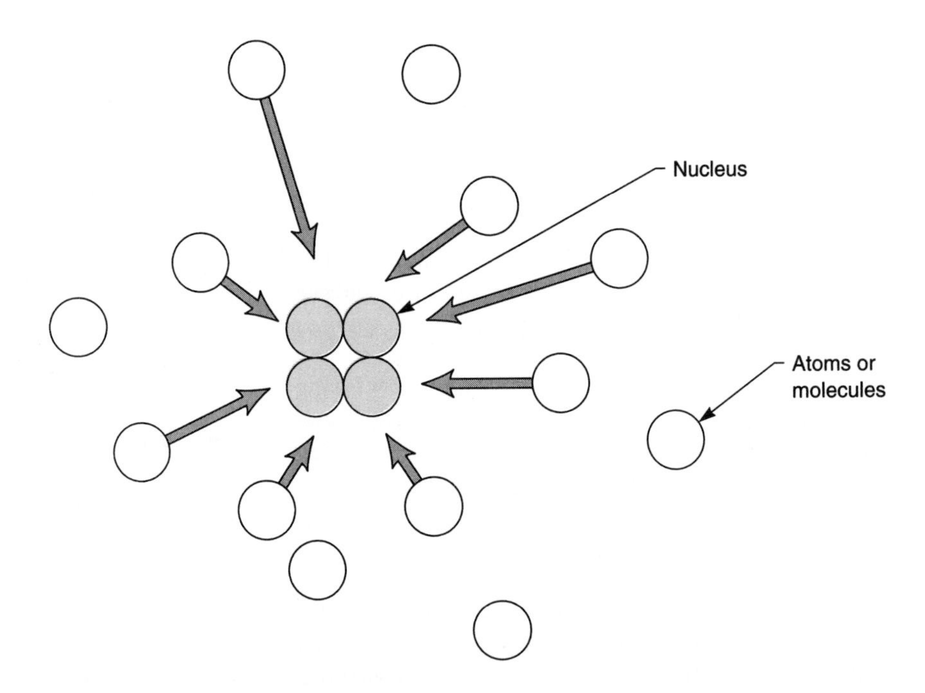

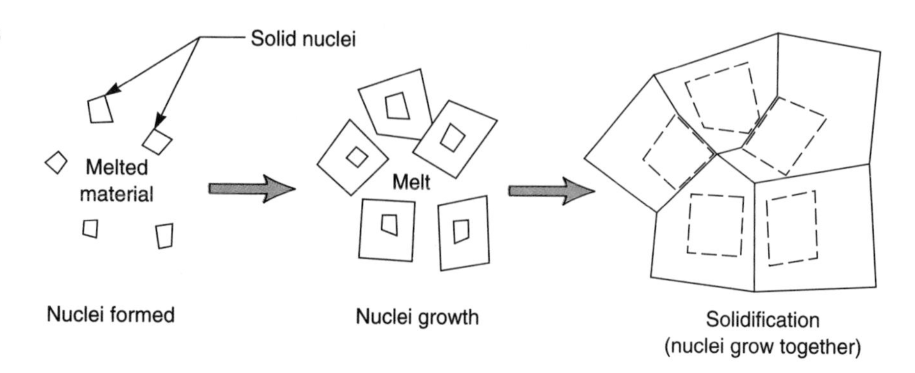

Figure 2-8. Formation of grains in a solid.

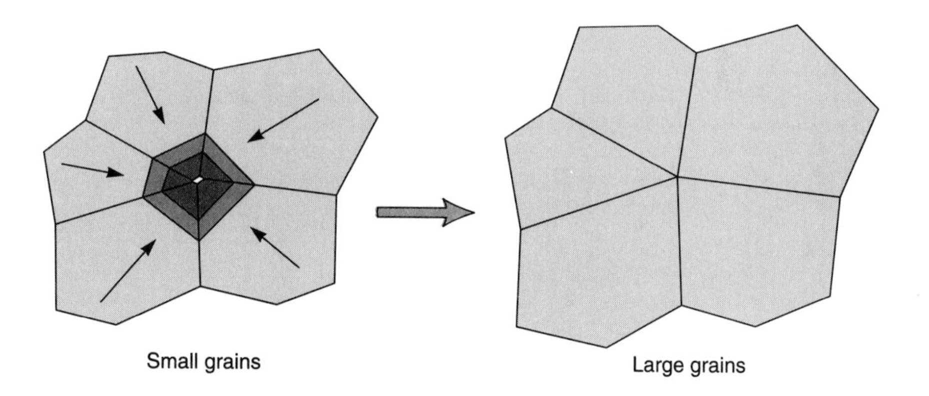

Figure 2-9. Grain growth.

Eventually, these individual crystals grow together, forming grains in the solid (Figure 2–8). Remember, each grain is a separate crystal.

After the grains have been nucleated and grown together to form a solid, the process of grain growth occurs. If the solid is maintained at temperatures just below the melting temperature, the larger grains will absorb the smaller grains in the solid (Figure 2–9). The result is that the slowly cooled solid will have large grains. However, if the solid is quickly cooled to room temperature, the grains will remain small.

Grains in most metals are extremely small and require magnification to be seen. Figure 2–10 shows grains in an annealed low-carbon steel magnified 500 times. A grain on this photograph, which measures 0.3 inch (7.6 mm) across, is actually only

$$0.3 \text{ in.}/500 = 0.0006 \text{ in. } (0.015 \text{ mm}) \text{ across}$$

This is about one-tenth the thickness of the paper in this book. Quenched or rapidly cooled steels have grain sizes much smaller than those shown in Figure 2–10.

Grains in metal coatings and certain alloys can grow quite large. Grains can easily be seen in tarnished brass door handles and galvanized (zinc-coated) products. Figure 2–11 shows grains over an inch in diameter on a galvanized steel telephone pole.

This entire process of melting and solidification is fundamental to the processes of casting and welding, which are discussed in later chapters.

Figure 2-10. Grains in steel (approximately 500X).
Photograph courtesy of USS Technical Center, Monroeville, PA.

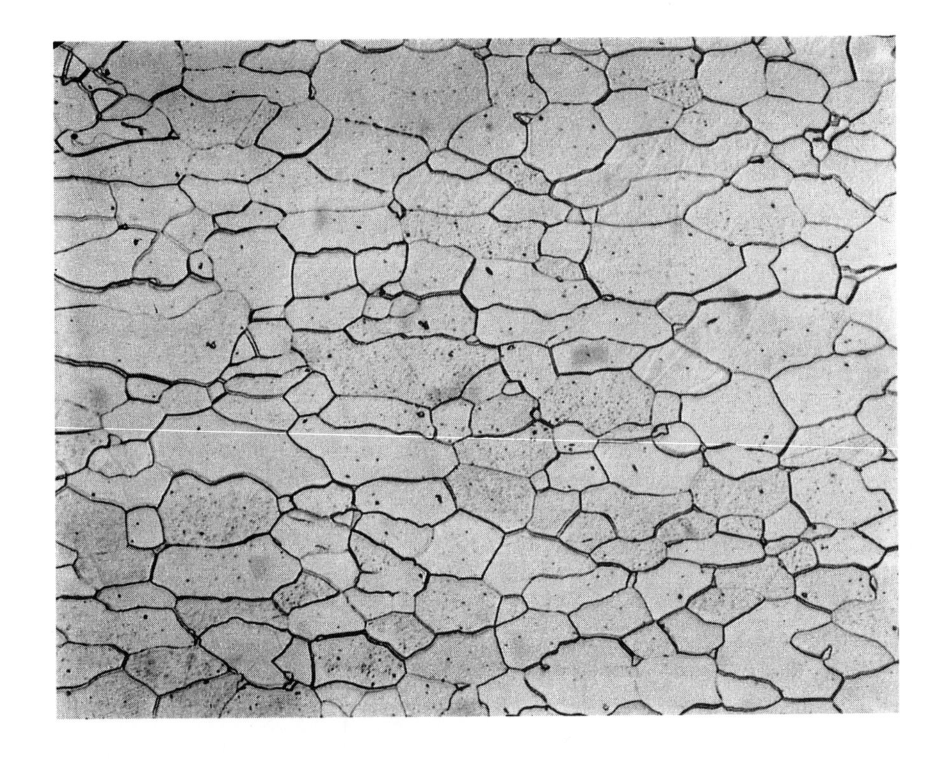

Figure 2-11. Zinc grains on galvanized steel (full size).

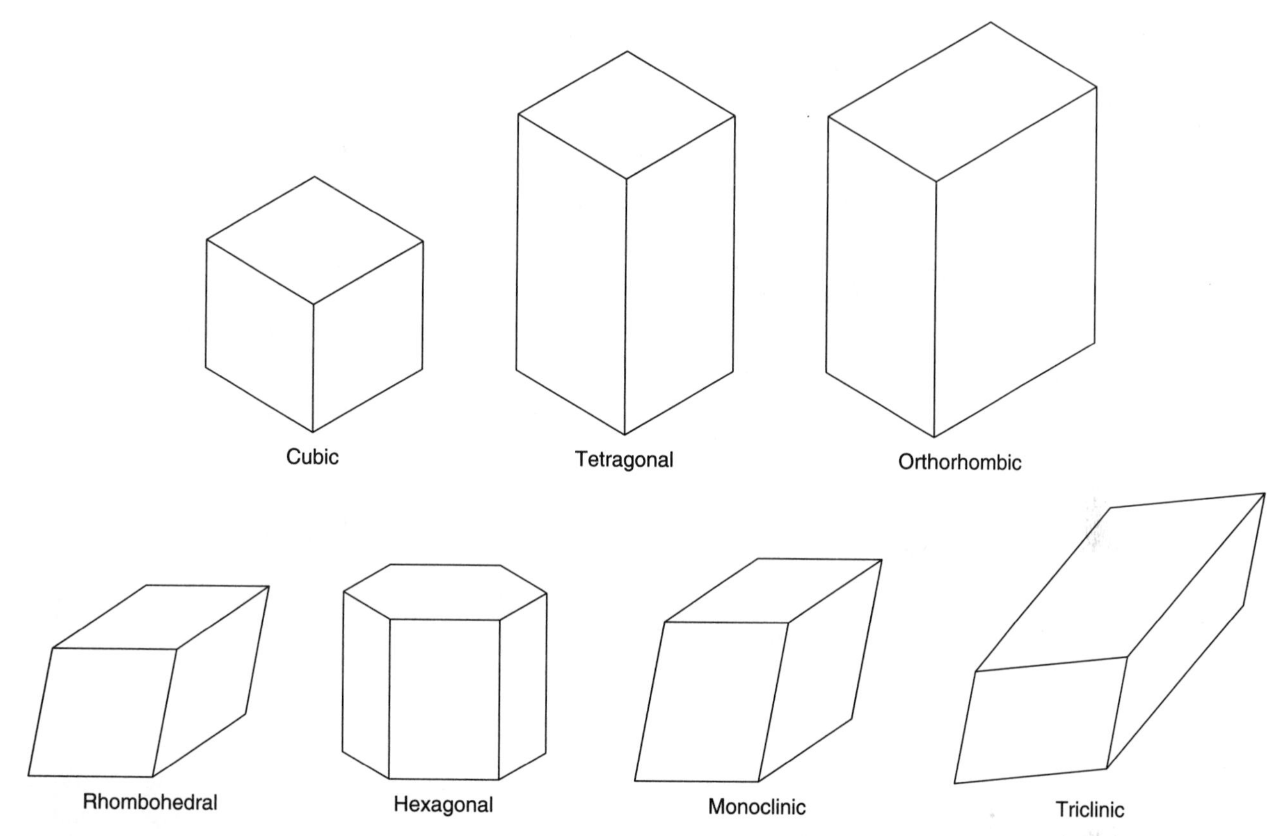

Figure 2-12. Crystal systems.

■ CRYSTAL STRUCTURE

Atoms, molecules, ions, or particles can arrange themselves in only seven different patterns or **crystal systems:**

- Cubic
- Tetragonal
- Orthorhombic
- Rhombohedral
- Hexagonal
- Monoclinic
- Triclinic

The basic shape of the systems is shown in Figure 2–12.

Fourteen modifications can occur in these seven crystal systems. These modifications are as follows:

- Cubic
  - Simple cubic
  - Body-centered cubic (BCC)
  - Face-centered cubic (FCC)
- Tetragonal
  - Simple tetragonal
  - Body-centered tetragonal (BCT)
- Orthorhombic
  - Simple orthorhombic
  - Body-centered orthorhombic
  - Base-centered orthorhombic
  - Face-centered orthorhombic

- Rhombohedral
  - Simple rhombohedral
- Hexagonal
  - Simple hexagonal
- Monoclinic
  - Simple monoclinic
  - Base-centered monoclinic
- Triclinic
  - Simple triclinic

These modified crystals are often designated in the literature by their initials. Body-centered cubics are simply noted as BCC, and face-centered cubic structures are listed as FCC.

The body-centered cubic crystal (BCC) simply has an extra particle inside the cube. The face-centered cubic crystal has extra particles in the center of the six faces of the cube. The base-centered structures have extra particles only in the parallel bases of the crystal.

Besides these crystal modifications, in some materials, the crystals overlap each other. For instance, the hexagonal structures often overlap to form the hexagonal close-packed (HCP) structure. Diamonds have a face-centered cubic structure that is overlapped five times to form what is often called the complex cubic structure. Figure 2–13 shows these structures.

Some crystals grow large enough to be seen by the naked eye. In most metals, however, the crystals or grains are only seen by means of a microscope. But regardless of how small they are, the crystals will always have the same shape. Consider Figure 2–14, which shows a simple cubic arrangement of atoms. If we tried

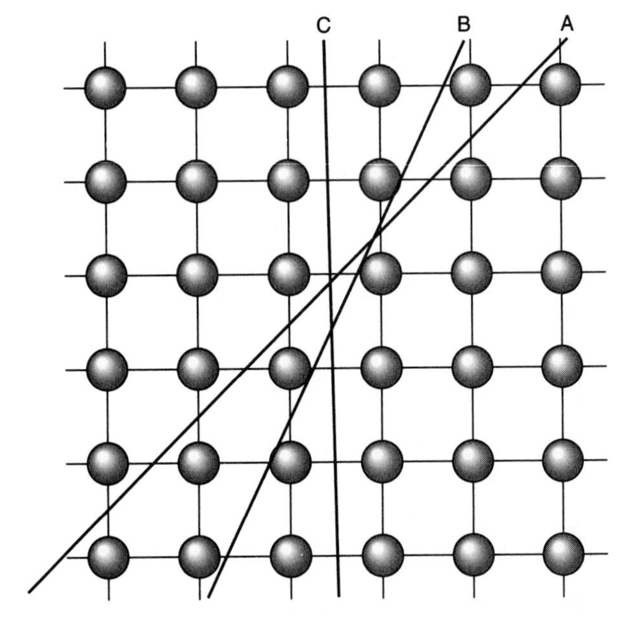

Figure 2-14. Fracture paths.

Figure 2-13. Close-packed crystals.

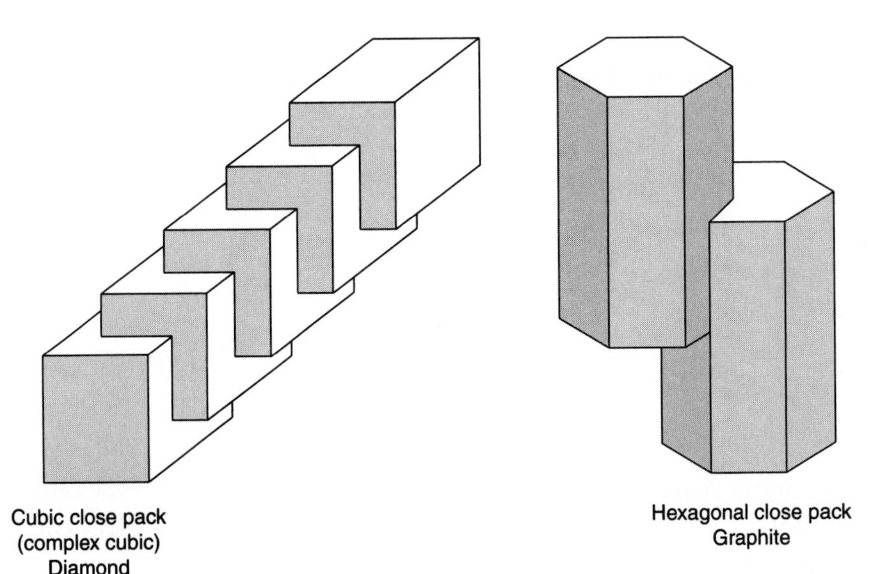

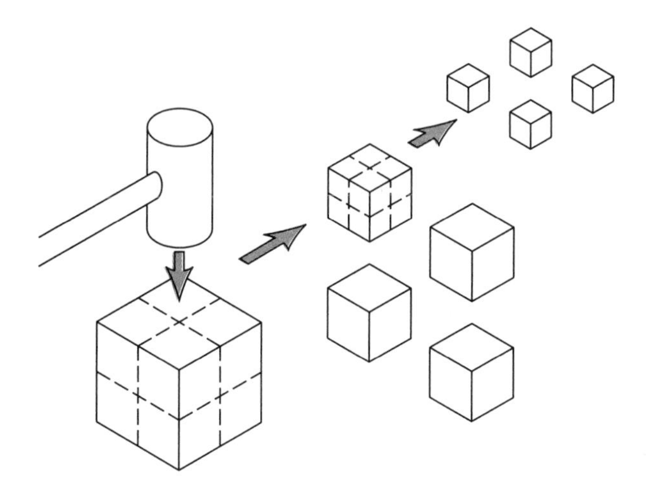

Figure 2-15. Crystal breakup.

to break the crystal along line A, 12 bonds would have to be broken. Path B requires only 8 bonds to be broken and path C requires only 6 broken bonds. Naturally, the material would always break along path C or one of equal weakness. Therefore, when large crystals are broken, they always break into smaller replicas of the large crystal, as illustrated in Figure 2–15.

In any crystal structure, there is a uniform distance between the particles. The atoms or particles align themselves into planes within each crystal. These planes can slide over each other in a mechanism called *slip.* The more methods of slip, or ways in which these planes can slip over each other, the more ductile the material becomes. Matter breaks when there is enough force to cause these planes to slip over each other. Slip is discussed further in the chapters in Section II.

Certain properties of a material are dependent on its crystal structure. *Density, ductility,* and *malleability* are examples of properties that depend on the crystal structure of the material. Face-centered and body-centered cubic materials are usually ductile, barring imperfections that can have more of an effect than the crystal structure itself. (Imperfections are discussed later.) Hexagonal close-packed materials are generally brittle in that they do not have as many *mechanisms of slip* as the cubic structures. One example of the brittleness of the hexagonal structure is seen in graphite. Slip always occurs between the hexagonal bases of the graphite crystal. Because the bonds between these planes are so much longer than those in the base, they are weaker. Therefore the base planes easily slide over each other. This principle of graphite makes it a very good dry lubricant.

STRENGTH PROPERTIES

Materials vary greatly in the magnitude of the forces they can withstand. Materials also differ in the way they resist forces applied in different directions or by different methods. It is necessary to understand the terminology in the discussion of the properties of materials.

Stress

Stress, in materials, is defined as the load per unit cross section of area. If a bar of any material is loaded by forces pulling it in opposite directions, the bar is said to be in *tension.* If the load or force pulling on the material is divided by the cross-sectional area of the bar, the result is the tensile stress applied to the sample. The equation for stress is

$$\text{stress} = \frac{\text{load}}{\text{area}}$$

or

$$S = \frac{P}{A}$$

From this equation, we can see that the stress in a sample with a 0.25-inch × 0.25-inch cross section, under a load of 1000 pounds would have a stress of

$$\begin{aligned}\text{stress} &= \frac{1000 \text{ lb}}{0.25 \text{ in.} \times 0.25 \text{ in.}} \\ &= \frac{1000 \text{ lb}}{0.0625 \text{ in.}^2} \\ &= 16{,}000 \text{ lb/in.2}\end{aligned}$$

The same stress would occur on a bar under a load of 4000 pounds that had a 0.5-inch × 0.5-inch cross section:

$$\begin{aligned}\text{stress} &= \frac{4000 \text{ lb}}{0.5 \text{ in.} \times 0.5 \text{ in.}} \\ &= \frac{4000 \text{ lb}}{0.25 \text{ in.}^2} \\ &= 16{,}000 \text{ lb/in.}^2\end{aligned}$$

Very often the units pounds/inch^2 (pounds per square inch) are abbreviated psi.

If the bar were cylindrical (that is, having a circular cross section), the calculations would be

$$\text{stress} = \text{load}/\pi r^2$$

A 0.5-inch-diameter bar under a load of 5000 pounds would have a stress of

$$\begin{aligned}\text{stress} &= \frac{5000\text{ lb}}{3.14 \times (0.25\text{ in.})^2}\\ &= \frac{5000\text{ lb}}{0.196\text{ in.}^2}\\ &= 25{,}500\text{ psi}\end{aligned}$$

Note that the calculated answer of 25,464 psi was rounded off to 25,500 psi. If measurements are read to only two significant figures (such as 2500 pounds or 0.50 inch), then there is no need to carry an answer to any more places. *The number of significant figures in the solution of a problem should never exceed the number of significant figures in the least accurate measurement.* Answers are rounded throughout this text.

In most countries, the metric system is used. The unit of stress in the metric system is the *pascal.* A pascal is defined as one newton per square metre. A newton is a kilogram-metre per second squared. Newtons are a unit of force, kilograms are a unit of mass. Kilograms can be converted to newtons by multiplying the number of kilograms by the acceleration of gravity in the metric system (9.8 metre/second^2). A mass of 10 kilograms can exert a force of

$$\begin{aligned}N &= \text{kg} \times 9.8\text{ m/sec}^2\\ &= 10\text{ kg} \times 9.8\text{ m/sec}^2\\ &= 98\text{ newtons}\end{aligned}$$

Consider a cylindrical bar 2 centimetres in diameter that holds up a 500-kilogram mass. The stress on the bar would be calculated as follows:

$$\begin{aligned}\text{area} &= \pi r^2\\ &= 3.14 \times \frac{(1\text{ cm})^2}{(100\text{ cm/m})^2}\\ &= 0.000314\text{ m}^2\\ N &= 500\text{ kg} \times 9.8\text{ m/sec}^2\\ &= 4900\text{ N}\end{aligned}$$

$$\begin{aligned}\text{stress} &= \frac{4900\text{ N}}{0.000314\text{ m}^2}\\ &= 15{,}600{,}000\text{ Pa}\end{aligned}$$

or

$$= 15.6\text{ MPa}$$

Stress can also be applied by pushing on the ends of a bar (or other shape). Compressive stress is calculated in the same manner as tensile stress. It is still "load per unit cross-sectional area."

Strain

A second term associated with materials properties is *strain.* Strain is the elongation of a specimen per unit of original length. Its equation is

$$\text{strain} = \frac{\text{elongation}}{\text{original length}}$$

or

$$\text{strain} = \frac{(\text{extended length} - \text{original length})}{\text{original length}}$$

or

$$e = \frac{z - z_0}{z_0}$$

If a 10.0-inch-long bar is stretched to a length of 10.1 inches, its strain is

$$\begin{aligned}\text{strain} &= \frac{10.1\text{ in.} - 10.0\text{ in.}}{10.0\text{ in.}}\\ &= \frac{0.1\text{ in.}}{10.0\text{ in.}}\\ &= 0.01\text{ in./in.}\end{aligned}$$

If we do a unit analysis of the last statement, it looks as if the "inch/inch" would divide out leaving the answer unitless. Indeed, this is often done but note that the numerator is inches of elongation, whereas the denominator is units of original length. This concept is sometimes used in engineering calculations in which stress should be left in inches per inch, metres per metre, and so on.

If a bar is placed under continuously increasing loads and the loads and elongations are measured at the same time at many different loads until the specimen breaks, the stresses and strains at each of those points can be calculated. A graph plotted from those stresses and strains for a typical material would be as shown in Figure 2–16. This is known as the *stress-strain curve* for the material. Note that the left-hand portion of the graph is a straight line, from the origin to point *P.* The line then starts to curve and reaches a maximum height at point *T.* From point *T,* the curve falls to point *R,* at which stress the bar breaks or ruptures.

If a bar has a stress applied that is less than that of point *P,* it will elongate but will return to its original length when the stress is released. Metals and other materials will stretch and rebound in much the same manner as a rubber band that has been stretched *if the applied stress is less than the elastic limit of the material.* A bar under greater stress than that at point *P* will be permanently stretched and will never return to its original length. The part of the curve up to point *P* is therefore called the *elastic region* of the curve. The maximum stress from which a bar will return to its original length is the material's *elastic limit,* sometimes called its *proportional limit.* The rest of the curve, to the right of the elastic limit, is the *plastic region.* The maximum stress that a bar will withstand before failing is its *tensile strength,* or *ultimate strength* and is shown as point *T* on the curve. The stress at which a bar breaks is its *breaking strength* or *rupture strength,* point *R* on Figure 2–16.

True Stress and Engineering Stress

If you were to stretch a rubber band, chewing gum, or Silly Putty, you would note that the cross-section diameter of the material, perpendicular to the direction of the pull, decreases. The same is true when a strain is placed on a metal bar. If the strain is in the elastic region, the decrease in the cross-section area is very small, but it exists. At strains larger than the tensile strength, the cross section decreases quite rapidly.

When calculating the stress in a material, the cross section should be measured at each load placed on the sample. If the load is divided by the actual cross section at that load, the *true stress* is calculated. However, because the decrease in area at loads within the elastic region is very small, and engineers seldom design structures or machines at stresses greater than the elastic limit, engineers usually divide all of the loads obtained in the tensile test by the *original* cross-section area (with no load applied). The results of these calculations produce the *engineering stress* in the sample. True stress is very seldom needed. Unless otherwise stated, engineering stress is used in most reports.

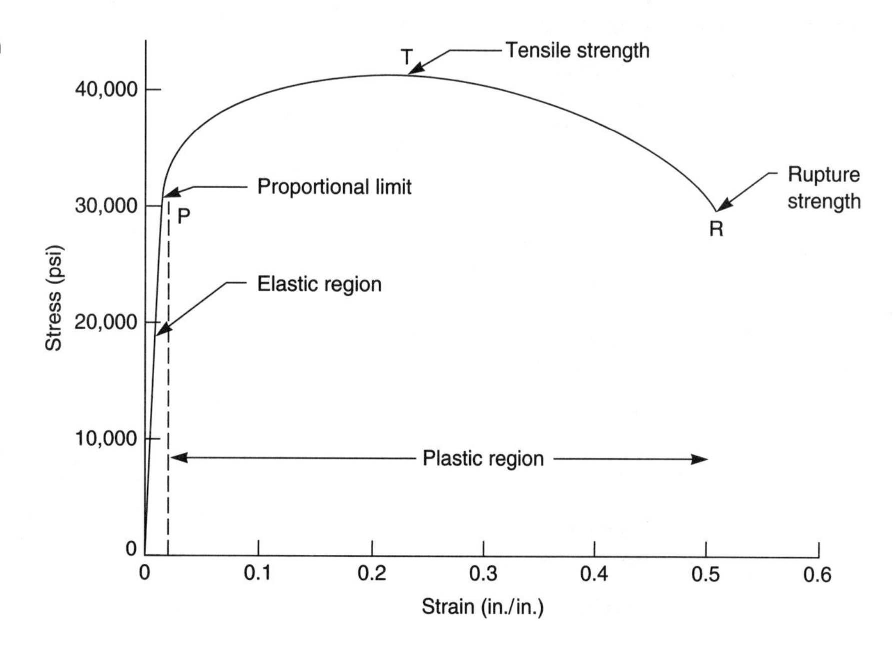

Figure 2-16. Typical stress-strain curve.

Yield Strength

Once the tensile strength of a material (point *T* on Figure 2–16) has been exceeded, the material starts to fail. Therefore, engineers and designers use a lower value as the limit of strength of the material. With the exception of shear pins and mechanical fuses, which are designed to fail at a given stress, most engineering design keeps the working stresses of a part in the elastic region of the material. Parts of machines, buildings, automobiles, aircraft, etc., are usually designed so that the expected stresses on them will not exceed the elastic limit. However, the elastic limit is difficult to pinpoint because it is nearly impossible to see exactly where the curve ceases to be a straight line. Therefore, an artificial standard is applied. If we measure a distance of 0.002 inch/inch (or metres per metre for that matter) on the strain axis, then draw a straight line parallel to the straight-line portion of the curve, the point at which that new line intersects the stress-strain curve is called the *yield point* or *yield strength* of the material. The yield point is the *engineering design strength* of the material (Figure 2–17).

In some materials the maximum stress that the material will take is also the breaking strength. That is, the tensile strength and the breaking strength are the same. These are defined as *brittle* materials. Cast iron and glass are examples of brittle materials. These materials will have a stress-strain curve similar to like that shown in Figure 2–18. If the breaking strength of a material is lower than the tensile strength, as shown in Figure 2–16, the material is considered *ductile.* Ductile materials can be stretched somewhat without breaking, brittle materials cannot. For brittle materials, the yield strength is often arbitrarily set as one-fourth the average of the tensile strength of the material.

Certain steels display another shape of the stress-strain curve. Steels often show a definite drop in strength after reaching the yield strength. The stress-strain curve for a typical low-carbon steel is shown in Figure 2–19.

Modulus of Elasticity

A factor on which the design of any part depends is the *stiffness* or *rigidity* of the material, that is, how much the material will give or stretch when a load is applied. This is found by determining the *modulus of elasticity* of the material. The modulus of elasticity is defined as the change in stress divided by the change in strain *in the elastic region of the curve.* Basically, it is

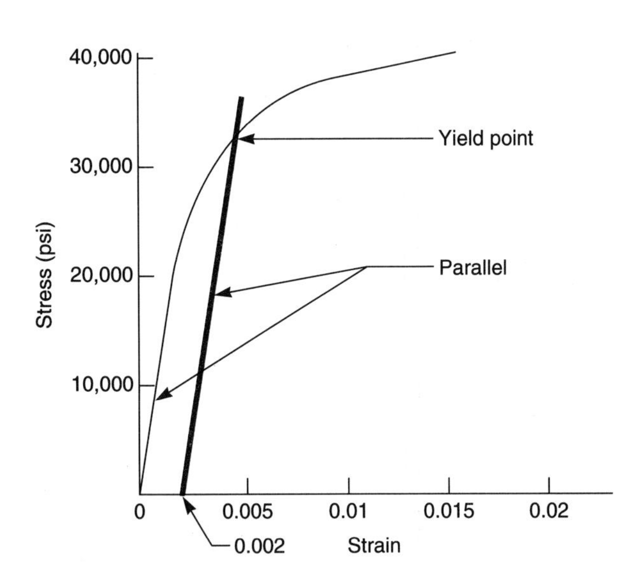

Figure 2-17. Yield point.

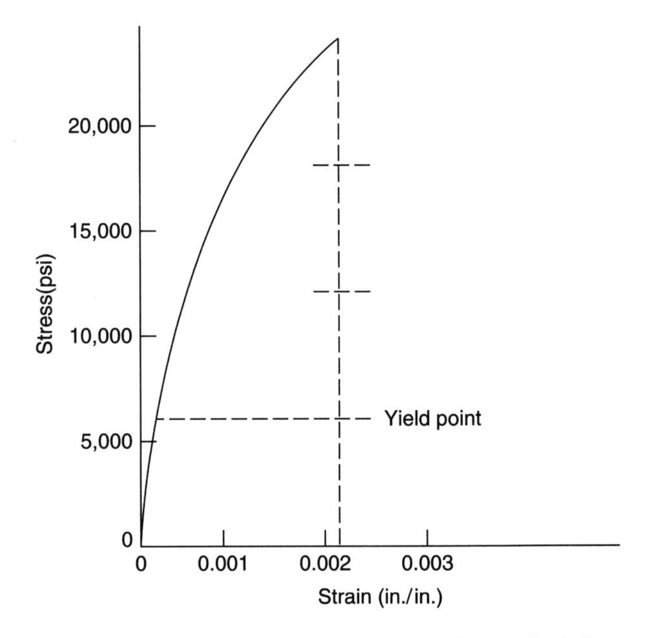

Figure 2-18. Stress-strain curve for brittle materials.

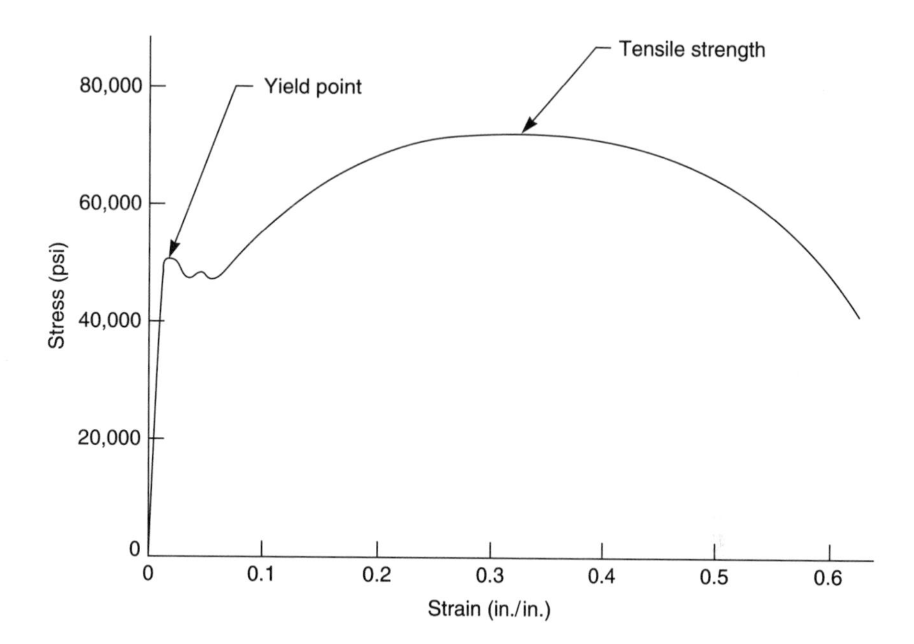

Figure 2–19. Stress-strain curve for low-carbon steel.

the slope of the linear portion of the curve. The formula for the modulus of elasticity is

$$E = \frac{\text{stress}}{\text{strain}}$$

or

$$E = \frac{\text{load/area}}{\text{elongation/original length}}$$

or

$$E = \frac{S}{e}$$

For example, if a steel test specimen having a diameter of 0.505 inch is elongated from 2.0000 to 2.0066 inches by a load of 2000 pounds, the modulus of elasticity is

$$E = \frac{2000\text{ lb}/(3.14[0.505/2\text{ in.}]^2)}{(2.0066–2.0000)\text{in.}/2.0000\text{ in.}}$$

$$= \frac{2000\text{ lb}/0.2002\text{ in.}}{0.0033}$$

$$= 30{,}300{,}000\text{ psi}$$

In other words, if a 1.000-inch-long bar of this steel could remain in the elastic region and stretch 1.000 inch (which it cannot), it would take 30,300,000 pounds to double its length. This is a measure of the rigidity of the metal and it is used in the design of everything from bicycles to automobiles and machinery to buildings. For reference purposes, Table 2–1 shows the accepted modulus of elasticity for several selected materials.

The modulus of elasticity, often called *Young's modulus,* allows a person to calculate the extent to which a material will deform under a given load. For example, if a tensile load of 10,000 pounds is applied

Table 2–1. Modulus of Elasticity

Material	Modulus of Elasticity (psi)
Aluminum	10.6×10^6
Copper	16×10^6
Magnesium	6.5×10^6
Nickel	30×10^6
Platinum	21×10^6
Silver	11×10^6
Steel	30×10^6
Tin	6×10^6
Titanium	16.8×10^6
Zinc	12×10^6

to a 5.000-foot-long steel bar having a diameter of 0.70 inch, by what length will it be elongated?

$$E = \frac{S}{e}$$

$$E = \frac{P/A}{(z - z_0)/z_0}$$

Then the change in length would be $z - z_0$.

$$(z - z_0) = \frac{P \times z_0}{A \times E}$$

$$(z - z_0) = \frac{10{,}000 \text{ lb} \times 5.000 \text{ ft} \times 12 \text{ in./ft}}{3.14 \times (0.72/2 \text{ in.})^2 \times 30 \times 10^6 \text{ lb/in.}^2}$$

$$= \frac{600{,}000 \text{ in.-b}}{12.2 \times 10^6 \text{ lb}}$$

$$= 0.049 \text{ in.}$$

Safety Factor

Still another limitation is placed on the engineering design. Because all design work is based on assumptions, and because there is always some doubt as to any piece of material's yield strength, a *safety factor* is applied. The safety factor is the number of times the engineering design maximum stress goes into the yield strength of the material. If an engineer designs a part so that the maximum stress it will ever undergo, under given design criteria, is 25,000 psi, and the yield strength of the material is 50,000 psi, the safety factor is 2. Safety factors of 2 to 4 are usually specified; however, for some cases higher or lower safety factors are used.

Compression

Just as stress, strain, tensile strength, breaking strength, and other properties can be found in tension, similar properties can be determined in *compression.* Compression is loading a specimen by squeezing the material (Figure 2–20). If a 6-inch-diameter cylinder of concrete withstands a load of 80,000 pounds before it breaks, its compressive stress (or compressive strength) is

$$\text{compressive stress} = \frac{80{,}000 \text{ lb}}{3.14 \times (3 \text{ in.})^2}$$

$$= 2800 \text{ psi}$$

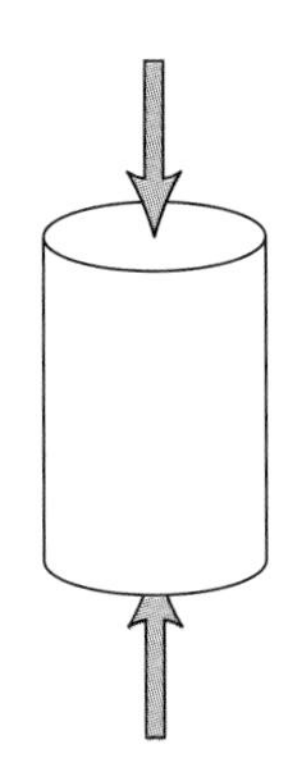

Figure 2–20. Compression.

If a 15-centimetre-diameter concrete test sample resisted a 35,000-kilogram load before it was crushed, the compressive strength of the concrete would be

$$\text{compressive stress} = \frac{35{,}000 \text{ kg} \times 9.8 \text{ m/s}^2}{3.14 \times (7.5 \text{ cm}/100 \text{ cm/m})^2}$$

or

$$= \frac{343{,}000 \text{ N}}{0.0177 \text{ m}^2}$$

$$= 19{,}400{,}000 \text{ Pa}$$

$$= 19.4 \text{ MPa}$$

which is equivalent to about 2800 psi.

Shear

Shear is defined as the application of opposing forces, slightly offset to each other (Figure 2–21). Shear forces result in a "cutting" action on a material. Shear stress is calculated by determining the load applied to the cut area. If a force of 9000 pounds is required to slice through a 3/8-inch- (0.375-inch-)diameter rivet, then the shear strength of the rivet can be calculated as

$$\text{shear stress} = \frac{\text{load}}{\text{total area cut}}$$

$$= \frac{P}{\pi r^2}$$

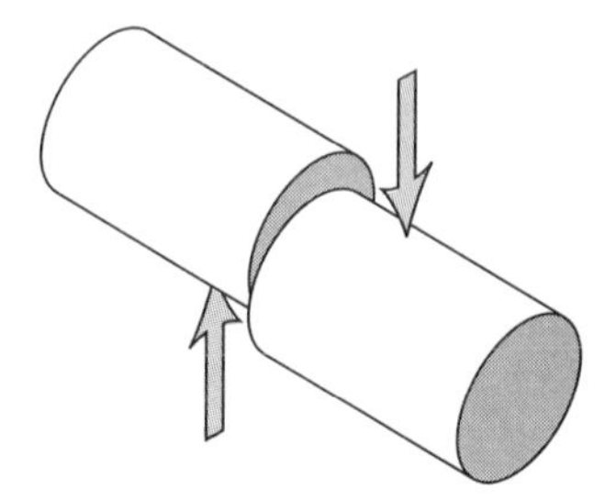

Figure 2-21. Shear.

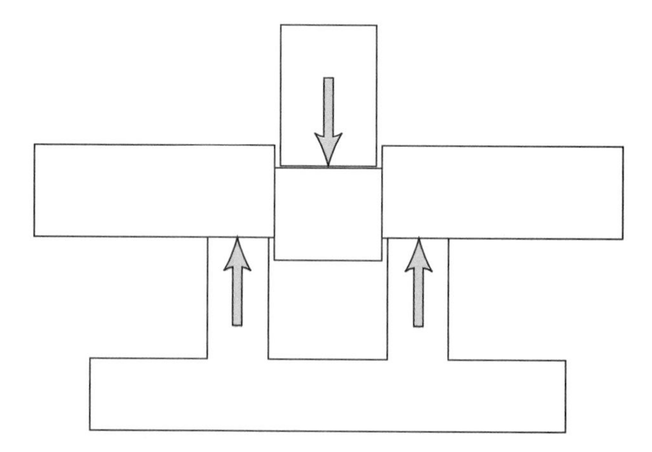

Figure 2-22. Double shear.

$$= \frac{9{,}000 \text{ lb}}{3.14 \times \left(\frac{0.375 \text{ in.}}{2}\right)^2}$$

$$= \frac{9000 \text{ lb}}{0.110 \text{ in.}^2}$$

$$= 81{,}000 \text{ psi}$$

Riveted sections are often designed with two plates riveted to a central one. The rivet would be cut or sheared at two places, resulting in *double shear* (Figure 2–22). Here twice the area would have to be cut in order for failure to occur. A rivet of the same strength, loaded in double shear, would take twice the stress in double shear as in single shear.

Torsion

Torsion is the twisting of an object (Figure 2–23). *Torsion strength* is the maximum twisting stress a material

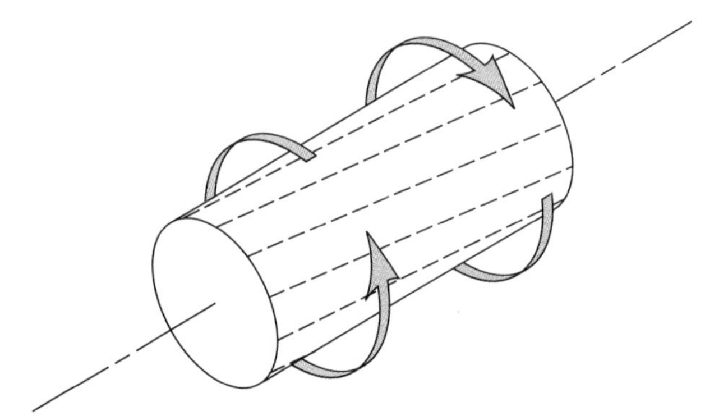

Figure 2-23. Torsion.

can withstand. Torsion is measured in foot-pounds, inch-pounds, or in the metric system, newton-metres. As seen in Figure 2–24, if a load of 100 pounds is applied to the end of a 12-inch wrench gripping a bolt head, pipe, or bar, then the torque or torsion stress would be

$$\text{torque} = 12 \text{ in.} \times 100 \text{ lb}$$

or

$$= 1200 \text{ in.-lb}$$

This could also be calculated as:

$$\text{torque} = 1 \text{ ft} \times 100 \text{ lb}$$
$$= 100 \text{ ft-lb}$$

Flexure

Any time a piece of material is "bent" it is said to be in *flexure.* To deal with the forces and the mathematics involved in flexure would require a course in the science of statics and therefore flexure is not discussed here. A few concepts concerning flexure, however, need to be covered.

The ability of a part to withstand flexure is a function of the geometry or shape of the part and the properties of the material. Referring to Figure 2–25, note that the material at the top of the beam is forced together, or *compressed,* during flexure. In hollow tubes or pipes that are bent or flexed, this often causes the tube to buckle or crimp if the radius of bending is too small. Conversely, the material on the bottom or outside of the bend is stretched or placed in *tension.* Brittle

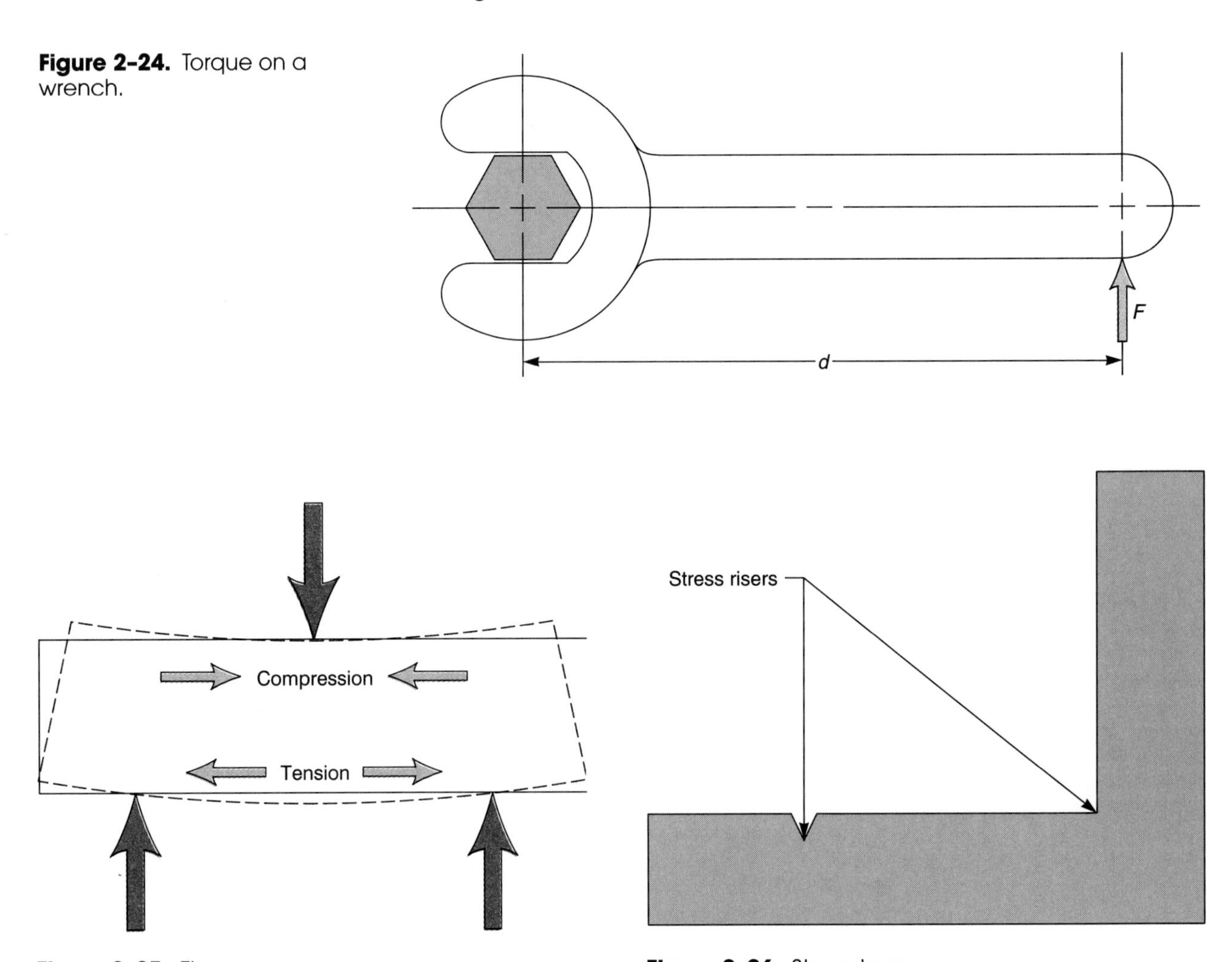

Figure 2–24. Torque on a wrench.

Figure 2–25. Flexure.

Figure 2–26. Stress risers.

materials, which cannot withstand tension, will crack and break at this point if the bending is very severe.

Whenever offset forces, no matter how small, are placed on a part, it will be flexed. In designing parts or members that must withstand flexure, a compromise must be made in selecting the material for the part. The part must be designed sufficiently large so that the yield point either in tension or compression is not exceeded. The material must be soft or ductile enough to withstand the tension on the outside of the curvature, yet strong enough in compression to withstand the tendency to buckle on the inner curvature of the part in flexure.

One further consideration should be met when considering flexure. Sharp corners concentrate the stress at a point. Such points are often called *stress risers* (Figure 2–26). Stress risers, under bending forces, usually lead to early failure of the part. No part should be designed with sharp internal corners. Corners should have relatively large radii corners instead of being sharp. These are known as *rounds* (external corners) and *fillets* (internal corners). Materials which are brittle or *notch sensitive* should be honed or polished to remove scratches that can act as stress risers.

PROBLEM SET 2-1

1. If a tensile force of 500 lb is placed on a 0.75-in.-diameter bar, what is the stress on the bar?
2. What is the tensile strength of a metal if a 0.505-in.-diameter bar withstands a load of 15,000 lb before breaking?
3. A cable in a motor hoist must lift a 700-lb engine. The steel cable is 0.375 in. in diameter. What is the stress in the cable?
4. If a steel cable is rated to take 800 lb and the steel has a yield strength of 90,000 psi, what is the diameter of the cable? (Ignore safety factor.)
5. If a tensile part in a machine is designed to hold 25,000 lb and the part is made from a material having a yield strength of 75,000 psi, what diameter must the part have?
6. If the part in Problem 5 must have a safety factor of 2, what diameter must the part have?
7. A tension link in a machine has a diameter of 0.325 in. The metal from which it is made has a tensile strength of 67,000 psi. Will this link be able to hold up a load of 20,000 lb?
8. A tensile part in a machine is designed to hold a load of 1800 lb, has a safety factor of 2.2, and is made of a metal having a yield strength of 55,000 psi. What is the diameter of the part?
9. If a compressive force of 2200 lb is applied to a concrete column having a diameter of 6 in., what is the stress on the column?
10. What diameter must a part have to support a load of 3000 lb (in compression) if the material from which it is made has a compressive strength of 2000 psi?
11. A table with four legs is made from a material with a compressive strength of 800 psi. It must support a load of 1200 lb placed in the middle of the top of the table. If each leg has a uniform square cross section, what are the dimensions of each square leg?
12. Consider a 200-lb man sitting on a three-legged milking stool. Each of the legs of the stool can withstand a compressive stress of 400 psi and is 1.5 in. in diameter. Will the stool collapse under the man?
13. An 18-in.-long torque wrench has a force of 80 lb applied to its end. What is the reading on the torque wrench in foot-pounds?
14. What force must be applied to the end of a 14-in. pipe wrench if a torque of 75 ft-lb is needed?
15. A torque of 150 ft-lb must be delivered to an axle but only 50 lb of force is available. What is the radius of the wheel driving the axle?
16. A shear force of 1800 lb is required to cut a bar having a diameter of 0.400 in. What is the shear strength of the material being cut?
17. A 0.25-in.-diameter rivet in double shear withstands a load of 5200 lb before breaking. What is the shear strength of the rivet in psi?
18. What diameter must a pin with a shear strength of 54,000 psi have to support a load of 15,000 lb in single shear?
19. What diameter must the pin have if the pin supporting the 15,000-lb load in Problem 18 is in double shear?
20. How many 0.25-in. rivets in single shear are needed to support a shear load of 125,000 lb if the rivet material has a shear strength of 75,000 psi?
21. If a 2.000-in.-long bar with a diameter of 0.505 in. stretches to a length of 2.008 in. under a load of 12,000 lb, what is the modulus of elasticity of the material?
22. If a 10-in.-long bar with a diameter of 0.750 in. stretches to a length of 10.030 in. under a load of 50,000 lb, what is the modulus of elasticity of the material?
23. A material has a modulus of elasticity of 20,000,000 psi. The diameter of the material is 1.25 in. and its original length is 5 ft (60 in.). How much will it stretch under a load of 2000 lb?
24. A material with a modulus of elasticity of 30,000,000 psi and an original length of 40 in. stretches to 40.125 in. under a load of 60,000 lb. What is the diameter of the bar if it stays in the elastic region?

Density

One of the more basic properties of a material, but a very important one in the design of many products, is *density.* Density is defined as the mass per unit volume of the material:

density = mass/volume

The mass of the material can be obtained simply by weighing it. The volume of complex shapes is sometimes difficult to calculate. According to Archimedes' principle, items submerged in water always displace their own volume. For small items a graduate or other volume calibrated container could be partially filled with water and the volume level noted. The item is then submerged, and the volume of liquid noted again. The difference in volume is the volume of the object.

In quoting the density of a material, the units must always be included. The density of iron, for instance, is

11.3 grams per cubic centimetre

or

6.5 ounces per cubic inch

or

705 pounds per cubic foot

Specific Gravity

Density is often confused with specific gravity. The *specific gravity* of a material is simply the density of the material divided by the density of water. The density of water, in the English system of measurements, is 62.4 pounds per cubic foot. In the metric system the density of water is 1 gram per millilitre. The specific gravity of lead, therefore, would be

$$\text{SG(lead)} = \frac{705\ \text{lb/cu ft.}}{62.4\ \text{lb/cu ft.}} = 11.3$$

or

$$\text{SG(lead)} = \frac{11.3\ \text{gm/cm}^3}{1\ \text{gm/cm}^3} = 11.3$$

The specific gravity of a material is the same in any measurement system. Specific gravity does not have any units. It would be improper to say that the density of lead is 11.3. The density would have to be 11.3 grams per cubic centimetre. It is correct to say that the specific gravity of lead is 11.3.

SURFACE PROPERTIES

Hardness

Hardness is a measure of a material's resistance to surface deformation. Hardness plays an important part in manufacturing. Cutting tools must always be harder than the materials that they cut. Hard materials will make cutting tools dull very quickly, which affects production rates and the costs of the product. Further, in products such as cutting tools, bearings, plow blades, and hammers, the hardness of the material from which they are made affects their quality. Therefore, a person involved in the design of manufactured products must understand the concepts of hardness of materials, how the hardness values are determined, and what the hardness numbers mean once they are established.

Rockwell Hardness

Many different hardness tests are commonly used in industry. One of the most common is the *Rockwell hardness test* (Figure 2–27). The Rockwell test makes use of three different indenters or points (Figure 2–28): a 1/16-inch steel ball, a 1/8-inch ball, and a black diamond conical or "brale" point. Rockwell testers are used primarily for testing metals.

In running the Rockwell test (Figure 2–29), the test sample is brought up against the indenter, and a "minor" load of 10 kilograms is applied. This minor load pushes the point through any surface dirt or corrosion and slightly into the material. The scale is then moved to the set mark and the major load is applied. The load is kept applied for several seconds in order for the test conditions to reach equilibrium. Then the major load is removed and the dial indicator is read. (Some newer Rockwell hardness testers use a digital readout.) The Rockwell tester uses three standard loads: 60, 100, or 150 kilograms. The three points combined with the three loads allows nine possible Rockwell scales, depending on which load is combined with which indenter. Table 2–2 shows these various scales.

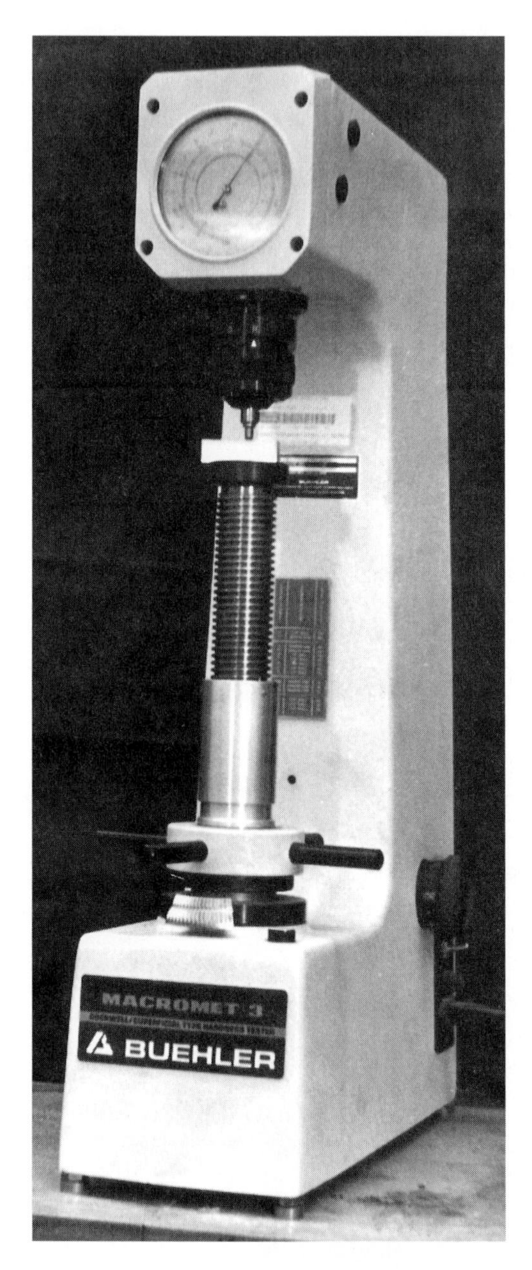

Figure 2-27. Rockwell hardness tester.

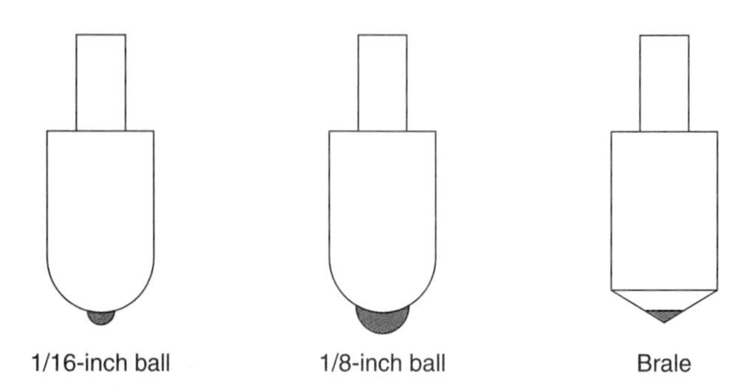

Figure 2-28. Indenters.

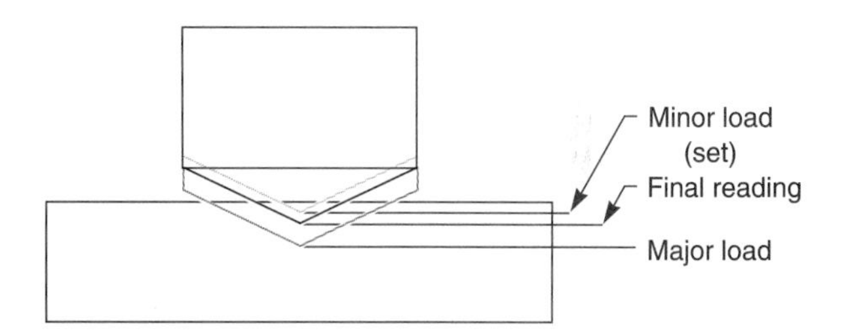

Figure 2-29. Rockwell schematic.

Table 2-2. Rockwell Scales

Scale	Load (kg)	Indenter
A	60	Brale
B	100	1/16-in. ball
C	150	Brale
D	100	Brale
E	100	1/8-in. ball
F	60	1/16-in. ball
G	150	1/16-in. ball

In reporting a Rockwell hardness number, the *scale* must be stated along with the hardness value. To say that a material has a Rockwell hardness of 50 is meaningless. A Rockwell C hardness of 50 (Rc50) is a very hard material, whereas a Rockwell B hardness of 50 (Rb50) is much softer. Conventional Rockwell hardness numbers range between 0 and 100. Note that no steel has a Rockwell C hardness of greater than 68.

The standard Rockwell test cannot be used on materials thinner than 1/8 inch because the indenter will punch through the specimen and merely test the anvil on which it rests. However, some Rockwell hardness testers are equipped with loads as little as 15 kilograms. These superficial testers are used on materials as thin as 0.006 inch and are reported under such scales as 15N, 15T, 30N, 45T, etc.

Brinell Hardness

A second common hardness test used to test metals is the *Brinell hardness test* (Figure 2–30). In the Brinell test, a 10-millimetre case-hardened steel ball is driven into the surface of the metal by one of three standard loads: 500, 1500, or 3000 kilograms. Once the ball is pushed into the material by the specified load, the diameter of the indentation left in the metal (Figure 2–31) measured in millimetres, is measured using a special Brinell microscope. The load and the diameter of the indentation are then entered into the following equation to determine the Brinell hardness number (BHN). For a complete derivation of this equation, see Appendix B.

$$\mathrm{BHN} = \frac{2P}{\pi D(D - \sqrt{D^2 - d^2})}$$

where P = load in kilograms, D = diameter of the ball (10 millimetres), and d = diameter of the indentation.

As an example, if a 3000-kilogram load left a 4-mm-diameter indentation, the BHN would be

$$\mathrm{BHN} = \frac{2 \times 3000}{3.14 \times 10(10 - \sqrt{10^2 - 4.0^2})}$$

$$\mathrm{BHN}_{3000} = 229$$

Here again it is important to state the load (3000, 1500, or 500 kilograms) at which the test was performed. Brinell hardness numbers usually vary from 100 to 600. If a Brinell hardness number falls outside this range, the test should be repeated using a different load. As is true with the Rockwell test, the Brinell test should not be used on soft or thin materials. In fact, the Brinell test can produce inaccurate results on materials less than 1/4 inch thick.

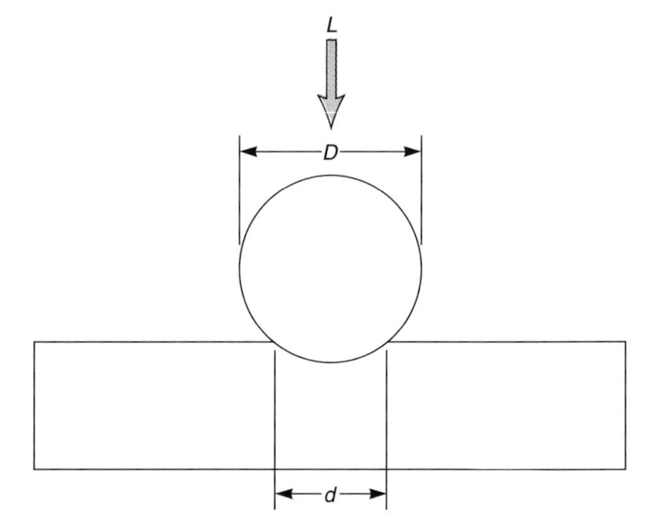

Figure 2–31. Schematic of Brinell test.

Figure 2–30. Brinell hardness tester.

Photograph courtesy of Tinius Olsen Testing Machine Co., Inc., Willow Grove, PA.

Vickers Hardness

Although the Rockwell and Brinell tests are the major hardness tests used in industry, several others that are also used should be mentioned. The *Vickers hardness test,* also known as the diamond pyramid hardness (DPH) test, uses a square pyramid point in much the same way as the Brinell tester uses the steel ball (Figure 2–32). The point is driven into the surface of the material using loads from 5 to 120 kilograms. The length of the diagonal of the square indentation is then measured and substituted into the following equation:

$$\text{DPH} = 1.8544 \times P/d^2$$

where P = load in kilograms and d = length of the diagonal in millimetres.

A 100-kilogram load that leaves an indentation of 0.80 millimetres would have a DPH number of

$$\begin{aligned}\text{DPH} &= (1.8544 \times 100)/0.80^2 \\ &= 290\end{aligned}$$

See Appendix B for the derivation of the DPH equation.

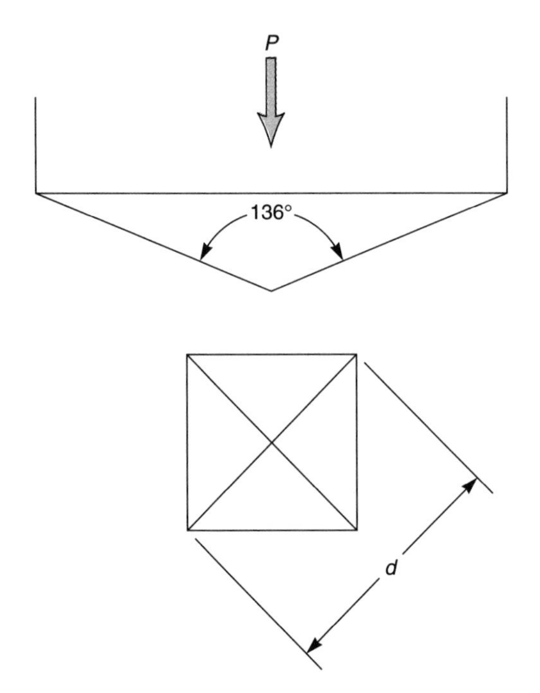

Figure 2–32. Schematic of Vickers hardness test.

Tukon Hardness

A test similar to the Vickers test is the *Tukon hardness test.* In the Tukon test, the indenter is an elongated pyramid rather than square. The indenter is known as the Knoop indenter (Figure 2–33). A load of 25 grams to 3.6 kilograms is lowered onto the indenter, forcing it into the surface of the metal. The length of the longest diagonal in the indentation is measured in millimetres. That length is then entered into the Knoop hardness number (KHN) equation:

$$\text{KHN} = 14.3P/L^2$$

where P = load in kilograms and L = length of the longer diagonal in millimetres.

Figure 2–33. Knoop indenter.

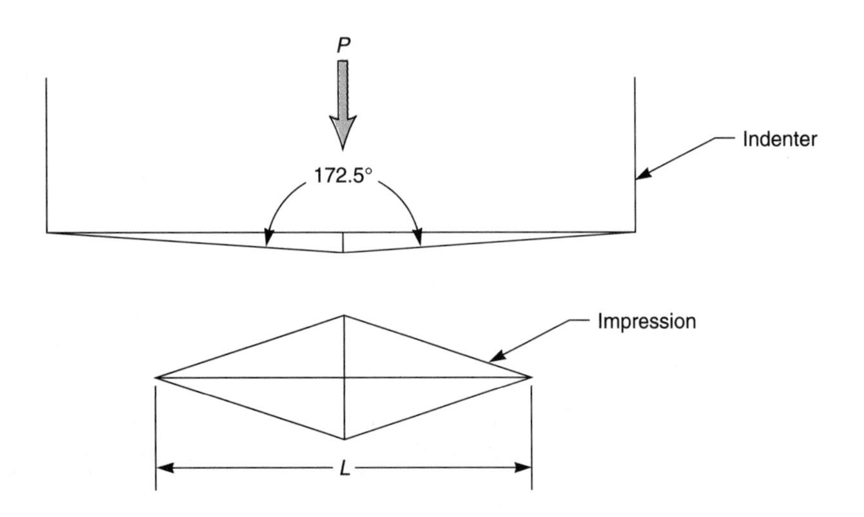

Suppose a load of 2.5 kilograms is used in the Tukon test, and the longer diagonal of the indentation is 0.29 millimetre. What is the Knoop hardness number?

$$\begin{aligned} \text{KHN} &= 14.3P/L^2 \\ &= (14.3 \times 2.5)/0.29^2 \\ &= 425 \end{aligned}$$

The Knoop hardness number will fall between 60 and 1000 for most metals. Appendix B contains the complete derivation of the KHN equation.

Scleroscope Hardness

The Rockwell, Brinell, Vickers, and Tukon hardness tests can be destructive tests. That is, they damage the tested specimen so much that it cannot be used for its intended purpose. The indentations left by these tests would ruin a ball bearing or drive shaft. The *Shore scleroscope* hardness test, however, is nondestructive. In this test a sapphire-tipped plummet is dropped on the surface of a metal from a fixed height. The plummet is in a glass tube, with a scale reading from 0 to 130 behind the tube. When the plummet hits the metal test sample, it bounces and the height on the scale is recorded as the scleroscope hardness (Figures 2–34 and 2–35).

Other Hardness Tests

There are many other hardness tests. Some are applicable only to metals, whereas others measure the hardness of plastics, rubber, wood, or other materials. For example, the *Mohs hardness test* is used to determine the hardness of rocks and minerals. The *Shore durometer* is used on plastics, rubber, and composites. The *Barcol impressor* is also applicable to plastics. Figure 2–36 shows a durometer and Figure 2–37 shows the Barcol impressor.

The engineering or technical student should become familiar with and develop a feeling for the different hardness ranges. A Rockwell C hardness greater than 40 is considered very hard. Materials having Rockwell B scale values of less than 60 are on the soft side. On the Brinell scale, steels and other materials having BHNs greater than 400 are very hard. Soft steels will fall below 200. Readings greater than 400 on the Vickers (DPH) and Tukon (KHN) scales are considered an indication of hard materials.

Problems in Measuring Hardness

There are problems involved in the measurement of the hardness of a material. First, the test applies to only a very small section of the surface of the material. The hardness at one point on a metal may not be the hardness a short distance away, or in the interior of the part. For this reason, several tests at different points on the part should be taken when using the Rockwell, sceleroscope, Vickers, or other hardness tests. Second, different hardness tests often measure different properties of the material and still refer to it as the "hardness" of the material. The Brinell, Rockwell, and Vickers testers leave a permanent indentation in the material and are thus measuring the resist-

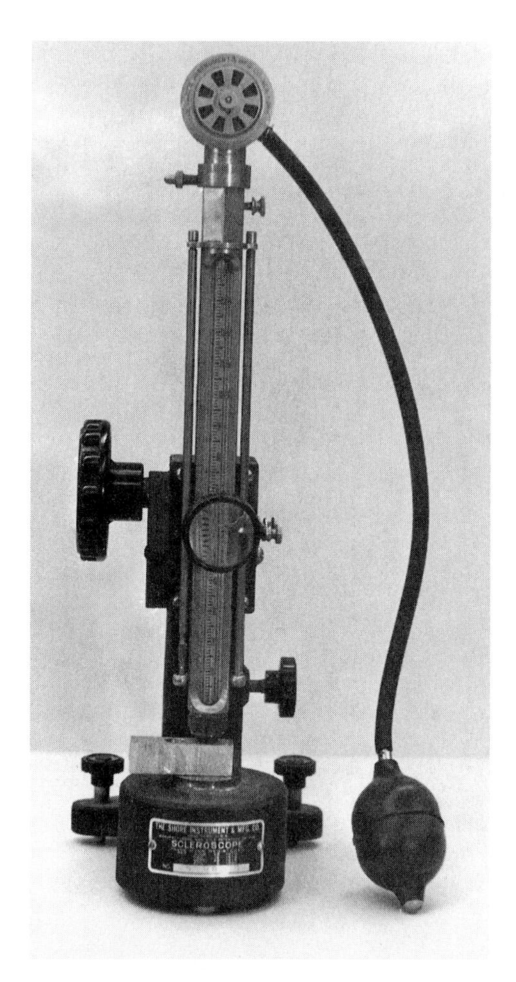

Figure 2–34. Shore scleroscope.

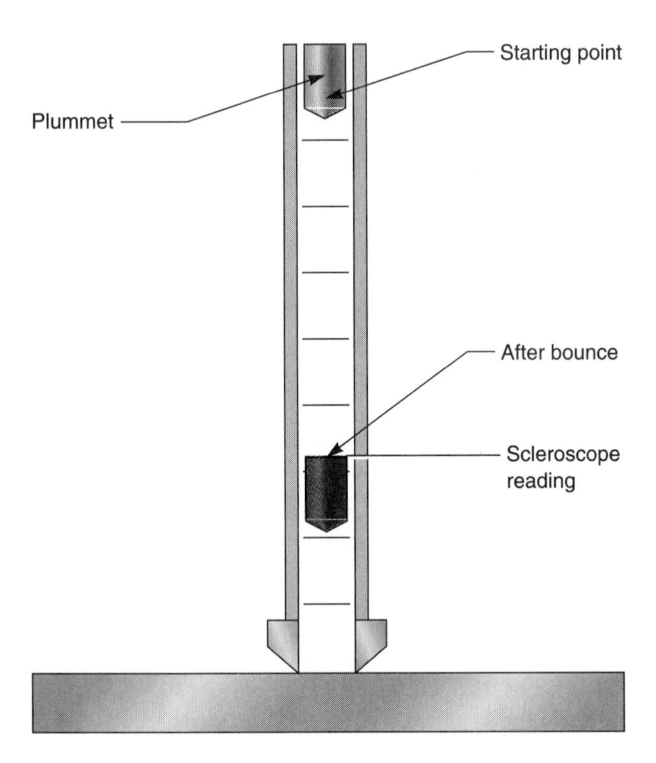

Figure 2-35. Schematic of scleroscope.

Figure 2-36. Durometer.

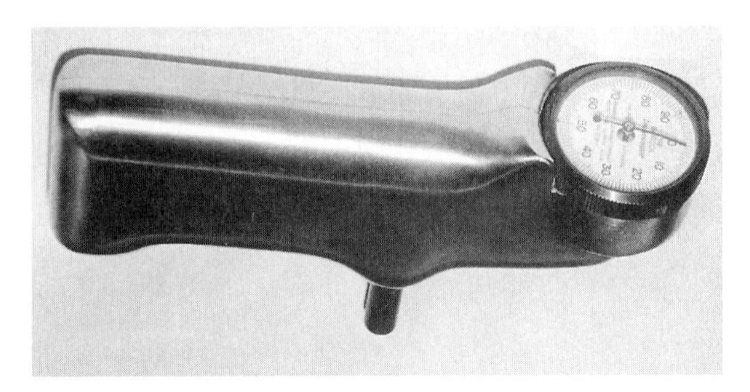

Figure 2-37. Barcol impressor.

ance to surface *plastic deformation.* The scleroscope, which leaves no dent or scratch on the test sample, measures the surface *elasticity* as an indication of hardness. The Mohs test refers to the resistance to surface *abrasion* as hardness.

These tests measure different properties of the material, yet we call all of them hardness tests. Conversion of a Rockwell to a Brinell hardness number is possible because they measure the same property of the material. Comparing a Vickers number to a scleroscope hardness number would make as much sense as comparing sweet potatoes and sonic booms—they have nothing in common.

PROBLEM SET 2-2

1. Calculate the missing values in the chart on the next page.
2. What is wrong with reporting the hardness of a knife blade as having a hardness of 58?
3. A 50-kg load on a Vickers test produces an indentation with a 0.790-mm diagonal. Is this considered a hard steel?
4. A Brinell 3000-kg test using a 10-mm-diameter ball produces an indentation of 2.6 mm in a steel test piece. Is this a hard steel or not?
5. If the Rockwell C hardness of a piece of steel is reported as 82, should it be accepted? Why or why not?
6. Should a Brinell hardness test be run on a 0.125-in.-thick piece of steel? Why or why not?

Chart for Problem 1.

	Test	Indenter size (mm)	Load	Indentation (mm)	Hardness
(a)	Brinell	10	3,000 kg	4.2	—
(b)	Brinell	10	500 kg	3.2	—
(c)	Brinell	10	1,500 kg	4.2	—
(d)	Brinell	10	3,000 kg	—	350
(e)	Vickers	DPH	10 kg	0.20	—
(f)	Vickers	DPH	100 kg	2.7	—
(g)	Vickers	DPH	25 gm	0.05	—
(h)	Vickers	DPH	—	3.2	225
(i)	Tukon	Knoop	500 kg	6.8	—
(j)	Tukon	Knoop	15 kg	—	500

7. Can a scleroscope hardness be related to a Brinell hardness test? Why or why not?
8. For what individual applications are the Brinell, Rockwell, Vickers, or scleroscope tests best suited?
9. Give at least three examples in which the hardness of a material should be known before a part is manufactured from it.
10. Which hardness test(s) would be suitable for measuring sheet metals?
11. What is the difference between density and specific gravity?
12. If a part weighs 22.4 g and displaces 3.5 mL of water, what is its density in grams per cubic centimetre? What is its specific gravity?
13. If a cylindrical bar has a diameter of 1.5 in., is 3.25 in. long, and weighs 0.87 lb, what is its density in pounds per cubic inch?
14. The specific gravity of nickel is 8.9. What is its density in grams per cubic centimetre? What is its density in pounds per cubic foot?
15. Consider the specific gravity of the following metals:

aluminum = 2.7

antimony = 6.7

cadmium = 8.6

nickel = 8.9

bismuth = 9.8

If a cube of one of these metals, 1 in. on a side, weighs 0.24 lb, what is the metal?

TIME-DEPENDENT PROPERTIES

Impact

Whereas the tensile strength, compressive strength, shear strength, and torsion strength are always determined in a *steady-state* test, impact strength is determined by a sudden blow to the material. In the steady-state test, the load is applied slowly and continuously until the material fails. Note, however, that the same material may fail under loads far less than its tensile strength if the maximum force is applied at one instant. A classic example of this is the children's toy Silly Putty. Silly Putty will stretch ductilely into a fine thread if gently pulled. If the Silly Putty is jerked, it will snap in a brittle fashion. Steel and other materials behave similarly. In impact, the entire load is applied in a few microseconds.

The speed at which the load is applied is known as the *strain rate* and is measured in inches per minute, metres per minute, millimetres per second, or similar units. The rate at which the material will stretch in the most ductile manner is the critical strain rate and depends on the temperature of the material. If steel is stretched at exactly the right rate, within 0.5°C of the critical temperature, it will stretch like a piece of taffy without breaking. On the other hand, the same steel will shatter in a brittle fashion if struck suddenly.

The impact strength of a metal can be determined by using one of three methods: the *Izod, Charpy,* and *tensile impact* tests. The purpose of these tests is to determine the energy that a piece of material can absorb in impact. *Energy* is defined as the ability to do work. *Work* is defined as a force acting through a distance and has the formula

$$W = F \times d$$

where W = work, F = force, and d = distance.

Work is measured in foot-pounds or newton-metres. If an object is dropped from a given height, the energy with which it hits the floor is given by the equation

$$E = w \times h$$

where E = energy in foot-pounds, w = weight in pounds, and h = the height in feet.

The energy a mass will impart to the floor depends solely on its height above the floor and not the path it takes to get to the floor. A 4-pound sledgehammer with the head 5 feet above the floor will strike the floor with an energy of 20 foot-pounds whether it drops vertically or is allowed to swing like a pendulum from a 5-foot-long handle. This principle allows us to design an impact test machine.

Refer to Figures 2–38 and 2–39 as you follow the derivation of the equation for the impact test. If a weight w is at the end of a pendulum having a length L, and starts its swing at an angle θ_1 with the vertical, then it will be a distance h_1 above its lowest

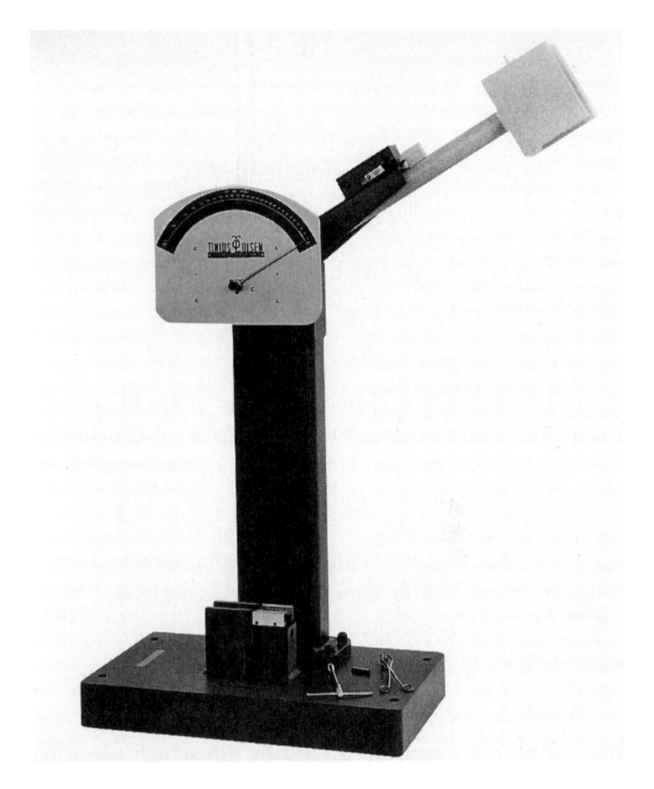

Figure 2-38. Impact test machine.
Photograph courtesy of Tinius Olsen Testing Machine Co., Inc., Willow Grove, PA.

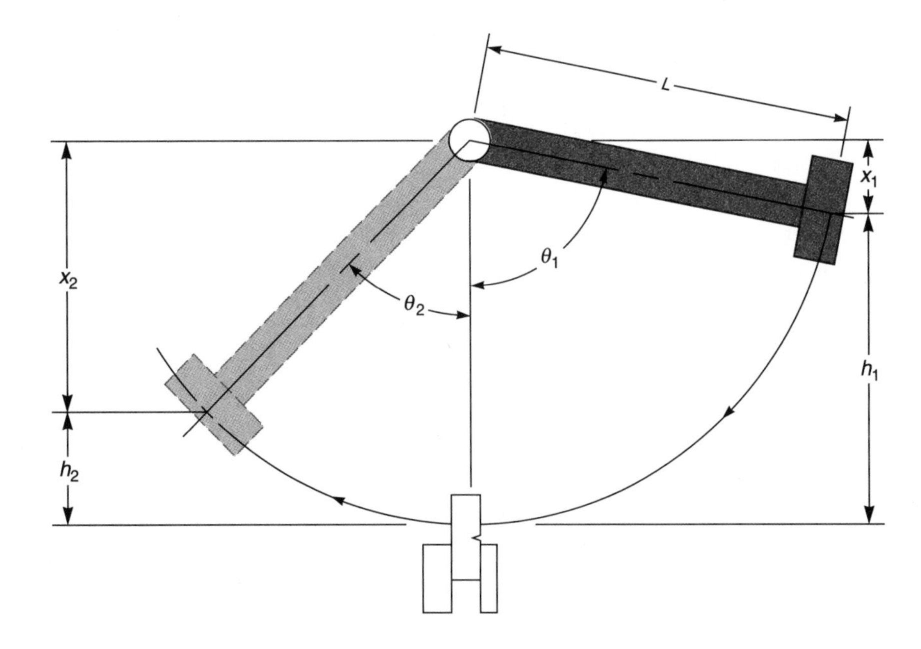

Figure 2-39. Schematic of an impact machine.

point. Since h_1 is the same as the distance $(L - x_1)$ and

$$x_1 = L\cos\theta_1$$

the vertical distance the pendulum falls is

$$h_1 = (L - L\cos\theta_1).$$

The energy E_1 of the hammer head at its lowest point of swing is

$$E_1 = wL\,(1 - \cos\theta_1)$$

If the pendulum is released from the angle θ_1, it will rise to the same angle on the other side of the swing (if the effect of friction is ignored). If the pendulum strikes a piece of material at the lowest point of its swing, the material will absorb some of the energy and the pendulum will rise only to the point at which the remaining energy is used. Call this angle θ_2. The energy E_2 left after impact can be calculated by the preceding equations as

$$E_2 = wL\,(1 - \cos\theta_2)$$

Therefore, the energy absorbed by the struck piece is $E_2 - E_1$ or

$$E_2 - E_1 = wL\,(1 - \cos\theta_2) - wL\,(1 - \cos\theta_1)$$

By means of algebra, this can be simplified to the equation

$$E = wL\,(\cos\theta_2 - \cos\theta_1)$$

This is the equation used in all of the impact tests.

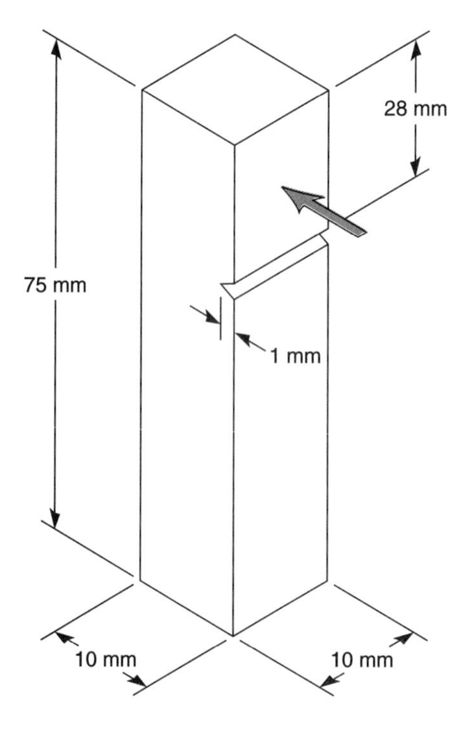

Figure 2-40. Izod test specimen.

Izod and Charpy Tests

Standard Izod and Charpy impact tests use a 60-pound hammer on a 35-inch (2.92-foot) pendulum arm. The original angle θ_1 is usually 135°. The original or set angle can be varied for use on plastics and weaker materials.

The difference between the Izod, Charpy, and tensile impact tests is in (1) the design of the test specimen "bars" and (2) the method by which the bars are held in the test machine. The Izod test bar is pictured in Figure 2–40 and the Charpy is shown in Figure 2–41. Both the Izod and Charpy test specimens have a sharp "V" notch cut into them. The notch acts as a stress riser and concentrates the impact force at that point. Without the notch, the test bar would often bend rather than break. The test would then measure only the energy a material could absorb in flexure, not the energy required to break it.

In the Izod test the test sample is held vertically by one end. The notch is toward the hammer and is struck on the top end. In the Charpy test the bar is held horizontally by both ends. The notch is away from the test hammer and is struck in the middle.

Tension Impact Test

In the *tension impact test,* the test specimen is held on one end and there is a large flange on the back end of the bar. As the hammer, which holds the test bar, is dropped, the flange catches the anvil at the bottom of the swing, grabbing the trailing edge of the bar and breaking it in tension (Figure 2–42).

We must know the impact test strength of a material to design any part or tool that must undergo sudden violent applications of force. Chisels, impact wrenches, jackhammer parts, and firearm parts are examples of impact applications.

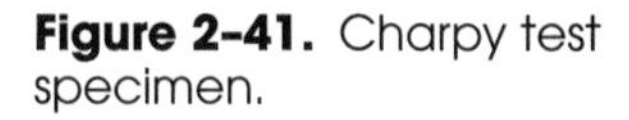

Figure 2-41. Charpy test specimen.

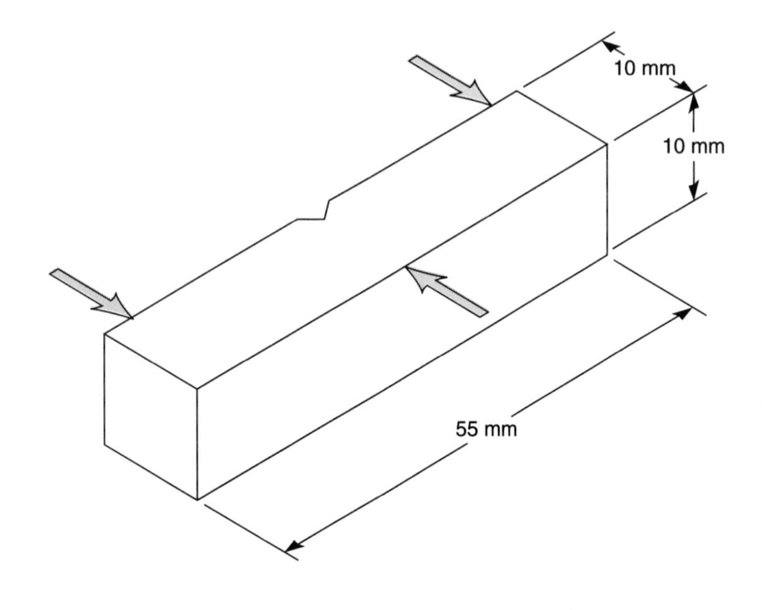

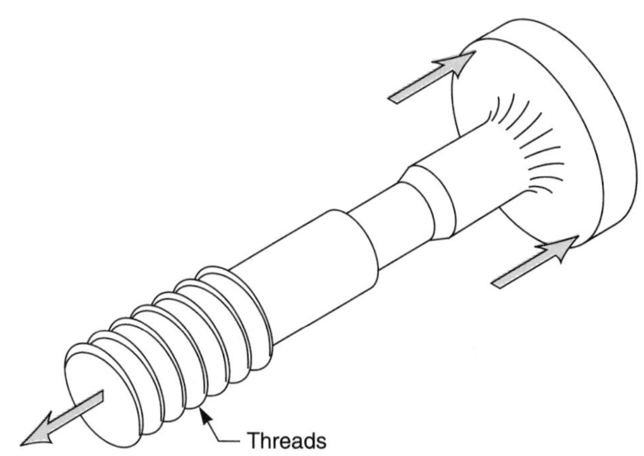

Figure 2-42. Tension impact test.

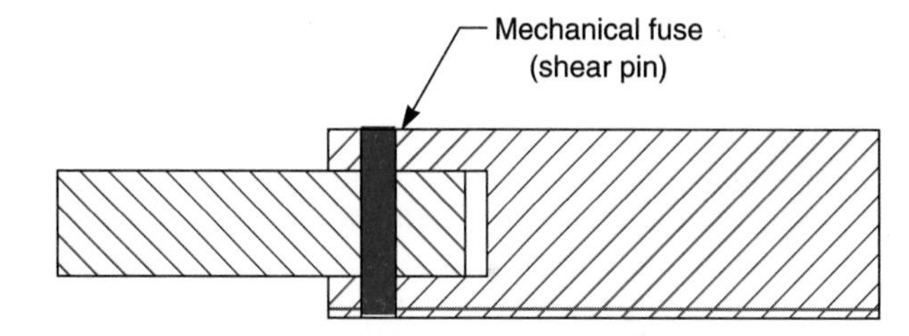

Figure 2-43. Mechanical fuse.

Any good mechanical design has a built-in *mechanical fuse.* These mechanical fuses are parts that are purposely designed to fail if the forces on the machine exceed its design limitations. The fuse should be a part that is easily replaced and relatively inexpensive. Gear lathes often have a soft brass gear in the gearbox as a mechanical fuse. If the gears jam or are otherwise overloaded, the brass gear shears, thus protecting the other gears from damage. In other words, a cheap gear is sacrificed to protect the rest of the expensive gear train. Shear pins on outboard motor propellers and drive shafts connected by shear pins are more examples of mechanical fuses (Figure 2–43). To design these mechanical fuses, we must know the tensile, compression, shear, torsion, or impact strengths as well as the hardness of the material. In the case of the brass gear, the shear strength of the brass determines the face area of the tooth needed to fail at a given load.

PROBLEM SET 2–3

1. What is the energy with which a steel shot put weighing 16 lb hits the floor when dropped from a height of 10 ft?
2. In a standard Izod impact test, a 60-lb hammer on a 35-in.-long pendulum arm drops from an angle of 135°. After striking a test specimen, the pendulum

rises to an angle of 105°. How much energy was absorbed by the specimen?

3. To test a plastic, the 60-lb impact hammer on a 35-in.-long arm is released from an angle of 70°. After impact, the hammer rises to an angle of 55°. How much energy was absorbed by the specimen?
4. In a standard Izod test (the 60-lb hammer on the 35-in.-long arm), the initial release angle is 135°. The test sample absorbs 35 ft-lb of energy. What is the angle to which the pendulum rises after impact?
5. In a standard Izod test (60-lb hammer, 35-in.-long arm, and 135° initial set angle), the angle after impact is 98°. If the same hammer is released from an angle of 90°, to what angle will it rise after impact with the same specimen?
6. A crude Izod impact tester is made to test plastics and other low-impact-strength materials by using a 4-lb sledgehammer that has a hole drilled through the handle at a distance of 36 in. from the center of the hammer head. The hammer is swung in a vertical plane about a nail through the hole in the end of the handle. The specimen is mounted in a vise directly under the nail as shown in the diagram. If the hammer is "dropped" from a horizontal position (90° from impact) and rises to an angle of 40° after impact, how much energy is absorbed by the specimen?
7. Give at least three examples for which the hardness of a material should be known prior to manufacturing a part from it.
8. Give at least three examples of mechanical fuses and determine whether they are shear, impact, tension, torsion, or compression protectors.
9. If you wish to produce a pair of pliers, which properties of the materials of construction are critical to the design?
10. What properties of a material of construction are necessary in the design of gears for a manual transmission for an automobile?

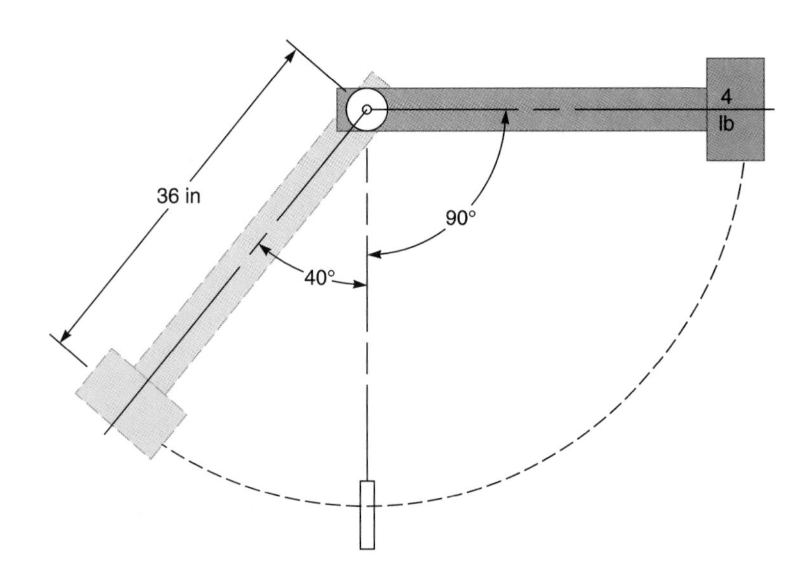

Creep

Any material undergoing continuous stress will, over a span of time, deform to some degree. The elongation caused by the steady and continuous application of a load over a long period of time is called *creep*. In impact, the load is applied in a few microseconds; in creep, the load is applied continuously for many months to many years. The amount of creep depends on the elasticity of the material, its yield strength, the stress applied, and temperature. In most metals, the higher the temperature, the greater the creep at a given load. This property allows engineers to determine roughly what amount of creep to expect by testing materials at several different elevated tempera-

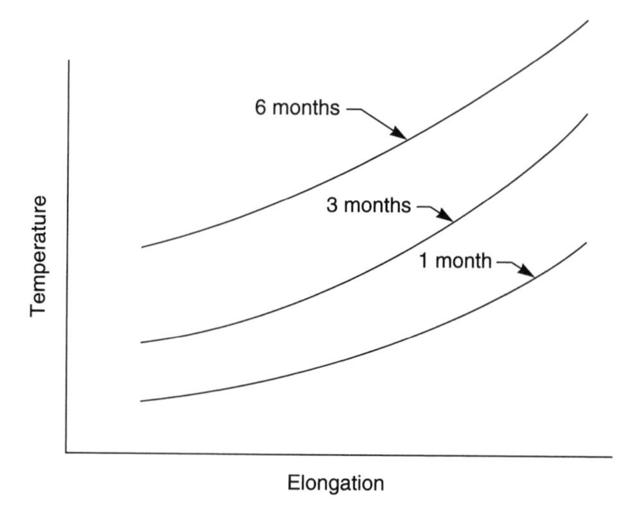

Figure 2-44. Creep versus temperature.

tures, then extrapolating that creep back to normal atmospheric temperatures (Figure 2–44). Generally creep is only taken into consideration in large static structures such as bridges and buildings.

Fatigue

Bending a wire back and forth several times at the same point will eventually break the wire. This failure of a material due to cyclic or repeated stresses is known as *fatigue.* To cause a material to break by fatigue, the yield point of the material need not be exceeded. Bending the metal only a few degrees, or in some cases a fraction of a degree, will eventually cause it to fail. Very often, fatigue in metals is caused by the vibration of the parts in a machine. The number of bending cycles required to fail a material depends on the severity of each bend. Figure 2–45 shows a sample graph of stress applied to a metal specimen versus the number of cycles required to fail the material. Note that this generalized curve consists of two rather straight lines. The extrapolated point at which these two lines meet is known as the *critical fatigue point.* If the strain through which a part is subjected is kept below this critical point, the effect of fatigue is minimized. Some materials are more fatigue sensitive than others. Usually brittle materials such as very high

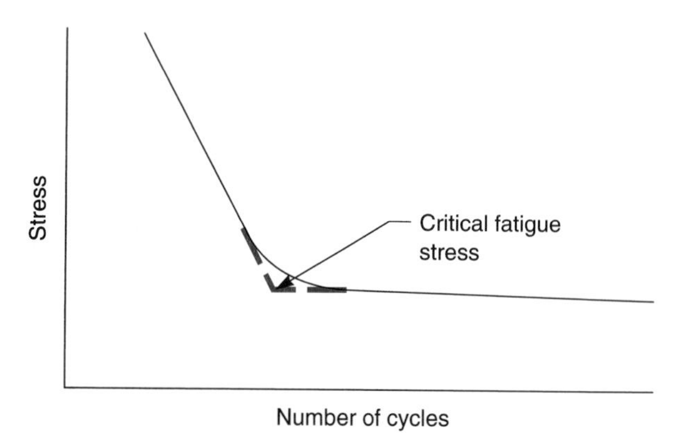

Figure 2-45. Fatigue.

strength steels, cast iron, age-hardened aluminum, glasses, and ceramics are more prone to fatigue than are softer steels and composites. Machine parts should be designed in a manner to minimize the cyclic or vibrational stresses.

PROBLEM SET 2-4

1. A 1.25-in.-diameter bar will take a maximum load in tension of 110,000 lb. What is the tensile strength of the material in pounds per square inch?
2. A bar has an original diameter of 0.625 in. and an original length of 10 in. At a load of 15,000 lb, the length has stretched to 0.631 in. What is the modulus of elasticity of the material?
3. What is the shear strength of a 0.5-in.-diameter bolt that needs 15,000 lb to cut the bolt in double shear?
4. A Brinell hardness test is performed on a block of metal. Under a 3000-kg load on a 10-mm steel ball, the indentation measures 4.55 mm. What is the Brinell hardness number of the material?
5. If a Rockwell hardness test uses a 1/16-in. ball and a 150-kg load, what scale is it?
6. In a Vickers hardness test, the diagonal of the indentation is 0.75 mm under a 50-kg load. What is the DPH number?

7. A product has a critical fatigue point at 500,000 cycles. Determine as many ways as possible of keeping the material from breaking under fatigue.
8. List as many ways as you can that will slow down the effects of creep.
9. List three engineering design situations where fatigue in metals must be considered.
10. If a dead weight such as an aircraft engine were suspended from a rod that had adequate strength to hold the weight, with no movement involved, would fatigue be a problem to be considered? Why or why not? Would creep be a problem to be considered? Why or why not?
11. Under a load of 3 kg, the longest diagonal made by the Knoop indenter in a Tukon hardness test is 0.25 mm. What is the KHN of the material? Is this a hard or soft material?
12. A Tukon hardness tester using a 20-kg load produces a KHN of 400. What is the length of the longest diagonal?

NONDESTRUCTIVE TESTING

A test is said to be destructive if it damages the part being tested to the point that it cannot be used for its intended purpose. Some tests may be destructive for certain applications but nondestructive for others. For instance, the indentation on the surface of a metal by a Rockwell hardness test may not damage a leaf spring for an automobile suspension to the point that the spring cannot be used. The Rockwell hardness test, in this case, would be a *nondestructive* test. The same Rockwell hardness test used to test the hardness of a ball or roller bearing might prohibit that particular bearing from being used. This hardness test, for this application, would then become a destructive test. Whenever possible nondestructive testing should be used.

There are manufactured items which can only be accurately tested by destructive means. For example, shotgun and rifle ammunition, photographic film, and the strength of shear pins, can only be accurately tested by destructive means. In these cases, the testing must be done by random sampling. In the case of shotgun shells, a small number of shells are selected at random from the production run and fired. If all, or the vast majority, of the shells work as they were designed, it is assumed that the rest of the shells in the production run are also good and the lot is accepted. If a significant number of the tested shells do not work properly, the entire lot is rejected.

There are problems associated with testing by sampling. How large, or what percentage of the production run must be removed for destructive testing to ensure that the test would be valid for the entire lot? If only one shell was tested out of every 10,000 shells, it is possible to pick the only bad one of the lot. Should the entire lot be rejected just because there was one bad shell? Then there is the problem of getting a truly random sample. There are many stories of biased samples being chosen to represent an entire lot. An example might be if three tomatoes taken from the top of the basket were chosen for testing. These tomatoes might be perfectly good while those at the bottom of the basket might be crushed and spoiled. Obtaining a truly random sample of a product is a science in itself. Another problem with testing by sampling is determining what percentage of the "test sample" must fail in order for the entire production run to be rejected. Here again, the science of statistics is used to make these calculations. People involved in manufacturing should take a course in statistics as part of their training.

There are many ways by which products can be tested nondestructively. In testing some products in which the property needed cannot be tested directly without destroying it, an *analog* or related property could be tested. For example, if the sugar content of a fruit had to be at a certain level for the product to be acceptable, one could analyze a sampling by destructive chemical means, but many defective pieces would still get through. If the density of the fruit was dependent on its sugar level, a "floatation" test could possibly be devised whereby the entire fruit production would be run through a tank containing a liquid that would allow the good fruit to sink and the defective pieces to float.

Other methods of testing involve looking for the defects themselves. Welds can be tested destructively by using tensile, bend, or impact tests, but how can welds be tested after they are placed *in situ* or on final product? Several nondestructive methods are available. *X-ray* examination can pinpoint holes or slag in the weld that could cause the weld to fail. Similarly, sound waves or *sonic testing* can be used to detect cracks in a casting or weld. *Dye penetrants* are made that are drawn into microscopic cracks in a part. *Electrical resistance* measurements between points on a part can often be used to determine if it is defective. For some applications, *magnetic tests* can be used to sort defective parts from acceptable ones.

Many parts are designed to pass or fail inspection simply based on their *measurements.* When possible, hand measurements are to be avoided due to the labor costs involved. Automated tests are preferred. Ball bearings, for instance, can theoretically be tested by rolling them over two trays containing holes. The first tray has holes the diameter of the smallest acceptable size. Bearings smaller than that limit fall through these holes and are discarded. The second tray has holes set at the largest acceptable size of bearing. All of the "good" bearings fall through these holes and get packaged for sale or used in further manufacture. The bearings that are too large do not fall through any of the holes and are rejected. Still another way of determining the acceptable size of some of the parts is to pass them over a scale or other weighing device.

Visual inspection is always a nondestructive test, but not always reliable. Workers can be trained to automatically reject parts that are obviously broken, flawed, or otherwise not acceptable.

Engineers are often required to devise test techniques that do not destroy a part, but still determine if it is defective.

■ DEFECTS AND IMPURITIES

Theoretically, the properties of a material are determined by the type of elements in it, the bonding structure, and the crystal structure. However, defects and impurities are seen in materials that upset and often override these theoretical considerations. Many types of defects are possible in a crystal; only a few of the more important ones are discussed here.

A *defect* in a material is defined as **anything that alters or upsets the orderly arrangement of particles in the crystal.** These defects can be classified as *point defects, line defects,* and *plane defects.*

Point Defects

Point defects are those that affect only a single point in the crystal. Point defects include *vacancies, interstitials, substitutionals, holes, electrons,* and others.

Vacancies are the absence of a nuclear particle at a place where it should be (see Figure 2–46). Because the particle is missing in the vacancy, the bonds are missing or weakened. A large number of vacancies in a material will greatly weaken the material. Further, vacancies allow the neighboring particles to move into the space left by the missing particles, at which time the vacancy simply trades places with the particle that moved into it. Vacancies also allow for other foreign particles to move into the empty space through a process known as *diffusion.* Carbon particles, for instance, can diffuse into the surface of iron then migrate through the vacancies further into the interior of the part. This is the principle on which *case hardening* works. Case hardening is discussed in Chapter 12.

Interstitials are particles located at positions other than the lattice points, or the normal positions of the particles. Figure 2–47 illustrates interstitials. Interstitials are often smaller particles than those of the parent matrix and they do upset the orderly array of particles. In the normal crystal, the planes of particles can slip over each other much like cards can slide over each other in a deck of cards. Interstitials can prevent these planes from slipping. It is similar to putting sand between the cards. When these planes are prevented from slipping, we say that the planes are *pinned.*

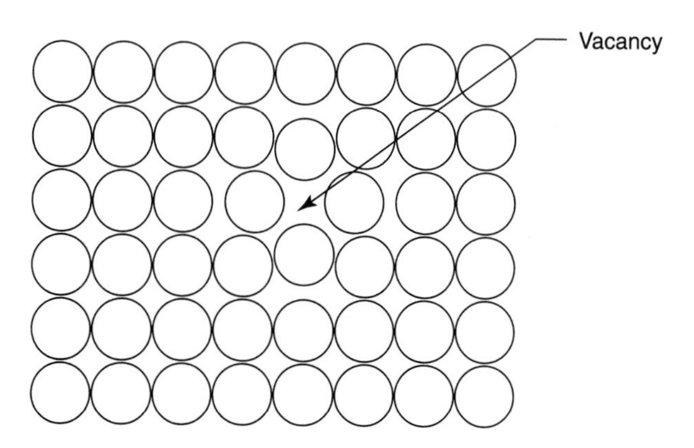

Figure 2–46. Vacancy.

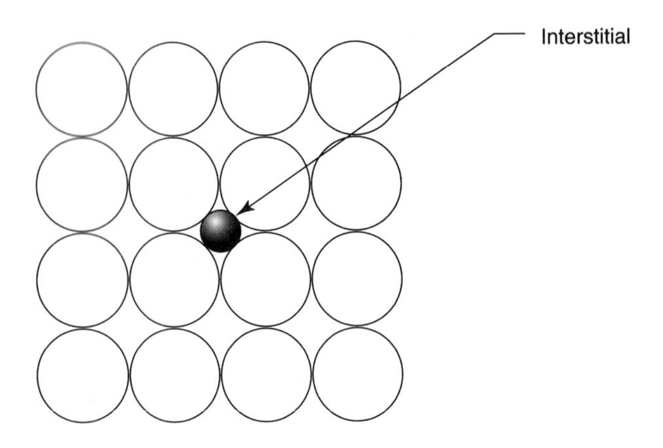

Figure 2-47. Interstitial.

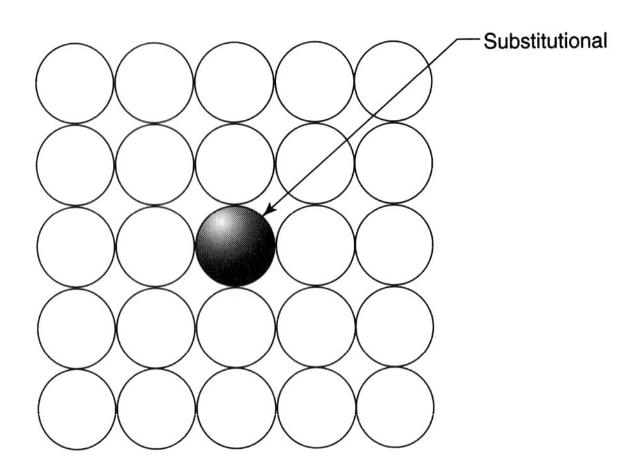

Figure 2-48. Substitutional.

In *steel,* which is an iron-carbon alloy having less than 2% carbon, the carbon is in the iron as interstitials. Each carbon atom pins the planes around it. When these planes are pinned, the material loses some ductility and become stronger. Pure iron, for example, has a tensile strength of less than 40,000 pounds per square inch. The addition of small amounts of carbon can increase that strength to more than 100,000 psi.

Substitutionals are foreign particles that take the place of particles of the parent material in the crystal lattice. If the substitutionals have a different diameter than the parent particles, they can distort the crystal structure and change the mechanical properties (hardness, tensile strength, ductility etc.) of the material. Figure 2–48 shows this concept. Substitutionals can also affect such properties as the corrosion resistance and electrical resistance of the material.

Holes and electrons affect the electrical properties of the material. Around every atom or other particle in the matrix there exist a fixed number of electrons. If substitutional atom has a greater number of electrons than the atoms of the parent matrix, a net negative charge will exist at that point. This extra **particle** electron creates the **imperfection electron.** (It is unfortunate that the particle and the imperfection have the same name, because this often leads to confusion.)

Conversely, if the substitutional has fewer electrons than the parent matrix, there is a net positive charge at that point. This absence of an electron, where it should be, is known as a *hole.* Do not confuse a hole with a vacancy. Vacancies indicate a missing nucleus, whereas a hole denotes a missing (particle) electron. Figure 2–49 illustrates both holes and electrons.

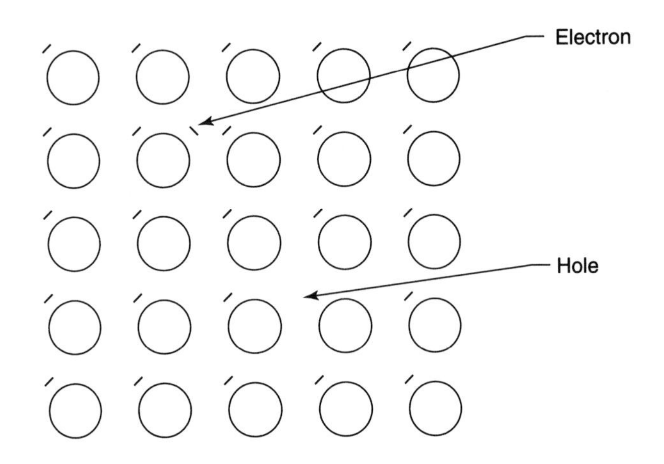

Figure 2-49. Holes and electrons.

Holes and electrons do not greatly affect the mechanical properties of a material, but they can significantly change its electrical properties. Materials containing holes and electrons are known as semiconductors and are used as the basis of all transistors and integrated circuits. Without them, computers, communications systems (including cell phones), many aircraft, satellites, and a host of other modern technological inventions would not be possible.

Line Defects

As strange as it may seem, line defects refer to a plane of particles in the crystal structure. The two major types are *line dislocations* and *screw dislocations.* A line dislocation is an extra partial plane of particles in the grain or crystal. As seen in Figure 2–50, this extra partial plane of atoms greatly distorts the crystal structure. Not only are the atoms about the dislocation forced outward, but a large empty space occurs at the end of the dislocation.

Line dislocations can be moved through the crystal. All that is needed is for the end of the dislocation to bond to the next adjacent atom or particle and join with the bottom part of that plane. This leaves the top part of the next plane of atoms as the dislocation (see Figure 2–51). However, if an interstitial or substitutional moves into the space at the end of the line dislocation, the end of that dislocation cannot move and the dislocation is pinned (Figure 2–52).

One method by which fatigue in metals occurs is through the generation of new dislocations by what is called a *Frank-Read source* (after its discoverers). Basically, the Frank-Read source generates dislocations that can eventually pile up, creating a crack in the material. Once a crack is formed, most of the strength of the material is gone, and the material quickly fails. Figure 2–53 tries to illustrate the process of crack formation.

Dislocations also play a role in such items as work hardening or cold working. Annealing of a metal, as

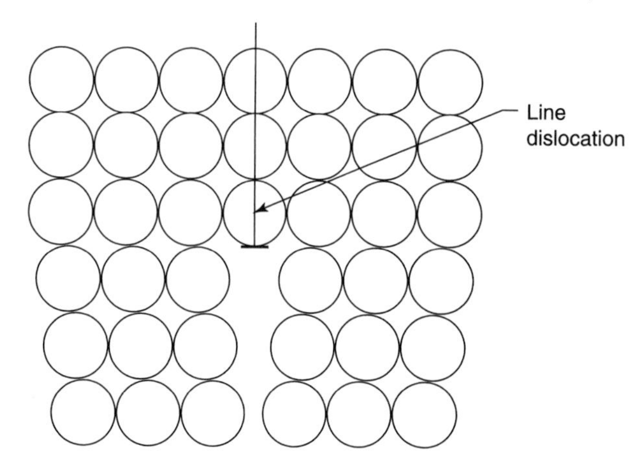

Figure 2–50. Line dislocation.

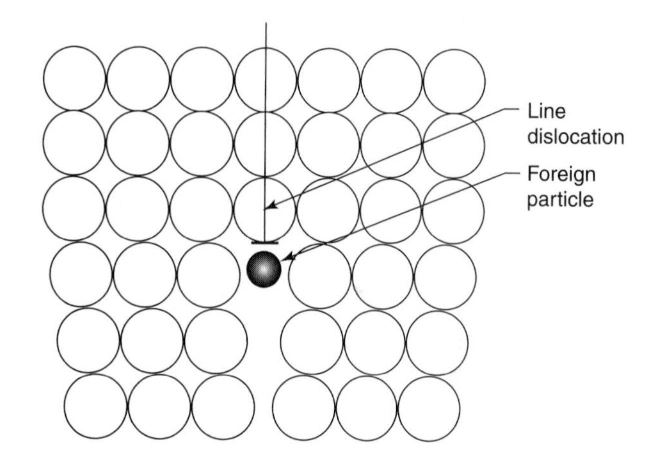

Figure 2–52. Pinned dislocation.

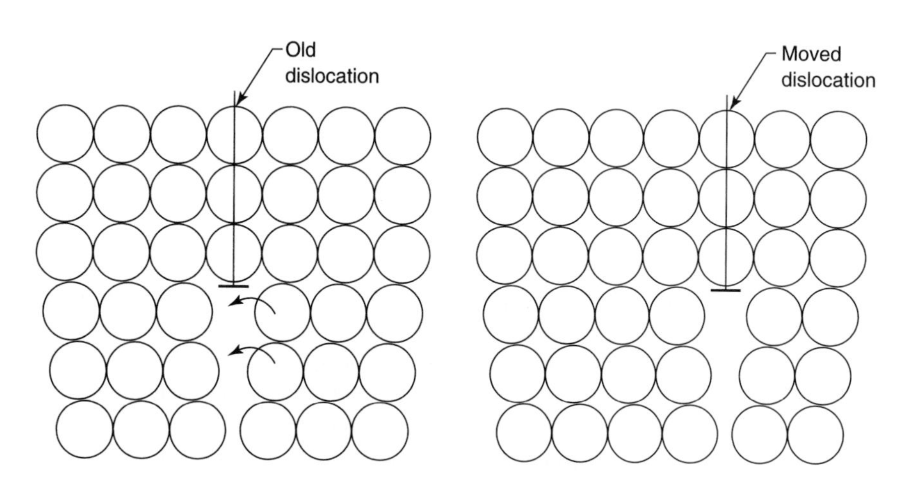

Figure 2–51. Movement of dislocation.

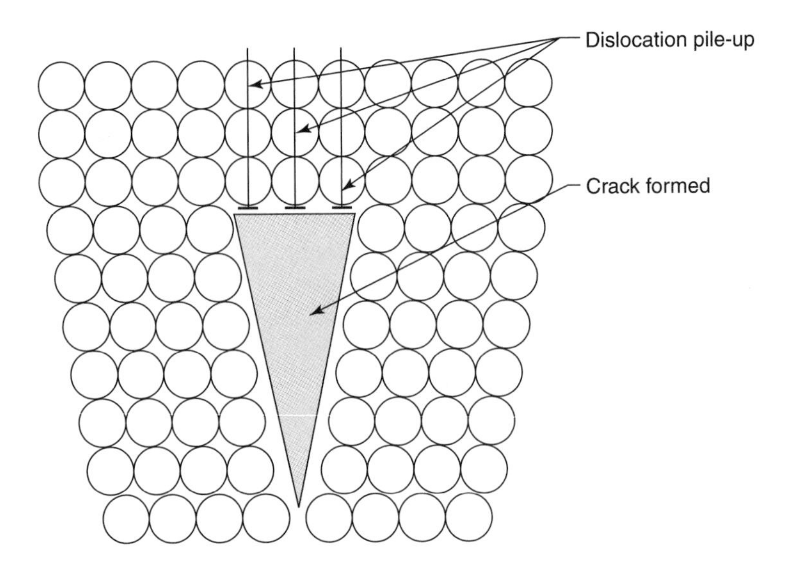

Figure 2–53. Crack formation.

discussed later in Chapter 12, works by allowing the crystals to reform and eliminate many of the dislocations. Thus the internal stresses in the material caused by dislocations are relieved and the material becomes softer, tougher, more workable, and less susceptible to failure.

Screw dislocations are the slight rotation of one plane of particles over its neighboring plane. This can be demonstrated by placing a slight rotation on a deck of cards. Figure 2–54 illustrates this idea. Screw dislocations occur during the solidification process in crystals. They tend to make the material more brittle in that they upset the slip planes in the matrix.

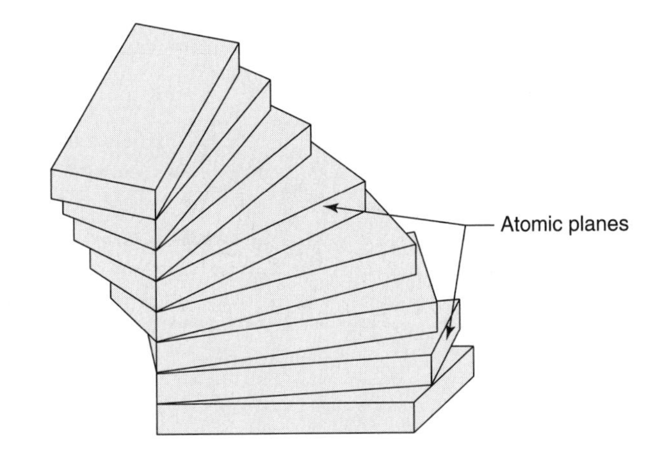

Figure 2–54. Screw dislocation.

Plane Defects

Just as line defects refer to planar imperfections in the crystal, plane defects involve three-dimensional imperfections. Among these are *twinning* and *inclusions.*

A twin in the crystal is a region within a grain that creates a mirror image of the parent crystal (see Figure 2–55). The analogy of the right-hand being a twin to the left-hand fits this case.

Twins can be formed in a crystal as the material cools and solidifies from the liquid state, or they can be formed by cold working the material. Those twins formed when the melt is solidifying are known as *thermal twins,* whereas those formed due to the mechanical deformation of the grain are called *mechanical twins.* Twins are often found in copper and other metals and can be seen as “streaks” across a single grain (see Figure 2–56).

Some minerals twin on a large scale and form quite large crystals. Quartz often twins as illustrated in Figure 2–56. The mineral “staurite” can twin at a 90° angle to form a perfect cross, which is frequently made into jewelry.

Microscopic twinning in a grain disrupts the slip planes and thus embrittles a material. Mechanical twins can usually be removed by annealing. Thermal twins usually remain in many crystals.

Inclusions are macroscopic portions of a foreign material in the matrix. Whereas the other imperfections

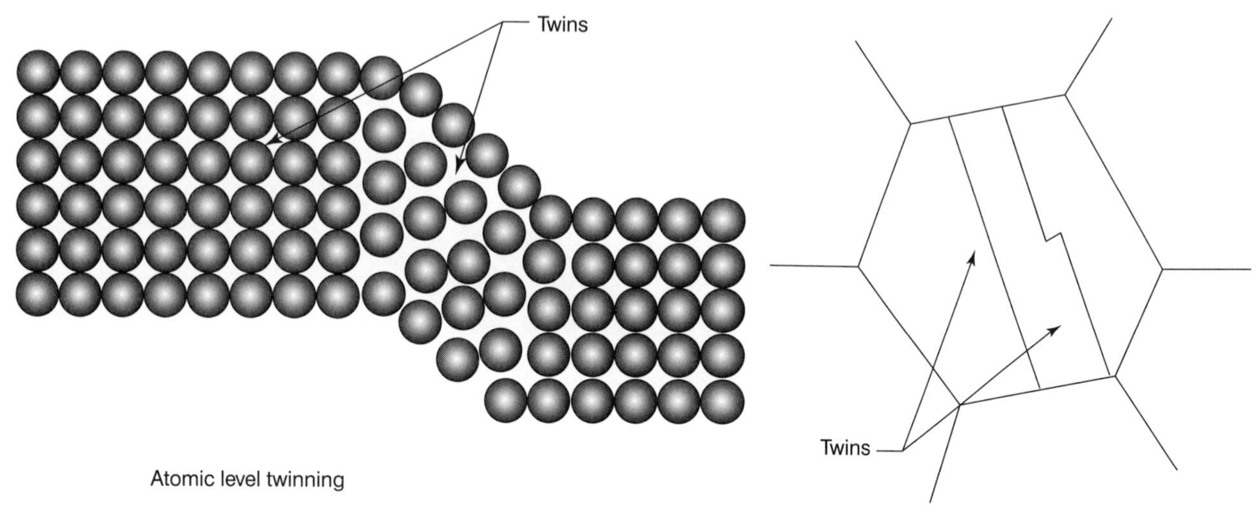

Figure 2-55. Twinning.

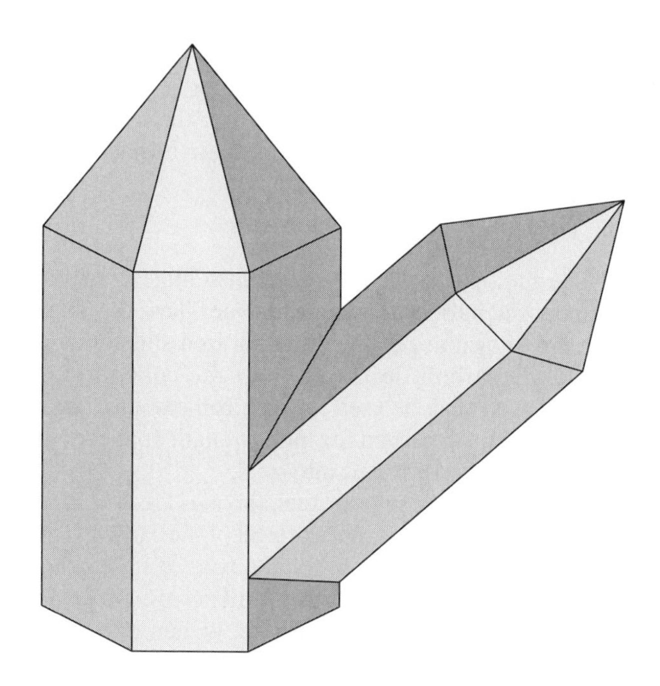

Figure 2-56. Twins.

just discussed occur on the atomic or molecular scale, inclusions involve a relatively large mass of material. Inclusions can usually be seen with an optical microscope. An air bubble in a piece of steel or weld is an inclusion. Likewise, a piece of slag embedded in a casting is an inclusion. Unlike some imperfections, which actually strengthen a material or produce other desirable properties in the material, inclusions are almost always detrimental to the performance of a material. Not only do they weaken the part themselves, but they can act as *stress risers* about which forces can concentrate. Inclusions often start cracks in the material. Aircraft, space, and other industries that must produce highly reliable products go to great expense to avoid inclusions in their manufacturing process. People's lives can depend on whether parts have inclusions in them.

Cast iron is an excellent example of a material with large inclusions. Steel has less than 2% carbon. Cast iron has more than 4% carbon. This extra carbon forms plate-like inclusions, which form small cracks in the iron matrix. This makes cast iron very brittle and weaker than steels. Cast iron is used for many casting applications instead of steel since its melting point is 500 degrees lower than steel. This makes it easier and cheaper to cast than steel, but it is not nearly as reliable. Figure 2–57 depicts inclusions.

Other types of imperfections also affect specific properties of materials, but these are left for a course in materials science to discuss.

Any of the imperfections discussed have an effect not only at the point of the defect, but they can also alter the properties of the material for many thousands of atoms around them. The area in which

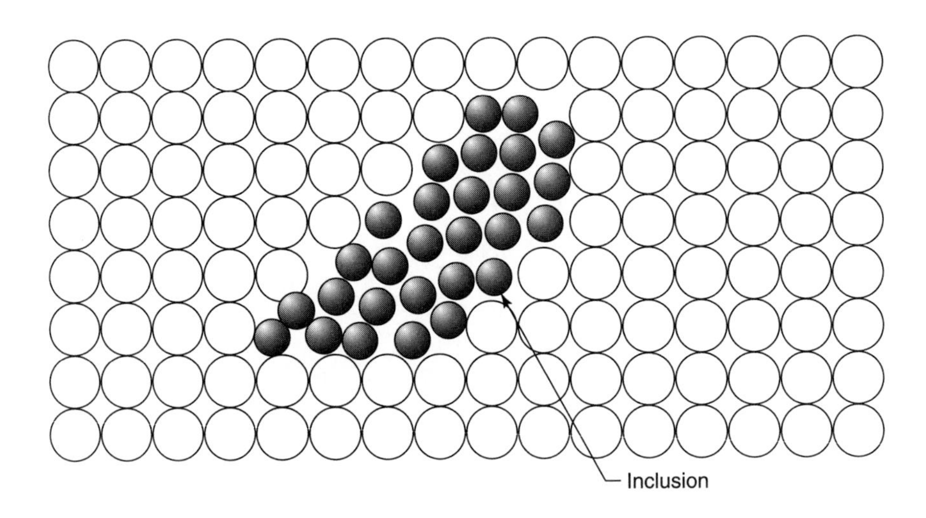

Figure 2–57. Inclusions.

an imperfection is "felt" is known as the *Gunier-Preston zone* or *G-P zone.* Many of the imperfections can be moved in a crystal, so the G-P zones often run into each other. When the G-P zones "collide," the effect can drastically change the properties of the material.

IRON AND STEEL

Many materials are used in manufacturing—metals, wood, plastics, composites—the list is endless. However, a few materials are used in so many applications that their production, and how they are made, warrants a short discussion here.

Steel is made from iron. Iron is found in many minerals in the earth. The most common, commercial ores from which iron is produced are **hematite, magnetite,** and **taconite.** Iron is also found in limonite, siderite, marcasite, and iron pyrites, but the concentration of iron in these ores is too low to make them commercially usable at this time. Magnetite and hematite are both oxides of iron. Iron pyrites is a sulfide ore that produces sulfur dioxide during the refining process. Sulfur dioxide produces acid rain which is detrimental to the environment.

Steel is produced from iron ore in two major steps: (1) the reduction of the ore to produce iron and (2) the conversion of the iron into steel.

To get the iron from the ore, the ore is crushed and mixed in layers with coke and limestone in a blast furnace (see Figure 2–58.) Air, heated to about 1100°F is blown through the mixture from the bottom of the blast furnace. At this temperature the carbon (from the coke) starts to react with the oxygen in the iron ore. When this reactions starts, the temperature increases to about 3000°F. After five or six hours, the carbon has removed the oxygen from the iron oxide leaving only iron and slag in the blast furnace. The chemical equation for the reaction is

$$(FeO \text{ and } Fe_2O_3) + C \rightarrow Fe + CO$$

The molten iron is drawn off into huge crucibles then poured into ingots known as *pigs.* These blocks of iron are known as *pig iron.* The pig iron still contains about 4% carbon, left over from the coke, at this stage. Pig iron can be used for cast iron castings, but it has far too much carbon for high-strength applications and it is usually converted into steel.

To convert pig iron into steel, the pigs are placed in a *converter,* which removes nearly all of the carbon. The desired amount of carbon, for the type of steel being made, and any other alloying elements are added to the purified iron. The molten steel is then drawn off and cast into ingots, blooms, billets, bars, or other forms. The steel forms are sent to rolling mills or other fabrication plants where they can be made into sheets, plates, wires, or beams of many sizes and shapes.

Several types of converters are available for converting pig iron into steel. The oldest type is the *open hearth converter.* Here the pig iron and recycled scrap are placed in a large, firebrick-lined "bowl" and air, heated to around 2800°F, is blown over the metal. The hot air melts the iron and burns off the carbon. Care

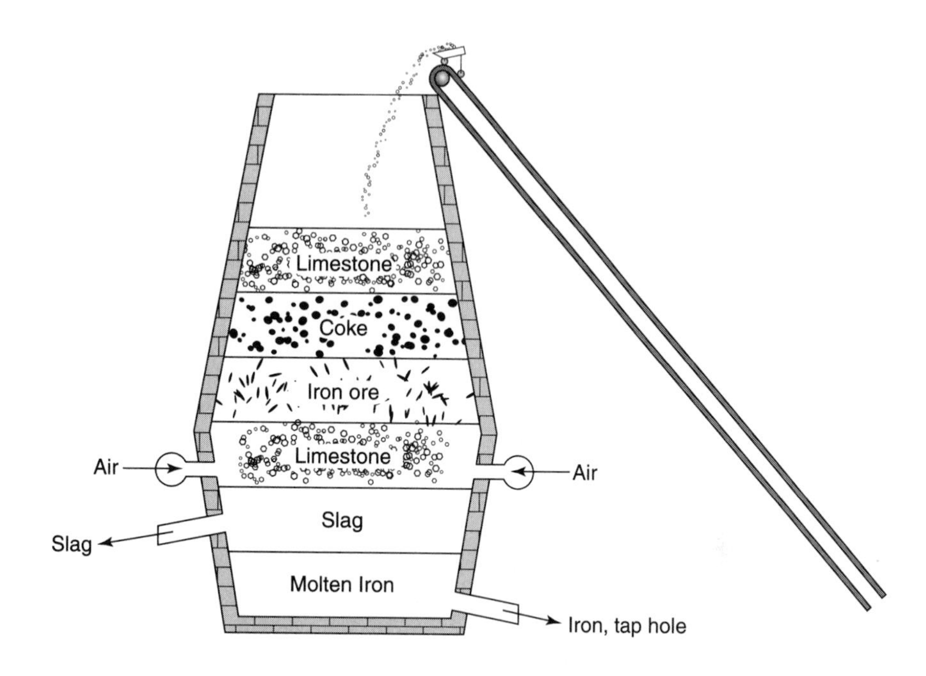

Figure 2–58. Blast furnace.

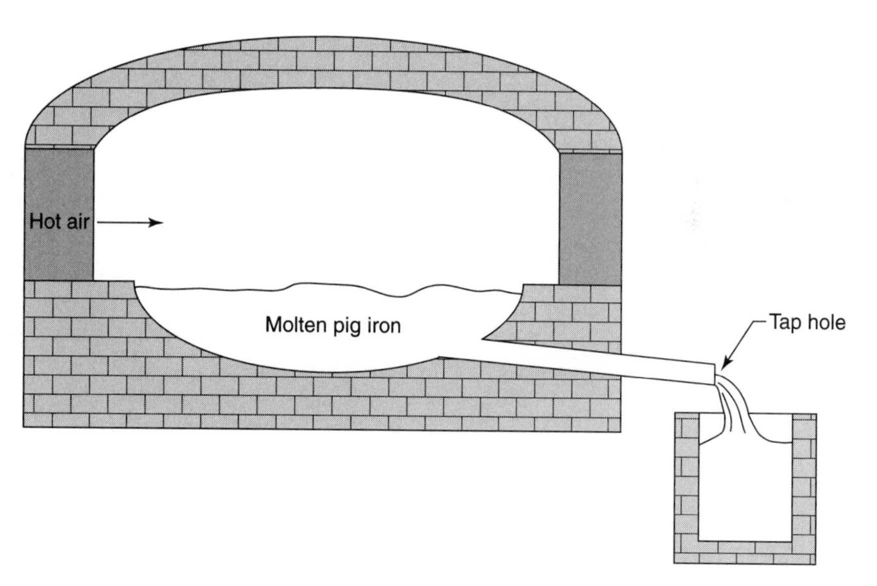

Figure 2–59. Open hearth converter.

must be taken to stop the process at the moment at which the carbon is completely burned off so as not to reoxidize the iron. This can be done by analyzing the color of the sparks produced. Carbon sparks have a yellow color, whereas iron burns with a bright blue-white spark. When the yellow color ceases, the iron is drawn off and cast into ingots (Figure 2–59).

A second converter is the *Bessemer converter* (Figure 2–60). In this converter, the pig iron and scrap are placed in a large firebrick-lined "crucible," which can be tipped on trunions for casting the finished product. In the Bessemer process, the hot air is blown upward, through the iron, burning out the carbon. The Bessemer converter does a better job than an open hearth converter of mixing the air with the metal, which speeds up the process, but it works on smaller batches. For economic reasons, the Bessemer process is declining in its use.

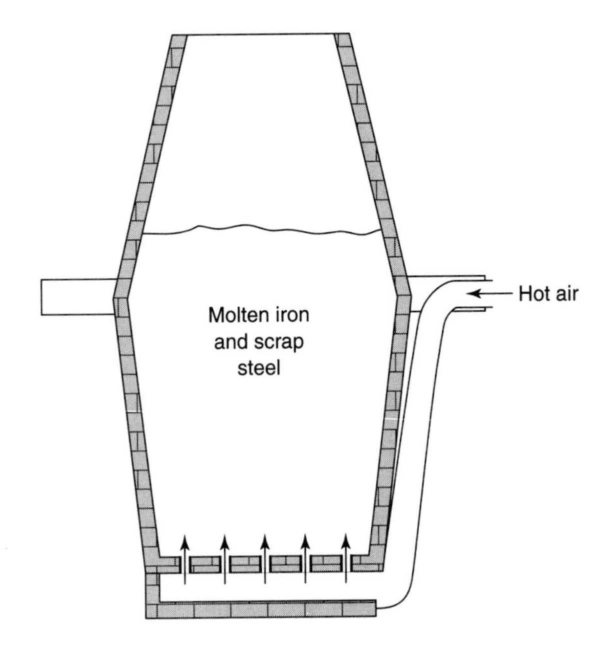

Figure 2-60. Bessemer converter.

A third type of converter is the *electric-arc converter.* In the electric-arc furnace, the pig iron is placed in the "bowl" and large carbon electrodes are brought into a position where an arc can be struck between the electrodes and the metal (see Figure 2–61). The arc burns out the carbon. Electric-arc steel is one of the "cleanest" steels made. It has very few inclusions, such as holes and slag, but it is expensive to produce. The electric power requirements are tremendous. The process requires about 400 kilowatts of electricity per ton of steel produced. (That is 200 watts per pound.) Applications requiring extremely high reliability often specify an electric-arc steel.

A fourth and relatively new (since 1954) conversion process is the *oxygen lance converter.* In this process molten iron is placed in the converter and a tube is lowered into the melt. Pure oxygen is then blown through the melt (see Figure 2–62). The principle is the same as for the other conversion processes, the carbon is burned out of the iron. The oxygen lance works on relatively small batches but produces almost as good a steel as the electric-arc converter.

Figure 2-61. Electric-arc converter.

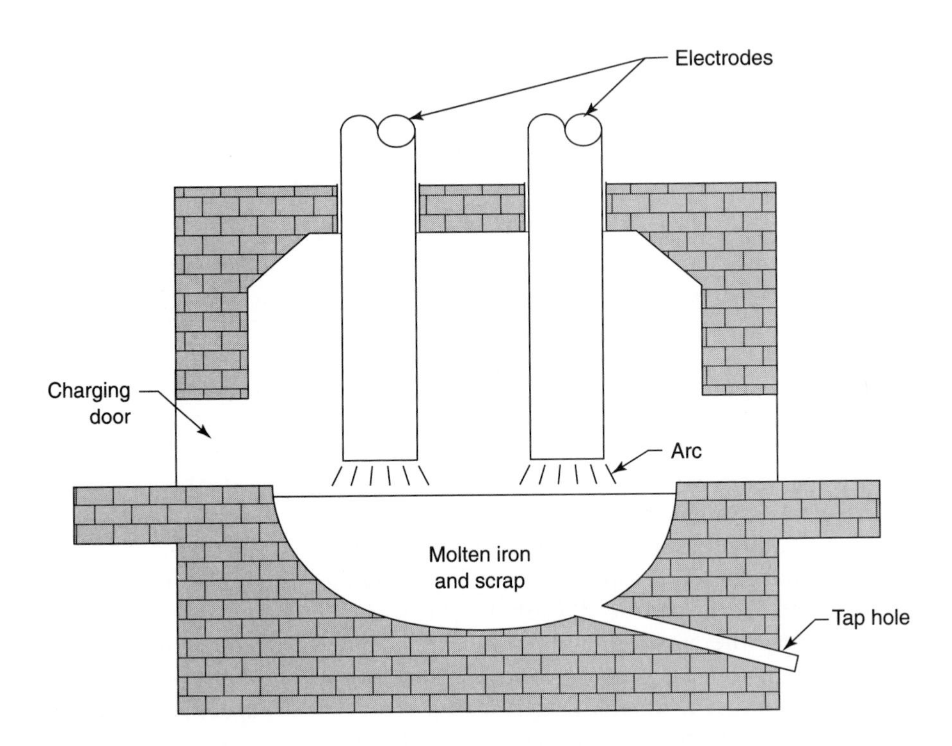

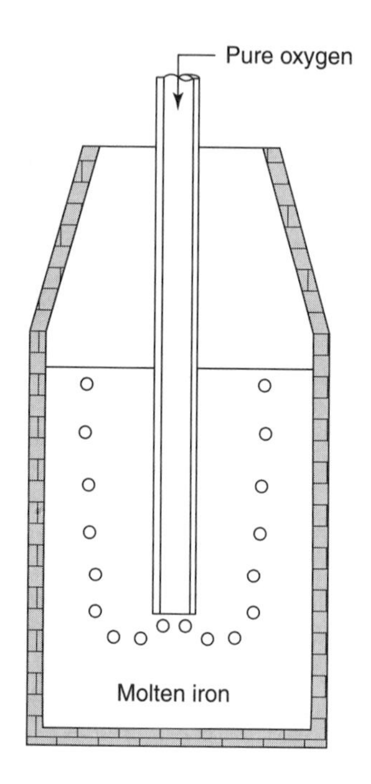

Figure 2-62. Oxygen lance converter.

Nomenclature of Steels

Steels differ by the amount of carbon and the alloying elements in them. Therefore, the nomenclature, or the way in which steels are named, must indicate the amount of carbon and the alloying elements present. Many companies have their own systems. For the compositions of these proprietary nomenclature systems you will have to consult the individual companies. However, the American Iron and Steel Institute (AISI) and the Society of Automotive Engineers (SAE) have developed a system that has become an adopted standard. The AISI/SAE system uses a four-number system. *The first two numbers of the designation are a code for the alloying elements present.* A 10xx steel is a plain carbon steel with only iron and carbon present. (Actually almost all steels have about 0.5% manganese alloyed with them to improve their ductility.) Table 2–3 gives the compositions for the various steels.

The last two numbers indicate the percent carbon in hundredths of a percent. Thus a 1040 steel is a plain carbon steel with 0.40% carbon. Likewise a 4120 is a chrome-molybdenum steel having 0.2% carbon. Steels are traditionally divided into three categories depending on carbon content:

1. Steels having less than about 0.25% carbon are considered low-carbon or mild steels.
2. Steels having between 0.25% and 0.55% carbon are the medium-carbon steels.
3. Steels having more than 0.55% carbon are high-carbon steels.

It must be admitted that these are arbitrary division points and are somewhat loosely used. Although steel can technically have up to 2% carbon, most commonly used steels have a carbon content of less than 1%. There are exceptions to this. In steels that have a five-number designation, the last three numbers indicate the percent carbon. These five-number steels have carbon contents of more than 1%.

Stainless steels have high concentrations of chromium and nickel and smaller amounts of copper, molybdenum, tungsten, and other elements. Stainless steels are divided into three categories: austenitic, martensitic, and ferritic stainless steels.

Austenitic stainless steels are face-centered cubic and cannot be hardened except by cold working. These steels withstand severe stresses at high temperatures but are not very resistant to corrosion by sulfurous gases. For instance, they would not be a good choice of steel for use in coal-fired boilers because of the high sulfur content of the coal. Austenitic stainless steels are nonmagnetic. These steels are designated by a three-digit numbering system and are the 200 and 300 series stainless steels.

Martensitic stainless steels are easily hardened by heat treatments. These steels have as much as 16% chromium with few other alloying elements. They are designated as 400 or 500 series stainless and are magnetic.

Ferritic stainless steels are designated the 405, 430, and 446 stainless steels. These are low-carbon steels (carbon content not more than 0.2%) and are magnetic. They are used particularly in furnace parts.

Many stainless steels are difficult to weld and are best welded by gas tungsten arc welding (GTAW).

Table 2–3. Steel Nomenclature

AISI/SAE Number	Type of Steel	Alloying Elements (%)
10xx	Plain carbon	None (0.4 manganese)
11xx	Free machining	0.7 manganese, 0.12 sulfur
13xx	High manganese	1.6–1.9 manganese
2xxx	Nickel steels	3.5–5.0 nickel
3xxx	Nickel-chromium	1.0–3.5 nickel, 0.5–1.75 chromium
40xx	Molybdenum	0.15–0.3 molybdenum
41xx	Chrome-molybdenum	0.8–1.1 chromium, 0.15–0.25 molybdenum
43xx	Nickel-chromium-molybdenum	1.65–2.0 nickel, 0.4–0.9 chromium, 0.2–0.3 molybdenum
46xx	Nickel-molybdenum	1.65 nickel, 1.65 molybdenum
5xxx	Chromium	0.4 chromium
61xx	Chromium-vanadium	0.5–1.1 chromium, 0.10–0.15 vanadium
81xx	Nickel-chromium-molybdenum	0.2–0.41 nickel, 0.3–0.55 chromium, 0.8–0.15 molybdenum
86xx	Nickel-chromium-molybdenum	0.4–0.7 nickel, 0.4–0.6 chromium, 0.15–0.25 molybdenum
92xx	Silicon	1.8–2.2 silicon

NONFERROUS METALS

Aluminum

Aluminum is one of the most abundant metals in the surface of the earth. It is estimated that 8.1% of the earth's crust is aluminum. All clay contains aluminum. The most used ore of aluminum is bauxite, which has the chemical formula AlO (OH).

Until a little over a century ago, extracting metallic aluminum from its ore was a very difficult, costly, and complicated process. It is recorded that until the late 19th century, aluminum cost as much as $540 per pound.

While Charles Martin Hall was still in college, he became determined to develop an inexpensive process whereby aluminum could be extracted from its ores. Within two years Hall perfected the technique. The **Hall process** involves dissolving the crushed bauxite in *cryolite,* which has the chemical formula Na_3AlF_6. The dissolved aluminum can then be separated from the ore by electrolysis. Large carbon electrodes are lowered into the mixture of cryolite and bauxite, which has been heated to 1800°F. A large direct current on the order of 100,000 amperes is passed through the solution to separate it from the nonmetallic elements. The molten aluminum can be drawn off in large ladles, then cast into ingots that weigh about 80 pounds each.

The process of purifying the ore has been modernized, to improve the yield and quality of aluminum, by what is called the **Bayer process,** which is applied before the Hall process is used. This process involves mixing the crushed and washed bauxite (or other) ores with lime (CaO) and soda ash (sodium carbonate, Na_2CO_3). This converts the aluminum ore to sodium aluminate ($Na_2Al_2O_4$), which is washed and filtered to remove the impurities, called the "red mud," and dried. Upon heating the sodium aluminate to about 2000°F it is converted into alumina (aluminum oxide or Al_2O_3). The Hall process can then be carried out to extract the aluminum.

The major drawback to the production of aluminum is the huge electric power requirement. Aluminum plants must be located near sources of cheap power. More than 20,000 kilowatts of electricity are required to produce a ton of aluminum.

The properties of aluminum make it an ideal metal for the fabrication of many objects. It has a specific gravity of 2.7, a tensile strength of between 10,000 and

80,000 psi depending on its alloying elements, a modulus of elasticity of 10.6 million psi, and an electrical conductivity that is about half that of copper. Although aluminum is extremely active chemically and oxidizes very rapidly in air, its oxide sticks to its surface, thus protecting the rest of the piece from further oxidation. Aluminum parts always have a dull sheen to them. That is the one-molecule-thick aluminum oxide protective layer on the surface. Aluminum parts should not be polished or shined; they will simply reoxidize very quickly to their dull finish.

Nomenclature of Aluminum

Aluminum is often used in the form of an alloyed metal. The alloying elements improve the strength, hardness, and other properties of the metal while not detracting significantly from its low density and other desirable characteristics. The Aluminum Association has developed a system for designating aluminum alloys. Aluminum alloys are usually designated by a four-digit system such as 6016 or 7075 aluminum. The first number indicates the major alloying element. A "1" indicates a 99% or more pure aluminum. A "2" as the first digit indicates copper. Table 2–4 lists the other alloying elements.

The second digit indicates the modifications of the original specification for the alloy. The last two digits are used to indicate the purity of the aluminum over 99% or to identify other alloying elements within the formula. Many handbooks of metallurgy have tables outlining the compositions of these specific alloys.

Table 2–4. Aluminum Alloy Designations

Designation	Major Alloying Element
1xxx	None, 99%+ pure aluminum
2xxx	Copper
3xxx	Manganese
4xxx	Silicon
5xxx	Magnesium
6xxx	Magnesium and silicon
7xxx	Zinc
8xxx	Other
9xxx	Unused as yet

Magnesium

Magnesium is noted for its light weight. It is the eighth most abundant element in the earth's crust. Magnesium is found in many minerals but its primary source for industrial purposes is **seawater.** Surprisingly, seawater contains 1272 grams of magnesium per metric ton. A little arithmetic will show that this converts to about a pound of magnesium per 100 gallons of seawater.

The recovery of the magnesium from seawater is simple in concept. The seawater is filtered through a bed of lime [calcium hydroxide or $Ca(OH)_2$] to convert the magnesium to magnesium hydroxide [$Mg(OH)_2$], which is insoluble in water and will form a precipitate. The precipitate can then be filtered out of the solution. The magnesium hydroxide is then reacted with hydrochloric acid to form magnesium chloride, which is soluble in water:

$$Mg(OH)_2 + 2\,HCl \rightarrow MgCl_2 + H_2O$$

The magnesium chloride solution is then thickened by evaporators and sent through electrolytic cells. The electrolysis produces the magnesium metal and chlorine gas, which can be reacted with water to form hydrochloric acid, which then can be recycled through the process.

The main problem with the production of magnesium is similar to that of aluminum: The process requires a tremendous amount of electricity. Further, magnesium plants need to be near the oceans.

Magnesium is so chemically reactive that it will ignite with an extremely high heat output when heated close to its melting point. For this reason, care must be taken when manufacturing parts from pure magnesium to prevent the magnesium from igniting. Large parts made of magnesium are difficult to ignite, but metal cuttings, dust, and even lathe turnings will ignite very easily. All manufacturing of magnesium requires that coolants be used on all parts being manufactured, that machinery be kept clean and clear of cuttings and dust, that personnel be trained in the proper handling of magnesium, and that stringent safety regulations be enforced. Even the building must be cleaned often to prevent particles of magnesium from collecting in hard-to-reach places. After all, magnesium is used in incendiary bombs.

Nomenclature of Magnesium

More than half of all magnesium products are made from magnesium alloys. The American Society for Testing and Materials (ASTM) and the Magnesium Association have developed a nomenclature for magnesium alloys. This system normally consists of two letters followed by two numbers then another letter. The first two letters represent the major alloying elements in descending percentage order. The numbers represent the percent of the alloying elements, and the last letter indicates the modification of the formula. An AZ81A magnesium would then indicate that the alloying elements are aluminum and zinc with 8% aluminum and 1% zinc. The A indicates that there have been no modifications of this formula. Again, metal handbooks usually include tables of the various magnesium alloys currently available.

Copper

Copper is one of the oldest metals used by humans. It can be found in its pure nugget form. It was discovered on the island of Cyprus or Kypros, from which it got its Latin name *cuprum.* Copper artifacts have been found dating to about 5400 B.C.

There are many copper ores but not many of them are commercially profitable. Further, many of the ores are sulfides that release large amounts of sulfur dioxide in their refining process. Therefore the oxide ores are more desirable. The most common commercially used ores are **cuprite, malachite,** and **azurite.** Chalcopyrite, bornite, chalconite, covelite, and chrysocolla can also produce copper. Turquoise contains copper but its copper content is not sufficient for economical production.

Copper is refined through a many-step process. After the ore is mined, it is concentrated to remove the soil, rock, and other undesirable elements, and then "roasted" either by allowing it to remain in the sun or by heating it in an oven to convert the sulfides to oxides. (This is the step that produces sulfur dioxide.) The ore is then put in a smelting furnace where it is heated to 2600°F. This drives off the oxygen, leaving the copper. The product of this step is called *matte copper,* which is only about 30% pure. This metal may contain metals such as gold, silver, and other impurities. The matte copper is then placed in a converter where air is blown through it to remove the impurities. The resultant product is cast into ingots. The ingots do contain air trapped on the surface. For this reason, the ingots are called *blister copper,* which is about 99% pure.

Copper is primarily used for four purposes: electrical conduction, heat conduction, alloying, and applications requiring a material that will not easily corrode. It must be very pure for these applications. The blister copper is therefore remelted and cast into copper anodes and placed in an electroplating cell. The copper is plated onto large cathodes as 99.9% pure copper. If the copper is to be used in electrical wire, it is purified even further by melting it in a reverberatory furnace in an oxidizing flame to prevent sulfur from being reabsorbed from the fuel. An old method was to melt the copper with a wood log or "pole" to keep the oxygen content low. Copper produced by this method is called *electrolytic tough pitch* or *ETP copper.*

Copper is the best commercially available electrical conductor. Therefore, much of the copper produced today is drawn into wire. Copper sheet is used for architectural decoration and roofing since it does not corrode easily. The electronics industry uses copper in printed circuit boards, heat sinks for transistors, and other applications. Copper is also alloyed with other metals to form bronze.

Bronze and Brass

Bronze was once defined as a copper-tin alloy. Today the definition has been broadened to "copper alloyed with any other metal," leading to aluminum bronzes, nickel bronzes, gold bronzes, beryllium bronzes, and so on. Beryllium bronzes are used in containers and mixing machines in the explosives industry, since they will not give off a spark if struck and have good corrosion resistance.

The particular bronze consisting of copper and zinc is known as *brass.* There are many brasses depending on the percent of zinc and other alloying elements. Naval and admiralty brasses are used for their corrosion resistance. Lead brass is often used in bushings and bearings in machinery. Table 2–5 shows the composition of some of the more commonly used brasses.

Table 2–5. Compositions of Brass

Type	% Copper	% Zinc	% Other
Red	85	15	
Low	80	20	
Yellow	65	35	
Muntz metal	60	40	
Lead brass	65	32	3 lead
Admiralty brass	70	29	1 tin
Cartridge brass	70	30	
Naval brass	60	39.25	0.75 tin
Free machining	60	39	0.75 tin 0.25 lead

Titanium

Copper is one of the oldest metals to be used by people; titanium is one of the newest. Titanium was discovered by an English clergyman, William Gregor, in 1701 but it was not purified until 1910. Even though titanium is the ninth most abundant element on earth, occurring in the minerals rutile, ilmenite, sphene, and many others, it has only been in the last half of the 20th century that it has been used significantly in manufacturing.

Rutile, the major titanium ore, is titanium oxide (TiO_2). It is not a simple process to purify the titanium from the ore. Rutile is finely ground and the titanium oxide then separated by electrostatic and gravitational separators. The titanium oxide is reacted with chlorine gas to produce titanium chloride:

$$TiO_2 + 2Cl_2 \rightarrow TiCl_4 + O_2$$

Titanium chloride reacts with moisture to produce a dense white cloud of "smoke," which can be used in skywriting or tracer bullets. To reduce titanium chloride to a pure metal, a stainless steel tank (called a "bomb" in the industry) is filled with sodium or magnesium metal and welded shut. The titanium chloride is run through the tank at about 800°F for approximately 48 hours. Because the magnesium or sodium is more chemically active than the titanium, it will replace the titanium in the chloride. The process is simply one of trading the magnesium or sodium for the titanium:

$$TiCl_4 + \text{Na or Mg} \rightarrow Ti + \text{NaCl or } MgCl_2$$

The titanium collects as a spongy metal on the inside of the stainless steel tank. After the reaction is completed, the tank is cut open and the titanium cleaned out of the tank. The spongy titanium is then remelted in a vacuum arc furnace and cast into ingots. Because this is a batch process, and must use either magnesium or sodium, which has already been refined, the recovery of titanium is an expensive process.

Titanium metal has some very unique properties that make it ideal for many applications. Titanium has a tensile strength of greater than 95,000 psi, which is as strong as high-quality steels, yet it has a specific gravity of only 4.54. Titanium is therefore roughly twice as heavy as aluminum. Like aluminum, titanium forms an oxide coating on its surface that protects it from further corrosion. Titanium will burn in nitrogen as well as oxygen.

Titanium alloys retain their strength at high temperatures where other metals lose strength. Titanium is a hard metal and is nonmagnetic. Some of the advances in technology would have been impossible without the use of titanium. The SR-71 Blackbird, which can fly at speeds over three times the speed of sound and at altitudes in excess of 85,000 feet depends on titanium to withstand the heat generated by its flight. The modern jet engine would be impossible without the titanium steel alloys used in the rotor blades. The supersonic transport plane, the Concord, uses several tons of titanium and titanium alloys. It is even ironic that the deep-diving submarine "Alvin," which was used to locate the wreckage of the *Titanic* was made largely of titanium. After all they both were named for the gods, "Titans."

Titanium also has many other uses. High-quality white paint or enamel uses titanium oxide as its pigment. Artificial bone transplants such as artificial hips, knees, and other joints are now being made of *sintered* titanium. Sintering is a process in which the metal is powdered then packed into a mold and the particles fused together. (See the chapter on Powder Metallurgy.) Sintered titanium is a very strong product and very light in weight. Because sintered metals have many voids between the particles, the bone will grow into the rough surface between the particles. This allows the bone to literally "weld" to the metal.

One interesting use of titanium is in the alloy **Nitinol.** Nitinol is 54.5% nickel and 45.5% titanium. It has a tensile strength of 110,000 psi. Moreover, it is a metal with a memory. A part can be bent into shape from a wire, sheet, or bar then heated to about 1000°F and slowly cooled. After cooling the Nitinol can be bent out of this shape, but when heated again, often in hot water, it will return to its original shape with a force of about 100,000 psi. New uses for Nitinol are still being discovered. Dentists are now using Nitinol wire in "braces" for straightening teeth. The possible uses of Nitinol are infinite and this promises to be a very exciting field in the future.

Titanium has a few drawbacks. It is more expensive than the more traditional metals. Further, titanium is difficult to machine. It is harder than many tool bits and dulls cutting tools very rapidly. In casting titanium parts, the metal must be melted in a vacuum often using remote control or robots to handle the molds. Titanium metal castings can be very brittle. Critical parts must be x-rayed to detect inclusions and other flaws. In spite of these flaws, titanium continues to be the "miracle metal" of the 20th century.

Several basic concepts are needed to understand the concepts of matter:

1. *Atom:* the smallest division of an element
2. *Proton:* a positively charged particle in the nucleus of an atom
3. *Neutron:* a neutral particle in the nucleus of an atom
4. *Electron:* a negatively charged particle in orbits around the nucleus of the atom
5. *Isotope:* atoms of an element with varying numbers of neutrons
6. *Chemical bonds:* the attractive forces between molecules that allow them to form molecules; there are ionic, covalent, metallic, and van der Waal bonds
7. *Gas:* that state of matter which has no structure
8. *Solid:* that state of matter which has a definite, long-range crystal structure
9. *Liquid:* that state of matter which has a random structure
10. *Crystal:* an orderly array of atoms, molecules, or ions
11. *Slip:* a direction in which atoms slide over each other
12. *Crystal systems:* the seven different ways in which atoms or molecules can arrange themselves

Several definitions are necessary to understand the properties of materials:

1. *Stress:* the load per unit cross-section area
2. *Strain:* the change in length divided by the original length
3. *Tension:* forces acting directly opposite and outwardly on a piece of material
4. *Compression:* forces acting directly opposite and inwardly on a piece of material
5. *Torsion:* forces twisting a material in opposite directions
6. *Shear:* forces slightly offset but parallel, acting on a piece of material in opposite directions
7. *Hardness:* the resistance to surface deformation

8. *Impact:* the sudden application of energy to a material
9. *Creep:* the deformation of a material under stress over a period of time
10. *Fatigue:* the failure of a material due to repeated or cyclic stresses

Materials can be tested by nondestructive tests. Nondestructive tests include:

Testing by sampling

Testing of analogs

X-ray scanning and photography

Sonic testing

Dye penetrants

Electrical resistance measurements

Magnetic tests

Visual inspection

To design and manufacture a part, we must know the properties of the materials used. Among those properties are the following:

1. *Tensile strength:* the maximum stress a material will withstand in tension
2. *Yield strength:* the engineering design strength of a material
3. *Compressive strength:* the maximum stress a material will withstand in compression
4. *Torsion strength:* the maximum stress a material will withstand in torsion
5. *Shear strength:* the maximum stress a material will resist in shear
6. *Modulus of elasticity:* the change in stress divided by the change in strain in the elastic region
7. *Density:* the mass per unit volume of a material
8. *Specific gravity:* The ratio of the density of a material to the density of water (in the same units)

The following hardness tests are commonly used in industry:

1. Rockwell
2. Brinell
3. Vickers
4. Tukon (Knoop)
5. Scleroscope

Hardness tests used on plastics, rubber, glass, and ceramics include the following:

1. Shore durometer
2. Barcol impressor

The time-dependent properties of a material are as follows:

1. Impact
2. Creep
3. Fatigue

Defects that can affect the properties of a material include:

1. *Vacancies:* a missing atom, ion, or molecule from a lattice
2. *Interstitial:* a particle located between the lattice points in a crystal
3. *Substitutional:* a different atom, ion, or molecule than the parent matrix
4. *Hole:* a missing electron in the crystal lattice
5. *Electron (imperfection):* an extra particle electron in the crystal lattice
6. *Line dislocation:* an extra partial plane of particles in the crystal or grain
7. *Screw dislocation:* the rotation of one crystal plane on top of another
8. *Twinning:* the formation of a mirror image of the parent crystal in a material
9. *Inclusion:* a macroscopic piece of foreign matter in the crystal

A few materials of manufacturing were discussed in this chapter. These included:

1. Iron and steel
2. Aluminum
3. Magnesium
4. Copper
5. Titanium

Chapter 3

Measurements in Manufacturing

Measurement has been important to people ever since they began to organize into societies. As business and trade increased, measurements were developed in order to determine the quantities of a product. Ownership of land demanded sophisticated measurements by which to define the boundaries of properties. Some of the early attempts at defining units of measurement are laughable by today's standards. Wheat, rice, and other grains were once sold in the marketplace by the *palm,* the amount of grain one could scoop up with the hand with the fingers and thumb closed. The ancient pyramids and other structures were built using the *cubit* as the unit of length. A cubit is the distance from one's elbow to the tip of the fingers. The *inch* was the length of three barley corns placed end to end. The heel-to-toe length was the *foot.* The *yard* was the length from the tip of one's fingers to the nose. The *span* was the length from fingertip to fingertip of one's outstretched arms. This length was adopted as the *fathom* for measuring the depth of water. The *rod* (now 16.5 feet), used to measure land, was set in the 16th century as "the length subtended by the heel to toe placement of the left feet of the first sixteen men as they left church on Whitsunday morning." Even the *mile* was developed by the ancient Romans as "millia passos" or one thousand paces. Since a pace (two steps) is roughly one's height, it is easy to see why a short Roman surveyor would place his "mile posts" approximately 5000 feet apart. During the reign of Queen Elizabeth I, the mile was fixed at 5280 feet. Weights were measured in *stones* (14 pounds equal one stone). Such units as the *hogshead* (63 gallons), *furlong* (1/8 of a mile, used at racetracks), and other old units are still used occasionally.

The shortcomings of these measurements are obvious. No two people could get the same results when measuring anything. It is remarkable that many of the structures built with these crude measurements are as symmetric and as square as they are.

Ever since the Industrial Revolution, which eventually brought about the use of interchangeable parts and mass production, the importance of accurate measurements with fixed and precise standards has been increasing in manufacturing. To accomplish the standardization of measurements, several governments set up organizations that were charged with maintaining the standards for all types of measurements. The world standard for metric measurements is in Paris, France. In the United States, the National

Institute of Standards and Technology (NIST), located in Gaithersburg, Maryland, just north of Washington, D.C., houses the standards for the country. The NIST was founded by Congress in 1901 as the National Bureau of Standards, but the name was changed in 1989.

The English (or inch) system of units is still used in the United States. We still use the inch, foot, yard, rod, mile, pound, pint, quart, gallon, etc., but they have been standardized by the NIST. The English system of weights, ounces, pounds, etc., is now derived from the metric standard of the international kilogram.

Several concepts about measurements need to be understood by all people connected with manufacturing. The definitions of the terms used in measurements, dimensional or unit analysis, the metric system of measurements, limits of accuracy, and some necessary statistical concepts are the focus of this chapter.

DEFINITIONS OF TERMS USED IN MEASUREMENTS

There is a difference between a "mistake" and an "error." *Mistakes* result from using incorrect methods, making faulty judgments, incorrect calculations, inattention to detail, carelessness, and so forth. Mistakes can be eliminated by repeating the operation correctly. Errors are inaccuracies in measurements due to the limitations of the measuring instruments. If one multiplies 3 times 6 and writes down 21, that is a mistake; it can be corrected. However, if you measure the thickness of a page in this book and get 0.0082 inch, and a second reading gives you 0.0078 inch, then you have an *error* in measurement. Further readings will also differ.

Nothing can be measured exactly. If one measures the length of a table with a yardstick, the smallest measurement that can be accurately read may be one-eighth of an inch. This limits the accuracy of the measurement. If a measurement demands finer units, then an instrument with finer divisions must be used. But even if a measuring instrument calibrated to one-millionth of an inch is used, there still could be an error of one-millionth of an inch. Errors can be minimized, but they cannot be eliminated. In fact, the law in science that prescribes the limits of accuracy in measurements is known as the Heisenberg uncertainty principle.*

Measurement (as the term will be used here) is simply the act of defining the measurable properties of some object. Measurable properties include diameter, length, hardness, flatness, roughness, electrical voltage and current, volume, weight, and many others.

Inspection is the examination of a part to determine whether or not it meets the requirements of the designer or engineer. Does it have the correct finish, hole size, shape, diameter, length, weight, etc.? Many of these determinations are found by taking measurements.

Testing is the determination of how a product will perform. Materials are tested to determine their tensile strengths, hardness, density, and other properties. Finished parts and products are tested to determine if they perform to design specifications.

Basic size is the size from which tolerances are measured. A hole in a piece of metal might have a specification of 1.875 ± 0.002 inch. The 1.875 is the basic size.

Nominal size is the size by which a measurement is usually called. For instance, a hole with a diameter of 1.875 ± 0.002 inch could be called a 1⅞-inch hole. In a lumber yard a finished board having a nominal size of 2 by 4 inches actually has a basic size of 1.5 by 3.5 inches.

Tolerance refers to the total range of acceptable values for a measurement. Parts used in manufactured products must be made to tolerances that will permit the interchanging of parts and allow the complete mechanism to work properly. For the hole specified as 1.875 ± 0.002 inches, the ± 0.002 inch is the specified tolerance for the hole.

* The Heisenberg uncertainty principle used in physics is based on the fact that electrons must change orbit in order to emit a photon or particle of "light" energy ("light" in this sense also includes everything from heat to radio waves to x-rays and gamma rays). This implies that the electron must be moving in order to see anything and therefore the limit of the accuracy of measurement would be the electron jump. Formally stated the Heisenberg uncertainty principle is "The product of the uncertainty in the position of a body at some instant and the uncertainty in its momentum at the same instant is equal to or greater than $h/2\pi$" (where h is Planck's constant = 6.63×10^{-34} joule-sec). We leave further discussion of this principle to a physics course.

Accuracy is how closely a measurement comes to the true value. It is often difficult to know what the true value of any measurement actually is. For instance, if one is trying to measure the speed of light, we can never know exactly what that speed is. All we can do is take an average of the results of many experiments used to determine the speed of light and use that average as an *accepted* value. It may or may not be accurate.

Precision is the term for how close together are the many readings of the same measurement. For example, if a baseball pitcher threw three pitches to exactly the same spot but outside the strike zone, he would have good precision but no accuracy (unless he was intentionally walking the batter). Conversely, if the pitcher threw three pitches, one high, one low, and one outside the strike zone, he might have good accuracy, in that he was near the strike zone, but no precision. To be successful, the pitcher needs to be able to throw the pitches all in the strike zone; that is, he needs both accuracy and precision.

A more mathematical example would be as follows: Suppose five people weigh the same electric motor and get readings of 4.25, 6.48, 5.37, 4.82, and 7.23 pounds. Certainly these readings are not very close together and therefore they are lacking in precision. On the other hand, if the readings came in at 5.30, 5.31, 5.29, 5.30, and 5.31 pounds, the readings would be much more precise.

Measurements may be accurate but not precise, or precise without being accurate. If five measurements of the length of a field were taken with an improperly calibrated distance meter, the readings might be very close together, yet totally wrong. Conversely, five different people could measure the same field with properly calibrated instruments, but differ significantly in their findings.

Validity simply means "does the measurement actually measure what it is supposed to be measuring?" For instance, using a yardstick to measure the length of a suitcase would be a valid measurement. A yardstick measures length. But often we cannot measure things directly. Determining the amount of light emitted by a light bulb by means of the amount of electric power (watts) it consumes may not be a valid measurement since much of the electric power is emitted as heat, not light.

The *reliability* of a measurement is its repeatability. Performing a speedometer check in an automobile by measuring the time required to travel between successive mileposts might not be reliable. The speed of the automobile as registered on the speedometer is an indirect measurement based on the revolutions per minute of the drive shaft or of some gear in the transmission. The indicated speed can change from day to day as the air pressure in the tires changes or treads wear down.

Measurements can be *objective* or *subjective.* In objective measurements, the person making the measurements does not influence the results obtained. Subjective measurements depend on the person making them. Measuring the diameter of a steel bar using a digital micrometer would be a much more objective test than measuring the amount of alcohol in a wine by tasting it. Every effort must be made to keep measurements as objective as possible.

DIMENSIONAL ANALYSIS

It is admitted that this is not a course in mathematics. However, individuals involved in manufacturing are often required to solve equations, use formulas, and perform mathematical calculations. There are a few mathematical techniques that, if properly learned and used, will help the individual to correctly solve these problems. One of these techniques is known as *dimensional analysis* or *unit analysis.*

With one exception, all numbers carry units with them. One buys 10 pounds of potatoes, one dozen eggs, or 5 gallons of fuel. A car can be driven at 55 miles per hour. A person's height can be measured either as 5 feet 10 inches, or 70 inches. In performing calculations, the units as well as the numbers must be consistent. In an equation, the units on one side of the equals sign must always be the same on the other side. In other words, the units can be treated just like numbers in an algebraic equation. Feet times feet equals square feet. Miles divided by hours gives miles per hour.

Therefore, when writing a number, always place the units with that number. The units can then be multiplied, divided, added, subtracted, raised to a power, etc., just the same as the numbers. For instance, if a room measured 10 feet by 12 feet, the floor area of the room would be

$$10 \text{ ft} \times 12 \text{ ft} = 120 \text{ ft}^2$$

If one makes the units on the left side of the equation the same as the units on the right side, then all that is left to do is the arithmetic on the calculator. Get in the habit of writing the units down with any number. Dimensional analysis will prevent a person from making mistakes such as dividing when one should be multiplying or adding unlike units such as gallons to pounds.

A few more examples of the use of dimensional analysis are in order. For instance, if one drives at an average speed of 55 miles per hour for 3.5 hours, the distance driven would be

$$55 \text{ mi/hr} \times 3.5 \text{ hr} = 192.5 \text{ mi}$$

The word *per* always means "divided by." Miles per hour means miles divided by hours. Since there are hours in the numerator of one factor and hours in the denominator of the other, the hours/hours divide out, leaving only miles.

One must also remember that only like units can be added or subtracted. Inches cannot be added to feet until either the feet are converted to inches or the inches converted to feet. To add 7 feet to 9 inches and get 16 is ridiculous. What are the units? The proper method would be

$$7 \text{ ft} \times 12 \text{ in./ft} + 9 \text{ in.} = 84 \text{ in.} + 9 \text{ in.} = 93 \text{ in.}$$

or

$$7 \text{ ft} + \frac{9 \text{ in.}}{12 \text{ in./ft}} = 7 \text{ ft} + 0.75 \text{ ft} = 7.75 \text{ ft}$$

Changing from one set of units to another becomes very simple if you use dimensional analysis. To change 35 "miles per hour" to "feet per second," use this formula:

$$\frac{35 \dfrac{\cancel{\text{mile}}}{\text{hr}} \times 5280 \dfrac{\text{ft}}{\cancel{\text{mile}}}}{60 \dfrac{\cancel{\text{min}}}{\text{hr}} \times 60 \dfrac{\text{sec}}{\cancel{\text{min}}}} = \frac{184{,}800 \dfrac{\text{ft}}{\cancel{\text{hr}}}}{3{,}600 \dfrac{\text{sec}}{\cancel{\text{hr}}}} = \frac{184{,}800 \text{ ft}}{3{,}600 \cancel{\text{hr}}} \times \frac{\cancel{\text{hr}}}{\text{sec}} = 51.3 \frac{\text{ft}}{\text{sec}}$$

Notice how the *units* divide out. Remember that in dividing by a fraction (such as minutes/hour) in the denominator, the denominator is inverted and the problem converts to a multiplication problem. Therefore, miles/hour divided by minutes/hour equals miles/minute.

Many times problems can be solved without equations just by working with the units of the definitions of the terms involved. For instance, one horsepower is defined as the power required to lift 550 pounds a height of one foot in one second. That's 550 foot-pounds/second. The power required, therefore, to lift a 1000-pound weight 10 feet in 15 seconds would be

$$\text{HP} = \frac{\dfrac{1000 \text{ lb} \times 10 \text{ ft}}{15 \text{ sec}}}{\dfrac{550 \text{ ft-lb}}{\text{sec/HP}}} = 1.2 \text{ HP}$$

The only exception to the rule that all numbers have units is for quantities such as π, $\sqrt{2}$, etc., which are often found in equations. These numbers are called *numerics* and have no units but must be included in the equations to produce the correct answer. For example, the area of a circle having a radius of 5 inches is

$$\begin{aligned} A &= \pi r^2 \\ &= 3.1416 \times (5 \text{ in.})^2 \\ &= 3.1416 \times 25 \text{ in.}^2 \\ &= 78.54 \text{ in.}^2 \end{aligned}$$

PROBLEM SET 3-1

1. If the tires on an automobile are checked with a standard tire gauge having graduations at 5-psi increments (psi = lb/in.^{2}), evaluate this gauge for accuracy, precision, reliability, validity, and subjectivity.
2. Evaluate an automobile speedometer for accuracy, precision, reliability, validity, and subjectivity.

Work the following problems using dimensional analysis. Make sure the units are made equal on both sides of the equation.

3. If an automobile is driven a distance of 180 miles in 2 hours and 30 minutes, what is the average speed of the car in miles per hour?
4. If a lathe is run at a speed of 500 rpm (revolutions per minute), how many revolutions will it make in 18 minutes?
5. If an automobile is driven at a speed of 55 miles per hour, what is the speed in feet per minute?
6. If a wire is being wound on a spool at 100 feet per minute, what is that rate in inches per second?
7. A feed rate on a saw is 30 inches per second. What is the feed rate in feet per minute?
8. Two men measure the length of a field, each starting from opposite ends and measuring toward the center. One man uses a yardstick and measures a distance of 32 yards, 2 feet, 3 inches. The other uses a tape measure and records 44.5 feet. What is the total distance in feet and inches?
9. There are 16 ounces in a pound. How many ounces would be in 44.5 pounds?
10. A rectangle measures 78 inches by 4.3 feet.
    a. What is the area in square feet?
    b. What is the area in square inches?
11. One horsepower is the ability to lift 550 pounds a height of one foot in one second or one horsepower is equal to 550 ft-lb/s. How many ft-lb/s are there in 32 horsepower?
12. If an elevator is to raise a total weight of 5000 lb a height of 30 ft in 20 sec, what horsepower motor would be required? (Refer to Problem 11.)
13. One horsepower is equal to 746 watts of electrical power. How many watts are required to run a 25-hp motor?
14. If 160 tons of coal are to be lifted 20 feet on a conveyor belt per hour, what horsepower motor is required to lift the coal? (Neglect the friction and other power requirements of running the belt.) How many kilowatts would be required? (1 kilowatt = 1000 watts)
15. If the cost of electricity is 9 cents per kilowatt-hour, how much would it cost to run a 0.5-hp pump motor per month if the motor is run 24 hours per day? (Refer to Problem 13.)
16. If it takes 6 sec for a 150-lb athlete to climb a 20-ft rope, how many horsepower does he generate?

USE OF FORMULAS

The technician must be acquainted with several geometric shapes, including the circle, triangle, rectangle, cylinder, and sphere. You should know how to calculate the areas, perimeters (distance around), surface areas, and volumes of these shapes. Formulas for simple shapes and other commonly used items should not only be memorized but also thoroughly understood. The key to the use of formulas is (1) to select the proper formula for its application and (2) to make sure the units are consistent on both sides of the equation. The formulas most needed are listed in Appendix C.

Example

Suppose a sheet of metal had two parallel sides but the other two sides were not parallel. The lengths of the two parallel sides were 78 and 92 inches, respectively, and the perpendicular distance between them was 54 inches. What is the area of the sheet?

Solution

The sheet is in the shape of a trapezoid. The formula for the area of a trapezoid is

$$A = \tfrac{1}{2}(L + l)h$$

where L is the length of the longer side, l is the length of the shorter side, and h is the height or distance between the parallel sides. Putting the numbers into the formula in the proper places, the area can be calculated as

$$\begin{aligned} A &= \tfrac{1}{2}(92 \text{ in.} + 78 \text{ in.})54 \text{ in.} \\ &= 4590 \text{ in.}^2 \end{aligned}$$

Notice that all of the units in this case were in inches, so the result is square inches. If the units are not the same, they must be converted to the same units prior to putting them into the equation.

PROBLEM SET 3-2

1. What is the area of a circle having a radius of 9 in.?

2. What is the circumference of a circle having a diameter of 12 in.?
3. What is the circumference of a circle having a radius of 8 in.?
4. What is the area of a circle having a diameter of 14 in.?
5. A shot put ring is 8 ft in diameter. What is its area?
6. What is the volume of a brick having dimensions of 2 in. by 3.5 in. by 12 in.?
7. What is the surface area of the brick in Problem 6?
8. What is the volume of a cylinder having a radius of 3 ft and a height of 8 ft?
9. How many square feet of sheet metal are used to make the cylinder in Problem 8?
10. If a spherical gas tank measures 16 in. in diameter, what is its volume?
11. What is the surface area of the gas tank in Problem 10?
12. A box has the dimensions of 15 in. by 8 in. by 24 in.
    a. What is the volume of the box in cubic inches?
    b. How much sheet metal in square inches would be required to make the box, neglecting any allowance for tabs or fold over?
13. A cylindrical can is 8 in. in diameter and 15 in. high.
    a. What is the volume of the can in cubic inches?
    b. What is the volume of the can in gallons? (*Hint:* There are 232 cubic inches per gallon.)
    c. What is the surface area of the can including top and bottom?
14. What would be the dimensions in inches of a cubic tank that would hold 1 gallon?
15. What would be the diameter in inches of a sphere that would hold 5 gallons?

■ SI METRIC SYSTEM

In 1960, at the Eleventh General Conference on Weights and Measures, held in Paris, a unified system of measurements was adopted. This standard is known as the *Système International* or *SI.* By 1975 Congress committed the United States to the "increasing use of, and voluntary conversion to, the metric system of measurement." Because many countries of the world with whom the United States has commerce will not accept products that are not in the metric system, it is believed by some that the United States will eventually go metric by default. Many industries that trade with foreign countries already use the metric system for their products. Many automobiles, for instance, use metric tools, bolts, and fixtures. For this reason students are encouraged to become well versed in, and be able to use, metric measurements.

The metric system is really much simpler than the inch system of units once you learn the system. In the inch system of measurements, there are 12 inches per foot, 36 inches per yard, and 5280 feet per mile. There is no common factor to the measurements. The word "inch" comes from an earlier word meaning "one-twelfth"; thus there are twelve inches in a foot. Why twelve? Because twelve can be divided into two, three, four, or six parts very easily. The inch can also be divided into halves, quarters, eighths, sixteenths, thirty-seconds, and sixty-fourths on tape measures and yardsticks. The decimally divided inch (i.e., divided into tenths, hundredths, etc.) was first used by Henry Ford in order to obtain the precision required to build the Model-T Ford cars. The foot is also divided decimally for work in surveying and civil engineering.

One nice thing about the metric system is that everything is 10 times or one-tenth the next measurement. The prefixes attached to each of the units of measurement are given in Table 3–1. Thus, ten metres is a dekametre, ten liters is a dekalitre, ten grams is a dekagram, and so on. A thousand grams is a kilogram, whereas one thousandth of a litre is a millilitre. To convert kilograms to grams just multiply by 1000 grams/kilogram:

$$16.2 \text{ kg} \times 1000 \text{ g/kg} = 16{,}200 \text{ g}$$

To convert millilitres to litres, multiply by 1 L/1000 mL:

$$8 \text{ mL} \times 1 \text{ L}/1000 \text{ mL} = 0.008 \text{ L}$$

Length Units

As in the inch system, the metric system has units of length, weight, mass, and volume. The fundamental

Table 3–1. Metric System (SI) Prefixes)

Multiplier			Prefix	Standard Symbol
1,000,000,000,000	or	10^{12}	tera	T
1,000,000,000	or	10^{9}	giga	G
1,000,000	or	10^{6}	mega	M
1,000	or	10^{3}	kilo	k
100	or	10^{2}	hecto	h
10	or	10^{1}	deka	da
1	or	10^{0}	—	—
0.1	or	10^{-1}	deci	d
0.01	or	10^{-2}	centi	c
0.001	or	10^{-3}	milli	m
0.000001	or	10^{-6}	micro	μ (Greek letter mu)
0.000000001	or	10^{-9}	nano	n
0.000000000001	or	10^{-12}	pico	p
0.000000000000001	or	10^{-15}	femto	f
0.000000000000000001	or	10^{-18}	atto	a

metric unit of length is the metre (not "meter"; that spelling is reserved for measuring instruments, such as a voltmeter). It is standardized as "1,650,763.73 times the wavelength of the $2p^{10}$ to the $5d^{5}$ jump of electrons in the krypton-80 orange-red line." For all practical purposes, it is about 39.37 inches. The abbreviation for metre is "m." Some years ago the inch and metric systems were brought into exact correspondence by redefining the length of the inch by a small amount. As a result the inch is exactly 25.4 millimetres.

This makes some conversions approximate, such as 1 millimetre = 0.03937 inch. These conversions are still usable since they are only off in the eighth decimal place.

Volume Units

The unit of volume is the *litre.* The litre is 1000 cubic centimetres. The volume of one gram of water is one millilitre, which equals approximately one cubic centimetre. The cubic centimetre and the millilitre differ only after the sixth significant figure. One quart is equal to 0.946 liters.

Force Units

Many important properties of engineering materials involve *force.* Force is treated differently in the inch and metric systems. In the inch system, the *pound* is the basic unit of force. For example, pressure is expressed in pounds per square inch (psi). In the metric system the *gram* is a unit of *mass.* Mass and weight are two different properties. Weight is a force, mass is not. Weight depends on amount of gravity pulling on an object, whereas mass does not. If this distinction is not clear, consider that, on the moon, an astronaut weighs about one-sixth of what he or she weighs on the earth. However, their mass remains the same since mass is a measure of the amount of "stuff" that is in an object. To make things more confusing, in the English (inch) system the unit for both mass and weight is the pound. We have pound force and pound mass, but people do not always distinguish between them. This confusion disappears in the metric system.

In the SI system, the unit of force is the *newton* (N). Therefore, in the SI system, pressure would be expressed in units of newtons per square millimetre, newtons per square centimetre, or newtons per square metre. One *newton per square metre* has been named a *pascal.* The stress discussed in Chapter 2 used the pascal.

Converting force units to mass units is very much like converting peaches to tomatoes. However, if the force of gravity can be assumed to be constant, as it is in most places on the surface of the earth, then a mass of one kilogram would weigh approximately 2.2046 pounds on the surface of the earth. This would make one pound equal to 0.4536 kilogram. In most calculations, we consider the kilogram to equal 2.2 pounds.

The relationship between force and mass is

$$F = ma$$

where F is the force, m is the mass, and a is the acceleration of gravity. The acceleration of gravity in the metric system is very nearly 9.8 metres per second squared. Therefore, a mass of 10 kilograms would have a force or weight of

$$\begin{aligned} F &= 10 \text{ kg} \times 9.8 \text{ m/s}^2 \\ &= 98 \text{ kg} - \text{m/s}^2 \\ &= 98 \text{ N} \end{aligned}$$

The most difficult part about the metric system is getting a "feel" for it. People raised with the English system can visualize a quart, gallon, pound, or yard. They know what a quart of milk looks like, but what about a litre of milk? We must learn to visualize the kilogram, litre, and metre. If you learn three conversion factors, you can do most calculations in the metric system:

$$\begin{aligned} &1 \text{ inch} &&= 2.54 \text{ centimetres} \\ &2.2 \text{ pounds} &&= 1 \text{ kilogram} \\ &1 \text{ quart} &&= 0.946 \text{ litres} \end{aligned}$$

Knowing the metric prefixes and the English equivalents, one can calculate anything else using dimensional analysis. For example, how many quarts are there in 5 litres?

$$\frac{5 \text{ L}}{0.946 \text{ L/qt}} = 5.3 \text{ qt}$$

How many metres are there in 3 miles?

$$\frac{3 \text{ mi} \times 5280 \text{ ft/mi} \times 12 \text{ in./ft} \times 2.54 \text{ cm/in.}}{100 \text{ cm/m}} = 4800 \text{ m}$$

Of course, one soon learns that there are approximately 1.6 kilometres to the mile, but that isn't necessary to do the conversion.

To convert 32 pounds to grams

$$\frac{32 \text{ lb} \times 1000 \text{ g/kg}}{2.2 \text{ lb/kg}} = 14{,}500 \text{ g}$$

By means of dimensional analysis, the student should now be able to convert any measurements in any units to any other units. If a rifle bullet was fired at 3000 feet per second, what is its speed in (a) miles per hour and (b) in metres per minute?

$$\frac{3000 \text{ ft/s} \times 60 \text{ s/min} \times 60 \text{ min/hr}}{5280 \text{ ft/mi}}$$
$$= 2050 \text{ mi/hr}$$

and

$$\frac{3000 \text{ ft/s} \times 12 \text{ in./ft} \times 2.54 \text{ cm/in.} \times 60 \text{ s/min}}{100 \text{ cm/m}}$$
$$= 55{,}000 \text{ m/min}$$

Some conversions from the English (inch) system to the metric system are a little more complex. For instance, if the density of iron is 495 pounds per cubic foot, calculate its density in grams per cubic centimetre. Here, both the numerator and the denominator must be converted:

$$\frac{495 \text{ lb}}{\text{ft}^3} \times \frac{\text{kg}}{2.2 \text{ lb}} \times \left(\frac{\text{ft}}{12 \text{ in.}}\right)^3 \times \frac{100 \text{ gm}}{\text{kg}} \times \left(\frac{\text{in.}}{2.54 \text{ cm}}\right)^3$$
$$= 7.95 \text{ g/cm}^3$$

It is equally important to be able to convert from the metric system to the English system. What would be the English system equivalent to a stress of 200 megapascals? Remember that a pascal is a newton per square metre, and a newton is a kilogram-metre per second squared. Using only the conversion factors listed above, the calculation is as follows:

$$200 \text{ MPa} \times \frac{(1 \times 10^6 \text{Pa})}{\text{MPa}} \times \frac{1\text{N}}{(\text{m}^2)(\text{Pa})} \times \frac{1 \text{ kg} - \text{m}}{\text{s}^2(\text{N})}$$
$$\times \frac{1 \text{ s}^2}{9.8 \text{ m}} \times \frac{2.2 \text{ lb}}{\text{kg}} \times \left(\frac{1 \text{ m}}{100 \text{ cm}}\right)^2 \times \left(\frac{2.54 \text{ cm}}{\text{in.}}\right)^2$$
$$= 29{,}000 \text{ lb/in.}^2$$

Of course, this problem would be much simpler if one remembered that the conversion factor from megapascals to pounds per square inch is 145 pounds/inch^2/MPa:

$$200 \text{ MPa} \times \frac{145 \text{ lb}}{\text{in.}^2\text{MPa}} = 29{,}000 \text{ lb/in.}^2$$

A summary of the metric system/inch system conversion equations follows:

1. 1 kg = 1000 g (exactly). (See Table 3–1 for more prefix conversions.)
2. 1 in. = 25.4 mm (exactly) = 2.54 cm (exactly) = 0.0254 m (exactly)
3. 1 kg = 2.2046 lb ≈ 2.2 lb
4. The weight of 1 kg ≈ 9.8 N
5. 1 qt = 0.946 L
6. 1 mi = 1.609344 km (exactly) ≈ 1.6 km
7. 1 Pa = 0.000145 lb/in.^{2} or 1 MPa = 145 lb/in.^{2}

PROBLEM SET 3–3

1. If a person weighs 155 pounds, what would he or she weigh in kilograms?
2. The person from Problem 1 is 5 feet, 8 inches tall. What is this height in centimetres?
3. How many pounds would 500 kilograms of tomatoes weigh?
4. How many kilograms are there in one ton of coal?
5. How many litres are there in 10 gallons of paint thinner?
6. Convert a speed of 75 kilometres per hour to miles per hour.
7. Convert a speed of 55 miles per hour to kilometres per hour.
8. Convert a speed of 120 feet per second to metres per minute.
9. How many square metres are in 10 acres of land? (*Hint:* One acre is equal to 43,560 square feet.)
10. The diameter of a shaft is 3.75 inches. What is this diameter in millimetres?
11. What would be the density in the metric system of concrete if its density is 145 lb/ft^3? (Give the answer in grams per cubic centimetre.)
12. Convert a pressure of 17.8 megapascals to pounds per square inch.
13. Convert a speed of 80 km/hr to feet per second.
14. A bandsaw blade runs at a speed of 56 inches per second. What is this rate in metres per minute?
15. A force of 175 pounds is equal to how many newtons?
16. The strength of a piece of steel is rated as 72,000 psi. What is the strength in pascals?
17. If the modulus of elasticity of an aluminum alloy is 16×10^6 psi, what is the modulus of elasticity in megapascals?
18. The air pressure in a tire was supposed to be 32 psi. Unfortunately, the tire gauge was graduated in units of kilopascals. If the gauge read 75 kilopascals, was the tire over- or underinflated?

LIMITS OF ACCURACY

As stated earlier in this chapter, nothing can be made exactly to size or be measured with absolute precision. Therefore, it is a common practice to put *limits* on dimensions. Sometimes these limits are stated explicitly. A dimension might be given as 35.40 ± 0.05 millimetres. Sometimes the limits are implied. The precision of a measurement is often implied by the number of digits stated. For example, if the measured value of a part is given as 8.52 inches, this implies that the part would be accepted if its actual dimension fell between 8.515 and 8.525 inches. Therefore, there is a difference between stating a measurement as 8.5 inches versus 8.500 inches. The former implies tolerances of 8.45 to 8.55 inches, whereas the latter has an implied precision of 8.4995 to 8.5005 inches. As straightforward as this may seem, there are times when it is difficult to state precision accurately with such a simple scheme. In those cases, do what seems to be reasonable. For example, the distance between the earth and the sun is 93,000,000 ± 700,000 miles (or $9.30 \pm 0.07 \times 10^7$ miles).

In today's industry, accuracy and precision are essential. However, precision is limited by the quality of the measuring instruments. It would be ridiculous to attempt to measure the diameter of an internal combustion engine bearing to the nearest thousandth of an inch using a foot ruler. Very often the measuring instruments are very delicate and very expensive. They should be treated with care and handled only by experienced people. The accuracy of a part is also a function of the quality of the tools and machinery used to make the part. One cannot turn a shaft to a ten-thousandth of an inch if the lathe has bearing play of a thousandth of an inch. Industry spends considerable time and money to maintain its production machinery in top condition.

Dimensions and tolerances are determined or derived primarily by the job or function the product or part has to perform. However, some manufacturing processes can provide closer tolerances than others. Sand casting, for example, is not as precise a process as machining. For that reason, sand-cast parts are usually made larger than they need to be, then machined to fit. Close tolerances should not be put on the drawings for a sand-cast part. With modern computer-controlled machining methods, the tolerances specified are often those that the properly maintained machine can hold. Wear in these automated cutting tools and machines can be detected when the tolerances are not being met. This is a change from the way it used to be done, which was to place tolerances on a drawing that the machinist had to meet.

However, good engineering and good production techniques require that the parts be manufactured only to the tolerances required for the part to perform successfully its intended job. Requiring parts to be made to finer tolerances than necessary increases the cost of the part. The increased cost might make the product uncompetitive and unmarketable. Parts produced to close tolerances result in products of higher quality. It is a delicate compromise to make a product that is economical yet of a quality sufficient to perform as expected. Safety should never be compromised. In military and commercial aircraft, reliability and safety demand the high precision, and the additional cost can be justified on that basis.

STATISTICS OF MEASUREMENT

Because no part can be made to an exact dimension, how do we determine if the part falls within acceptable tolerances? Answer: take repeated measurements. Then two questions become very important. The first is *where is the middle of the measurements?* The second is *how much are the measurements spread out?* These questions are answered by statistical techniques. Therefore, the technician should understand some of these statistical terms.

Measures of Central Tendency

If several, seemingly identical parts are made and each measured, there will always be slight differences in measurements between them. The middle of the measurements is called its *average.* In statistics the average is a measure of *central tendency.* However, the average can be calculated by and interpreted in several different ways. These include the *mean,* the *median,* and the *mode.*

Mean

The mean (or arithmetic mean) can be found by adding the measurements of the same dimension of all of the parts, then dividing by the number of parts measured. For instance, 10 pistons for an internal combustion engine have these diameters (in inches):

3.501 3.498 3.502 3.500 3.497
3.499 3.498 3.497 3.496 3.498

The sum of these measurements is 34.986 inches. Therefore, the mean would be

$$34.986 \text{ in.}/10 = 3.4986 \text{ in.}$$

The formula for the arithmetic mean is

$$M = \text{sum values/number of values}$$

Always check to see if the mean is somewhere between the top and the bottom values measured.

Median

Another measure of central tendency often used is the median. The median is the middle value obtained. To get the median, place the values in order from the top to bottom (or bottom to top, it doesn't matter) and count down halfway from either end. Arranging the values given previously in order they would be

3.502 3.501 3.500 3.499 3.498
3.498 3.498 3.497 3.497 3.496

Because there is an even number of values (10), the middle score or median would fall halfway between the fifth and sixth from the bottom or the top. The median in this case would be 3.498 inches.

In small groups (less than 30), the median is often more significant than the mean. If one number differs greatly from the rest, it would affect the mean greatly, but not affect the median at all. For instance, if the top value of the example were 3.600 inches instead of the 3.502, the median would stay the same, but the mean would be changed to 3.508, which is larger than 9 of the 10 values. Such a value is misleading about the majority of values obtained. For larger groups, the mean and the median tend to come very close to the same value.

Mode

Mode is the name given to the most-often-recorded reading. In the preceding example, it would be 3.498. Often the data are "grouped" or put in uniform increments and the mode determined by which *group* has the most readings. In sample sizes of more than 60, a mode will usually exist. Sometimes there may be two or more modes. In small samples, the measurements may be evenly distributed with no apparent mode. The mode can easily be seen in a graph in which the number of measurements of each value is plotted against the values themselves (Figure 3–1).

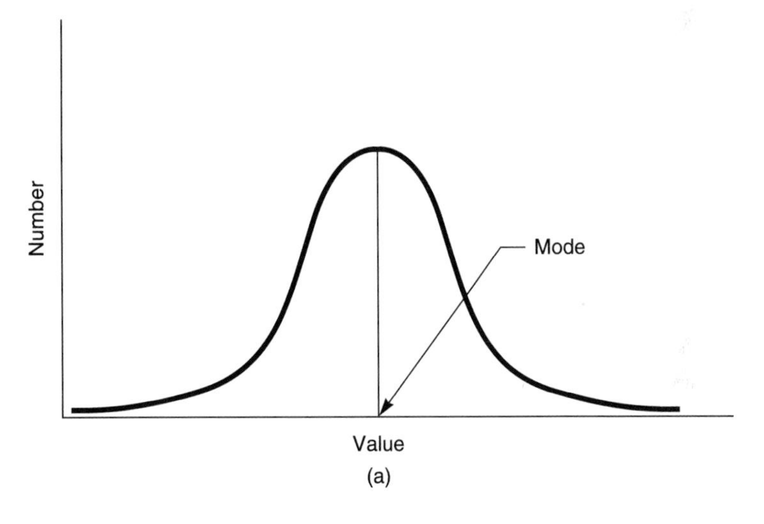

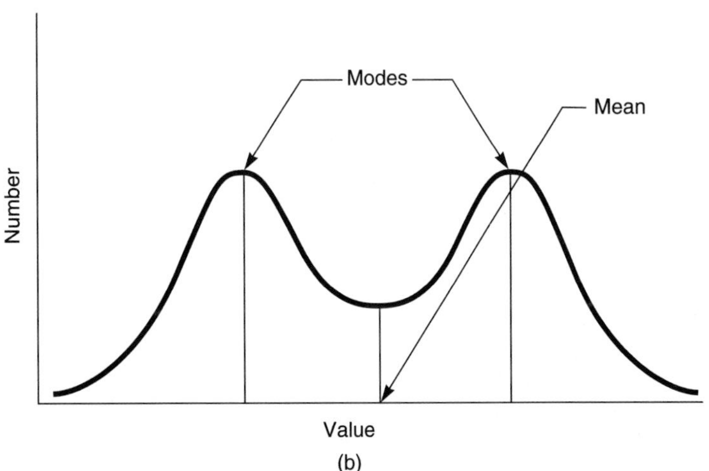

Figure 3–1. Modes. **a.** Normal distribution. **b.** Bimodal distribution.

Measures of Dispersion

Range

The second question is "how far are the values spread out?" Several methods are used including the *range* and *standard deviation.* The range is simply the maximum less the minimum values. For the previous example, the range would be

$$3.502 - 3.496 = 0.006 \text{ in.}$$

Standard Deviation

The standard deviation is found by the formula:

$$\text{SD}\sqrt{\frac{\Sigma(X-M)^2}{N}}$$

where SD = standard deviation, X = individual score, M = mean, and N = number of scores. Note that the Greek letter sigma (Σ) means to sum the values after it.

To calculate the standard deviation, it is often helpful to make a chart as follows:

X	$X - M$	$(X - M)^2$
3.502	0.0034	0.00001156
3.501	0.0024	0.00000576
3.500	0.0014	0.00000196
3.499	0.0004	0.00000016
3.498	−0.0006	0.00000036
3.498	−0.0006	0.00000036
3.498	−0.0006	0.00000036
3.497	−0.0016	0.00000256
3.497	−0.0016	0.00000256
3.496	−0.0026	0.00000676
		$\Sigma(X - M)^2 = 0.00003240$

Therefore, the standard deviation would be

$$\text{SD} = \sqrt{\frac{0.0000324}{10}}$$
$$= \sqrt{0.00000324}$$
$$= 0.0018$$

On the normal distribution curve, there is a point of inflection where the curve changes from a concave to a convex shape. The distance between the mean and this "point of inflection" represents the standard deviation

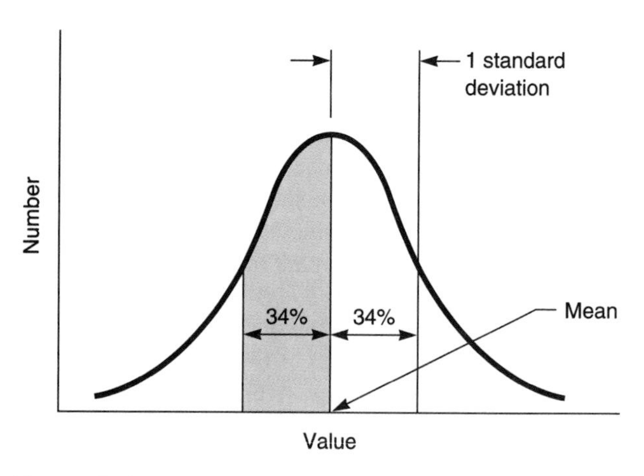

Figure 3-2. Standard deviation.

(Figure 3–2). In a normal distribution, 34% of the scores or values will fall between the mean and one standard deviation above or below the mean. In the example shown earlier, this would imply that 68% of the scores would fall between one standard deviation above the mean and one standard deviation below the mean.

$$3.4986 \text{ in.} + 0.0018 \text{ in.} = 3.5004 \text{ in.}$$

and

$$3.499 - 0.0018 = 3.4968 \text{ in. in diameter}$$

As can be seen by these data, 7 of the 10 values lie within one standard deviation above and below the mean. This is very close to the predicted 68%, which means that the data were normally distributed.

Since the standard deviation is a measure of dispersion from the mean, it can be used for many purposes. The more precise measurements become, the smaller the standard deviation. Tests such as the one just illustrated can show whether a machine is getting worn out or if a machinist is not producing parts that meet standards. If the mean, median, mode, or midrange is not sufficiently close to the design criteria, and if the standard deviation does not lie within the design tolerances, a problem is indicated that should be investigated.

Variance

A similar term sometimes used in statistics as a measure of dispersion from the mean is *variance.* Variance is simply the square of the standard deviation, or

$$v = \frac{\sum(x-M)^2}{N}$$

Variance can be used in the same manner as the standard deviation.

■ PRODUCTION TESTS

In industry, parts are continuously being inspected and tested. Individual parts may be inspected or tested as they come from the machines. The entire product is tested as it leaves the final assembly line. The inspection and testing of a product as it leaves the factory is called *quality control.* The purpose of quality control is to make sure the product is working properly and to prevent, if possible, that product from having to be returned to the factory.

Many industries also inspect and test products that they purchase from other companies. This is called *quality assurance.* Ensuring that the purchased parts will function properly allows them to be used with confidence in the purchasing company's products.

Every product can be inspected and tested by some means, as discussed in Chapter 2. In some tests, some of the product must be sacrificed. The best way of testing rifle bullets, for instance, is to fire them. Of course, the product is destroyed. So in this case, the testing is done by checking a small sample. If all of the small sample works properly, it is assumed that the rest of the product will too. (This assumption, however, is not always valid.) Electric wire, on the other hand, can be tested by measuring its resistance. Since resistance can be measured without damaging the wire, all wire can be tested before it is used. Statistics plays an important part in these sampling techniques.

Physical measurements are often made with micrometers, gauge blocks, dial indicators, and optical comparators (Figure 3–3). The proper use of these instruments requires a certain amount of training but it is essential that technicians learn to use them properly.

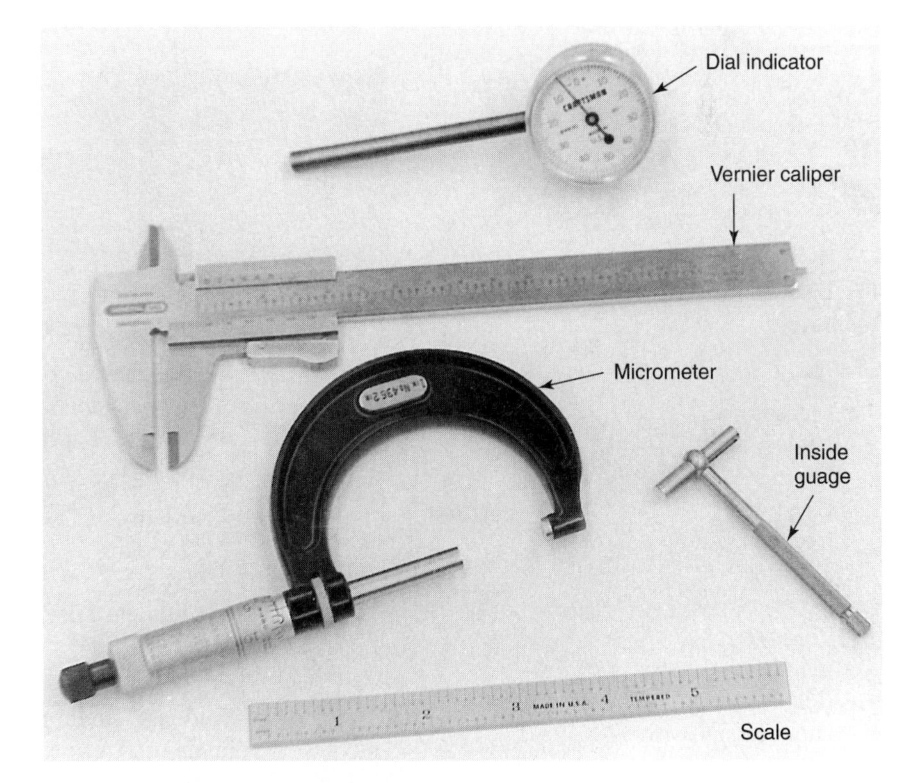

Figure 3–3. Measuring instruments.

PROBLEM SET 3–4

The following data are given for the measurements of the tensile strengths of a piece of metal. All values are in pounds per square inch.

50,750	51,200	55,100	53,850	52,000	54,100
51,000	55,500	53,400	52,600	52,550	53,800

1. What is the mean of the data above?
2. What is the median of the above data?
3. Group the data above in 1000-psi increments (i.e., 50,000–50,999, 51,000–51,999, etc.) In which group is the mode?
4. What is the standard deviation of these data?
5. What is the variance of these data?
6. What is the range of these data?
7. What are the values of one standard deviation above the mean and one standard deviation below the mean?
8. If an inspector was told to reject all parts more than one standard deviation away from the mean, would the diameters listed be accepted or rejected compared against the following standards?

   Mean 5.675 in. SD 0.003 in.

   a. 5.657 in.
   b. 5.673 in.
   c. 5.677 in.
   d. 5.670 in.
   e. 5.679 in.
9. You are given the following weights of people (in pounds).

152	175	110	147	156	165
225	125	130	187	205	114

   a. How much is the mean weight?
   b. How much is the median weight?
   c. How much is the range of weights?
   d. How much is the standard deviation?
10. If one measures the length of a room, what are the limits of accuracy of the measurement?

SUMMARY

In this chapter the following subjects have been covered:

- Definitions of:

  Measurement
  Inspection
  Testing
  Basic size
  Nominal size
  Tolerance
  Accuracy
  Precision
  Validity
  Reliability
- Dimensional analysis
- Use of formulas
- Metric (SI) system

  Metric (SI) system prefixes
  Conversion factors
- Limits of accuracy
- Statistics of measurements

  Mean
  Median
  Mode
  Standard deviation
  Variance
  Range

Students should now be familiar with measurements in manufacturing.

PHOTO ESSAY 3–1

Reading the Micrometer

METRIC MICROMETER

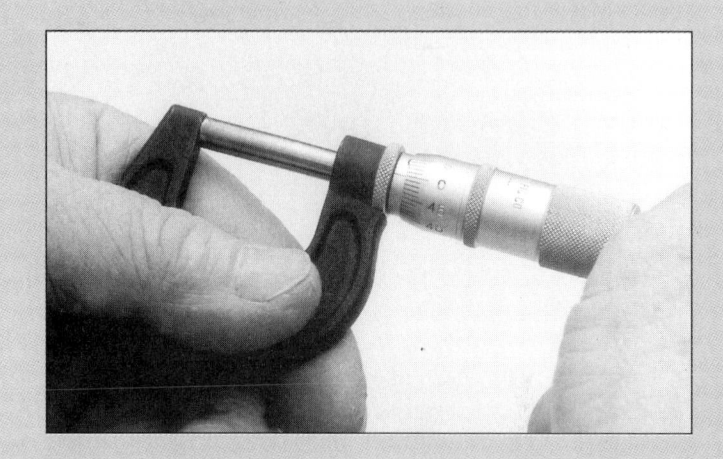

STEP 1 Clean the faces of the anvil and the quill. Turn the friction knob to screw the faces together. *Do not tighten the faces tighter than that allowed by the friction knob.* Overtightening can damage the micrometer. Make sure the micrometer is reading zero. If it does not read zero, adjust the micrometer. (See instructions for the individual micrometer for adjusting procedures.)

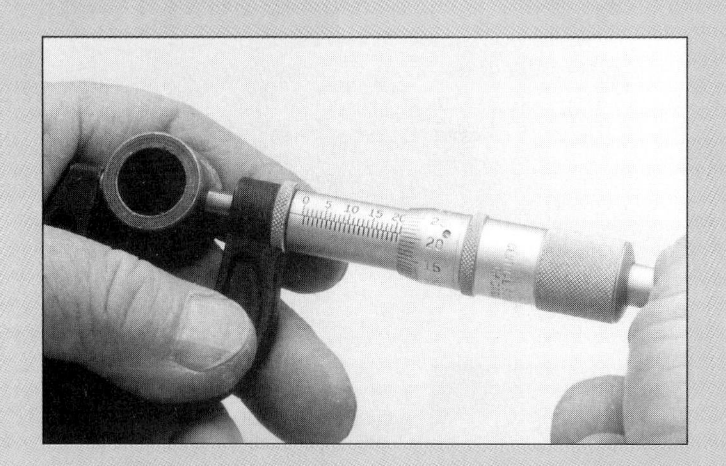

STEP 2 Open the screw of the micrometer and place the object to be measured between the anvil and the quill. Tighten the micrometer against the object using only the friction knob.

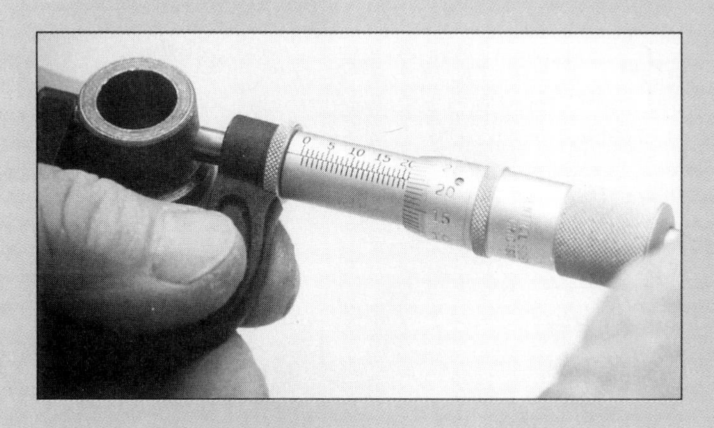

STEP 3 Read the micrometer. Note that each mark on the drum of the micrometer is one (1) millimetre or 0.1 centimetre. Each mark on the barrel is 0.01 millimetre or 0.001 centimetre. The reading here is therefore 21.21 millimetres or 2.121 centimetres.

INCH SYSTEM MICROMETER

STEP 1 Clean, close the micrometer, and zero or adjust the micrometer as shown in Step 1 for the metric micrometer.

STEP 2 Place the object to be measured in the micrometer and tighten the micrometer about the object *using only the friction knob.*

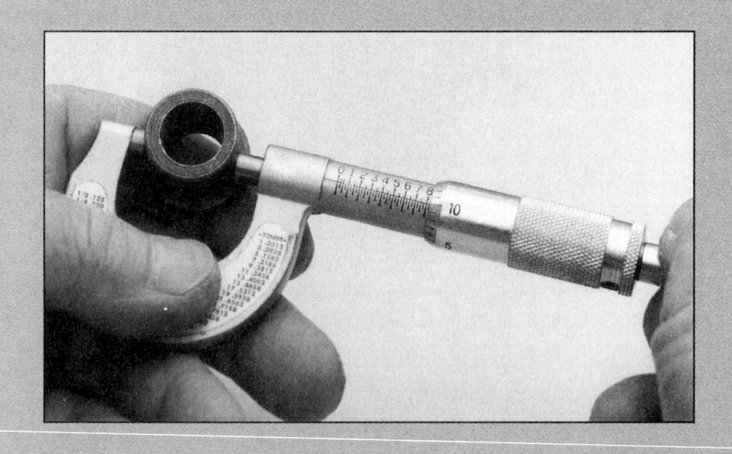

STEP 3 Read the micrometer. Note that each number on the drum is 0.1 inch. Because there are four marks between each number, each mark is 0.025 inch. Rotating the barrel once moves it back one mark or 0.025 inch. The reading here is back one mark beyond the 0.8 inch number and the barrel is indicating 10. The total reading is therefore 0.8 + 0.025 + 0.010 or 0.835 inch.

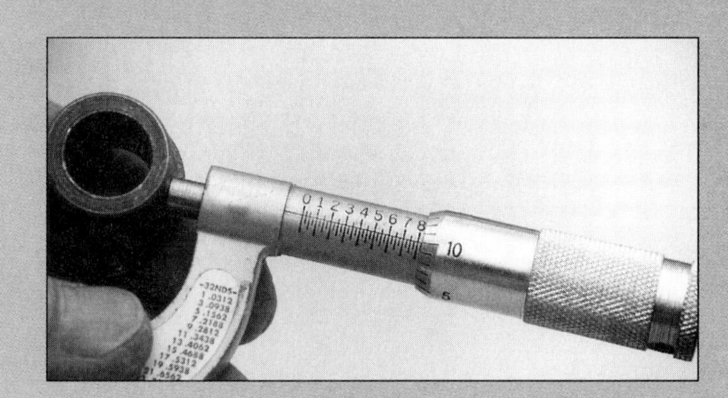

PHOTO ESSAY 3-2

Using the Dial Micrometer

STEP 1 Clean the faces of the jaws, close the jaws completely using the thumb wheel, and set the dial to zero.

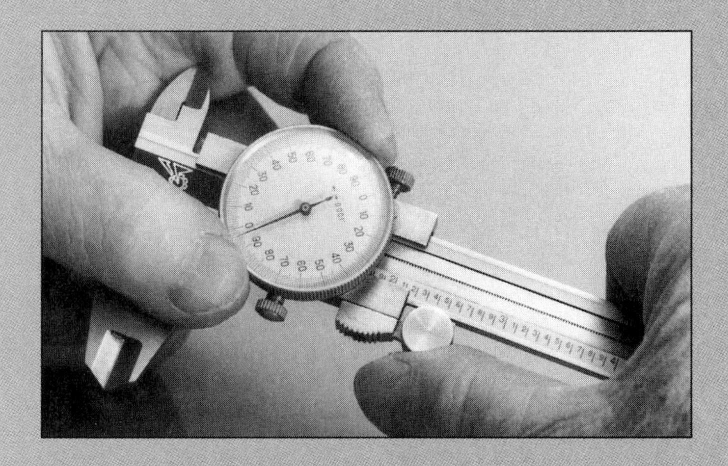

STEP 2 Tighten the set screw to lock the dial in place.

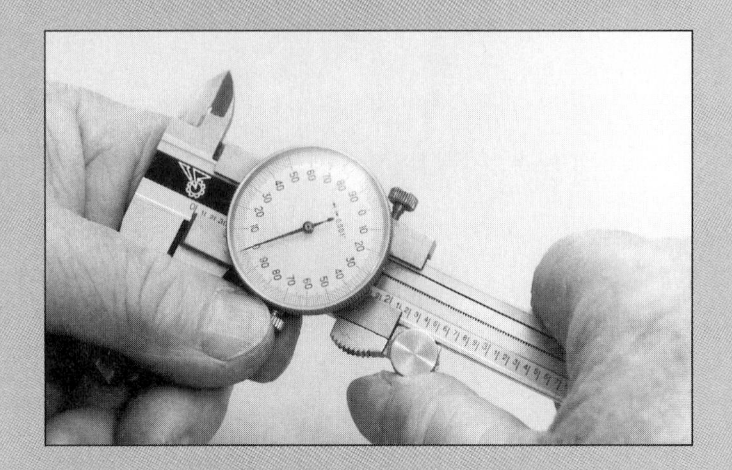

STEP 3 Place the object to be measured between the jaws and gently close the jaws over the object using the thumb wheel. Do not force the jaws. Excessive force could bend the micrometer and ruin it.

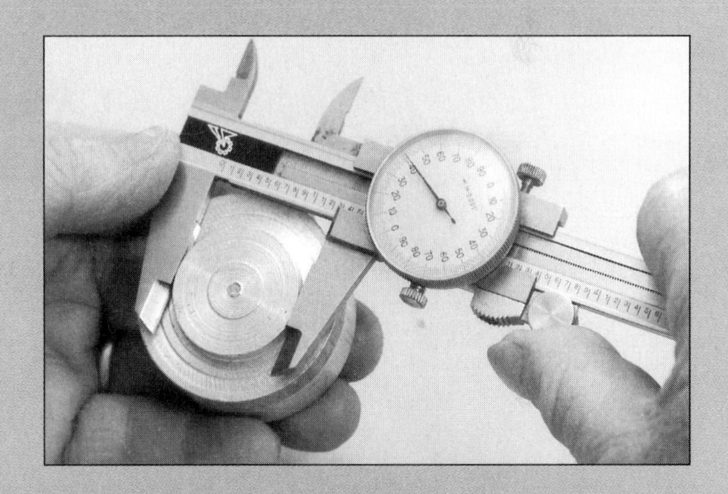

STEP 4 Read the slide and dial. In this example the slide is resting between 1.2 and 1.3. The dial is reading 41 or 0.041. The total reading is therefore 1.241 inches.

PHOTO ESSAY 3–3

Using the Vernier Caliper (Inside Measurement)

STEP 1 Close the caliper to make sure it is reading zero to start. If it does not read zero, note the reading that must be subtracted (added for negative readings) from the final reading. Place the object to be measured over the blades of the caliper. Slide the blades back with the slide wheel to make firm, complete contact with the object.

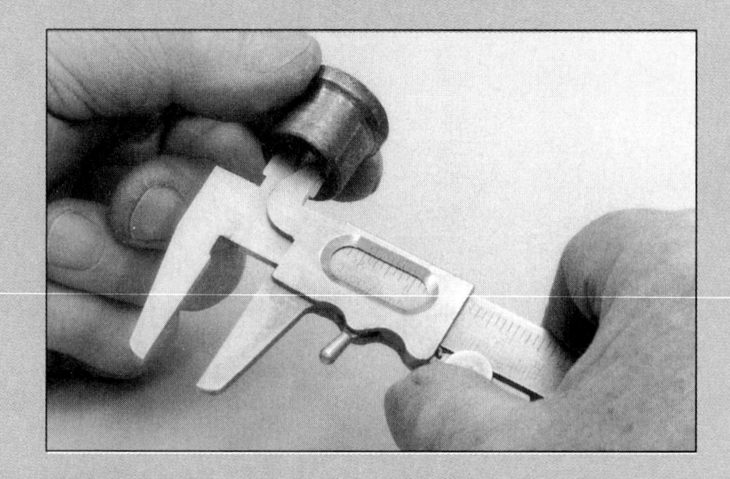

STEP 2 Take a reading. This vernier caliper is calibrated in inches on the top scale and centimetres on the bottom scale. On the inch scale, each mark is 1/16 inch. In the measurement shown here, the index mark is between the 1/2- and 9/16-inch marks. Count the number of lines on the slide scale to find the line that lines up with a line on the caliper. (In this example, it is the fifth line.) This is the number of 1/128 inch to be added to the initial reading. The total reading is therefore 1/2 + 5/128 or 64/128 + 5/128 or 69/128 inch. On the metric scale, the index is between 1.3 and 1.4 centimetres. The seventh line on the slide matches a line on the caliper. The reading is therefore 1.30 + .07 or 1.37 centimetres.

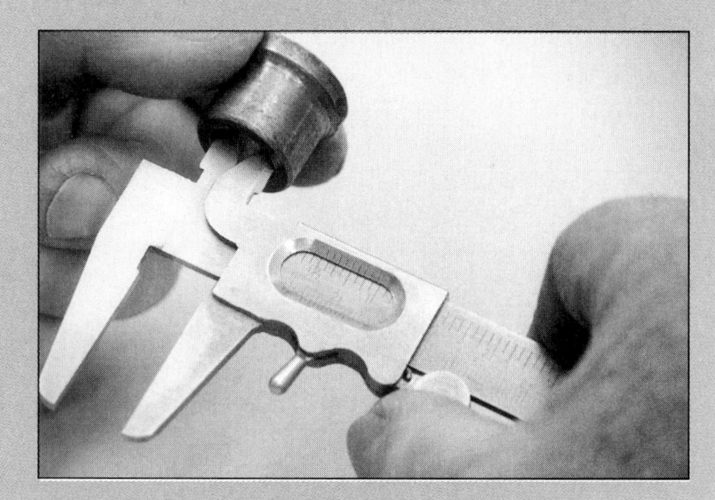

SECTION

II

Material Removal

■ INTRODUCTION

The manufacture of a part often requires that some of its material be removed. This can be done in many ways. Mechanical methods, such as sawing, turning, milling, grinding, and drilling are commonly used. However, material can also be removed by burning, by chemically dissolving, or by electrical methods. Many of the methods of removing material used in industry are discussed in this section.

Chapter

4

Mechanical Methods of Material Removal

■ FUNDAMENTAL CONCEPTS OF MATERIAL REMOVAL

Materials fail or break in several ways. Brittle materials break into large chunks, whereas ductile ones may stretch before breaking. Cutting or breaking a piece of material implies that somehow the bonds between the atoms must be severed. The ease with which these bonds can be broken and, therefore, the methods used to break them often depend on several factors:

- The properties of the material being cut
- The properties of the cutting tool
- The speed at which the material is cut

The failure of a material begins at the microstructure level. If a steady force is slowly applied to a ductile solid, the planes between the atoms slide over each other (Figure 4–1). In ductile slip, bonds are continuously broken and reformed, but this takes time in the form of a few milliseconds.

No cutting tool, no matter how sharply the edge is formed, is sharp enough to cut a single atomic bond. Even if it were, it would quickly dull. Therefore, *the cutting action of a tool actually "pushes" the material ahead of the cutting tool.* The "pushed" material breaks away from the rest of the stock in little pieces at the slip planes. Figure 4–2 shows how these slipped pieces (called *chips*) come from a piece being turned. Chips are similarly formed by other cutting tools. These slip planes can easily be seen by microscopic examination of the chips coming from a lathe turning (Figure 4–3). Chips from a ductile material will be long strings of metal, whereas the chips from brittle materials come off in little chunks (Figure 4–4).

If a brittle material is struck too rapidly, all of the bonds are broken at once and there is not sufficient time to reform them. The result is impact as described in Chapter 2. In brittle materials the slip planes are "pinned" in such a way that they cannot slide, and the bonds break all at once in a brittle fracture through the easiest path (Figure 4–5).

In some materials such as cast iron there are so many flaws that the material breaks through the flaws. In cast iron these flaws are a result of an excess of carbon in the iron. This excess carbon creates flat plates, which have no strength and are, in effect, cracks in the iron. This is why cast iron is brittle and has an unpredictable strength (Figure 4–6).

Figure 4–1. Steady-state slip between planes.

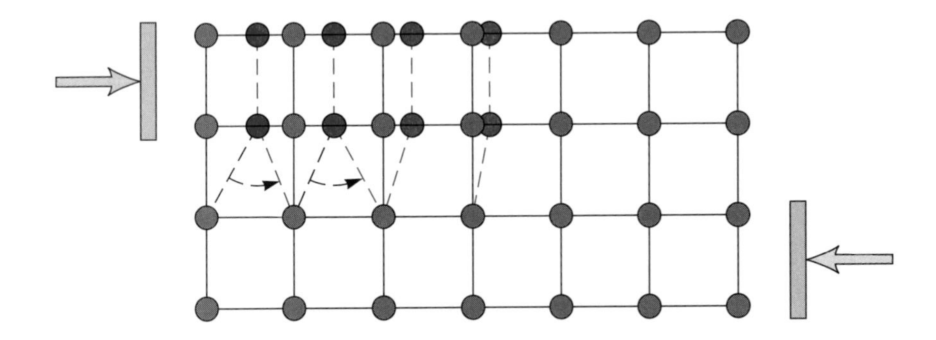

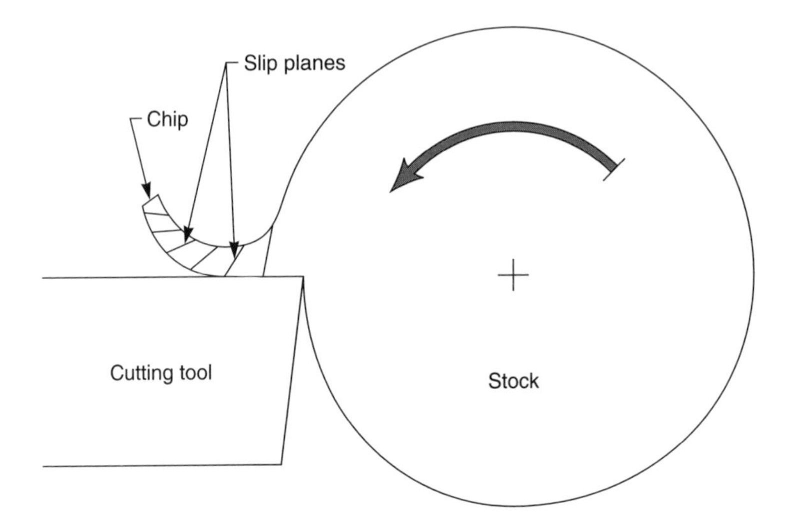

Figure 4–2. Cutting action of lathe.

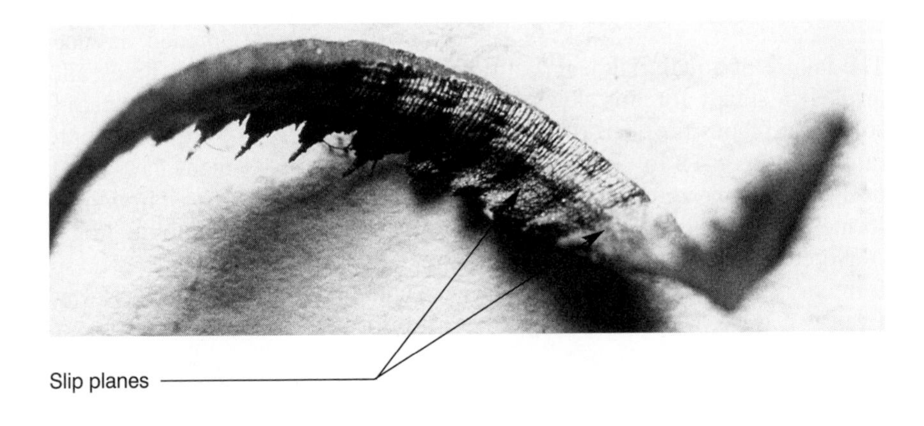

Figure 4–3. Slip planes in turnings.

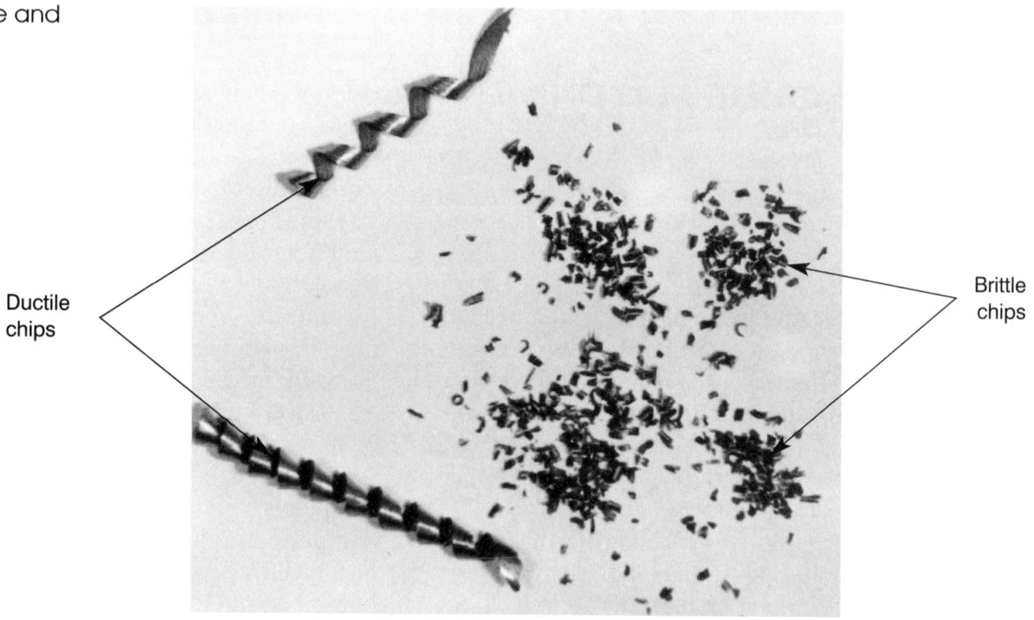

Figure 4–4. Ductile and brittle chips.

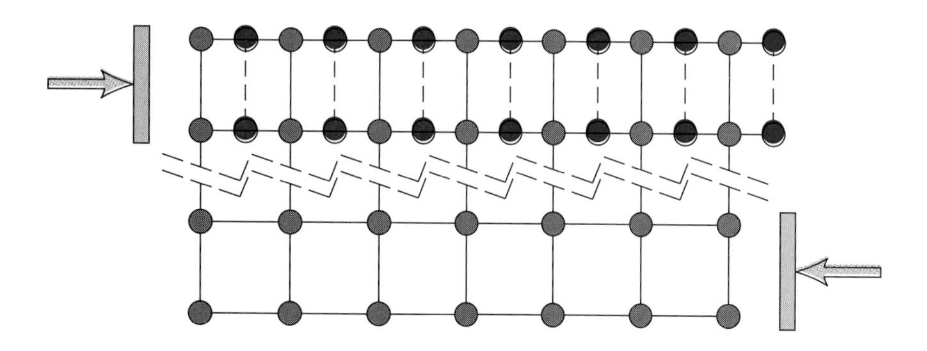

Figure 4–5. Brittle slip.

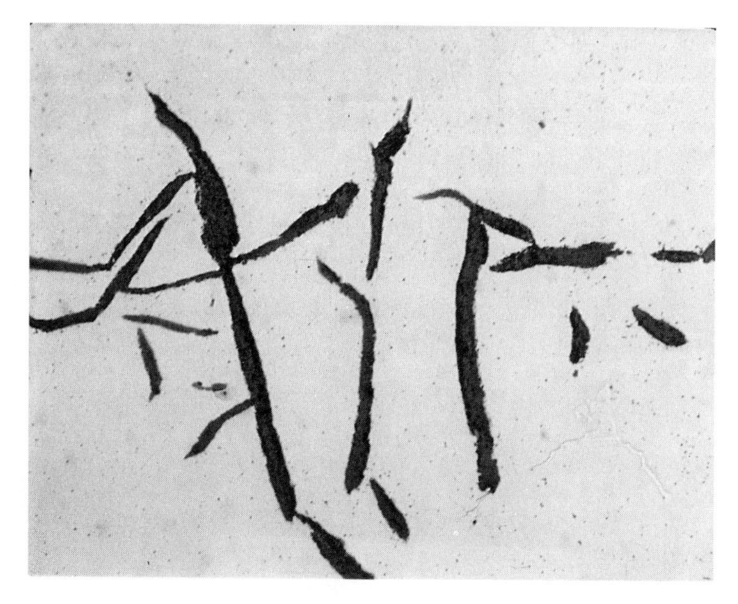

Figure 4–6. Cast iron (approximately 500X).
Photograph courtesy of USS Technical Center, Monroeville, PA.

■ METHODS OF MATERIAL REMOVAL

Handwork

Handwork is the oldest form of material removal. People have been removing material ever since they found they could strip the bark off of a tree or chip flakes of flint from a stone to make a spear point. Since humans are limited in the amount of force they can apply to a tool, and since their control of the cutting tools is not consistent, handwork is slow and imprecise when compared to power tools. Nonetheless it still has its place in industry and is still used.

Just about any definition of handwork that exists does not hold universally. At one time, handwork was defined as "any operation or procedure whereby the only force applied to a tool is provided by the operator of that tool." Today, many of these so-called "hand tools" such as saws, drills, planes, etc., have motors attached to them. Still, they are handheld. If we use the definition of handwork as "the use of any tool whereby the tool is held and guided by the worker's hand," then what about tools that are guided by the worker but supported by a hoist or crane? Let us just say that handwork involves the use of tools manually by a person.

Handwork is commonly used in making one-of-a-kind items for which it would be impractical to devise a tool to make only one piece. Handwork is also used for delicate operations where individual attention must be paid to each part. Many parts are still finished by hand. Because the labor costs for handwork are expensive, as much as possible of the material should be removed by faster methods.

Even though handwork is an old art, many of the hand tools used today have been improved to the point that persons skilled in their use can make products of high quality. Tools such as planes, saws, drills, punches, hammers, pliers, wrenches, paintbrushes, squares, torches, chisels, and a host of other items are but a sampling of hand tools still used today. Figure 4–7 shows a few of the old hand tools used before small, portable, and powerful electric motors were developed. Many modern tools are just improvements on these relics.

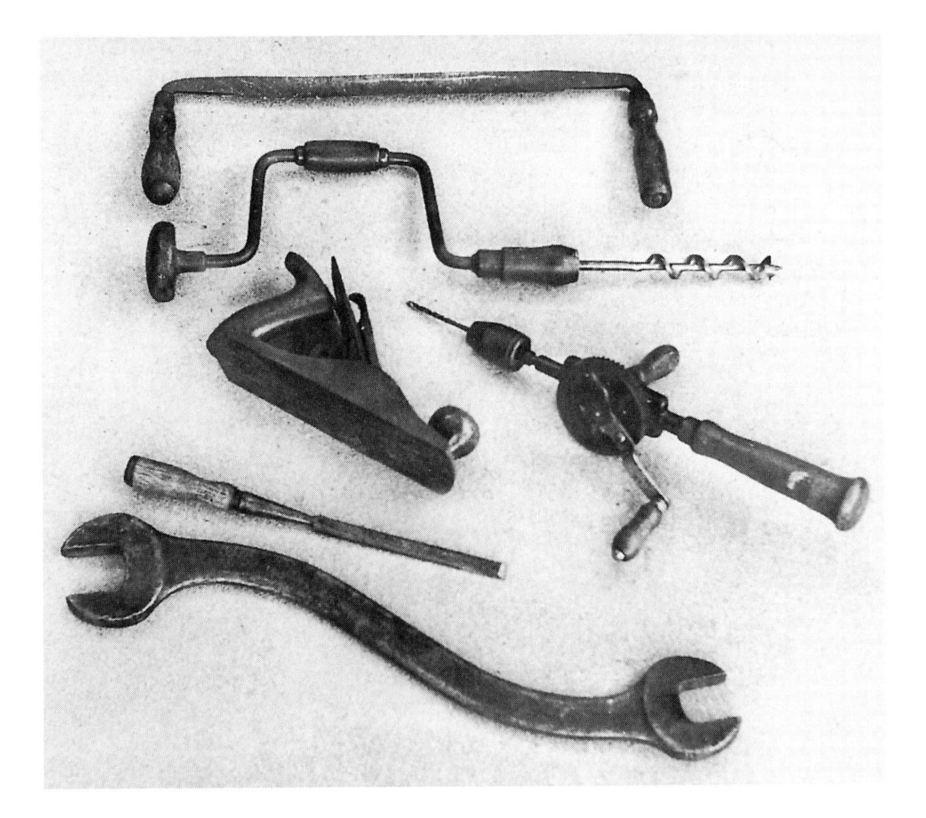

Figure 4–7. Old hand tools.

Figure 4-8. Lathe.

■ POWERED MECHANICAL METHODS OF MATERIAL REMOVAL

Many power tools are available that can remove material from a piece of stock. Probably the two most versatile machines are the *lathe* and the *vertical milling machine.* By using a lathe to remove material to make cylindrical, conical, spherical, or even threaded shapes, and a vertical mill to remove material for flat surfaces, a machinist can make just about any part needed. Machinists often brag that with these two machines, they could even make another lathe or milling machine.

Turning

Turning involves the use of a lathe, and is used primarily to produce conical or cylindrical parts (Figure 4–8). With common attachments, flat faces, curved surfaces, grinding, and boring can be done with a lathe. It is one of the most versatile machines in industry. Lathes come in all sizes—from very small jewelers' lathes used to produce watch parts to lathes that produce huge drive shafts for oil tankers. The principles on which they operate are all the same. A piece of stock is held in a *chuck* or *collet* and rotated against a cutting tool. The cutting tool removes the material (Figure 4–9).

The use of the lathe involves several steps. After the stock is placed in the chuck or collet in the *headstock,* a conical-shaped hole is bored into the other end by holding a specially shaped *center-hole drill* in the *tailstock.* Once the hole is made, the drill is replaced with a *dead-center* or *live-center* (Figure 4–10) to support the other end of the stock. (The difference between a dead-center and a live-center is that the live-center is mounted on ball bearings so it can rotate with the stock. The dead-center does not rotate and the stock merely spins about it.) Parts that are being turned at their end or "faced" (or if the stock protrudes from the chuck no more than the diameter of the stock) can be turned without using a live-center or dead-center. Parts that are long and thin must be supported at both ends because the stock will bend away from the cutting tool, resulting in an uneven cut.

The cutting tool is mounted in the tool post holder and is forced into the rotating stock. The stock turns against the cutting tool and the material is removed.

As with many machines, the parts of a lathe have been given specific names. Figure 4–11 gives the names of the more visible parts.

Figure 4-9. Cutting tool.

Figure 4-10. Live-center.

Figure 4-11. Lathe nomenclature.

Lathes have been designed for specific applications. One type of lathe is a *plumber's lathe,* used only to cut and thread pipe. Lathes that perform many functions to produce a single part are *turret lathes.* Several different tools can be held in the turret located on the end of the bed opposite the headstock. Each tool is designed to perform a given task in a series of steps. After the first tool's task is completed, the turret is rotated to the next tool and that tool does its job. The turret is then rotated again and the process continues (Figure 4–12).

Most lathes hold the stock horizontally and rotate it about a horizontal axis. However, in some heavy industry, the weight of the stock is too much to hold at both ends. In these instances, *vertical lathes* are used, and the stock is rotated about a vertical axis while the cutting tools move up and down against the stock. Figure 4–13 shows a vertical lathe.

Flat surfaces, too, can be produced by a lathe on the end of the stock by moving the cutting tool across the end of the stock. This operation is called *facing* the stock (Figure 4–14).

On a metal lathe, many of the controls are power driven. The *feed* is the amount the tool is moved hori-

Figure 4–12. Turret head lathe.

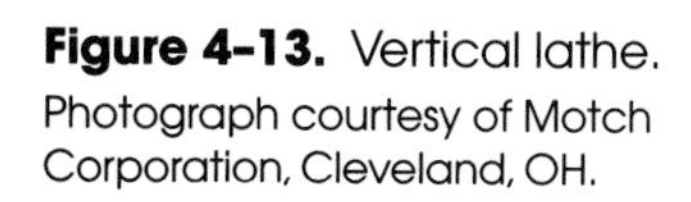

Figure 4–13. Vertical lathe. Photograph courtesy of Motch Corporation, Cleveland, OH.

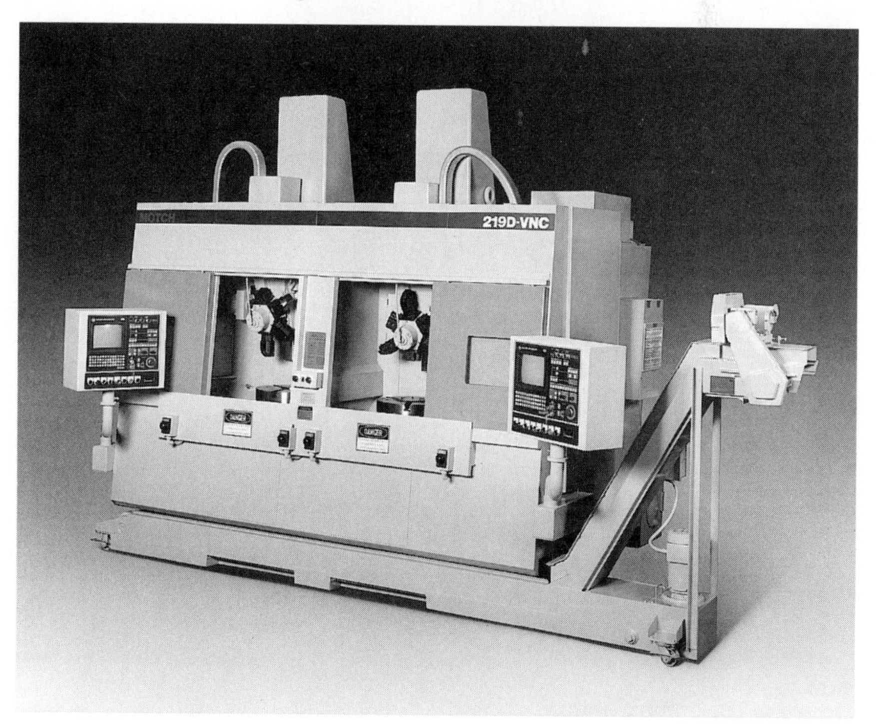

Figure 4–14. Facing process.

zontally as the stock turns. Feed is usually set in inches per turn (revolution) or millimetres per turn. Feed rates are set by a gear box mounted on the lathe. This gear box turns a long *worm gear (lead screw),* which moves the tool carriage at an even rate. The use of this gearing also permits the turning of *threads* on the lathe. The lathe has a *chasing dial,* which allows the gearing to be engaged at the same point on a revolution so that cuts can be duplicated many times to make the thread. These controls are shown in Figure 4–15.

Knurling does not involve the removal of material, but it is another operation that can be done on a lathe. A special knurling tool consisting of rollers with vee-type groves in them is held in the tool post and pressed against the workpiece. The grooves on the

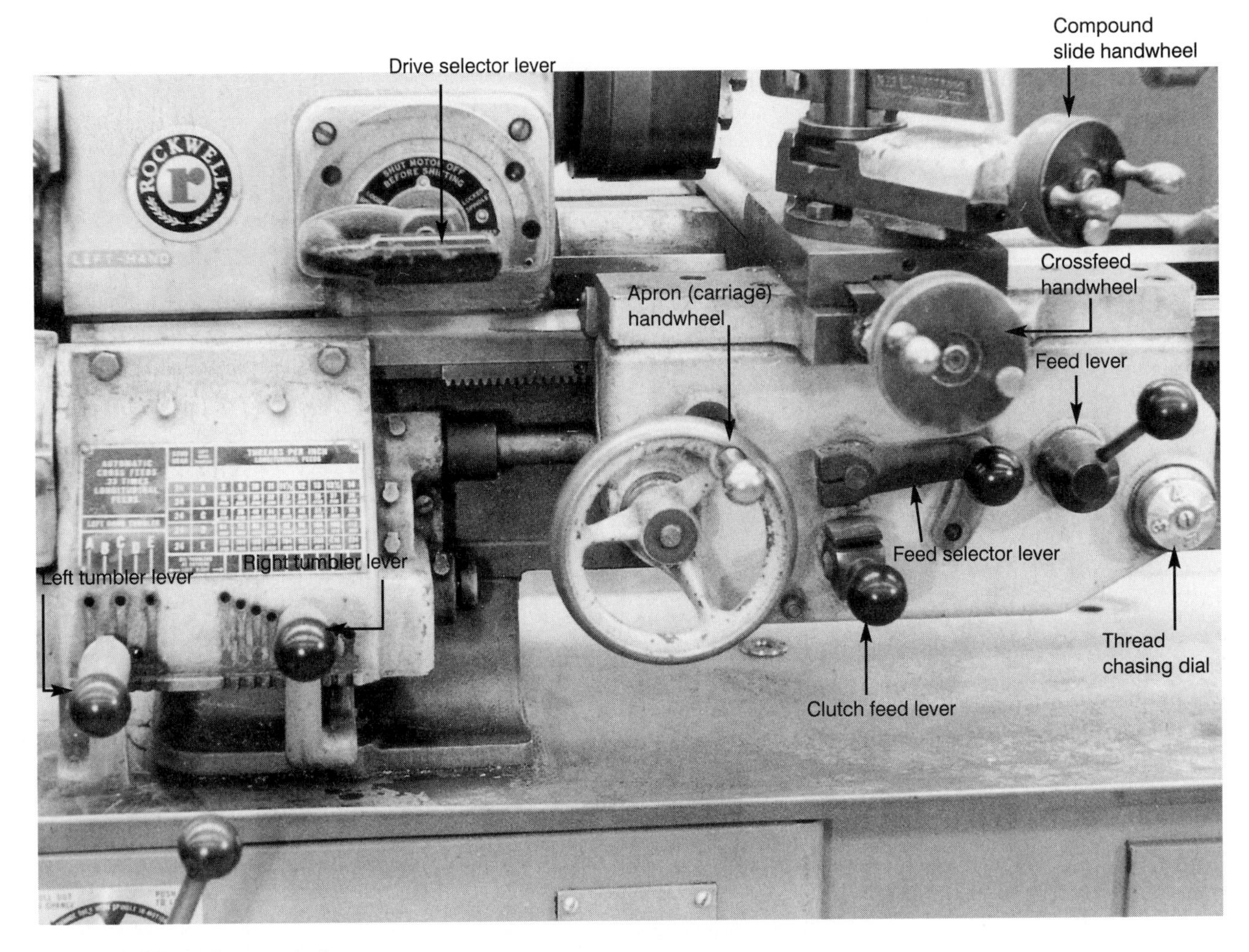

Figure 4–15. Lathe controls.

Figure 4–16. Knurling.

knurling tool press against the metal and raise ridges on the workpiece. Besides being decorative, knurling is often used to enlarge the diameter of the stock slightly and to increase the friction of a surface that is to be mated with another surface. Several different sizes and types of knurling tools are available including several different sizes of diamond knurls and parallel knurls. Figure 4–16 shows a knurling tool.

Milling

Milling is used to produce flat surfaces on a piece of stock. Mills are generally divided into two types, horizontal and vertical. In the *horizontal mill,* the stock is clamped in a vise or a fixture (or clamped to the table) and moved against a circular cutting tool. The cutting tool is rotated about a horizontal axis and the stock is moved against the moving teeth of the tool. The shape of the cut depends on the shape of the cutting tool. If a flat cut is desired, the cutting tool will have flat teeth. A cylindrical cut is made by rounding the teeth of the cutting tool. Figure 4–17 shows a horizontal mill.

Vertical mills are the more versatile of the mills, but are generally not as big nor can they take as large a stock as the horizontal ones. In the vertical mill, the tool is held vertically and rotated about a vertical axis. The stock is still held in a vise and moved horizontally against the cutting tool. The cutting surface of the tool can be the end of the tool (*end milling*) or the side of the tool (*side milling*). If necessary, the head of the milling machine can be rotated through a horizontal axis or a vertical axis. This allows the cutting tool to be used to produce "vee" cuts, or to cut a desired angle on the stock. The freedom to rotate the head also allows for proper alignment of the cutting tool, but it also makes it more difficult to keep the tool in alignment since these adjustments can slip if not carefully set and anchored. Figures 4–18 and 4–19 show a vertical mill.

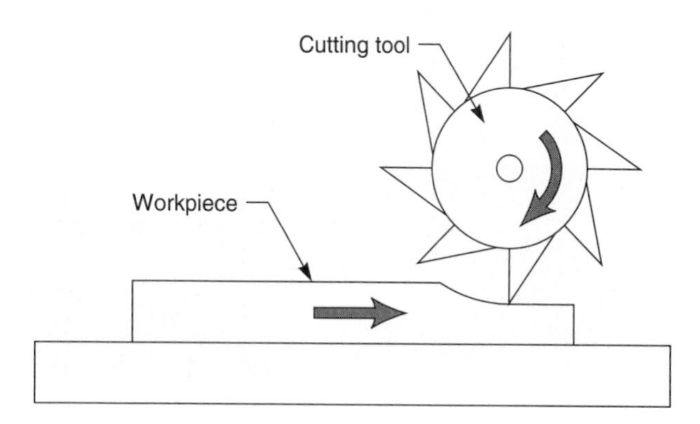

Figure 4–17. Horizontal mill.

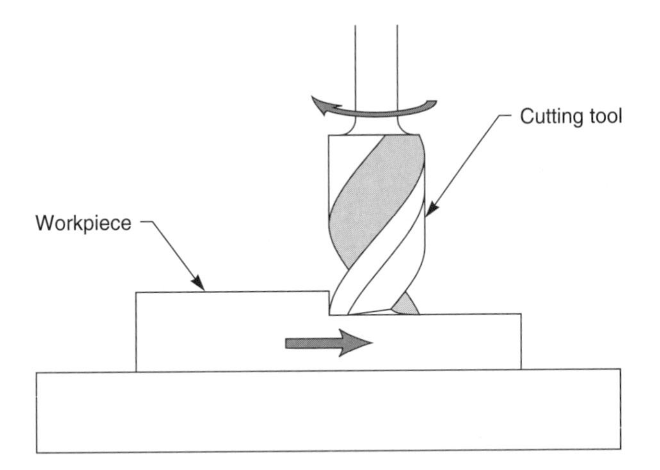

Figure 4–18. Vertical mill.

Figure 4–19. Vertical mill.

Both vertical and horizontal mills come in many sizes and can be manually, computer, or robotically controlled.

Shaping and Planing

Shaping and *planing* both produce flat surfaces and longitudinally curved surfaces on the stock. In shaping, the stock is held stationary while the cutting tool moves horizontally above it. In planing the cutting tool is held in one place while the stock is passed over it. The cutting tool in either the shaper or planer may be stationary or rotating. Different shapes are produced by the shapes of the cutting tool and the depth of successive passes over the stock. Both planers and shapers can be used on metals, wood, composites, or plastics. The machine in Figure 4–20 is a shaper, whereas Figure 4–21 shows a planer.

Routing

A close relative of the shaper and the planer is the *router.* Routers use specially shaped cutting tool to remove material in a definite geometry. Routing can be done on a milling machine by means of specially shaped tools. Curved edges on table tops, curved surfaces on plastics, and even ogives made from metals can be made with routers. Routers can be handheld

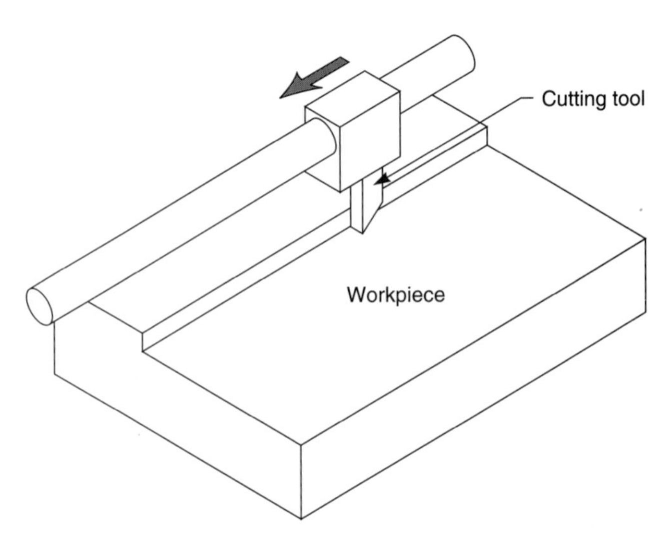

Figure 4–20. Shaper.

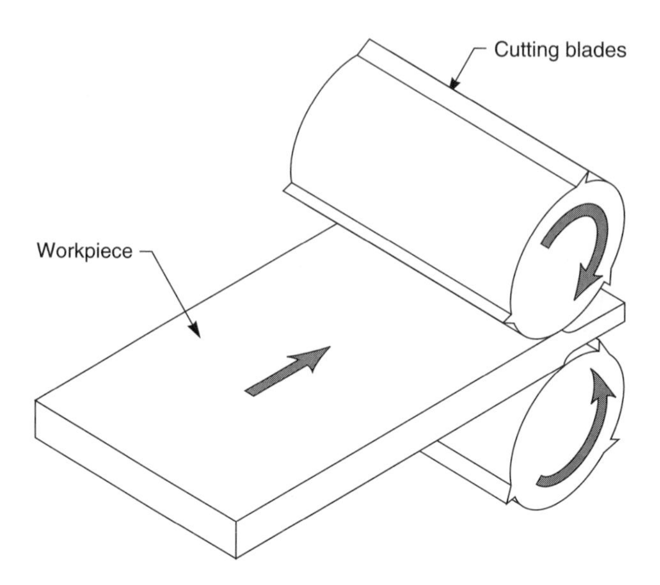

Figure 4–21. Planer.

Figure 4–22. Router.

and the tool moved over the stock, or mounted in a fixed machine and the stock moved against the tool. Although routers are most commonly used in woodworking, they do have their applications with plastics, composites, and metals. Figure 4–22 shows a router.

Broaching

How can a triangular hole be produced in a piece of metal? It can be done by hand by drilling a round hole and filing out the corners using a triangular file. This can also be done by means of a similar method that takes advantage of machine tools. This method is called *broaching.* A broach is a cutting tool with teeth much like a file. If a triangular hole is desired, a triangular broach is used. The tool has a slight taper from a small end to a larger end. Once a hole that will admit the smaller end of the broach is cut, the broach is placed in the hole and pulled through to cut the triangular hole. Just about any shape and size hole can be made by broaching. Figure 4–23 depicts a broaching tool.

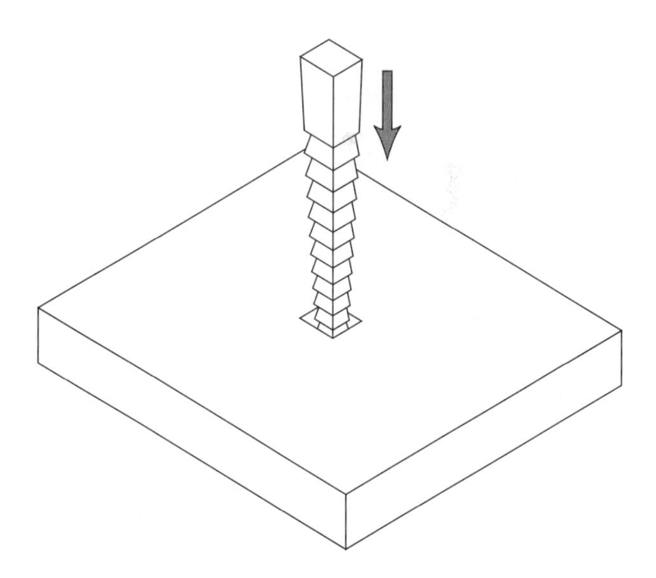

Figure 4–23. Broaching tool.

Drilling and Boring

What is the difference between drilling a hole and boring a hole? Both are used to make cylindrical holes in materials. In the operation of *drilling,* the stock is held stationary and the cutting tool (the drill) is rotated. In *boring,* however, the cutting tool is held stationary and the stock is rotated. Holes are bored on a lathe, but drilled on a drill press. Figures 4–24(a)

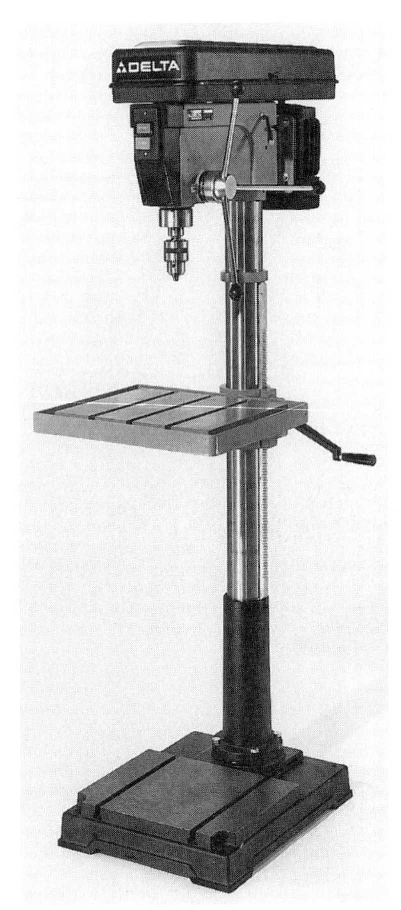

a.

b.

Figure 4-24. **a.** Small drill press. **b.** Large drill press.
Photograph 4-24 (a) courtesy of Delta International Machinery Corporation, Pittsburgh, PA.

and (b) shows a drilling operation, and Figure 4–25 shows a boring bar in a lathe.

The common method of drilling involves the use of a *twist drill.* A twist drill, however, has two problems. The flat point on the tip of the drill (the web) has poor cutting action and, therefore, the drill point tends to "wander" when pressure is applied to it. The second problem with a twist drill is that the drill itself does not have a uniform cross section. This allows the drill to "wobble" while cutting the hole (Figure 4–26). As with many common tools, many parts of a drill have names. Figure 4–27 shows the nomenclature for a twist drill bit.

Drills cannot be used to produce holes with accurate and precise tolerances. To make sure the hole is located properly, a center punch must always be used before drilling the hole. The center punch "dimple" need only be sufficiently deep to keep the center of the drill in the right location. Further, large-sized drilled holes must have a *pilot hole* of smaller diameter drilled, which is followed by the larger drill. When very close tolerances are needed, a hole should be drilled undersized then *reamed, honed, ballized,* or *bearized* to the tolerances. Reaming and honing are discussed later in this chapter and the other techniques are covered in later chapters.

Proper shaping of the drill bit ends is essential for good cutting. The rake angle is used to prevent the back side of the twist from interfering with the cutting surface. The softer the material to be cut, the sharper the lip angle should be. Most general-purpose drills have a 59° lip angle with about a 10° rake angle.

In boring a hole with the lathe, a pilot hole large enough to permit the boring bar to pass through is made in the stock. The boring bar with its cutting

Figure 4-25. Boring bar.

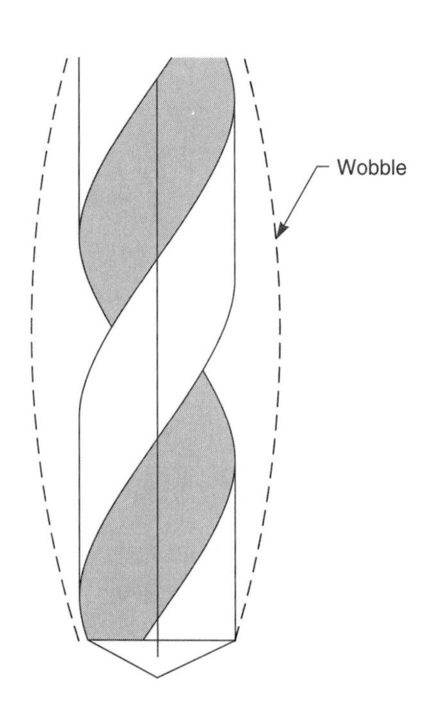

Figure 4-26. Drill action.

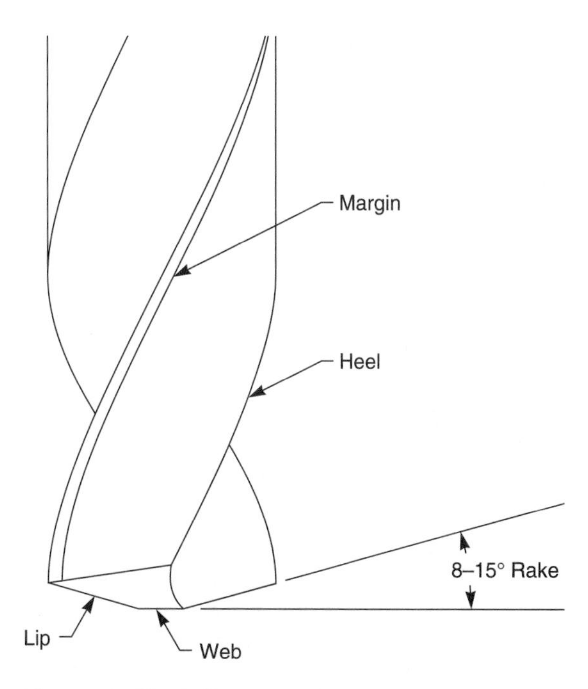

Figure 4-27. Drill nomenclature.

surface is then placed inside the stock and the hole enlarged to meet the specified tolerances. Usually the boring bar is large and rigid enough to keep the cut straight and meet tolerances. However, in some cases in which the hole is very large and long, the pilot hole is filled with sulfur or some plastic. A pilot hole is then drilled in this plastic and the boring bar inserted in the hole. The boring bar with the cutting tool is then pulled through the workpiece while both the plastic and the workpiece are removed by the cutting action of the tool. The sulfur or plastic serves as a bearing for the boring tool for the entire cut. This provides very close tolerances. Note that this type of operation is only done for the final cut. Figure 4–28 illustrates this procedure.

Reaming and Honing

Unlike a drill, a *ream* has straight cutting edges (or "flutes"). The ream can be made to hold close tolerances. The ream might have a tapered section to allow for slight enlargement of holes. It might also have a cylindrical section at the top to finish holes to very close tolerances. The major problem with the ream is that it often leaves a slightly rough surface on the inside of the hole. This surface roughness is of no consequence if the hole is just for a bolt or fastener, but if the hole is to be a bearing surface, the roughness may cause unnecessary wear on the shaft it holds. In some brittle materials, the surface roughness inside the hole may be a point at which *stress-corrosion cracking* or fatigue can start. It is essential that this surface roughness be removed in these materials.

To avoid this hole roughness, the hole is often honed. Hones are small grindstones, or polishing stones, which are attached to a shaft and rotated within the hole (Figure 4–29). If very fine tolerances are desired, or if very smooth surfaces are needed, hones are usually employed. The classic example of the use of hones is in the preparation of cylinder walls in automobile engines and automobile brake cylinders.

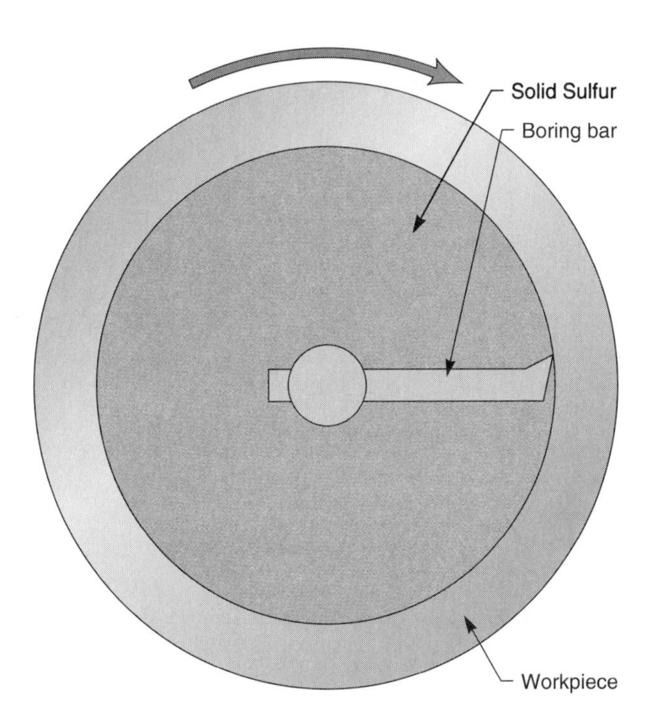

Figure 4–28. Final cut boring.

Sawing

Sawing involves the separation of two pieces of material by removing the material between them. Sawing is applicable to wood, metals, plastics, composites, and other materials and can be accomplished in any of a number of ways because many types of saws exist. A list of a few of the saws would include the following:

Circular saws
Reciprocating saws
Band saws
Chain saws
Abrasive saws
Wire saws
Water saws

As with any operation, sawing has its advantages and its limitations. It is a quick and cheap method of removing material. It is much easier to cut off a large piece of unwanted material than to turn, grind, or mill it off. However, sawing leaves a rough surface on both sides of the cut. It may pull fibers if the stock is of a fibrous nature. It may also leave severe burrs or ragged edges on the workpiece.

Set in Teeth

Another problem of sawing arises from the fact that the saw blade is between the two cut surfaces while

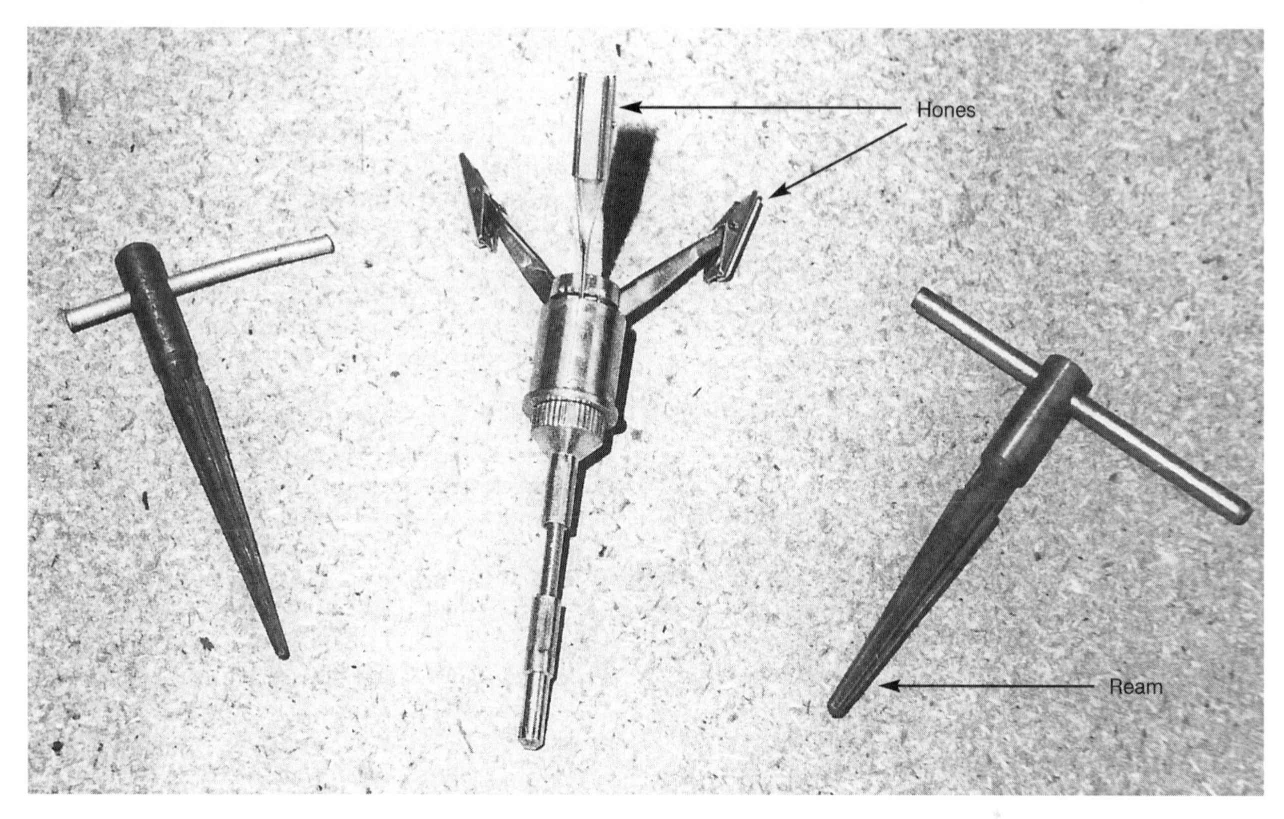

Figure 4–29. Hones and reamers.

the cut is being made. For this reason, the teeth of the saw must cut a path wider than the rest of the saw to allow it to pass through the material without binding the blade, causing friction and generating heat. This is done by putting a *set* on the teeth (Figure 4–30). Teeth on the saw blade are bent slightly in the opposite directions. Further, the outwardly bent teeth are sharpened on the outside to cut the surface. Saws are "set" in several different ways. In straight set saws, every other tooth is bent in the opposite direction. In raker set saw blades, one or more teeth which have no set are placed between the outwardly bent teeth. A wavy set is used on fine-tooth saw blades (such as hacksaw blades) since the teeth are too small for individual sets.

Blade selection for sawing depends on the hardness, ductility, rigidity, and other internal properties of the material being cut. The general rule is that the harder the material, the finer and closer the teeth must be. In cutting a steel, a blade may have from 14 to 30 teeth per inch depending on the hardness of the steel.

Since aluminum is a very "gummy" material and tends to stick to the saw teeth, a coarser saw of 8 to 12 teeth per inch may be required. Ripping saws for wood may have 4 to 8 teeth per inch. (*Ripping* refers to cutting the wood parallel to the grain. Crosscut blades are used to cut perpendicular to the grain.) Crosscut saws generally have 8 to 12 teeth per inch. When sawed, plywood tends to tear and leave a ragged edge. Plywood blades, therefore, have very little set and are generally very closely set teeth.

Circular Saws

Circular saws include bench saws and power hand saws. Circle saws are commonly used in cutting

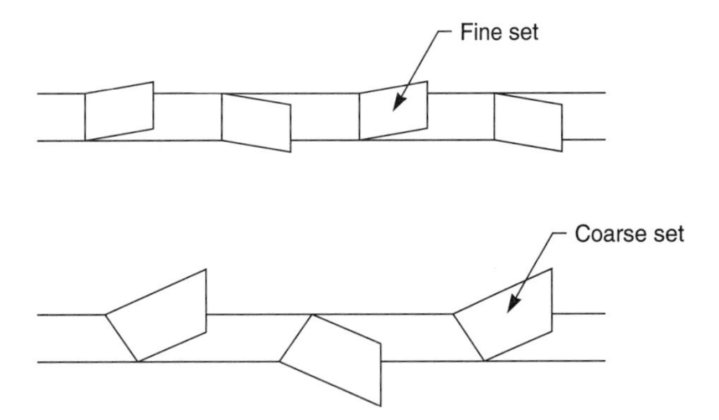

Figure 4-30. Set in saw teeth.

wood, but they are not necessarily limited to that material. Plastics, composites, ceramics, and even some metals can be cut with circle saws if the proper blade is chosen. These saws come in various standard sizes. Some have tilting beds or arbors to cut angles. *Radial arm saws* (sometimes called *overhead saws*) move the cutting blade over the material. This allows the blade to be moved and the workpiece to be held stationary. Radial arm saws are generally used for "rough" work since they are faster than the table saws. (There is no need to move the stock.) In a table saw, the motor and blade are more rigidly held and, therefore, allow for more precise work. Circular saws, in general, cannot cut a curved surface. Figures 4–31 and 4–32 are examples of circle saws. Figure 4–33 is a radial arm saw.

Jigsaws, Hacksaws, and Band Saws

Reciprocating saws such as the *jigsaw, hacksaw,* and *power hacksaw* usually require more time to cut a piece of stock since the blade cuts only half the time (on either the forward or the backward part of the stroke, but not both). Their advantage is that they do not heat up the workpiece as much as a band saw does, and usually more delicate work can be performed. Jigsaws can be used for internal cuts and can be used to cut curved surfaces. A variation of the jigsaw is the handheld saber saw. (See Figures 4–34 and 4–35.)

Band saws use an endless blade and cut continuously. As a result, they are very fast, cheap, and easy to use. Band saws can be operated vertically as in Figure 4–36, or horizontally, in a cut-off position as

Figure 4-31. Table saw.
Photograph courtesy of Delta International Machinery Corporation, Pittsburgh, PA.

Figure 4-32. Hand circle saw.

Figure 4-33. Radial arm saw.
Photograph courtesy of Delta International Machinery Corporation, Pittsburgh, PA.

Figure 4-34. Jigsaw.
Photograph courtesy of Delta International Machinery Corporation, Pittsburgh, PA.

Figure 4-35. Saber saw.

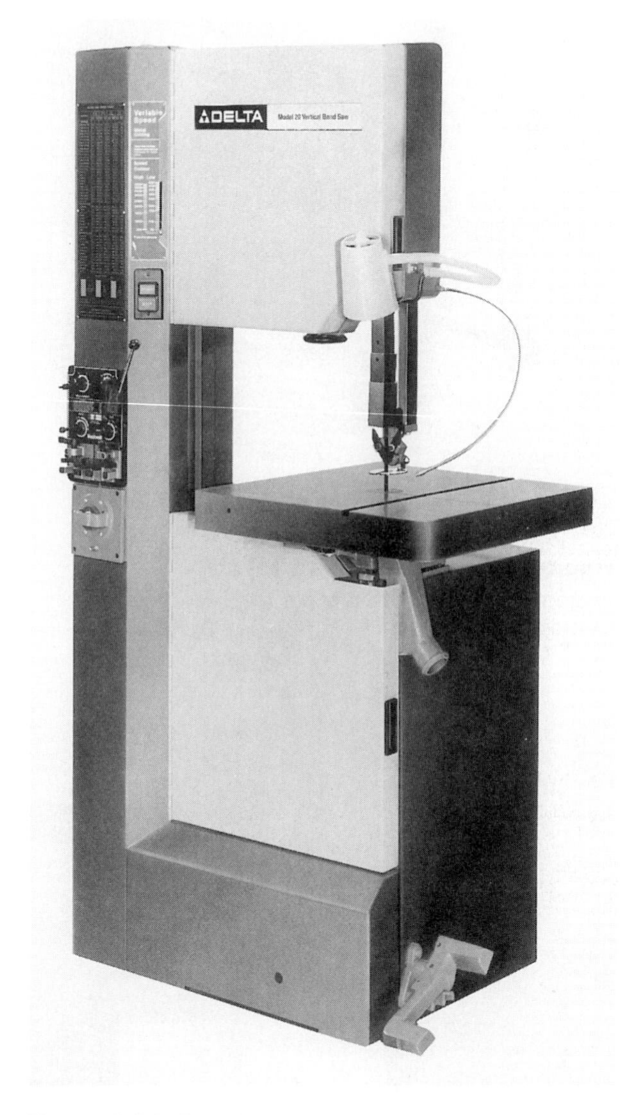

Figure 4-36. Band saw.
Photograph courtesy of Delta International Machinery Corporation, Pittsburgh, PA.

shown in Figure 4–37. They do generate more heat than other saws but they can be cooled with water or cutting oil. When using band saws or power hacksaws, cutting should always occur on the side of the stock that allows the most teeth to rest on it at a time. Cut the flat side not the end (Figure 4–38). Since each tooth takes a deeper cut as it passes over the stock, the more teeth that are resting on the stock, the more material removed per minute and the faster the cut. Second, if the edge of the stock is cut, fewer teeth come in contact with the stock, resulting in more force per tooth, and a greater chance of breaking a tooth out of the blade. Further, cuts made on the narrow end set up vibrations in the material in much the same manner as a bow being drawn over a violin string. This vibration produces a screech, known as "chatter," which is not only unpleasant, but leaves rough, uneven surfaces on the cut stock.

Abrasive Saws

A major principle of cutting is "the cutting tool must always be harder than the material being cut." In cutting high-carbon, hardened steels, there is very little difference in hardness between the steel blades and the stock. Steel blades just won't work in this case. Very often an *abrasive saw* must be used. These can be blades made of corundum, emery, garnets, or diamonds imbedded in the metal matrix. These blades do not cut their way through the workpiece as much as they grind their way through. Do not cut soft materials on an abrasive saw. The soft material gums up the blade and it just quits cutting. Figure 4–39 shows an abrasive saw.

The action by which abrasive saws work generates considerable heat. This heat can change the hardness of the material near the cut surface. This change of hardness is of little consequence if the material is being prepared for welding because the welding process heats the material anyway. These saws should not be used where the heat would damage the material. Abrasive saws can be liquid cooled to the point where they do not damage the cut surface, but this slows down the cutting action considerably.

Chain Saws

A variation of the band saw is the *chain saw* (Figure 4–40). This saw uses a linked chain, similar to a bicycle chain, which is run around a grooved bar of steel. The chain contains hook-shaped teeth for cutting. The teeth cut on either side of the chain, leaving a sufficiently wide path for the bar to clear. Although the chain saw is used mainly for logging and rough lumber work, it has been used by sculptors and others. The concept of the chain saw can be applied to other manufacturing situations. It is just another method of removing material.

Figure 4–37. Band saw in cut-off position.

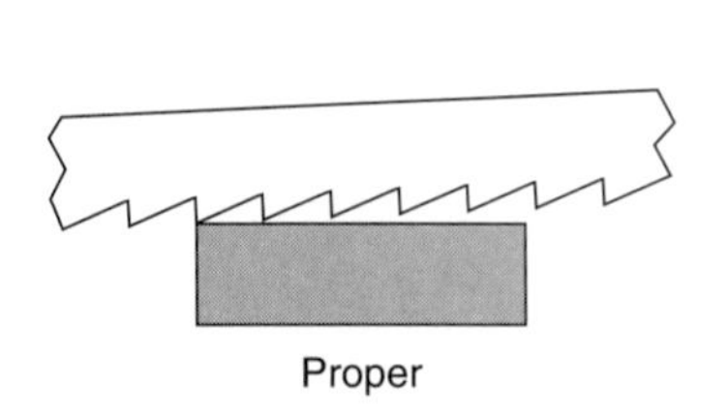

Figure 4–38. Proper stock placement.

Figure 4–39. Abrasive saw.

Figure 4–40. Chain saw.

Hole Saws

Cylindrical saws that will cut circular pieces from a stock are called *hole saws* (Figure 4–41). Hole saws come in several sizes and can be used on drill presses, vertical mills, or handheld power drills. They are not accurate ways of making a hole but are used for cutting large-diameter holes for which drilling and boring are not feasible.

Shearing and Punching

As defined in Chapter 2, *shearing* is the application of parallel but slightly offset forces. These offset forces can be used to cut materials. Common scissors are often called shears. The scaled-up principle can also be used

Figure 4-41. Hole saw.
Photograph courtesy of Black & Decker (U.S.), Inc., Towson, MD.

to cut sheets of steel, aluminum, composites, and a host of other materials. Often these shearing machines are huge and can cut steel plates up to 0.5-inch thick.

Punching is simply the shearing of any shaped hole in a sheet or plate of material. Shearing and punching are the fastest and cheapest method of cutting but the tolerances that result are not acceptable for very fine work. Metal parts are often cut by shearing or punching in a fraction of a second. Figure 4–42 shows sheet metal shears. Figure 4–43 shows a punch.

Grinding

Grinding is the removal of material by abrasion (Figure 4–44). A harder material is simply rubbed against a softer material and the softer material is worn away. It is a principle as old as sharpening a blade on a rock. Grinding usually generates considerable heat, and care must be used not to change the internal physical properties of the stock by this heat. Excessive heat can soften a hard steel. Parts being ground are often cooled by oil or water either continuously or intermittently to prevent this softening from occurring.

Grinding wheels that have been properly "dressed" can produce surfaces that have very close tolerances. Dressing a wheel keeps the wheel round and the face flat. Keeping a grinding wheel properly maintained is very important. Grinding wheels that are rounded, grooved, or uneven will not produce good cuts. Figure 4–45 shows one of several types of grinding wheel dressers. These dressing tools are made either from very high carbon steel "wheels," which rotate against the uneven grinding wheels and knock off the high points, or from diamond-tipped tools, which simply cut away the uneven parts of the wheel. Grinding is often a slower technique of material removal than the others already mentioned so generally it is used as a finishing operation. One caution about grinders. Soft materials such as aluminum, zinc, lead, and pot metals

a.

b.

Figure 4-42. **a.** Large sheet metal shear. **b.** Small sheet metal shear.

Figure 4-43. Punch.

Figure 4-44. Grinder.
Photograph courtesy of Baldor Electric Company, Fort Smith, AR.

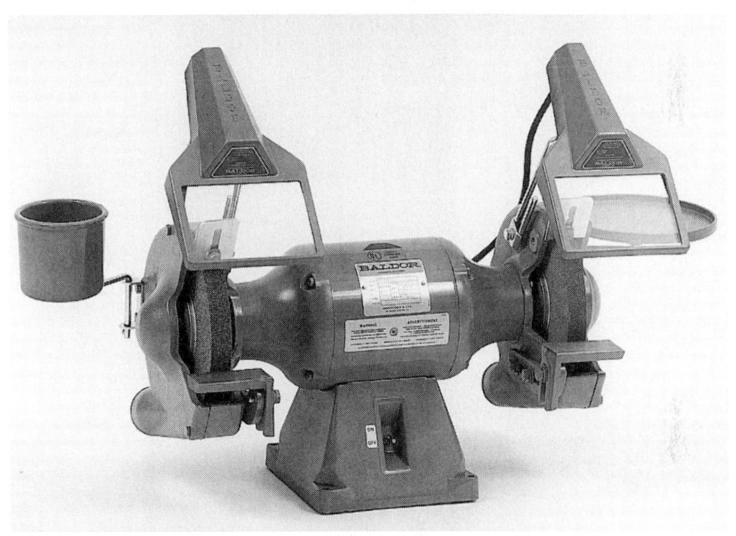

Figure 4-45. Grinding wheel dresser.
Photograph courtesy of Baldor Electric Company, Fort Smith, AR.

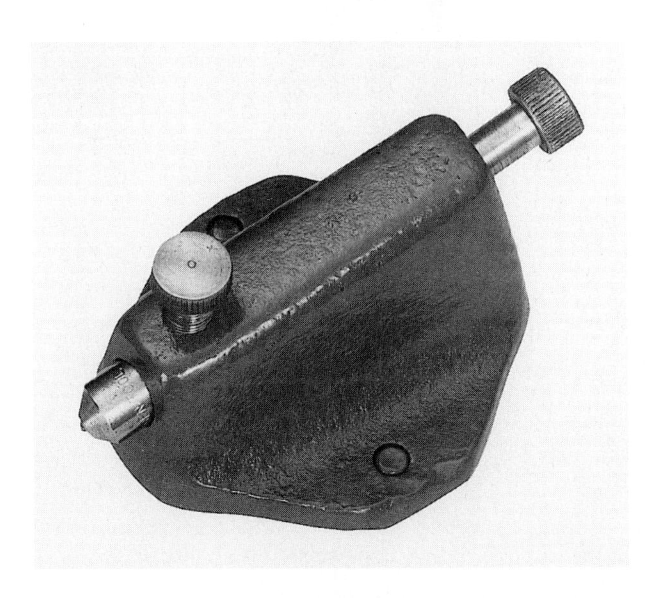

should not be placed on grinding wheels. These metals simply gum up the wheels and prevent them from cutting.

Tool Post Grinder

Cylindrical surfaces can be ground on a lathe by the use of a *tool post grinder* (Figure 4–46).

Blanchard Grinding

Flat surfaces are finished by a process called *Blanchard grinding* in which the grinding wheel is moved across the surface of the stock in a machine. Figure 4–47 shows a Blanchard grinding machine used in industry.

Figure 4–46. Tool post grinder.

Figure 4–47. Blanchard grinder.

Grit Size

Grinding wheels and abrasive papers and cloths come in many different *grits.* The grit of a paper or grinding wheel refers to the size of abrasive particles. For instance, a "100 grit" sandpaper has abrasive particles of uniform size that will just pass through a "Number 100 sieve." The number 100 refers to the number of openings per linear inch. A 100 sieve will have 100 openings per linear inch (Figure 4–48). The higher the number the finer the grit.

Selection of Grinding Wheels

Grinding wheels are made in many sizes and of several different materials. The selection of the proper type of grinding wheel depends on the material being ground, the speed of operation of the wheel, and the type of ground surface desired. Many grinding wheels use high-hardness, naturally occurring minerals such as the following:

- Diamonds
- Corundum (aluminum oxide)
- Emery
- Garnet
- Quartz (flint)
- Pumice
- Diatomaceous silica (infusorial earth or tripoli)
- Rouge and crocus (finely powdered iron oxide used for polishing)

Some of the abrasives used in grinding wheels are synthetically made and include the following:

- Crystalline alumina (trade names: Borolon, Aloxite, Lionite, and Alundum)
- Silicon carbide (trade names: Carborundum, Carbolon, and Crystolon)
- Boron carbide (trade name: Norbide)
- Crushed steel
- Diamonds (Most industrial diamonds are synthetic.)

Further, new types of "superabrasives" are now appearing on the market. These materials last longer than conventional abrasives, but are more expensive.

In making the wheel, these abrasives are bonded by means of some cementing agent. Wheels are made by *vitrification,* the *silicate process,* or *organic bonding.* In the vitrification process, the abrasive grains are mixed with a glass or porcelain, molded into a wheel, then fired at a temperature of 2500°F. Very large wheels can be made by this method.

In the silicate process, sodium silicate ("water glass") and fillers are used as the bonding agent. The mixture is tamped in a mold and baked at a moderate temperature.

Organically bonded wheels use a mixture of abrasive with such organic materials as shellac, rubber, or synthetic resins. Synthetic-resin-bonded wheels can be used in very high speed grinding wheels and are used in the majority of high-speed wheels used in foundries, welding shops, and cutoff wheels. While the vitrified wheels are limited to about 6500 surface feet per minute, resin-based wheels can be operated as high as 9500 surface feet per minute. A surface foot

Figure 4–48. Sieve screens.

per minute is the speed at which a single cutting point passes the surface of the material being cut. This is discussed later in this chapter.

Grinding wheels are specified by their grit size and their grade. The grade of the wheel is its relative strength and hardness of the wheel. The grade of the grit depends on the type of abrasive and the bonding material. Grit sizes range among standardized sizes of 8 to 900. (See the section on grit size earlier in this chapter.) Finer size wheels also carry the notation F, FF, FFF, and XF. The hardness of the wheel is given by a letter notation from A to Z with A being the softest wheel and Z being the hardest.

Grinding wheels vary in size and thickness from a diameter of a fraction of an inch (used for making small instruments such as watches) to more than 60 inches. When ordering a grinding wheel, the diameter, thickness, arbor or hole size, grit, and grade must all be specified.

Sanding

Sanding and grinding are often lumped together as one operation. Both of them remove material by abrasion, a very small piece at a time. The differences are subtle. In grinding, the wheel is made by fusing together the abrasive particles. In sanding, the abrasive particles are glued to a cloth or paper. The paper or cloth backing for the abrasive allows much more flexibility in the shape of the grinding surface. While grinders have a circular, conical, or cylindrical wheel, which makes flat surfaces difficult to grind, sanders can have belts or disks that can be run over flat surfaces. Figure 4–49 shows a dual sander that can use either belts or disks. Abrasive papers can be purchased from about a 40 grit to 600 grit. Although grits are sometimes specified up to 1000, these generally are used for polishing rather than for sanding or grinding.

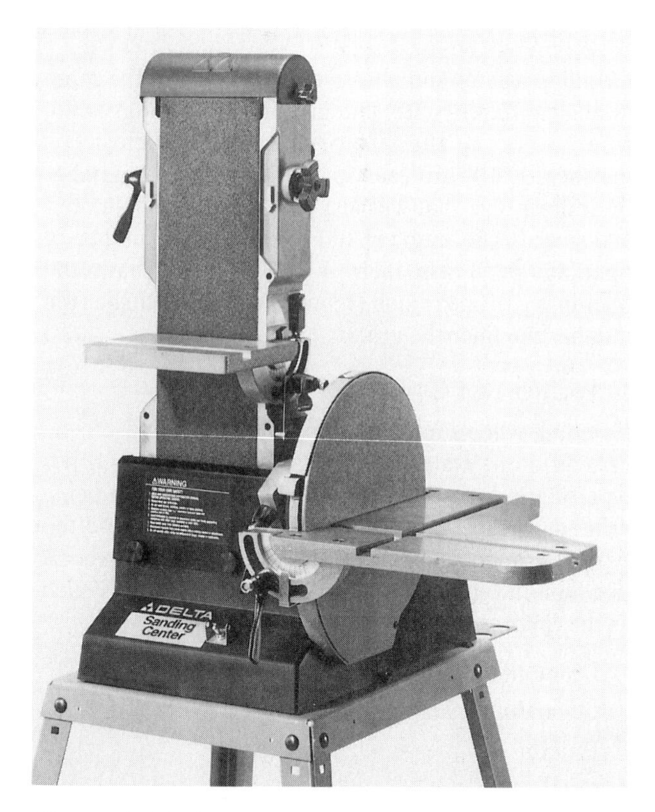

Figure 4–49. Sander.
Photograph courtesy of Delta International Machinery Corporation, Pittsburgh, PA.

Although metals, woods, plastics, composites, and even glass can be sanded or ground, grinding is generally used on the harder surfaces, whereas sanding is used on softer ones. Grinders are better for harder metals because the grinding wheels last longer and the abrasive surfaces can be shaped more precisely to produce the tolerances the metals require. Sanding disks, belts, and other types are cheaper, but wear out faster than grinding stones. Both grinders and sanders can be used for fine or coarse work.

Polishing

There is a difference between sanding or grinding and *polishing*. In sanding and grinding, the material is actually removed from the surface of a piece of material; in polishing, the high points on the surface are simply moved over into the low areas to produce a very smooth and shiny surface. The heat generated due to friction by the polishing wheel or cloth helps in this material movement. Polishing is discussed further in Section VII on finishing.

CUTTING TOOL SHAPING

Cutting tools are custom shaped for the job they are to do. If the plans call for a sharp "vee" to be cut into a piece of stock, then a vee-shaped cutting tool is used. Round cuts require round tools and so on. But besides the shape of the cut, other factors also control the shape of the cutting tools. The material being cut, the quality of cut desired (i.e., roughing versus a finish cut), the feed rate, and feed depth all affect the shape of the cutting tool.

The various surfaces of the cutting tool have been named. To discuss the types of tools, this nomenclature should be learned. Figure 4–50 depicts a common single-point cutting tool for a lathe, planer, or shaper, with the surfaces named. Figure 4–51 breaks the various parts of the cutting tool into one surface at a time for ease of discussion.

The side, back, and end rake angles are determined by the materials being cut and the type of cut being made. Hard materials require very little side or back rake angle. Materials such as aluminum, which tend to stick to the tool and pile up on the end, require a larger rake angle. The pileup of cut material on the tool creates a dull edge and the tool will not cut smoothly.

The *back rake angle* usually controls the direction of chip flow. The side and back rake angles can also act as chip breakers. Small rake angles bend the chips at greater angles and tend to break them up in small pieces. This is desired for roughing of stock but the cut is not as smooth as provided with larger rake angles.

The *side* and*end relief angles* are put on the cutting tool simply to keep the bottom of the cutting tool out of the way and keep the part of the tool *behind* the cutting edges from touching the stock. The relief angles should not be very great because they tend to weaken the cutting tool. Relief angles that are too large could cause the tool to break if given a sudden blow. The larger the stock being turned, the smaller the relief angles can be. Also "roughing" tools used for making deep cuts and removing large amounts of material quickly should have very shallow relief angles.

The *end* and *side edge angles* are dependent on the length of chip contact and the feed rate. In general, large cutting edge angles (especially the end cutting edge angle) tend to weaken the tool point. This would require the tool to be sharpened more often and,

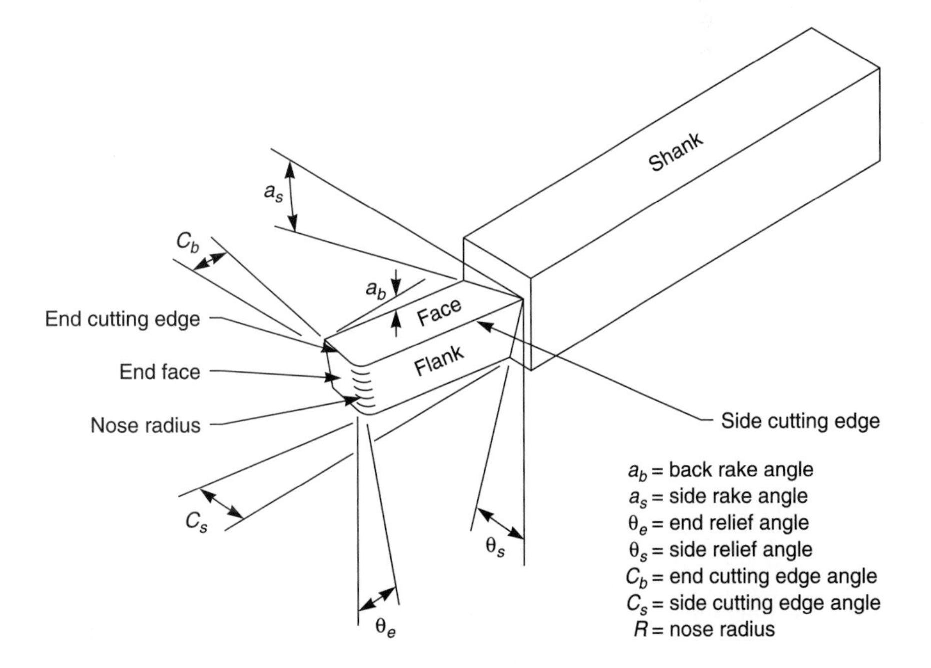

Figure 4–50. Cutting tool nomenclature.

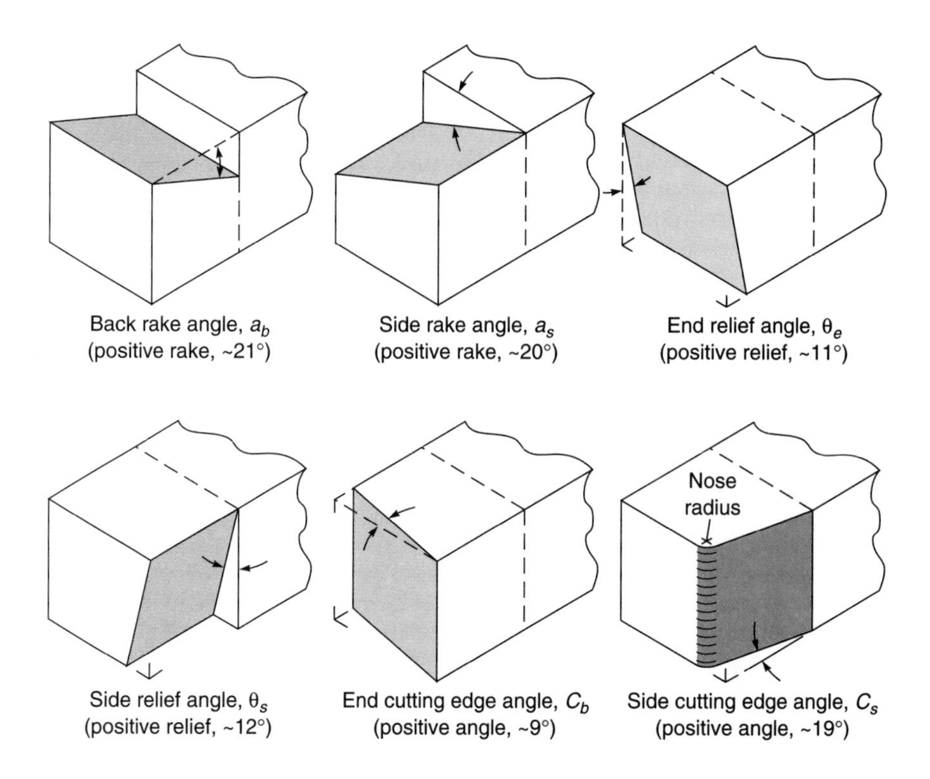

Figure 4–51. Cutting tool angles.

therefore, reduce production. However, an end cutting edge angle that is too small causes excessive force perpendicular to the workpiece. These excessive forces require more horsepower in the motor to get the same depth of cut.

A large nose radius, resulting in a more rounded tip, makes the tool stronger and is generally used for roughing cuts. However, the larger radii can cause tool chatter. A small nose radius is preferred for small-diameter workpieces and for finishing cuts.

A good starting geometry for cutting tools for work on mild steel stocks, aluminum, and hard plastics is listed in Table 4–1.

Cutting tools are usually specified by designating the angles in the following order:

$$a_b, a_s, \theta_e, \theta_s, C_b, C_s, R$$

Thus, a tool specification for a steel might read:

5,10,8,6,5,15,3/64

while a cutting tool for aluminum might be:

20,15,12,12,6,15,1/8

Cutting tools are made from specially formulated and alloyed steels known as *high-speed steels* (HSSs). These steels are generally high-carbon steels containing vanadium, molybdenum, and chromium. Other blades and cutting tools have pieces of tungsten carbide fused to the cutting surface. Carbide inserts can also be gripped in specially designed shanks. Several of the standard carbide inserts are shown in Figure 4–52. The use of ceramic cutting tools for cutting metals is now being adopted by industry. For roughing purposes, the HSS cutting tools are the best choice. They are inexpensive, can be easily resharpened, and are not extremely brittle. The HSS tools will take considerable shock. Their drawback is that they tend to dull faster, especially in the cutting of harder metals.

Carbide tips will cut harder steels, but they are brittle and should not be used for roughing purposes. These tips are more difficult to resharpen than the HSS variety. Because they will take heat better than the HSS, they can be run at higher revolutions per minute with a resultant increase in production. Carbide-tipped tools can produce closer tolerances and

Table 4–1. Recommended Cutting Tool Geometry

Angle	Abbreviation	Steel	Aluminum	Plastics
Back rake angle	a_b	0–20°	20–30°	0°
Side rake angle	a_s	5–15°	15–18°	0°
End relief angle	θ_e	5–12°	10°	15°
Side relief angle	θ_s	5–12°	8°	10°
End cutting edge angle	C_b	5–6°	5°	5°
Side cutting edge angle	C_s	15°	15°	15°
Nose radius	R	3/64 in.	1/8 in.	1/8 in.

Figure 4–52. Carbide cutting inserts.

better finishes than the HSS tools can. Often a roughing cut will be made with a HSS tool and the machinist will then use a carbide tool to finish the work. Carbide tips are more expensive than HSS tips, but the increase in production often makes their use cost effective.

Ceramic tools are not affected by heat, and can be operated at extremely high revolutions per minute. However, these tools are similar to glass in brittleness. Ceramic tools are generally used only for the final, very light cut on very hard steels. Titanium and other extremely hard metals are also machined with ceramic cutting tools. Ceramic tools can give extremely fine finishes to the surface of a material since the lathe can be operated at maximum rotational speed.

FEEDS AND SPEEDS

A few quick definitions are in order here: *Cutting speed* is the velocity of the surface of a workpiece as it passes the cutting tool. In machines in which the workpiece is held stationary, the cutting speed is the velocity at which the cutting tool passes over the stock. Speed is usually given in surface feet per minute (SFPM). *Spindle speed* is the rotational speed in revolutions per minute at which the lathe, milling machine, saw, grinder, or drill press is running.

Feed is the rate of advance of the cutting tool per revolution. *Depth of cut* is the distance to which the cutting tool enters the workpiece. The width of the removed chip is equivalent to the depth of the cut.

Heat is the biggest enemy of machining. Heat is produced by friction between the cutting tool and the stock. The higher the speed at which the tool passes the stock, the more heat is generated. It doesn't matter whether the stock is turning against a stationary cutting tool or the tool is moving against the stationary stock. The result is the same. Heat is concentrated in the very small cutting tip of the tool. At excessive speeds, the cutting edges can melt, requiring the tool to be resharpened or a new one inserted. This poses a dilemma. For high production rates, the drill, lathe, milling machine, or saw should be run at high speed. Yet high speeds reduce tool life, which slows production rates. Graphing tool life and production rate against the speed of the machine would produce a graph similar to Figure 4–53. From this graph, it is evident that the optimum cutting speed for maximum production would be at the point of intersection of the two lines. This optimum cutting speed depends on three factors: (1) the material being cut, (2) the material of the cutting tool, and (3) the diameter of the stock being turned on a lathe (or the diameter of the drill or milling tool). In practice, the actual production rate that can be achieved depends also on the sharpness of the tool, the uniformity of the material being cut, the actual grain structure of the workpiece, and the type and condition of the coolant fluid. Therefore, it is the responsibility of the machine operator to be sensitive to changes in these conditions. The optimization of production often depends on the machine operator adjusting quickly to these changes.

However, it *is* possible to give some idea of an appropriate starting point at which to set feeds and speeds for cutting materials. Table 4–2 gives a few selected recommended cutting speeds for different materials using HSS or carbide cutting tools. Once the machining operation is started, these values should be modified to optimize tool life and productivity.

Note that the recommended cutting speeds are given in SFPM or surface feet per minute; that is the number of feet of stock surface that passes the cutting tip per minute. To determine the SFPM, multiply the circumference of the stock or the circumference of the cutting tool (whichever is moving) by the revolutions per minute. The circumference can be found by multiplying the diameter by pi (3.1416). The equation for the recommended turning rate for optimum production is therefore:

$$N = C_s \pi D$$

where N = spindle speed (rpm), C_s = recommended SFPM, and D = diameter (ft) of the rotating piece. If a 3-inch-diameter piece of low carbon steel was being turned on the lathe and cut with a HSS cutting tool, the recommended rpm would be

$$C_s \text{ (from Table 4–2)} = 160 \text{ SFPM}$$

$$N = \frac{160 \text{ SFPM} \times 12 \text{ in./ft}}{3.1416 \times 3 \text{ in.}}$$

$$= 204 \text{ rpm}$$

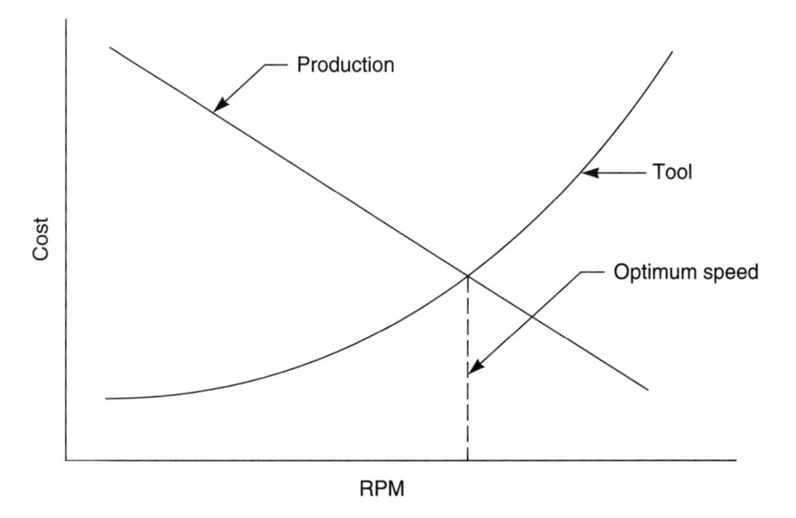

Figure 4–53. Optimum cutting speed.

Table 4–2. Recommended Cutting Speeds and Feeds

Workpiece Material	Turning				Milling				Drilling[1]
	HSS		Cemented Carbide		HSS		Cemented Carbide		HSS
	Speed (SFPM)	Feed (IPR)[2]	Speed (SFPM)	Feed (IPR)[2]	Speed (SFPM)	Feed (IPM)[3]	Speed (SFPM)	Feed (IPM)[3]	Speed (SFPM)
Aluminum	300	0.030	1200	0.031	750	20	1200	80	550
Brass	300	0.030	600	0.108	150	20	350	34	300
Copper	300	0.030	750	0.090	150&20	1000	60	200	
Iron, cast (medium)	75	0.015	260	0.062	95	24	225	24	150
Magnesium	900	0.030	1200	0.030	600	20	1200	80	500
Steel[4]									
Cold rolled	150	0.015	375	0.035	90	18	175	19	160
Low carbon (annealed)	160	0.015	550	0.035	125	20	350	40	220
Stainless (316)	100	0.015	350	0.030	70&20	250	30	90	100

Notes:

1. Feeds same for all materials:

Drills diameter (in.)	0.125	0.250	0.500	0.750	1.000
Feed (IPR)	0.004	0.006	0.010	0.012	0.014

2. Inches per revolution.
3. Inches per minute. (Some sources specify starting feed per tooth (inches).)
4. Speeds and feeds for steel are a function of Brinell number; these are average values; optimum values may differ significantly from the table values.

Of course, no lathe could be set at exactly 204 rpm so the probable setting would be 200 rpm. The lathe could be run faster, but the tool would get dull and wear out faster, or the lathe could be run slower with a resulting reduction in production rate.

If a piece of aluminum was being milled with a 1.5-inch-diameter milling tool, then the recommended speed on the mill would be

$$C_s \text{ (from Table 4–2)} = 750 \text{ SFPM}$$

$$N = \frac{750 \text{ SFPM} \times 12 \text{ in./ft.}}{3.1416 \times 1.5 \text{ in.}}$$

$$= 1910 \text{ rpm}$$

This may be faster than many mills can run, so the machine would be set at maximum rpm. A nomograph that solves this cutting speed equation is found in Appendix B.

Feed and Depth of Cut

The feed and depth of cut are limited by the forces placed on the cutting tool. Once the speed is set, the feed needs to be selected to maximize the rate of production without overheating the tool. If the carriage is operated by hand, gradually increasing the feed will increase the temperature of the tool. On a steel workpiece, the chips will discolor as the temperature increases. Therefore, the feed can be increased to the point where the chips begin to have a bluish cast, which represents the maximum tool temperature HSS tools can safely endure. No such convenient color indicator is available when machining aluminum, magnesium, or other materials.

Remarkably, the temperature will not rise very much if the *depth of cut* is increased. This is because the length of the tool edge in contact with the hot chip will increase exactly as the depth of cut increases, keeping the rate of heat production per *inch of cutting edge* a constant. However, depth of cut *does* directly affect the power required as is shown later.

For a lathe operation in which the carriage is being advanced by the lead screw, it is natural to specify the feed as "so many inches per revolution." This amount is easily determined from the setting of the lead-screw gear box:

$$\text{lead (inches per revolution)} = 1/\text{gear box reading (threads per inch)}$$

Tool life can be improved by reducing the feed and retaining the production rate by increasing the depth of cut. However, increasing the speed and/or the feed and/or the depth of cut will increase the power requirement that the motor must provide. It will also increase the forces on the workpiece, and increase the stress on all parts of the lathe. These increased forces may present problems of chip removal (long continuous chips are often formed) and cause "chattering" and tool failure.

The calculated rpm is used as a starting point. Most machinists usually run the lathe a bit slower while making roughing cuts to save tool life. However, finishing cuts generally have a slower feed but higher rpm to leave the surface of the metal in a smoother condition.

■ COOLANTS FOR CUTTING TOOLS

The greatest cause of tool cutting-edge failure is overheating. Therefore, tool cutting-edge temperature generally is the limiting factor in the production rate of the cutting tool. Although cutting operations can be done "dry," a "coolant" is generally considered to be a basic part of any cutting or grinding operation.

A coolant sprayed onto or running through the point of chip formation can accomplish the following:

1. Significantly reduce tool temperature by carrying away heat.
2. Reduce heat generation by lubricating the points of contact between the chip and the tool face and also between the cut working surface and the tool flank.
3. Carry away chips that could interfere with a smooth cut on the workpiece.
4. Suppress corrosion both on the material being cut and the machinery. Some materials, such as magnesium, tend to corrode very easily and even catch fire in some cases. Appropriate cutting fluids reduce this effect considerably.

A variety of liquid coolants is available. They can be classified into the following categories:

1. *Chemical:* simply a water-based solution
2. *Emulsions* (soluble oils): oil with an emulsifying agent to make it dissolve in water
3. *Semichemical fluids* (semisynthetics): less water and a greater variety of additives
4. *Cutting and grinding oils:* a combination of animal, mineral, and vegetable oils or fats, with some additives

The choice of coolant type for a specific application is based on the competing considerations of cooling, lubrication, corrosion prevention, bacterial-growth control, operator safety, wetting ability, compatibility with other components of the system, recyclability, and cost.

The coolant typically costs about 2.5% of the manufacturing cost of a product. The great majority of the remaining product cost is labor and tools. It is clear that if changing to a more expensive cooling fluid can reduce both labor and tooling costs by 10% (a reasonable figure), then using the more expensive cutting fluid may save money overall.

Coolants degrade with use by, among other things, evaporation, contamination by lubricating or hydraulic oils from the machine, zinc from funnels or other galvanized containers, hard water mineral, bacteria, machine paint fragments, dirt, and minute particles of removed material. Disposal of waste cutting oils is a problem and some of them must be removed to hazardous waste dumps. Others may be reprocessed economically and reused almost indefinitely, thereby reducing both production costs and degradation of the environment.

POWER REQUIREMENTS FOR MACHINING

In selecting a lathe to perform a given task, several factors must be considered. Is the lathe large enough to handle the size of stock required? The size of the headstock, the swing or clearance over the bed, and the length of the bed must be considered. Another factor is the horsepower requirement needed in the motor to meet the designed production rate. The steps in computing the horsepower requirements are as follows.

First the material removal rate must be calculated. How many cubic inches per minute of material are to be removed from the stock? Think of the material removed as a cylinder with a thickness equal to the depth of cut and a length equal to the length of the travel of the cutting tool in 1 minute. In a lathe this would amount to multiplying the depth of cut (in inches) by the tool feed rate in inches per revolution (IPR) and the recommended cutting speed. Since the recommended cutting speed is in surface feet per minute, this must also be multiplied by 12 to get cubic inches. The formula for the material removal rate (MRR) would be:

$$\text{MRR} = 12fdC_s = fd\pi DN$$

where f = the tool feed rate (inches/revolution), d = the depth of cut (inches), C_s = the cutting speed (SFPM), D = the workpiece diameter in inches, and N = the spindle speed in rpm. The tool feed is often set on a lathe as the number of cuts per inch. This is done to make turning of threads on the lathe easier. For instance, if 13 threads per inch are to be turned, the controls would be set on that number. The tool feed rate f for the preceding equation would then be the reciprocal of that number. For instance, if the lathe was set to cut 60 threads per inch (TPI), then the tool feed rate f would be

$$f = 1/60 \text{ TPI} = 0.0167 \text{ in./revolution}$$

Consider the problem of a lathe cutting 60 threads per inch, a depth of feed of 1/8th inch, and a recommended cutting speed for the material being cut of 300 SFPM. The material removal rate would then be:

$$\begin{aligned}\text{MRR} &= 12 \text{ in./ft} \times 0.0167 \text{ in./revolution} \times 0.125 \text{ in.} \\ &\quad \times 300 \text{ ft/min} \\ &= 7.5 \text{ in.}^3\text{/min}\end{aligned}$$

(The units of "revolutions" are often omitted since they do not measure length or arc.)

Power is the rate at which energy is used per unit of time. As discussed in Chapter 3, one horsepower is defined as the power required to raise 550 pounds one foot in one second. The horsepower requirements for a machine tool depend on the cutting force on the tool and the cutting speed. The total force on the cutting tool is the vector sum of the downward force on the cutting tool F_c due to the rotation of the stock against it, and the force driving the cutting tool along the axis of the stock F_f. Because the tool moves a very small fraction of the distance that the stock is turning against the tool, F_f is very small and can be ignored in these calculations (Figure 4–54). The horsepower requirement at the cutter is then found by the equation

$$P_c = F_c C/33{,}000$$

The F_c is a function of the hardness and toughness of the material being cut, the cutting tool geometry, and the friction generated in cutting the chip. It is therefore very difficult to calculate. Usually tables are used that show the unit horsepower requirements or the horsepower required to remove one cubic inch of material per minute (P_u). The unit horsepower requirements of a few selected materials are given in Table 4–3.

The power requirement to run the machine is then calculated by multiplying the unit horsepower requirements by the material removal rate or

$$P = P_u \times \text{MRR}$$

Thus if a stainless steel shaft with a Brinell hardness number of 200 were turned at a MRR of 7.5 inch^3/minute, the power needed at the tool would be:

$$\begin{aligned} P &= 1.7(\text{hp/in.}^3/\text{min}) \times 7.5(\text{in.}^3/\text{min}) \\ &= 12.75 \text{ hp} \end{aligned}$$

Of course, this is assuming a 100% efficiency in the motor and lathe. If a machinery loss factor is applied, the required horsepower should be divided by the efficiency of the machine. If the machine just discussed was 70% efficient, the power of the motor should be:

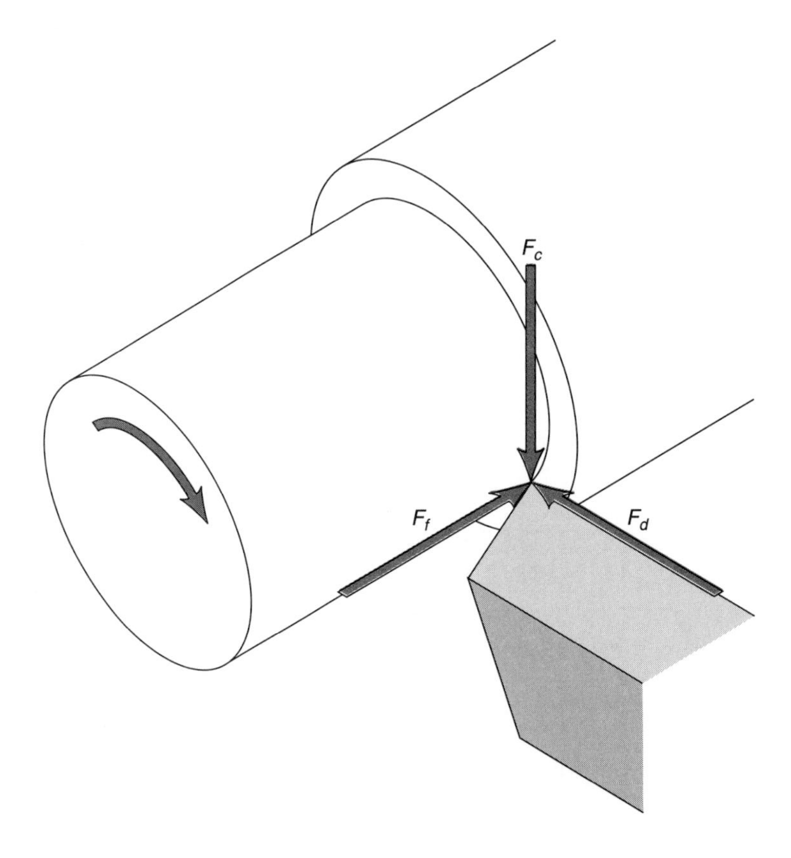

Figure 4–54. Forces of a lathe cutting tool.

Table 4–3. Unit Horsepower Requirements

Material (BHN)	Hardness (P_u)	Unit Power (hp/in.^{3}/min)
Steel	120	1.2
	200	1.4
	300	1.8
400	2.0	
Stainless steel	200	1.7
	400	2.3
Cast iron	200	1.1
Magnesium	50	0.2
Aluminum	100	0.3
Copper	150	1.2
Titanium	250	1.2

$$P = 12.75 \text{ hp}/0.7 = 18.2 \text{ hp}$$

An 18- to 20-hp motor would probably be installed, depending on which could be bought most economically.

The power requirements for milling, drilling, and other machines are calculated in the same way. The unit horsepower requirements for the various cutting tools are very similar to those just given.

PROBLEM SET 4–1

1. A 4-in.-diameter piece of mild, low-carbon steel is to be turned on a lathe using a carbide cutting tool. What is the optimum speed of the lathe? (*Hint:* Use Table 4–2.)
2. A 0.5-in.-diameter hole is to be drilled in a piece of 316 stainless steel with a HSS drill. At what rpm should the drill press be set?
3. A lathe has a maximum speed of 1500 rpm. Could it be run at this maximum rpm using a carbide-tipped tool to cut a 2-in.-diameter piece of aluminum?
4. Consider a lathe operated at 280 rpm with a feed rate of 120 TPI. To take a 0.010-in.-deep cut on a 3-in.-diameter piece of brass, what is the material removal rate?
5. We want to operate a lathe to cut a 4-in.-diameter piece of mild steel (BHN of 120) at 200 rpm. The depth of cut will be 0.05 in. and the feed rate will be 60 threads per inch. What horsepower will be needed at the tool?
6. We want to cut a piece of pipe to be welded to a steel plate. List three methods by which the pipe could be cut. Which would be the best method? Why?
7. What cutting tool material would be best suited for mass production of aluminum parts? Why?
8. At what rpm should a drill press be set to drill a 0.25-in. hole in a piece of stainless steel? Work from Table 4–2.
9. Would a carbide-tipped cutting tool be a wise choice for cutting a piece of copper? Defend your answer.
10. You have available a power hacksaw, a vertical milling machine, and a grinder to remove 0.25 in. of material. Under what conditions would each be a good choice of tool?

This chapter has discussed the advantages and the limitations of several mechanical methods of material removal. Material can be removed by

- Handwork
- Turning
- Milling
- Shaping or planing
- Routing
- Broaching
- Drilling and boring
- Reaming and honing
- Sawing
- Shearing and punching
- Grinding
- Sanding

For good cutting, the shape as well as the material of the cutting tools is important. The rate at which the tool is fed against the workpiece helps establish the rate of production, the tool life, and the optimum speed for cutting any material. Coolants on the cutting tool help increase the optimum cutting speed and thereby increase production.

The power required at the cutting tool can be approximated by multiplying the material removal rate and a factor called the *unit horsepower requirement* (from a table). Then the cutting-tool horsepower is divided by the assumed efficiency of the machine to calculate the required motor horsepower.

PHOTO ESSAY 4-1

Setting Up the Lathe for Manual Control

STEP 1 Place the stock in the chuck of the lathe and tighten it firmly.

Although three jaw chucks generally center the stock, center alignment should be checked periodically. With four jaw chucks, the center alignment must be checked for each piece. This is especially true if the jaws are independently adjusted. (*Note:* For safety reasons, never leave a tee-wrench in the chuck. Notice that the spring on the tee-wrench will pop the wrench out of the chuck if left unattended.)

STEP 2 With the stock protruding slightly from the headstock, bore the hole for the tailstock center. This is done with a special countersinking tool held in a chuck in the tailstock.

STEP 3 Move the stock out of the headstock the desired length and retighten the chuck or collet. Place the tail center in the hole just bored. (*Note:* If the tailstock does not rotate with the stock, it is called a *dead-center*. If it rotates with the stock, it is a *live-center*.)

STEP 4 Place the proper cutting tool in the tool holder. Place the tool holder in the tool post and adjust the cutting tool to meet the stock just slightly above the centerline of the stock. Keep the length of the tool protruding from the tool holder and the length of the tool holder from the tool post as short as possible to prevent excessive vibration and bending of the tool.

STEP 5 Set the feed rate. This is done differently on different types and makes of lathes. Here it is done by setting gears for the number of cuts per inch.

STEP 6 Adjust the speed of the lathe. This is a variable-speed lathe so the lathe must be turned on first. With some lathes the speed is adjusted by belts or gears while the lathe is turned off.

STEP 7 Set the depth of cut. This is done by adjusting the cross-feed dials.

STEP 8 Engage the transverse feed and make the cut.

STEP 9 At the end of the cut, stop the carriage by disengaging the worm gear. It is usually a good procedure to stop the carriage a few millimetres before the end of the cut and finish the cut by moving the hand feed WHEEL.

STEP 10 Back the cross-feed out so the cutting tool does not touch the stock. Move the carriage back to the starting point and repeat Steps 7 through 10 as many times as necessary to complete the part.

Stop the lathe periodically to measure the diameters to determine the remaining material to be removed. Cutting tools may have to be changed for the last or finishing cuts. Special shapes may also require special cutting tools.

Note that in modern automated and computer-controlled turning operations, most, if not all, of the preceding steps are done automatically or by robots. Still many small machine shops use manually controlled lathes.

Photo Essay 4–2 Drilling

STEP 1 Mark the location of the center of the hole with a sharp scribe.

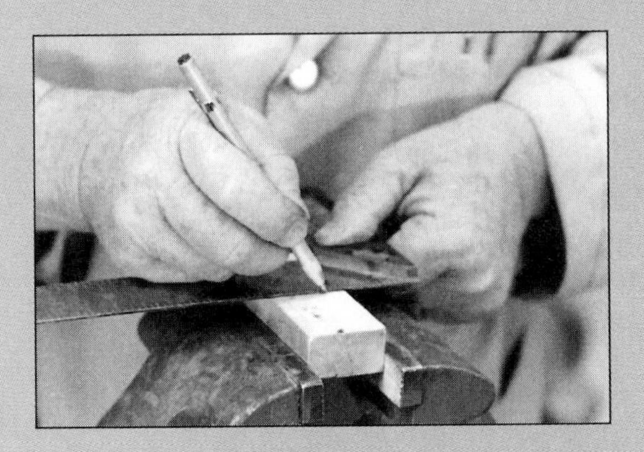

STEP 2 Carefully punch the center point into the metal with a center punch.

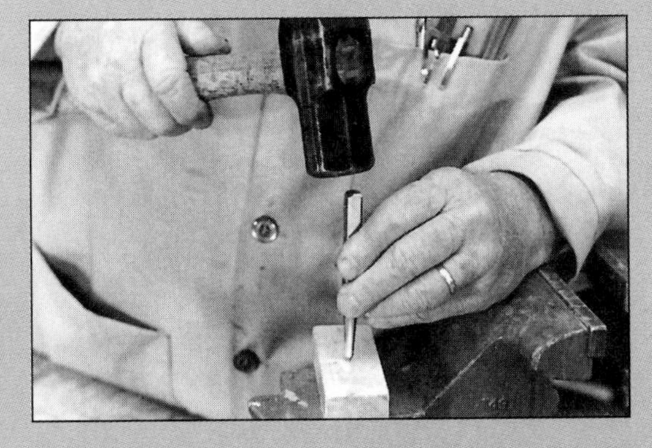

STEP 3 Select the proper drill and insert it into the chuck of the drill press. Make sure the drill is in the center of the chuck. (Small drills can easily be placed off center in the chuck.) Large holes may require a small pilot hole to be drilled prior to using the large drill.

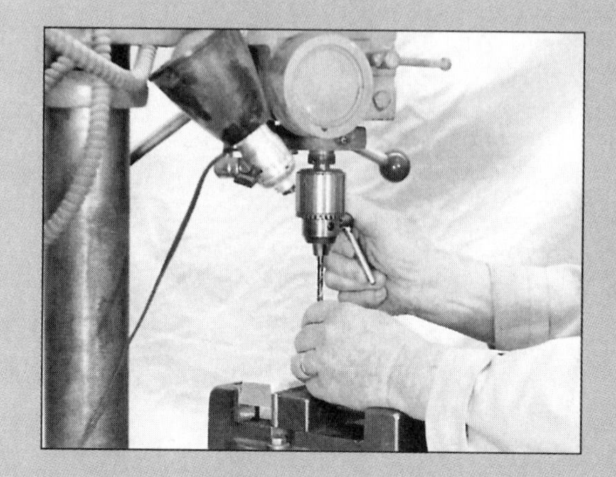

STEP 4 Place the workpiece in the vice, bring the tip of the drill down to the workpiece, and move the vise to locate the center punch mark directly under the tip of the drill.

STEP 5 Securely lock the vice in place on the bed of the drill press. *Do not try to hand hold a metal workpiece while drilling or an accident may result.*

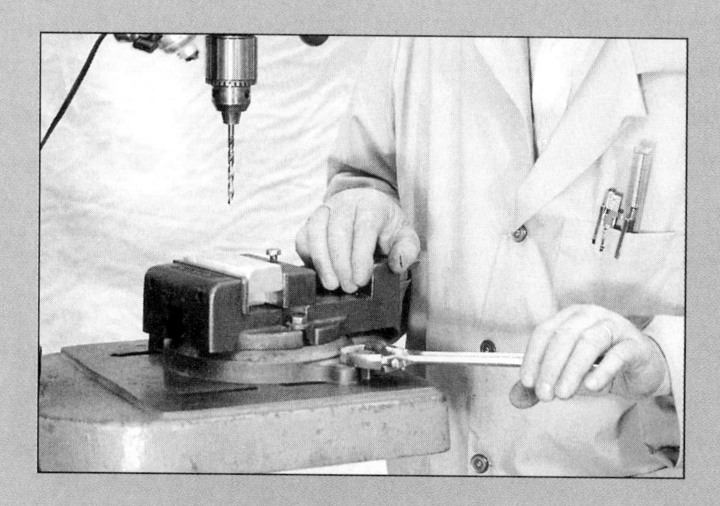

STEP 6 If the drill is not to go completely through the workpiece, set the depth of the drill using the stops on the press.

STEP 7 Set the proper speed (rpms) of the drill press. The recommended speed depends on the size of the drill, the material being cut, and the material of the drill. (*Note:* This is a variable-speed drill press, which allows the drill speed to be set by a dial handle. Many drill presses are belt or gear driven, which requires belts, pulleys, or gears to be shifted to set the drill speed.)

STEP 8 Turn on the press and drill the hole. Apply a firm, steady, but not excessive force to the drill. Let the drill do the cutting. Forcing the drill too fast generates excessive heat, which dulls the drill and causes "drill wobble," resulting in an inaccurate hole.

STEP 9 For large holes, the small pilot drill can be removed and replaced with the larger drill. Do not move the workpiece. The larger drill will require a different depth setting (Step 6) and a lower rpm for the press (Step 7).

Photo Essay 4-3 Milling

STEP 1 Select the proper cutting tool for the job. This includes the size of the tool, the shape of the tool, and the number of cutting edges on the tool. Insert the tool into the milling machine.

STEP 2 Calculate or look up the proper cutting speed for the size and material of the cutting tool and the material being cut. Adjust the milling machine drive mechanisms to the proper rpm setting.

STEP 3 Adjust the vise or fixture for the proper angle of the cut.

STEP 4 Insert the workpiece in the vise, jig, or fixture and firmly tighten.

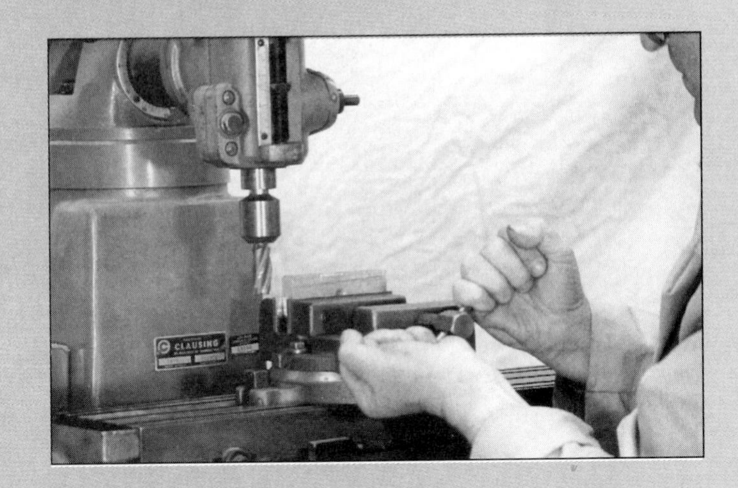

STEP 5 Set the depth of cut by either lowering the cutting tool or raising the bed of the mill.

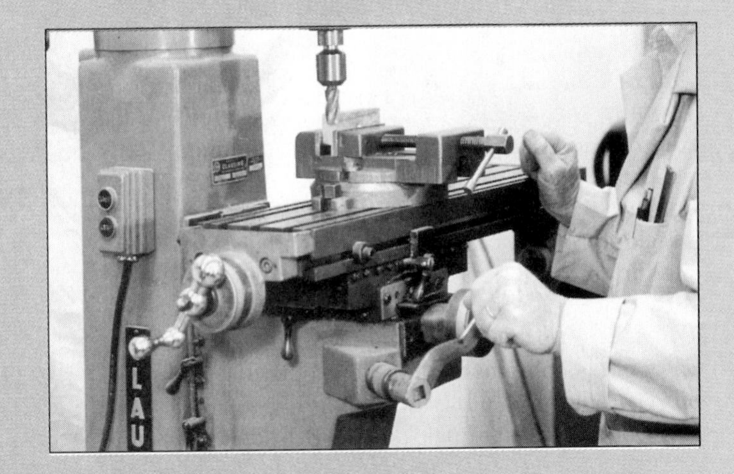

STEP 6 Make the cut. Always make the cut against the rotation of the cutting tool. Cutting in the direction of the rotation of the cutting results in a rough and inaccurate cut. (Note the chips flying from the material. Always use eye and face protection.)

Photo Essay 4-4 Grinding

STEP 1 Dress the wheel. Using a wheel dresser, make sure the grinding wheel is flat across the front and completely circular. The wheel shouldn't vibrate or wobble.

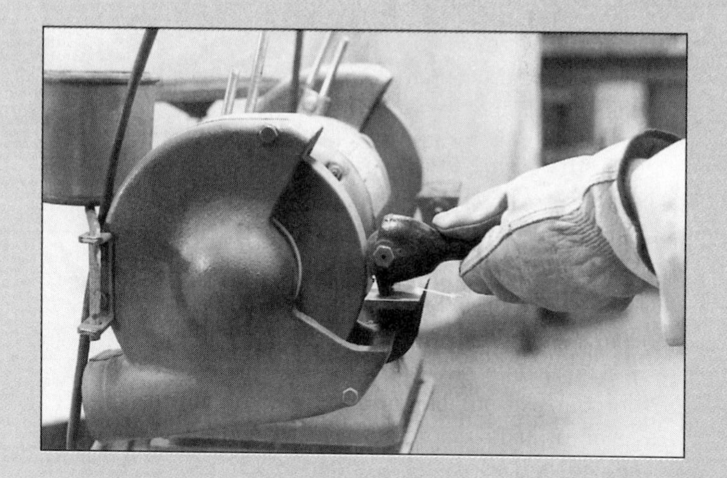

STEP 2 Adjust the rest so that it is no more than 1/8th inch (3 millimetres) between the rest and the face of the grinding wheel. (Greater distances can allow the workpiece to become trapped between the rest and the wheel and cause accidents.) If an angle is to be ground on the workpiece, the rest can be tilted to that angle.

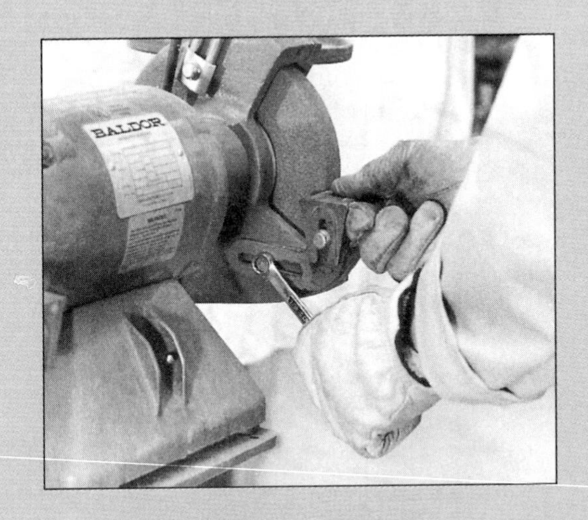

STEP 3 Grind the workpiece to the desired shape. The workpiece should be held firmly on the rest and against the wheel.

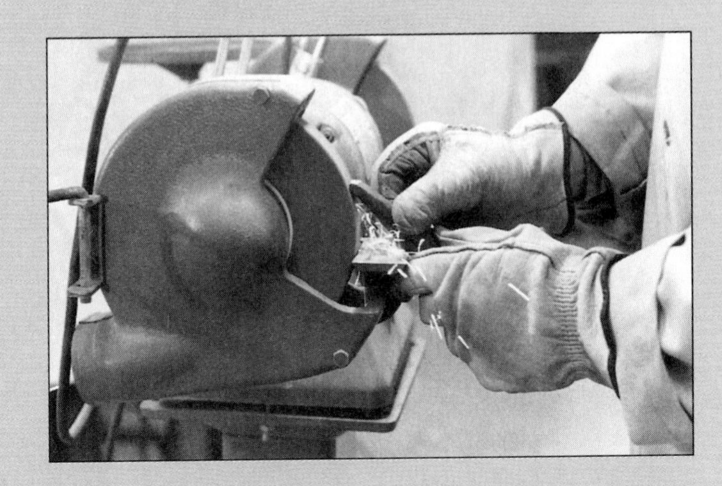

STEP 4 Keep the workpiece cool. It may have to be dipped into water at frequent intervals. If the workpiece is allowed to get too hot, the hardness of the metal may be affected.

Chapter

5

Electrical Methods of Material Removal

■ ELECTRICAL DISCHARGE MACHINING

In electric-arc welding, the electric arc melts the metal where the arc hits it. If that melted metal could be removed continuously, a "separating" process would occur. The process of removing metal with an electric arc is called *electric discharge machining (EDM)*. One of the newer manufacturing processes, EDM allows us to perform some operations that are nearly impossible to do any other way.

The word *discharge* appears in the name because electrical energy is stored in a capacitor and then discharged to create an electric arc (or spark) between the workpiece and an "electrode" many times a second. In the mid-1960s, power supplies became available that were able to produce strings of pulses without the use of storage capacitors, but the name *EDM* is still used. The arc is caused by electrons being drawn from one conductor, through the air or other fluid, such that they hit another conductor. As the electrons pass through the fluid, they knock electrons from the atoms in the fluid making "ions" in the fluid. It is the ionized fluid that causes the spark. As the electrons hit the second conductor, they do so with such force that the atoms in the conductor are melted free from their surrounding atoms.

In EDM, the electrode and the workpiece are generally immersed in a moving insulating fluid, called the *dielectric* (such as kerosene), which is an important part of the process (Figure 5–1). Each arc creates a small glob of melted metal, surrounded by a bubble of gas. Once the current (or electrons) stops flowing, the bubble collapses and the moving fluid carries away the solidified bit of metal. The verb used to describe this action is *erodes.*

If that sounds like a slow process, it is, even though the arcs may be repeated several thousand times a second. However, EDM has many advantages over faster methods of material removal, such as the following:

1. The hardness of the metal is no consideration. That means that (a) extremely hard materials (such as those used in dies) can be formed using EDM and (b) any object that was formed of several materials with different hardness can be machined easily using EDM.
2. Fragile parts are not deformed by the forces usually associated with machining operations. EDM is a gentle process.

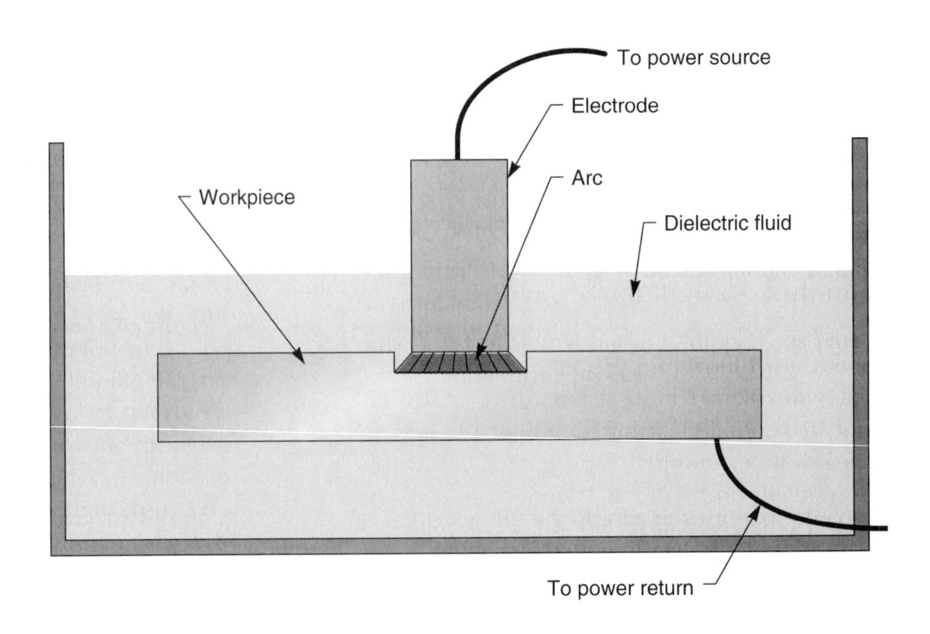

Figure 5-1. Electric discharge machining.

3. Holes with extremely complex shapes can be made using EDM.
4. EDM is easy to automate. It is not unusual to start an automated EDM machine and then come back the next day to remove the finished part.
5. EDM can produce surfaces to very fine dimensional and roughness tolerances.

GENERAL OPERATING PRINCIPLES OF EDM

In EDM, the electric arc jumps between an electrode and the metal workpiece. The design of the electrode is critical to the success of EDM. The first electrodes were generally made of graphite, because graphite carries electricity well and is easy to form, even into complex shapes. That proved to be so successful that the most popular electrode material today is still graphite, although brass, copper, and alloys are used in special applications.

Once the electrode and the workpiece are immersed in the moving fluid and are close enough for the spark to jump between them, erosion begins. The erosion action continues until all of the workpiece material closest to the electrode has eroded and the spark gets weaker. An automatic control senses this condition and moves the electrode closer to the workpiece to keep the sparking current constant. As this process continues, the cavity that forms in the workpiece gradually assumes the shape of the end of the electrode, regardless of how complicated a shape the electrode has.

To be precise, there is always a small gap between the electrode and the workpiece so that the cavity is always slightly larger than the electrode. However, the size of this *overcut* can be controlled accurately so that the electrode can be made undersized by precisely the necessary amount to compensate for this effect.

The speed of any material removal process is generally specified by the rate at which material is removed. The units of material removal are generally cubic inches per minute or cubic centimetres per minute. For EDM it is more appropriate to use cubic inches (or cubic centimetres) per hour. Material removal rates in the range of 0.005 to 1.5 inch^3/hour (0.08 to 24 centimetre^3/hour) are common. Note that where values are as rough and as subject to change as these are, inches-to-metric conversions will be only approximate.

The material removal rate is influenced by the melting temperature of the workpiece material and is faster for materials of lower melting temperature.

Experience has shown that the removal rate is proportional to (melting temperature)$^{-1.24}$; in particular,

$$V = 440T^{-1.24}\ \text{in.}^3\text{/hr-A}$$

or

$$V = 7200T^{-1.24}\ \text{cm.}^3\text{/hr-A}$$

where T is the melting temperature in degrees Celsius. Study the following example. To compare the material removal rates possible for aluminum, steel, and tungsten, use the following numbers:

	Aluminum	Steel	Tungsten
Melting temperature (°C)	660	1530	3410
Temperature$^{-1.24}$	0.000319	0.0001125	0.0000416

To get a rough comparison, multiply by 100,000 and round off

Aluminum	Steel	Tungsten
32	11	4

It is worth emphasizing that it takes eight times as long to remove a volume of tungsten as to remove the same volume of aluminum by EDM because tungsten's *melting point* is higher, not because it is harder.

Material removal rates calculated from the above equations are

	Aluminum	Steel	Tungsten
in.^{3}*/hr-A*	0.14	0.05	0.018
cm^3*/hr-A*	2.3	0.81	0.30

The material removal rate is also proportional to the sparking current; the actual relationship being roughly 0.05 inch^3/hour-ampere (0.05 inch^3/hour-ampere = 0.8 centimetre^3/hour-ampere). That means that a current of 1 A is capable of removing:

$$(0.05\ \text{in.}^3\text{/hr-A}) \times 1\ \text{A} = 0.05\ \text{in.}^3 \text{ of steel per hour}$$

or

$$(0.8\ \text{cm}^3\text{/hr-A}) \times 1\ \text{A} = 0.8\ \text{cm}^3$$

of steel per hour. A current of 20 A would remove approximately:

$$(0.05\ \text{in.}^3\text{/hr-A}) \times 20\ \text{A} = 1\ \text{in.}^3\text{/hr}$$

or

$$(0.8\ \text{cm}^3\text{/hr-A}) \times 20\ \text{A} = 16\ \text{cm}^3\text{/hr}$$

This translates to 24 inch^3 in a 24-hour day.

Another restriction is the maximum current that the electrode can handle without local overheating (and consequent electrode disintegration). For graphite, the maximum design current density for the working surface is 50 amperes per square inch (50 A/in.^{2} or 8 A/cm^2 = 0.08 A/mm^2). This means that if an electrode is to erode a hole into a steel block, and if the cross section of the hole is to be 0.5 inch^2, the maximum current would be

$$(50\ \text{A/in.}^2) \times (0.5\ \text{in.}^2) = 25\ \text{A}$$

Using the numbers from the preceding paragraph, the material removal rate would be:

$$(0.05\ \text{in.}^3\text{/hr-A}) \times 25\ \text{A} = 1.25\ \text{in.}^3\text{/hr}$$

Since the area of the hole is 0.5 inch^2, then the hole will be eroded ("drilled") at a maximum rate of

$$(1.25\ \text{in.}^3\text{/hr})/(0.5\ \text{in.}^2) = 2.5\ \text{in./hr}$$

Surface roughness increases with the sparking current. Therefore, the operator should choose the largest practical current that would create the *desired surface quality.* It is a compromise between producing the necessary surface quality and increasing the current to the maximum limit of the electrode to increase productivity.

Besides the material removal rate for the workpiece, the rate at which the electrode is eroded must also be considered. The *wear ratio* is defined as the workpiece volume removed divided by the electrode volume removed. Wear ratios from 3:1 to 100:1 are common, depending on the combination of materials used. For example, consider the following: The melting temperature of graphite is 3500°C. Following the earlier example, the material removal rate for graphite is approximately 0.018 inch^3/hour-ampere, which is the same as for tungsten. Therefore, the predicted wear ratio for tungsten/graphite is approximately 1:1. For steel/graphite it is 3:1 and for aluminum/graphite it is 8:1. Since this is an oversimplified calculation, actual values may vary considerably from these values. Nevertheless, they illustrate how the wear ratio can vary considerably depending (at least partly) on the materials being worked.

Normally it is difficult to machine any part made of materials of different hardness. For example, a drill bit will wander into the area of softest material rather than go straight. Just try drilling out a hardened steel bolt that has broken off in a piece of aluminum. Not so with EDM. Erosion rates are not affected by material hardness so that holes are truer if made by EDM. Hole length-to-diameter ratios of 20:1 are possible with EDM; otherwise, the limit for a hole made by a conventional twist drill is about 10:1 before drill "wander" makes the hole exceed tolerances.

Manufacturing processes that use force to shear or punch holes in thin stock generally distort the stock and leave a "burr," which then needs to be removed by another machining operation. Again, this is not so with EDM. With EDM it is possible to make small holes in honeycomb, thin-wall tubing, or other fragile structures or parts. The edges of a hole made by EDM are smooth and free of burrs, and the stock is not distorted in any way.

Ram EDM Machines

The earliest EDM equipment was designed to make the cavities in dies by letting the electrode work its way straight down into the metal block until the desired depth was achieved. These early machines were therefore called *die-sinking* or *plunge EDM* machines. Since the vertical position of the electrode was controlled by a hydraulic ram, they were (and are) also called *ram EDM* machines (Figure 5–2).

A die that is used for forging or punching operations must be made out of very hard material such as tool steel, exotic alloys, or carbides. Even then it wears down over the course of a day and needs to have the details "sharpened" before it can be used again. This takes a skilled tool-and-die maker a day or more to do. Therefore, before the invention of ram EDM, it was necessary to have several dies available for the same operation so that one was being used while the others were being reworked. EDM changed all that.

With ram EDM, it is a simple matter to mount the proper electrode in the EDM machine, clamp the dull die in place, and turn the machine on before closing for the night. In the morning, the die is all ready to go for another day! Barring accidents, only one die is needed, which results in considerable cost savings.

With some modifications, ram EDM machines can do far more than straight die-sinking. Two modifications to the basic design are orbiting and rotation of the electrode about an axis perpendicular to the direction of ram motion. *Orbiting* refers to a small circular motion of the electrode. The circular path of any point on the electrode is in a plane parallel to the surface of the workpiece; equivalently, the orbiting is around an axis perpendicular to the surface of the workpiece. The effect of orbiting can best be visualized by imagining that a hole has been sunk into the workpiece and the electrode is being withdrawn. Before the electrode is quite clear of the hole, however, a small orbiting motion is added, increasing the size of the hole. If the size of the orbiting is increased

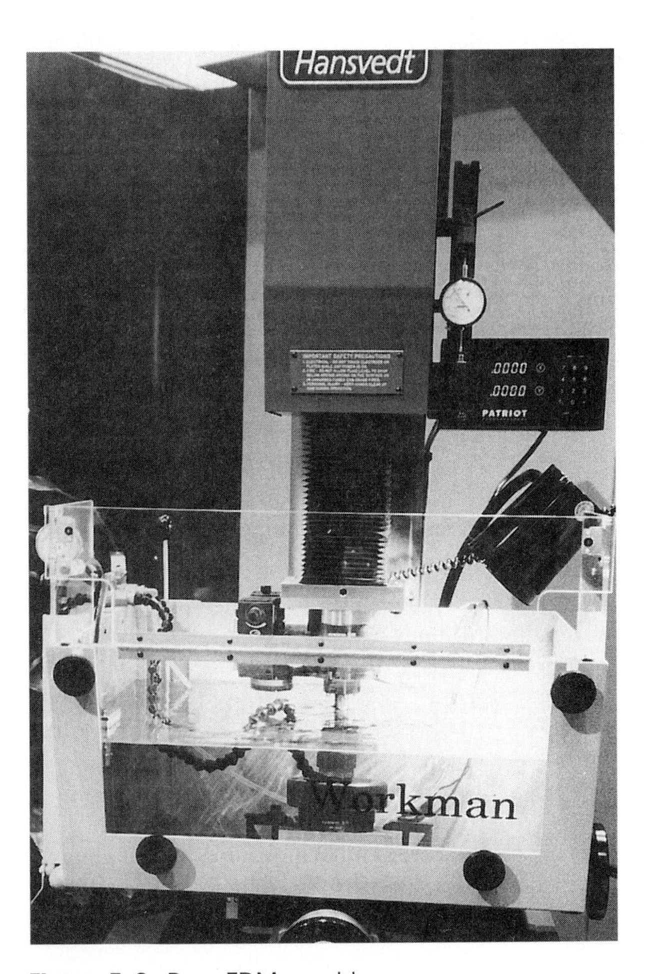

Figure 5–2. Ram EDM machine.

as the electrode is withdrawn from the hole, then the entrance to the hole becomes tapered in a completely controlled way.

Another use of orbiting can be visualized if an electrode having the shape of the head of a nail were to be used. Then, once the "head" has eroded a hole of some depth, if the downward motion is replaced with an orbiting motion, an "undercut" hole is created.

Rotation of the electrode is of value if the head of the electrode is not rotationally symmetric. For example, suppose that an L-shaped electrode is being used. Then, once a hole of some depth has been eroded, the electrode could be revolved slowly about a vertical axis, producing undercutting of a complex shape. Alternatively, threads could be eroded into very hard material by using an electrode in the shape of a tap that is rotated like a conventional tap to cut threads. The possibilities are virtually endless!

Wire EDM Machines

Later forms of EDM machines use a moving wire as the electrode, rather like how one would use a jigsaw. The wire may be made of copper, brass, molybdenum, or tungsten, and it typically has a diameter of 0.002 to 0.010 inch (0.5 to 0.25 millimetre). Wear ratio is not a concern because each part of the wire is "used once," feeding off the supply spool at speeds between 2.5 and 150 millimetres/second (0.1 to 6 inches/second) (Figures 5–3, 5–4, and 5–5).

Since the wire spools can be moved around at several angles, wire EDM machines are capable of making much more complex shapes than are the ram EDM machines. Moreover, they produce parts with a much greater repeatability than can be made by conventional machining techniques. One example is a hole through a block where the outline of the hole is the numeral "3" on one face of the block and the letter "D" on the opposite face (Figure 5–6). It's difficult to imagine how such a hole could be produced by any method other than wire EDM. Note that holes that vary in cross section are not all that rare in manufacturing. Any "nozzle" through which softened material is extruded must have such a shape and be made of very hard material to withstand the large forces involved in the extrusion process. Figure 5–7 shows some other shapes made by means of wire EDM.

With ram EDM, material removal rates are in units of volume divided by time (as one would measure a milling operation). Wire EDM is more like a band saw or a jigsaw: The material removal rate is the "cut" surface area divided by time. Typical values for wire

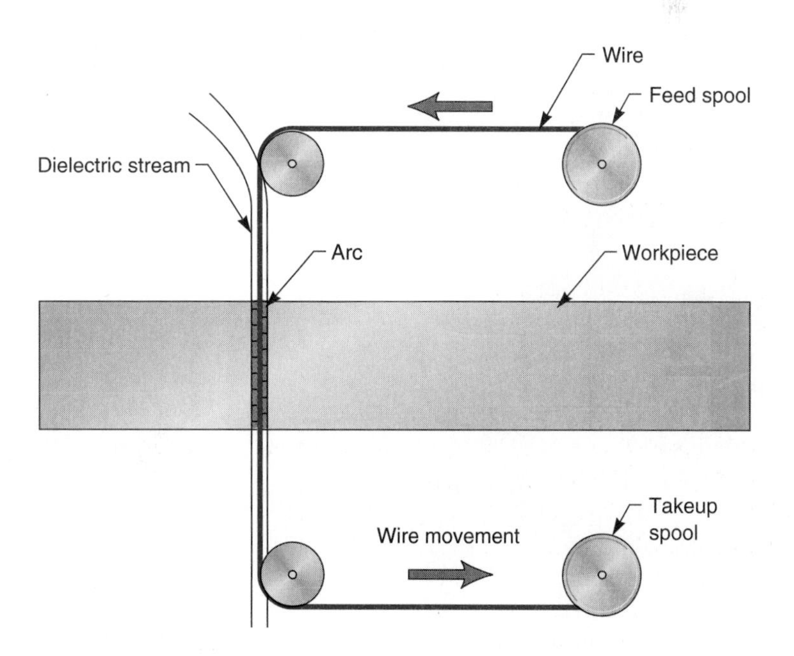

Figure 5–3. Wire EDM.

Figure 5-4. Wire EDM.

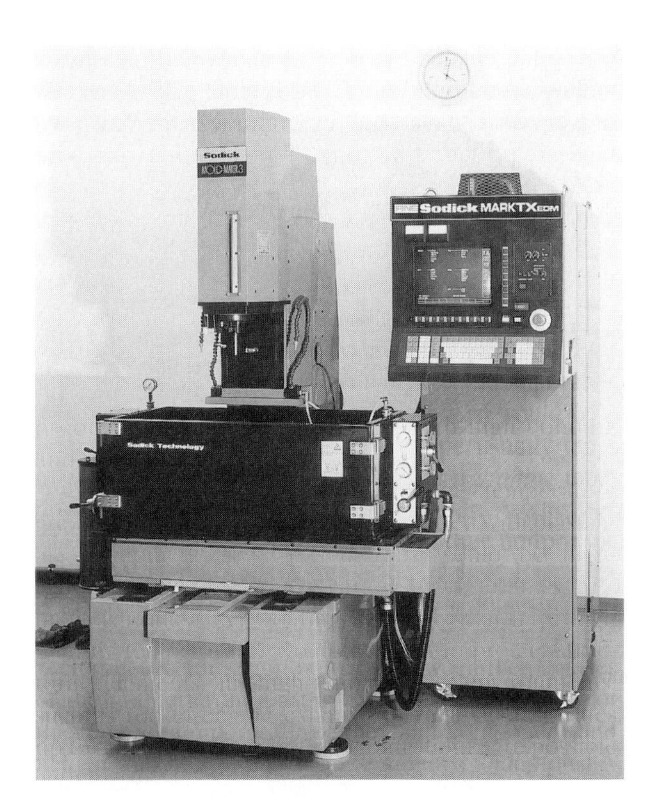

Figure 5-5. Wire EDM.

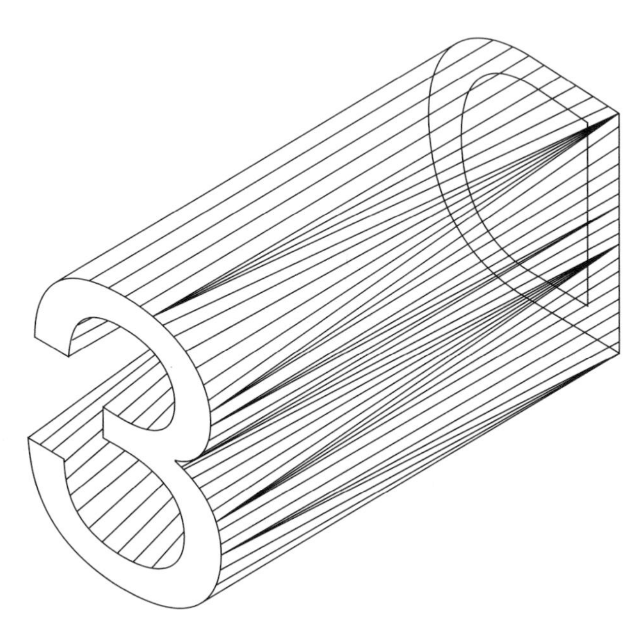

Figure 5-6. Complex shape made by wire EDM.

Figure 5-7. **a.** EDM shapes. **b.** Gear made by wire EDM.

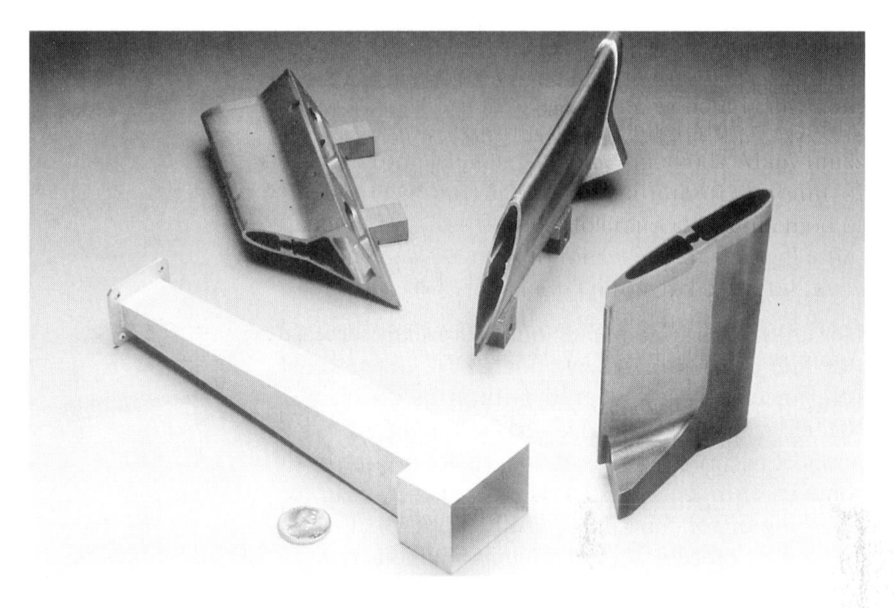

a.

b.

EDM range up to 4.0 inches^2/hour or 43 millimetres^2/minute. For example, if a piece of stock is 16 mm (0.63 inch) thick, then the speed at which a slot could be cut by wire EDM would be:

$$(43\ \text{mm}^2/\text{min})/16\ \text{mm} = 2.7\ \text{mm/min}$$

or

$$(4.0\ \text{in.}^2/\text{hr})/0.63\ \text{in.} = 6.3\ \text{in./hr}$$

Newer machines are coming out that greatly increase this rate of cutting. When using EDM, dimensional tolerances of two hundred millionths of an inch (0.0002 inch) are common; surface roughness tolerances of twenty millionths of an inch are possible with some care. In fact, wire EDM is frequently used for polishing or for final dressing to dimensional tolerances. This process is called *skimming* in industry.

Inside cuts with wire EDM can easily be made. A small hole is drilled through the workpiece and the wire fed through the hole. This can be done by the machine itself since the wire can be caused to follow a stream of the kerosene or insulating fluid through the hole and be caught by the machine on the other side of the hole.

Surface roughness values are commonly given in *microns* (μm). This is an abbreviation for micrometres, not microinches. For conversion, 1 μm = 30.37 microinches, or 1 μm = 0.00004 inch approximately.

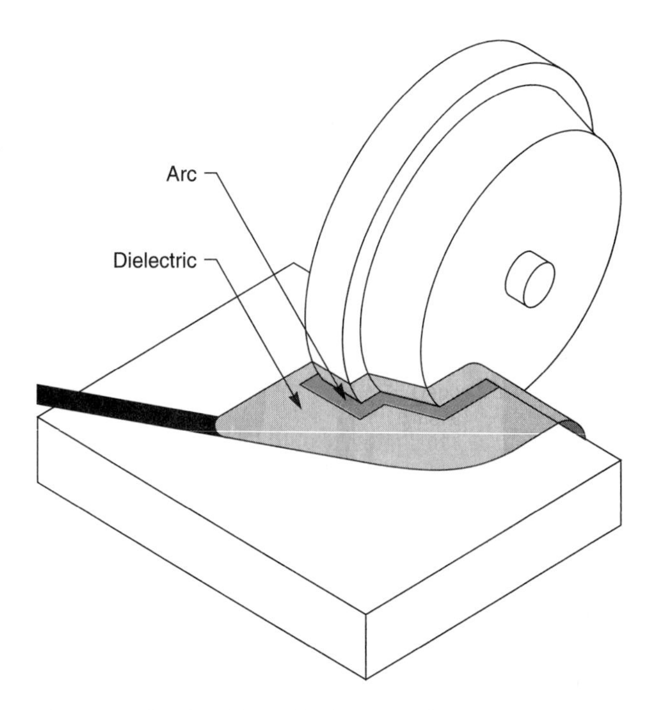

Figure 5-8. Electrical discharge grinding.

Electric Discharge Grinding

An *electrical discharge grinder (EDG)* is essentially a bench grinder with the grinding wheel replaced by an EDM electrode of the same shape. EDG has essentially the same advantages and disadvantages over conventional machining methods as does EDM. Hardness of material is no problem, fragile parts are not deformed, and EDG produces parts to very fine tolerances. Not only that, the cross section of the EDM "grinding wheel" can be fairly complex, thereby producing a surface that is fairly complicated (Figure 5–8).

Electrochemical Grinding

There is another process that only superficially resembles EDG: *electrochemical grinding (ECG).* In it, the surface of the work is treated with a chemical that loosens the grains of the metal. This action is accelerated by passing an electric current through the solution. The loosened grains are removed by a conventional grinding wheel. The advantage of this process is that, compared to conventional grinding, far less force is required to remove the grains, resulting in less heat and less distortion of fragile workpieces.

■ DISADVANTAGES

There are some limitations to EDM, EDG, and ECG:

1. They are slow methods of material removal with a resulting higher cost per pound of material removed.
2. The machines require elaborate filtration systems to eliminate the tiny metal particles that are removed from the workpiece.
3. Eventually, the spent fluid must be removed to a waste dump, which adds to the cost of operation of the machines.

PROBLEM SET 5-1

1. We noted earlier that 1 micron (one millionth of a metre) nearly equals 39.37 microinches (one millionth of an inch). Start with 1 micron and do a units-conversion analysis to confirm the figure of 39.37 microinches.
2. Give the alternative names for "ram EDM." Distinguish between ram EDM and wire EDM.
3. Describe a typical manufacturing job where EDG would be an ideal process to use.
4. List the advantages and disadvantages of EDM over conventional material removal techniques.
5. Describe one object that would be a good candidate to be made by ram EDM and another object that would be a good candidate to be made by wire EDM. What makes the difference?
6. Explain why it is important to use conventional machining methods to "rough out" a cavity before using EDM.
7. Why should the dielectric fluid be kept moving through the gap of an EDM operation? (That is, what would likely happen if it didn't?)
8. Which physical property of the workpiece material determines the material removal rate? Why is this property the determining factor?
9. Define the following terms: dielectric, erode, overcut, undercut, wear ratio.
10. Explain why the units for material removal rate are volume/time for ram EDM but area/time for wire EDM.
11. Explain why the wear ratio is not a concern when cutting a through-hole with ram EDM.
12. What current would be required to remove 0.15 in.^{3}/min of steel? Use the maximum current density for graphite to calculate the minimum surface area for the electrode.
13. What is the shortest time that it would take to remove 20 cm^3 of steel with a graphite electrode whose working area is 2.5 cm^2?
14. Suppose that we want to create a crescent-shaped hole in a block of metal, and that the available processes are ram EDM and broaching. Compare ram EDM and broaching (as candidate processes to create the hole) with respect to similarities and differences, and advantages and limitations.
15. We want to make a clover-shaped hole in a piece of soft steel. How could this be done?

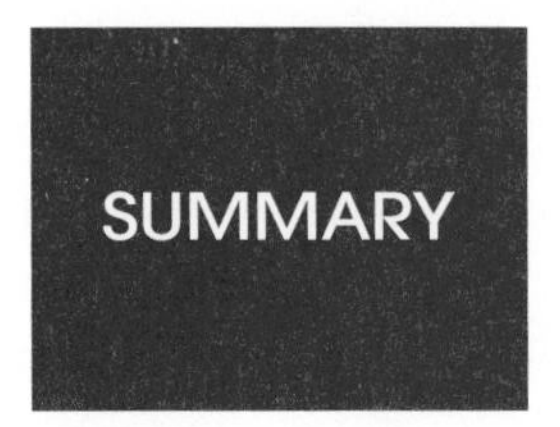

SUMMARY

Electrical means of material removal include the processes of electrical discharge machining (ram EDM, wire EDM, and electric discharge grinding) and electrochemical grinding. Though slower than mechanical methods of material removal, the electrical methods can make parts that would be impossible to make by mechanical means.

Chapter

6

Chemical Methods of Material Removal

■ CHEMICAL MACHINING

Acids dissolve metals; alcohols and acetone dissolve certain plastics; water dissolves sugars and table salt. It is possible to use this dissolving action in carefully controlled situations to remove material. Among these techniques are *chemical machining* and *electrochemical machining.*

The cutting of odd-shaped holes in metals always poses a problem. The hole could be broached, sawed, or perhaps punched. But both broaching and sawing would involve a two-step operation (drill a hole first then broach or saw the rectangle), and punching cannot be done easily on very thick pieces. However, holes can be made easily by chemical machining.

Suppose that we needed an odd-shaped hole in a piece of magnesium. The simplest method of chemical machining would be to coat the magnesium stock with a layer of plastic or wax, remove the wax from the area to be removed, then dip the block in an acid. The parts of the metal still covered with wax are protected and the acid cannot act. The acid does, however, attack the unprotected metal to dissolve it (Figure 6–1).

The crude method just outlined can be refined to do very delicate work. In applications where weight is critical, we might calculate that a metal web only 0.005 inch thick is needed. To mill or otherwise machine a part leaving a web of metal only 0.005 inch thick would be difficult because the metal web would probably tear under the force of the cutting tool. With chemical machining, however, this is a very simple process: Just take the part out of the acid when it has dissolved down to the desired thickness.

Chemical machining does have some unique problems associated with it. Once the surface of the metal is eroded, the chemical can attack under the protec-

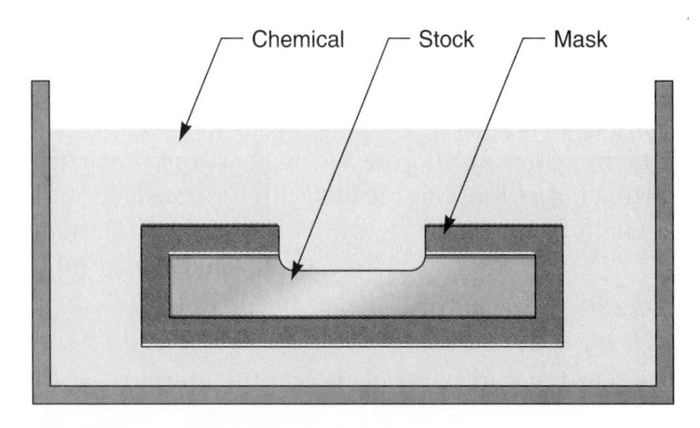

Figure 6–1. Chemical machining.

tive coating and thus enlarge the hole. To counter this action, the hole can be "undersized" on the mask so that by the time the chemicals have eroded the metal to the desired depth, the hole will be the proper dimensions (Figure 6–2). A second technique is to remove periodically the item being chemically machined, recoat the surface, *along with the interior surface,* then proceed with the chemical machining (Figure 6–3).

The fact that chemical machining does involve "undercutting" can be put to practical use. For instance, suppose a curved hole had to be cut under a piece of metal (Figure 6–4). This could be accomplished by chemical machining simply by controlling the recoating process so that the chemicals undercut the top part.

Further refinements of chemical machining involve the use of photographically laid masks on the metal. By coating the surface of the metal with a light-sensitive emulsion, a design can be projected onto the surface and developed. The parts of the surface that are to be chemically machined are then removed, by another chemical, while the places that are to remain unetched remain covered and protected against chemical attack. The part can then be dipped into the corrosive solution and the uncovered places removed. Printed circuit boards for electronic applications are often completed using this technique. (Figure 6–5)

Since images can be photographically reduced, even projected through a microscope onto a very small area, and since material only a few atoms thick can be removed by chemical machining, chemical machining is often used in the design of integrated circuits for electronic applications. These integrated circuits, so small that they are almost invisible to the naked eye, are now used in everything from calculators, hearing aids, and heart pacemakers to advanced-technology radar systems.

Chemical machining has many advantages over mechanical methods. It can be used for work too delicate for mechanical methods. Chemical machining will often work on materials on which mechanical tools do not work well. Soft metals such as aluminum, copper, and magnesium often tear and cannot be machined to extremely fine tolerances by mechanical means. Machining titanium with a cutting tool generates so much heat that it simply dulls the tool. Chemical machining presents no problems in these respects. One big advantage of chemical machining is its ability

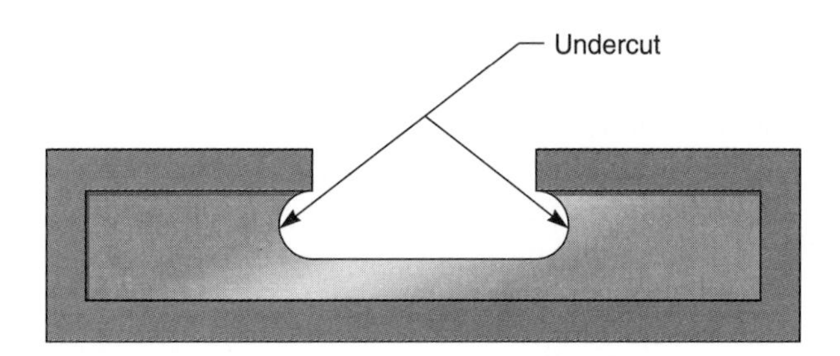

Figure 6–2. Undercutting.

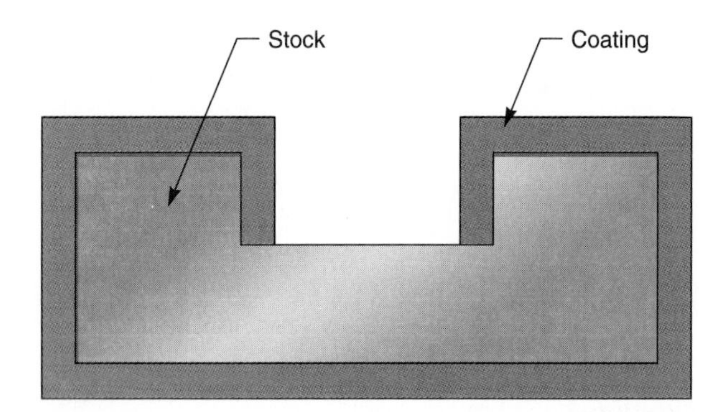

Figure 6–3. Recoating.

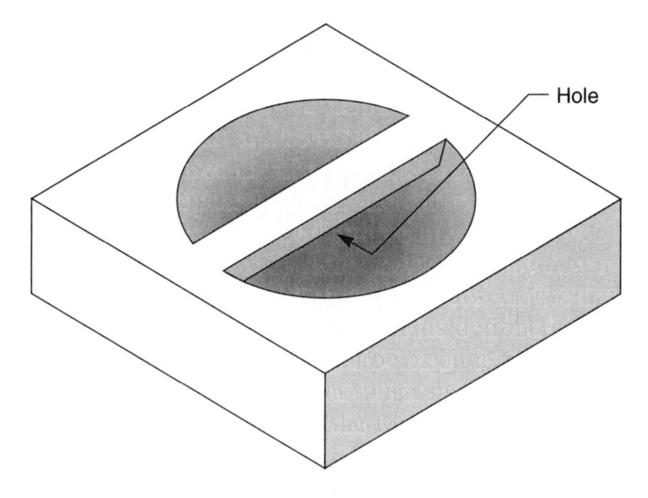

Figure 6–4. Cutting a hidden hole by means of chemical machining.

Figure 6-5. Chemically machined printed circuit board.

to remove very small amounts of metal (a few atoms if necessary). As a result, chemical machining is a manufacturing process that will often work when no other process will.

As with any material removal technique, chemical machining also has its weaknesses. Internal surfaces might be relatively rough compared to honed surfaces. Chemical machining is a relatively slow process. The underlying principle of chemical machining is that of corrosion. To remove a pound of metal by *mechanical* machining would require only a few seconds to a very few minutes. To remove a pound of metal by chemical methods might require several hours. Chemical machining requires quite a bit of skill and training on the part of the operator and it requires constant watching, measuring, and recoating of surfaces to finish the product. Therefore, chemical machining becomes an expensive method of material removal on a per-pound basis.

Another disadvantage of chemical machining lies in the use of acids or other chemicals in the process. Often these chemicals are hazardous to the health of the people using them. Elaborate venting and scrubbing systems must be employed to protect the operators and the environment. Further, the used chemicals must be taken to a toxic waste dump for disposal. Some cities and localities have enacted laws banning chemical machining. As a result, some companies have been forced to move their chemical machining facilities into remote areas, away from population centers. This creates problems when transporting raw materials and other supplies, as well as the finished products, to and from the parent plants. Chemical machining is strictly regulated everywhere in the United States.

■ ELECTROCHEMICAL MACHINING

Although corrosion and chemical attack on metals by acids is considered to be a chemical process, it is actually the result of an electronic action. The process is similar to that of energy storage in batteries and to electroplating. For a metal to be dissolved by an acid and go into solution, it must give up electrons. Not only does the metal give up electrons, but it also does so with considerable force. The force by which the electrons are released from the metal can be measured in volts. Each different metal will give a different voltage to the electrons when dissolving. If two different metals are placed in a liquid that conducts electricity (such as an acid or even seawater), one of the metals will give up its electrons with greater force than the other and a net voltage difference between the metals will result. If the two metals are connected by a wire, the voltage difference causes an electron current flow, which can be used to light light bulbs,

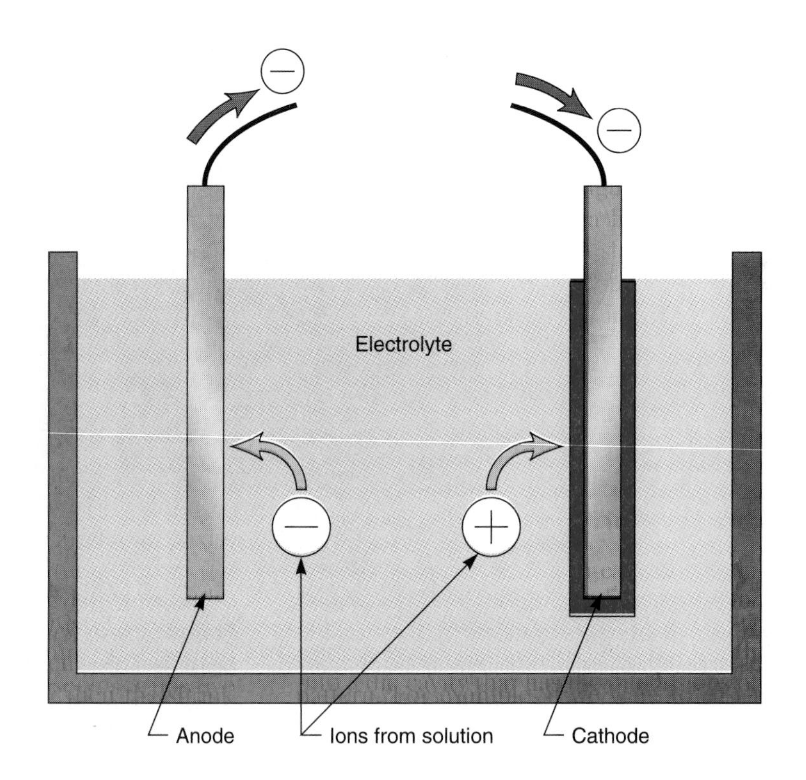

Figure 6–6. Electric cell.

start engines, or for any other purpose requiring direct current. This is the principle of the electric cell (Figure 6–6).

In the electric cell, the easier the metal goes into solution, the higher the voltage. This metal is called the *anode.* The less active metal is the *cathode.* However, if voltage is applied to the anode, the process can be reversed and the more active metal will pull ions out of the solution and the less active metal will go into solution. This is the principle of electroplating, which will be discussed in more detail in the chapter on material addition. In the plating cell, the metal being built up is called the cathode and the metal going into solution is the anode.

Metal is being removed from the anode in the plating cell. The removal of this metal can be carefully controlled by masking the area that should not be removed and allowing the unmasked area to go into solution. This technique, which uses an electrical current to assist the corrosion process, is called *electrochemical machining* (Figure 6–7).

Electrochemical machining is much faster than chemical machining and will work on metals such as copper, which will not easily dissolve in acids. However, electrochemical machining has the same drawbacks as chemical machining. It is expensive and poses environmental problems. But it will work in situations for which no other method of material removal will work. It is a valuable tool in industry.

PROBLEM SET 6–1

1. Under what conditions might chemical machining be the best method of material removal?
2. Would electrochemical machining be a good choice for a material removal method for producing low-priced automobiles? Why or why not? Defend your answer.
3. Could chemical machining be used to make parts out of ceramics? Why or why not? Defend your answer.

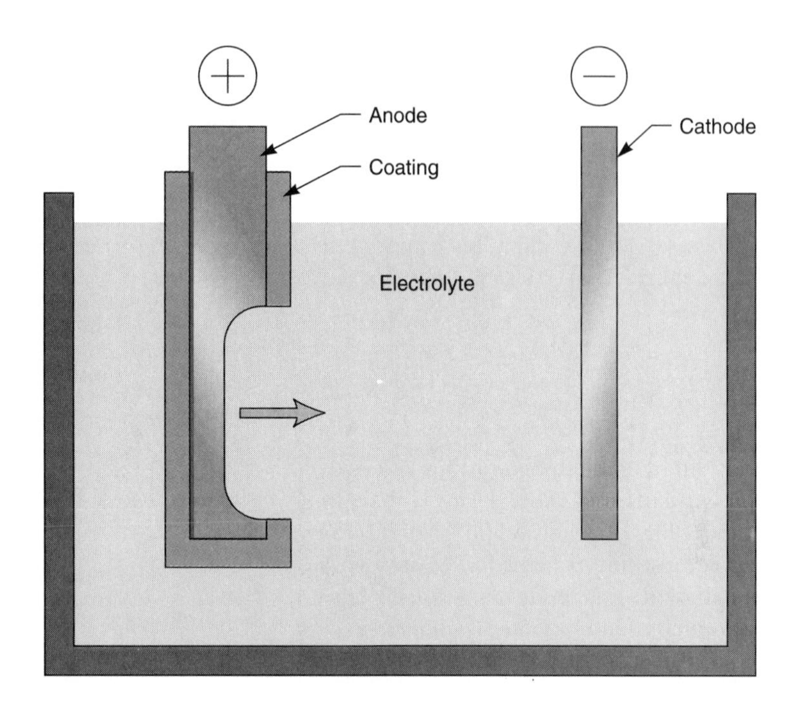

Figure 6–7. Electrochemical machining.

4. What are the environmental concerns of using chemical or electrochemical machining?
5. What is the difference between electroplating and electrochemical machining?
6. What is the difference between chemical machining and electrochemical machining?
7. What are the advantages of chemical machining?
8. What are the advantages of electrochemical machining?
9. We want to make a padlock "key" by chemical machining. Outline a process whereby this could be done.
10. Besides the environmental concerns, what other drawbacks are there to chemical and electrochemical machining?

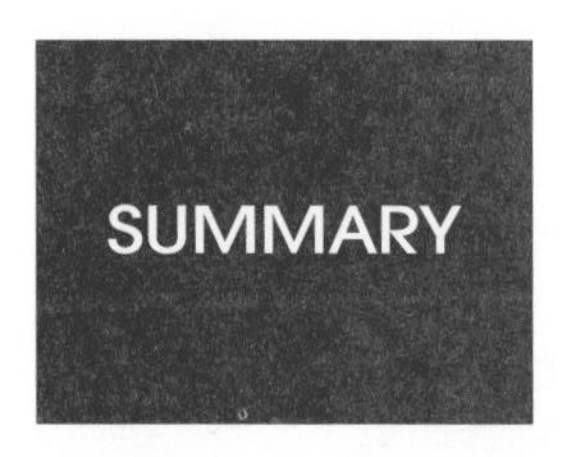

Chemical and electrochemical machining allow parts to be made that could not be made by other techniques. Chemical machining can be used to manufacture parts that are too delicate to be made by mechanical means. Chemical machining can "reach" areas not accessible to other forms of material removal, and it can also be used with a photographic masking process.

Electrochemical machining is somewhat faster than chemical machining, but it is a little more limited in the scope of parts that can be manufactured by it. Both chemical and electrochemical machining are relatively slow processes when compared to mechanical methods of material removal. As such, they are more expensive processes.

Both chemical and electrochemical processes pose environmental problems. The disposal of used chemicals is an expensive proposition. Care must also be used in the handling of the acids and other chemicals. Areas in which the chemicals and electrochemical plating cells are used must be properly ventilated and workers must be protected from the chemicals.

Chapter 7

Thermal and Other Methods of Material Removal

Braided nylon rope is often "burned" through rather than cut so as to fuse the ends of the strands and therefore prevent unraveling of the cut ends. The actual process, however, is for the most part a melting process rather than a burning process. Burning destroys the material by oxidation; melting simply changes it into a liquid. But melting is a process that is used frequently as a material-removal process for certain types of materials.

■ HOT WIRE CUTTING

Certain kinds of plastic stock may be cut with a hot wire or a hot soldering gun. When used on foamed thermoplastic material, this is a fairly fast operation that leaves a surface that is smooth enough for many applications. But when used on dense stock, the melted plastic may resolidify behind the moving hot wire, thereby rejoining the two parts again. In such cases, a hot air jet to blow away the melted plastic may be required. Use of a thicker wire is not generally a practical solution because the amount of power required increases rapidly with increasing wire diameter. Figures 7–1 and 7–2 depict a hot wire cutter.

Although most of the action is simple melting, some of the plastic is burned, producing potentially harmful gases. Plastics such as polyvinyl chloride (PVC) and "Saran" liberate deadly chlorine gas when burned. Acrylonitriles contain the cyanide radical, which is toxic. Appropriate safety precautions must be taken and might require the use of ventilating hoods or respirators.

Hot wire cutting of plastics will not work on thermosetting high polymers, which only get harder when heated. Neither will it work on some polymers that are good insulating materials. Polyurethane foam is difficult to cut with a hot wire because it sticks to the wire and thermally insulates it. The heat from the wire cannot get through the coating of polyurethane to melt the rest of the foam and the cutting action stops.

■ OXY-FUEL GAS CUTTING

Separation by melting is also accomplished by means of a gas cutting torch that is used to make rough cuts in steel plates. The typical oxy-fuel (oxygen-acetylene or oxygen-hydrogen) cutting torch (Figures 7–3 and 7–4) has a circle of oxygen gas jets that heat the metal

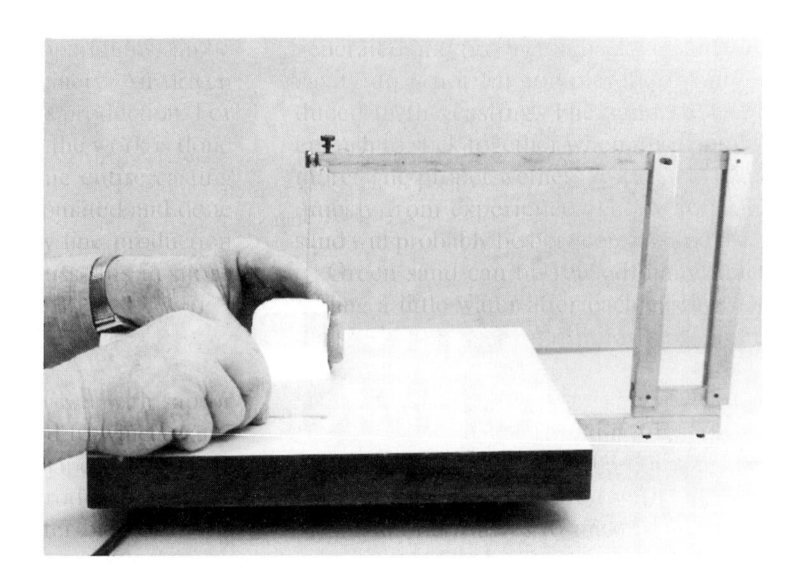

Figure 7-1. Hot wire cutter.

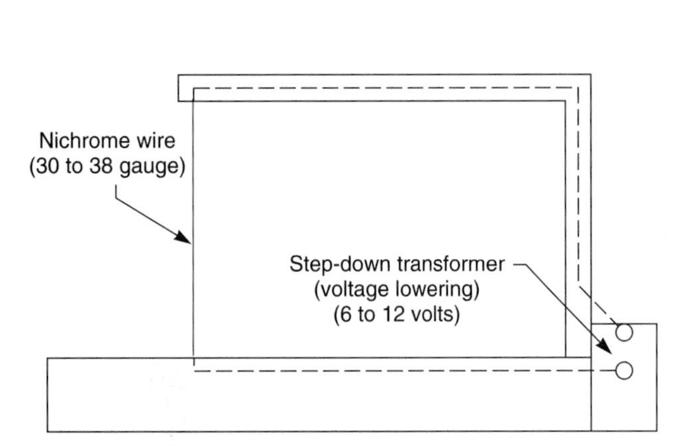

Figure 7-2. Hot wire cutter schematic.

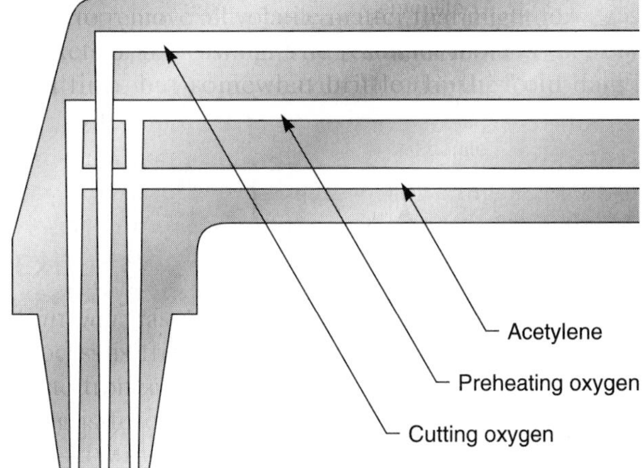

Figure 7-3. Cutting torch schematic.

to melting temperature. These surround a central oxygen jet. Once the workpiece material is hot, the central jet is turned on, which oxidizes the melted metal. The excess oxygen has two actions: It oxidizes (burns) the metal and blows the molten metal away, ensuring that the molten metal will not reharden in place to fuse the stock back together.

Cutting torches are either operated by hand to do rough cutting or operated by automatic machines to produce remarkably smooth cuts in heavy stock (Figure 7–5). Tolerances on the order of ± 0.04 inch (± 1 millimetre) are common. It is a fast, inexpensive process that requires fairly little operator training. Plates as thick as 12 inches (300 millimetres) may be cut with an oxy-acetylene gas mixture. Thicker plates can be cut with oxy-hydrogen or other special gas mixtures.

In industry, the machines used for torch cutting are often computer controlled and have multiple cutting torches. Multiple parts can be cut simultaneously in this manner. Figure 7–6 shows such a machine.

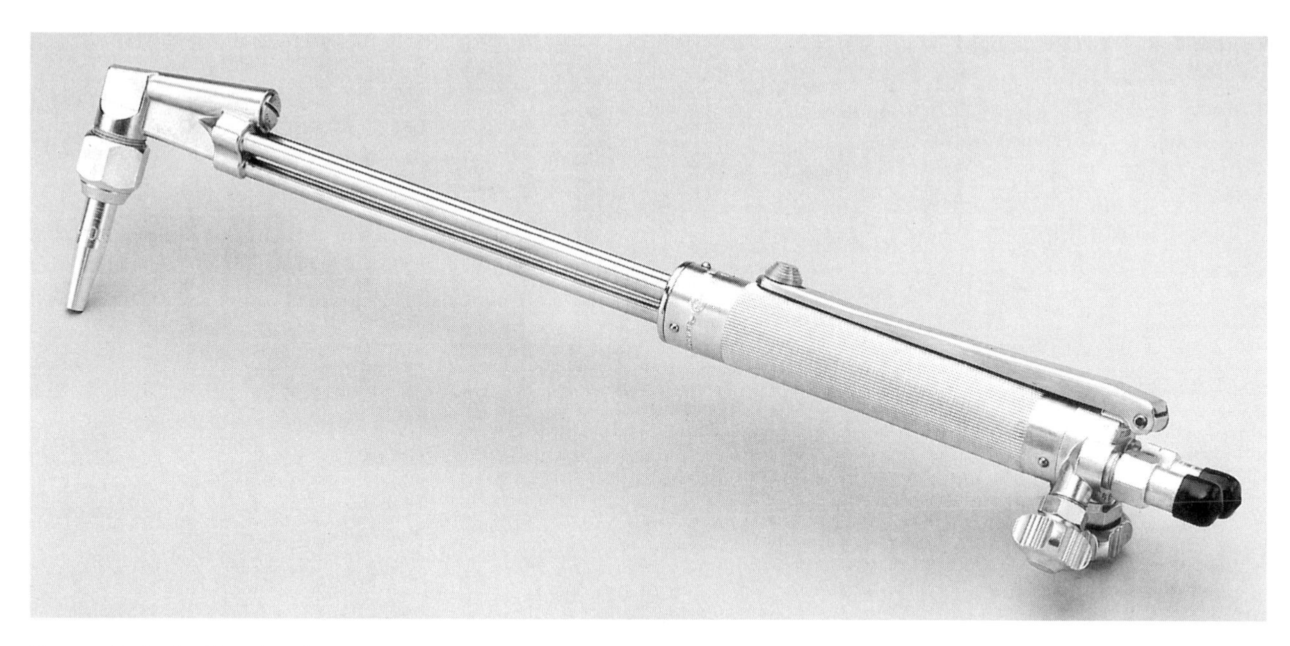

Figure 7-4. Cutting torch.
Courtesy of The Lincoln Electric Company, Cleveland, Ohio.

Figure 7-5. Cutting with a torch by hand.

Figure 7-6. Multiple cutting torches.

■ LASER CUTTING

A third, thermally related cutting application involves *vaporizing* the material: *laser* cutting. Laser stands for "*l*ight *a*mplification by *s*timulated *e*lectromagnetic *r*adiation." A magnifying glass and sunlight can be used to burn a piece of paper. Lasers work in much the same way. Lasers produce light, which is energy, of a single frequency. Since the light is of one frequency, it can be focused to a very fine point and its energy concentrated to a very high degree. Extremely high temperatures can be produced by focused laser beams. Lasers can be used to cut nearly any material: fabric, wood, plastics, and metals. In fact, the ability of the first high-powered lasers to "burn" through one or more razor blades gave rise to a trend to rate laser power in "Gillettes"—a practice that mercifully died out quickly.

One advantage of laser cutting is that the heat can be precisely localized so that nearby material is not disturbed in any way. Thick material can be cut partway through if that's desired. Laser cutting also leaves a surface finish that may be smooth enough that no further finishing process is required.

The effectiveness of laser cutting depends critically on three properties of the material being cut:

1. The surface property of *absorptance,* which is the opposite of reflectivity. If the workpiece reflects the laser energy instead of absorbing it, no heating of the workpiece can occur.
2. The bulk property of *thermal conductivity,* combined with the thickness of the workpiece, determines how well heat is conducted away from the target site. If it is conducted away efficiently, then the temperature of the workpiece will never become high enough to melt or vaporize the workpiece material.
3. The bulk property of *heat of fusion* is the amount of heat that must be added to a material that is already heated to its melting temperature *to change it from a hot solid to a hot liquid.* If this requirement is too high, the workpiece may become very hot but will not melt.

As for material removal rates, the laser beam "burns" through the workpiece, producing a narrow opening or "kerf" similar to that of wire EDM or jigsaw blades. Therefore, the appropriate unit of measure for material removal rate is area per unit time.

A very powerful laser (say, 6 kilowatts) can cut through 3/8-inch steel at a rate of 45 inches/minute:

$$(3/8 \text{ in.}) \times (45 \text{ in./min}) \times 60 \text{ min/hr} = 1000 \text{ in.}^2\text{/hr}$$

Contrast that to 4 inch^2/hour for wire EDM!

Lasers produce a very clean cut because no gas or other material is involved in the cutting action.

OTHER METHODS OF MATERIAL REMOVAL

Energy is required to remove material. That energy can come from many sources. In recent years, several exotic techniques that make use of "unusual" sources of energy to remove material have been developed. A few of them are discussed here.

Water Saw

Water is even being used to cut materials. It has long been known in hydraulic or placer mining that the force of water can carry away matter. This principle is now used in the *water saw.* A fine abrasive is added to water and the water forced through a fine nozzle, under very high pressure, against the surface to be cut. The stream of water may be only a millimetre wide and pressures of up to 50,000 psi are used. The actual saw is not large, but the hydraulic pump that drives the water is often several times the size of the saw. The material is held in a moving bed and moved against the stream. Titanium and composites, materials that are not easily cut by conventional techniques, are easily cut by water saws (Figure 7–7).

Wire Saw

Another recent innovation often used in the composites industries is the *wire saw.* Fine diamond fragments are embedded on a fine wire by means of running the steel wire through a liquid copper bath, which has the diamond powder in suspension. These wires are about 100 feet long and are wound on a spool. In a machine similar to a jigsaw, the wire runs from a spool above the workpiece to one below it. The entire length of

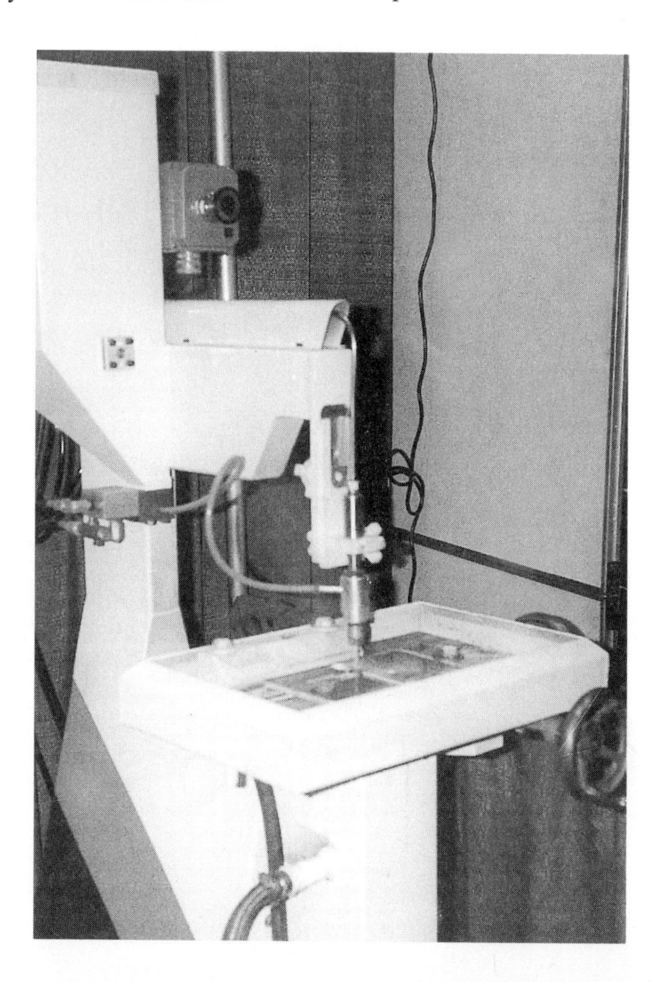

Figure 7–7. Water saw.

wire will run from one spool to the other then the action is reversed. It generally takes about 3 seconds for the wire to run its length. The fine wire can cut straight or sharply curved surfaces equally well. Again, the bed holding the workpiece is moved against the saw blade. Figures 7–8 and 7–9 depict a wire saw and Figure 7–10 shows the microstructure of the wire.

Electron-Beam Machining

Thomas A. Edison found that heating a wire makes the wire give off electrons. Electrons carry a negative charge, which will be attracted toward a positive charge. The higher the positive charge, the faster the electrons will move. Electrons are tiny bits of matter. As is true in everything from bowling balls to hydraulic mining, matter under motion can knock other matter out of the way. To see the evidence that electrons can move matter around, one simply needs to observe the effect of a spark on the surface of a piece of metal. Look at the mechanism of a worn-out switch to see the scarring of the surface by electrons. Electron-beam machining makes use of the momentum of these electrons to cut materials.

In electron-beam machining, electrons are generated at a cathode, then focused and driven toward the workpiece by electrostatic fields. The action is similar to focusing sunlight rays through a lens. This pinpoint beam of electrons can be used to cut matter. Figure 7–11 is a schematic of electron-beam machining.

The atoms of the air are many times bigger than electrons. As such, the air will absorb the energy of electrons flowing through it. Air and other gases can be ionized in this manner and caused to glow. (This is the principle of the fluorescent light.) Therefore, in electron-beam machining, the workpiece must be placed in a vacuum.

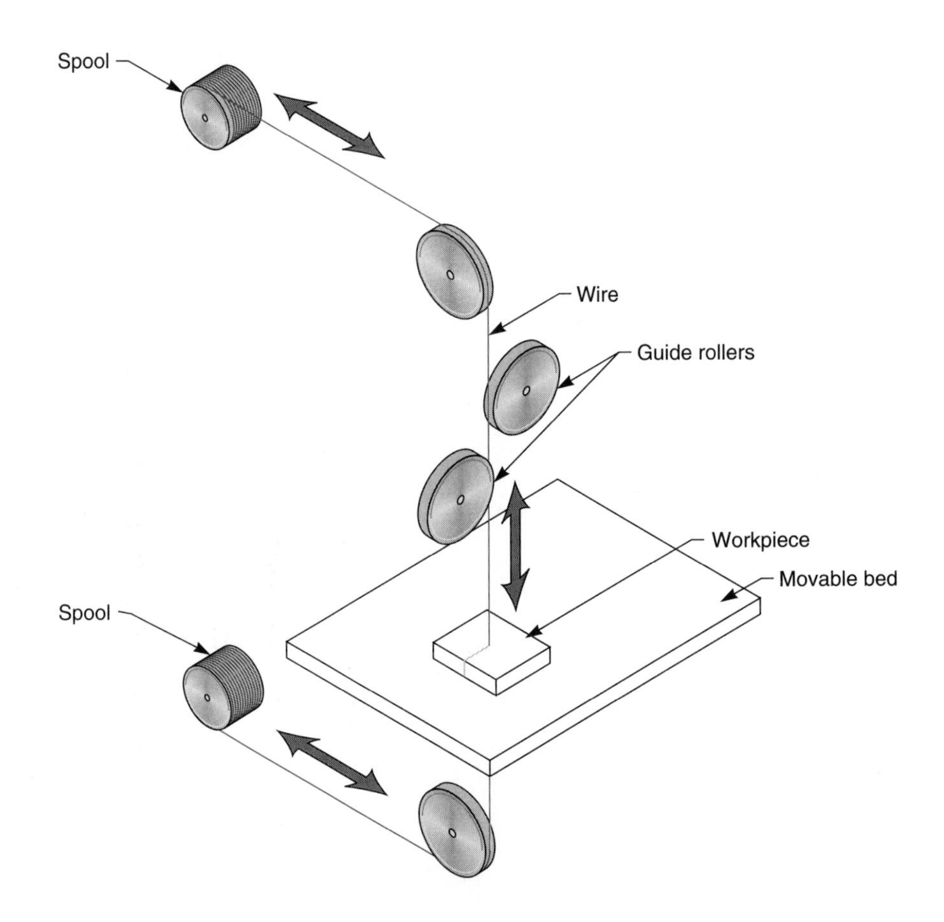

Figure 7–8. Wire saw schematic.

Figure 7-9. Wire saw.

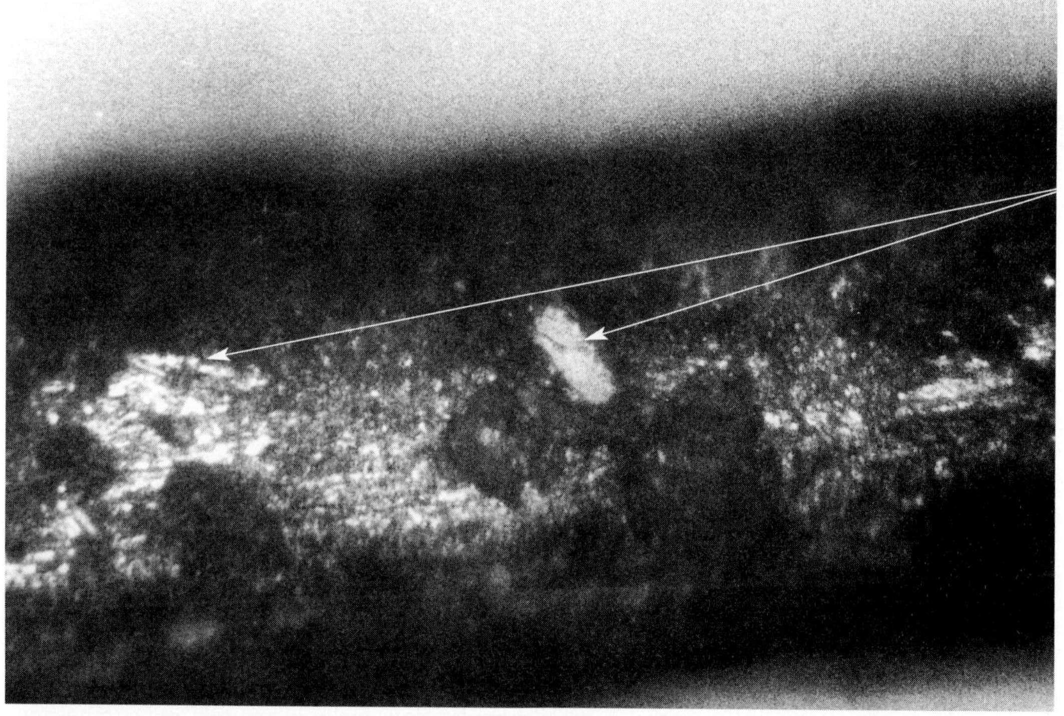

Figure 7-10. Enlarged view of the wire saw blade.

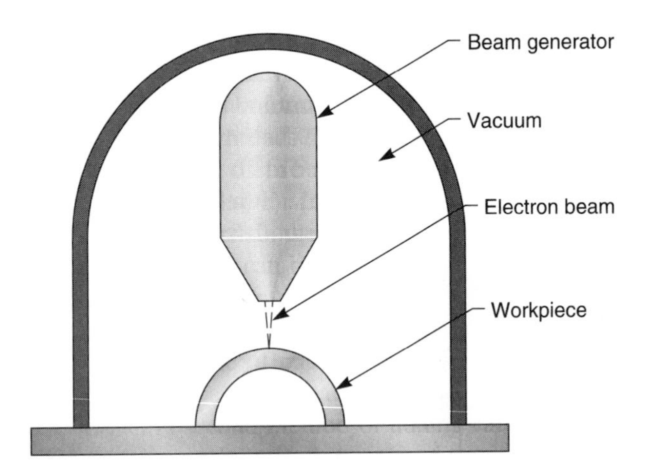

Figure 7-11. Electron-beam machining.

Electron-beam machining is the cleanest of all methods of material removal. It is a relatively slow process because each part must be inserted into a chamber and a vacuum established before the process can start.

Ultrasonic Machining

Sound is energy. Properly directed, sound can be used to remove material from a piece of stock. Many ultrasonic cleaners are commercially available. Higher powered versions of these ultrasonic cleaners, which make use of focusing chambers, can also knock atoms or particles from a workpiece. In ultrasonic application, the frequency of the sound is sometimes important.

PROBLEM SET 7-1

1. List three examples for which hot wire cutting might be a good method of material removal.
2. List three examples for which a cutting torch would be the best method of material cutting.
3. List three examples for which laser cutting might be a good method of material removal.
4. List the major advantages and limitations of laser cutting.
5. List the major advantages and limitations of oxy-acetylene cutting.
6. List three applications for which water saw cutting would be a good method of material removal.
7. Would the use of a water saw to cut up automobile tires be a good method to do so?
8. Which method of cutting would be better for titanium—a water saw or band saw? Why?
9. List three applications for which electron-beam machining might be a good method of cutting.
10. Given the processes of oxy-acetylene cutting, water saw cutting, and electron-beam cutting, summarize or list the advantages and limitations of each.

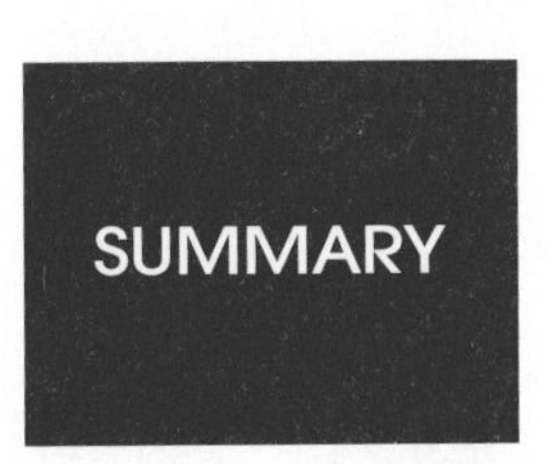

SUMMARY

This chapter has discussed the advantages and the limitations of several forms of material removal as well as the methods by which they work. These included the thermal methods of cutting with a hot wire, cutting torch, and lasers. Some newer methods of material removal include the water saw, wire saw, electron-beam cutting, and ultrasonic cutting.

Many of the methods discussed in this chapter have specific applications including metals and composites.

PHOTO ESSAY 7-1

Oxy-Fuel Gas Cutting

STEP 1 Attach the cutting torch to the welder, open the tank valves, and set the gas pressures as outlined in Photo Essay 14-1. Ignite the fuel and adjust the fuel (acetylene, red hose) for a flame that "feathers" out almost back to the torch. With the cutting lever valve closed (note the position of the left thumb in the picture), adjust the oxygen for a neutral flame. (The small cones of blue should be clear back to the torch.)

STEP 2 When the cutting lever is closed, the flame should become an oxidizing flame (see Figure 14-9).

STEP 3 Preheat the metal to be cut to a red or orange color, leaving the cutting valve lever in the closed (up) position.

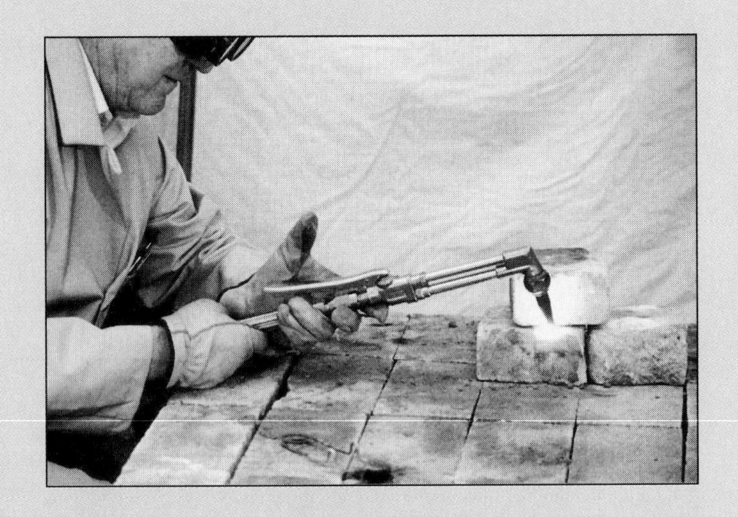

STEP 4 Once the metal is red hot, depress the cutting valve lever. The excess oxygen will burn through the metal. Move the torch along the line of cut steadily and evenly to get a smooth cut. Shut off the torch and shut off the tanks in the manner described in Photo Essay 14–1.

SECTION

III

Material Addition

Chapter 8
Material Addition

INTRODUCTION

Instead of *removing* material, we often need to reverse the process to *build up* a workpiece or *add* material to it. The techniques of **material addition** include the following:

- Electroplating by various methods
- Electroforming
- Chemical plating
- Dipping
- Hardfacing
- Metallizing (including wire and powder methods)
- Vapor-deposited coating

Chapter

8

Material Addition

■ ELECTROPLATING

The electric cell was discussed in Chapter 6 in the section on electrochemical machining. *Electroplating* also involves the use of an electric cell. To add a metal to a workpiece, the workpiece is placed on the cathode (negative pole) of a direct-current power source. The cathodic workpiece is then immersed into a solution containing the metal ion that is to be added to the workpiece. The negative charge on the cathode attracts the positively charged metallic ions to the workpiece. When the ions touch the workpiece, they pick up an electron from the negatively charged metal and become a metal atom and stick to the workpiece. Given enough time, the workpiece will be covered or *plated* with the metal atoms from the solution.

Of course, the solution will be depleted of its metal ions. However, if the same metal that is being plated out of solution is used as the anode, or positively charged pole, then the negative ions from the solution will remove electrons from the anode, causing the metal atoms to become ions and go into solution to replace the plated-out atoms. Remember, the anode gets smaller while the cathode gets larger. Figure 8–1 illustrates this action.

Electroplating is simple in theory, but to produce a good plated part—one on which the plated metal will permanently adhere, be of even thickness, and have the specified properties of the finished product—several factors must be carefully controlled. These include, but are not necessarily limited to the following:

- Concentration of the electroplating solution or electrolyte
- Roughness of the cathode
- Preparation and cleanliness of the cathode or part being plated
- Voltage and current applied to the electrodes
- Temperature of the electroplating solution
- pH, or acidity, of the electroplating solution
- Masking or rotation of the cathode during the plating process
- Compatibility of the workpiece and the metal being plated

A first, simple attempt at electroplating something is usually enough to convince a person that electroplating is as much an "art" as it is a science.

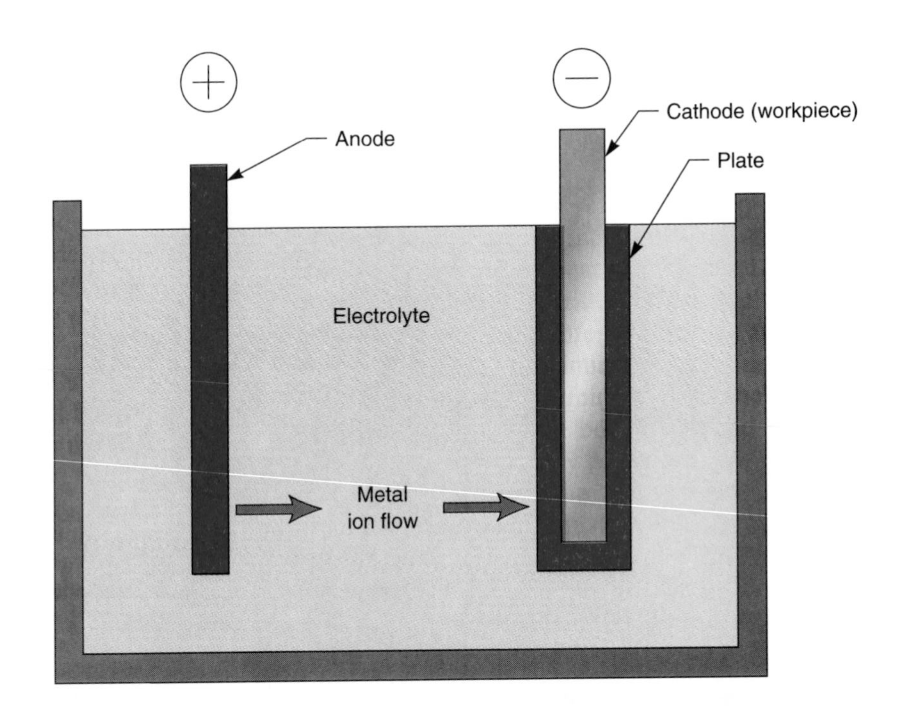

Figure 8-1. Plating cell.

The plating solution must contain the metal ion that is to be deposited on the workpiece. For copper plating, the electrolyte is usually copper sulfate ($CuSO_4$). Nickel plating uses nickel chloride ($NiCl_2$). Silver plating can be accomplished with silver nitrate ($AgNO_3$), but silver cyanide ($AgCN$), silver ammonium cyanide ($AgNH_3CN$), and other soluble silver salts are often used. Chrome plating uses chromic acid in the form of chromic oxide (Cr_2O_3) dissolved in an acidic solution. Remember, the electrolyte must contain the metal to be plated out and must be soluble in water.

The concentration of the plating solution varies with the metal being plated. If the concentration of the metal ion is too low, there is little to be plated out and the electroplating cell may start to generate hydrogen bubble from the electrolytic decomposition of the water. If the concentration is too high, all of the metal-carrying compound may not dissolve in the water and will instead settle out at the bottom. The solubility of the plating material is also dependent on the temperature.

The voltage used in plating is usually very low, around 2 to 6 volts. If the voltage is too high, the water in the electrolyte will decompose into hydrogen and oxygen, forming bubbles on the workpiece (cathode). This produces a pitted surface and often prevents the plated material from sticking to the workpiece. Too high a voltage will also produce a "spongy" surface, which makes the plating useless.

For the plated metal to adhere to the surface of the cathode, the cathode must be carefully prepared. The workpiece is first "pickled" or dipped into an etching solution (often dilute nitric acid), which dissolves a little of the grain boundaries in the metal. This provides a slightly roughened surface to which the plated metal can mechanically bond. After pickling, the surface of the workpiece is "stripped" or placed in a cleaning solvent to remove all traces of grease, including fingerprints. The workpiece is then washed and carefully placed in the plating cell.

The amount of electric current (in amperes) is determined by the distance between the electrodes, the surface area of the electrodes, and the concentration of the plating solution. Different types of plating require different currents. These currents are usually specified in amperes per square foot of surface area being plated. The concentration of the metallic ions near the cathode is constantly being depleted by the deposition of the metal onto the workpiece. If the concentration of the electrolyte near the anode gets too low, the plating action stops. To prevent this, the

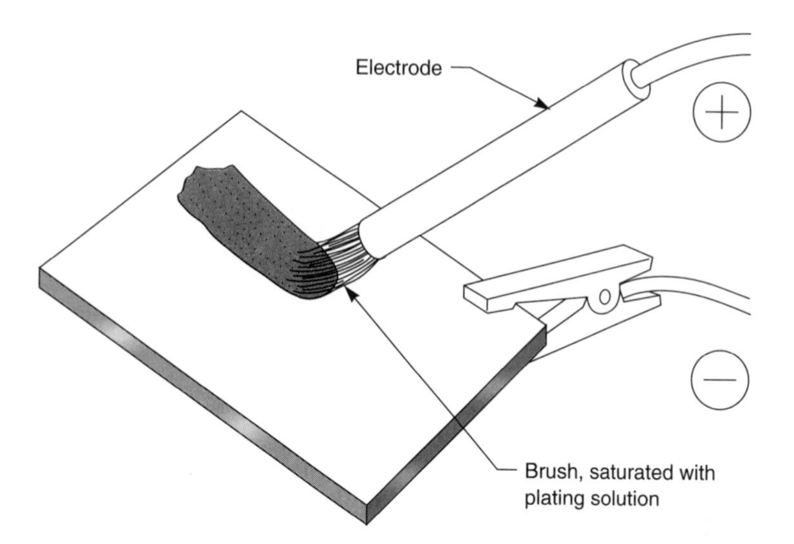

Figure 8-2. Brush plating.

solution must be stirred constantly during the plating process.

The temperature of the plating solution is another factor that must be controlled. The passage of an electric current through the plating solution will heat it. Plating often is done at temperatures above 100°F, so care must be taken not to boil the solution.

The thickness of the plate depends on the time the workpiece is left in the solution, the currents used, and other factors. Plate thicknesses from a few atoms to a few inches are possible.

Another factor in electroplating is the compatibility of the plated metal and the cathodic metal. For instance, chromium will not stick to steel. Copper, however, adheres very well to steel. Chromium will not stick to copper either, but it *will* stick to nickel and nickel will stick to copper. Therefore, to apply a chrome plate to a piece of steel, the steel must first be copper plated, then nickel plated, and then it can be chrome plated to the desired thickness. The copper and nickel plates need to be only a few atoms thick, less than a thousandth of an inch each. This process is often referred to as *triple plating.*

All in all, considerable skill and practice are required to produce the beautiful shiny electroplated objects we have become used to seeing.

Beside providing a shiny or corrosion-resistant surface to a metal part, electroplating can be used to increase the size of a part. Sometimes worn engine parts such as drive shafts are ground undersize, then electroplated to increase the diameter to slightly larger than specified, then reground to tolerance. This process can provide a wear-resistant surface at a lower cost than making a new part.

Brush Plating

A variation of electroplating is *brush plating.* In this process, a properly prepared (pickled and stripped) workpiece is connected to the negative pole of a battery or other source of direct-current electricity. A paintbrush that has a wire embedded in it is connected to the positive pole. The brush is dipped into the electrolyte and the solution is painted onto the workpiece. The current flows from the brush through the electrolyte and the metal is plated on the stock. This process is often used to repair chipped or damaged plating and for small parts (Figure 8–2).

■ ELECTROFORMING

Many irregular-shaped objects that are very difficult to make by mechanical means are easier to *electroform.* The part shown in Figure 8–3, a 1-millimetre-thick seamless expansion joint in the shape of an elliptical bellows, is an example.

In electroforming, a pattern is made of the part. In the case of the elliptical bellows, the pattern could be made out of clay or plaster of paris with the outside dimensions of the pattern being the inside dimensions of the bellows. The pattern is then coated with an electrical conducting metal either by vacuum deposition or spraying. This metallic coating need not be very thick, just thick enough to conduct the current required for plating. The coated pattern is then placed in a plating tank and electroplated to the desired thickness. On removal from the tank, it is washed and the pattern chipped or dissolved out of the inside of the bellows. The part is then ready for finishing and use.

It sounds very simple, but again, a lot of "technique" is involved in electroforming. Since the side of the workpiece nearest the anode would have a higher current density than the other side, the nearer side would have a thicker plate than the opposite side. To prevent this, the cathode must be rotated constantly. Different thicknesses can be purposely achieved by "baffling" the cathode. This is done by placing obstructions between the anode and cathode (Figure 8–4).

As with any chemical process, care must be taken to protect the workers from fumes and chemical spills. Further, electroplating and electroforming chemicals can only be used for a fixed time before they must be replaced. Disposal of the waste chemicals is a problem. Current law requires them to be transported to a toxic waste dump. Much time, effort, and money are being expended to discover ways of rendering these chemicals inert so that they can be disposed safely, easily, and cheaply.

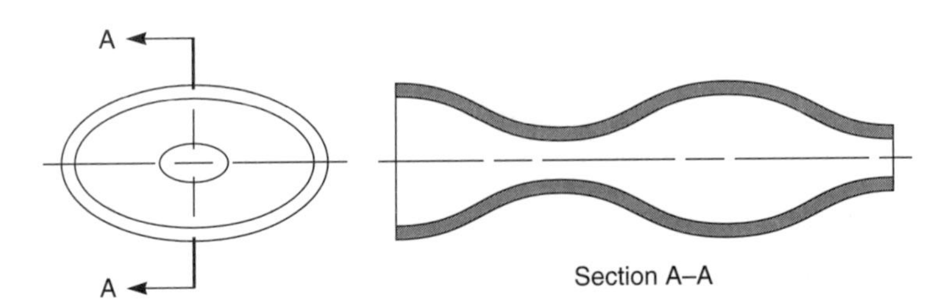

Figure 8–3. Elliptical bellows.

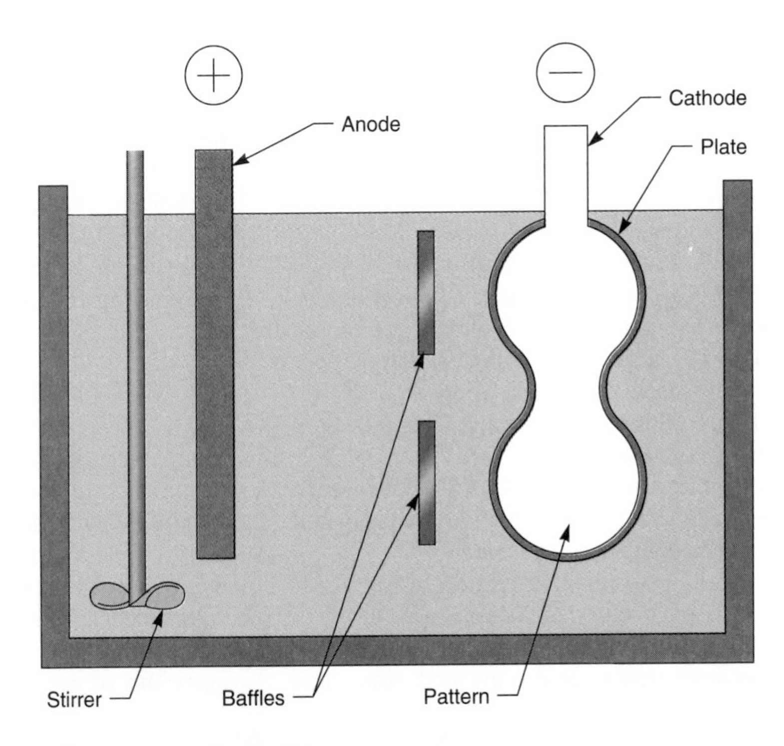

Figure 8–4. Electroforming cell.

■ CHEMICAL PLATING

One of the principles of chemistry is that some metals are more chemically active than others. Lithium, sodium, and potassium are very active metals that dissolve very quickly in water. Conversely, gold, silver, platinum, and copper are very inactive metals, to the point where they even resist attack by acids. A list of the metals from the most active to the least active is called the *electromotive series* of the elements. A partial listing of the electromotive series is shown in Table 8–1.

If a metal of higher chemical activity is placed in a solution containing the metallic ions of a lower activity, the more active metal will go into solution, replace the lower activity ions in the solution, and the metal of lesser activity will plate out as a free metal. This principle of chemistry has led to a practice called *chemical plating.*

If a steel rod is placed into a bath of copper sulfate, a little of the iron on the surface of the rod will go into solution, while the copper in solution will be deposited on the surface of the rod. If carefully done, this plating will adhere to the surface of the steel just as it did in electroplating. Again, the concentration of the solution as well as its temperature and acidity must be strictly controlled. Further, the plated material must be removed periodically and washed to remove any loosely plated metal on the surface. Metals such as copper, gold, silver, and platinum can be chemically plated with relative ease. Figure 8–5 illustrates chemical plating.

Table 8–1. Electromotive Series

MOST ACTIVE
Lithium
Potassium
Barium
Strontium
Calcium
Sodium
Magnesium
Aluminum
Zinc
Chromium
Iron
Cadmium
Cobalt
Nickel
Tin
Lead
Hydrogen (neutral)
Antimony
Bismuth
Copper
Silver
Platinum
Gold
LEAST ACTIVE

Figure 8–5. Chemical plating.

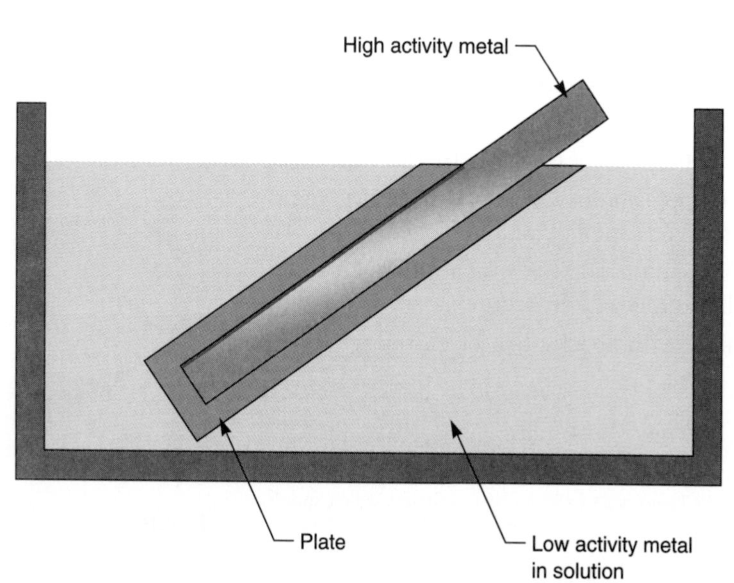

DIPPING

Dipping is not a new method of manufacture. Homemade candles have been made by dipping for centuries. In recent years, the dipping of metals has been used as a method of material addition.

A part can simply be immersed into a molten metal of a lower melting point. After removal, the molten metal solidifies on the surface of the stock, thus adding to its thickness. A method used to protect iron and steel for the last century involves dipping the steel into molten zinc. The zinc-coated steel is referred to as *galvanized steel.* Not only does the zinc provide a mechanical barrier to protect the steel, but since zinc is higher on the electromotive series than iron, when the zinc-coated steel is placed in water or acid, the electrons from the zinc are given to the steel, making the steel cathodic. It is the anode that goes into solution. The cathodic steel will be protected against corrosion as long as there is any zinc left on the surface of the steel. Only after all of the zinc is dissolved will the steel be corroded. This process is often referred to as *cathodic protection.*

PROBLEM SET 8–1 *

1. What are the differences between electroplating and dipping?
2. List the steps involved in electroplating a piece of steel.
3. Which electrode (anode or cathode) must the part being plated be connected to? Why?
4. What variables must be controlled in electroplating?
5. List five applications where brush plating would be the preferred method of plating.
6. List three situations for which electroforming would be a preferred procedure.
7. What are the drawbacks to electroforming?
8. What is the difference between electroplating and chemical plating?
9. Could a piece of copper be chemically plated with zinc? Why or why not?
10. Could a piece of lead be coated with copper by dipping? Why or why not?
11. Make a list of five items in an automobile that are electroplated.
12. Make a list of items that are "dipped" as part of their manufacturing process.

* Some data may need to be obtained from sources such as *The Handbook of Chemistry and Physics* or from technical engineering handbooks.

HARDFACING

Hardfacing refers to any of a large variety of processes that add a significant coating of material to the surface of an object (Figure 8–6). The purpose generally is to provide a surface that is harder than that of the underlying object in order to improve the resistance to wear caused by corrosion, heat, oxidation, abrasion, impact, etc. However, the term "hardfacing" is not always an accurate description, because some coating materials (such as brass, silver, or gold) may be softer than the original piece onto which they are deposited. In that case, they may be added to improve electrical conductivity, reflectance, or other nonmechanical properties.

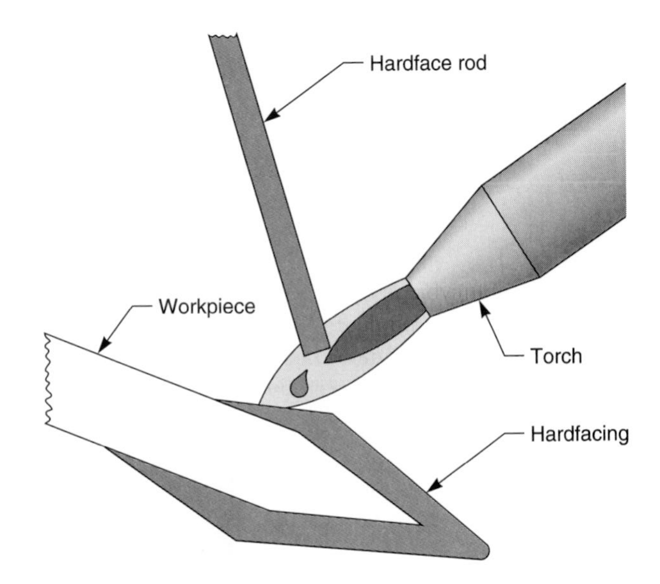

Figure 8–6. Hardfacing.

Hardfacing can be done with oxy-fuel gas, arc, or plasma-arc welding apparatus. (See Chapter 14 for an explanation of these processes.) As the melted coating material is added to the already melted surface of the object, an alloy forms that has the desired mechanical or chemical properties. For example, a hard, brittle, tough alloy may be welded onto the tines of a rototiller or the "buckets" of a ditch digging machine to reduce wear on the cutting surface while retaining the shock-handling capacity of the supporting metal.

METALLIZING

In *metallizing,* melted metal or ceramic is sprayed (from a distance of between 4 and 10 inches, or 100 to 250 millimetres) onto some existing object so as to build up a layer of the sprayed material for whatever reason. The metal may be powdered or in rod or wire form, whereas the ceramic is only in powdered form. The coating material is generally melted in an oxy-fuel flame (or in an electric arc or plasma arc) and blown against the object as a fine spray of melted coating material.

Some sense of the process of metallizing can be gained by looking at the alternative names for this process: "metal spraying" or "thermal spraying." Metallizing is a process quite similar to spray painting; in fact, the equipment used is referred to as a "gun" just as in spray painting. Also, just as in spray painting, the gun may be either hand held or operated by an industrial robot or other piece of equipment.

Bonding Problems

The process of making candles by dipping was mentioned earlier as an example of material addition. When it is done correctly, the outer layer (at each stage) is softened by immersion into the melted wax so that a secure bond is formed between the old wax and the new layer. It's easy to imagine how poor the bond would be if, instead, the candle were taken from a freezer and dipped only briefly into the melted wax. But that's very nearly what is done in several "metallizing" processes; obviously the issue of the strength of the bond must be considered for each of the metallizing processes discussed in the next section. The strength of the bond can be influenced by the choices of three variables: (1) the temperature of the original workpiece, (2) the droplet velocity, and (3) flame temperature.

In the processes of welding and some soldering applications, the original piece is heated to its melting point and held there while the added material is melted and mixed with the melted portion of the original piece. (A "cold solder joint" occurs when the melted materials are not hot enough to flow together and, therefore, do not form a mechanical bond.)

In other processes (brazing and "tinning," for example) the original piece is heated in the presence of a "flux" so that the melted material to be added "wets" the surface, creating a strong bond. However, in this case, the original material remains solid because the temperature is not raised high enough to melt it.

In metallizing, however, melted droplets of metal or ceramic are sprayed against a metallic (or ceramic or plastic) surface, solidifying on contact; little heating of the original piece is possible so that the droplets solidify on the surface, resulting in a weak bond and low material density. Proper preparation of the surface is essential so that the bond will be as strong as possible. In fact, it is common to cut or sandblast fine grooves into the surface of the object and then peen over the ridges between the grooves so that the sprayed material is held in place mechanically rather than by chemical bonds (Figure 8–7).

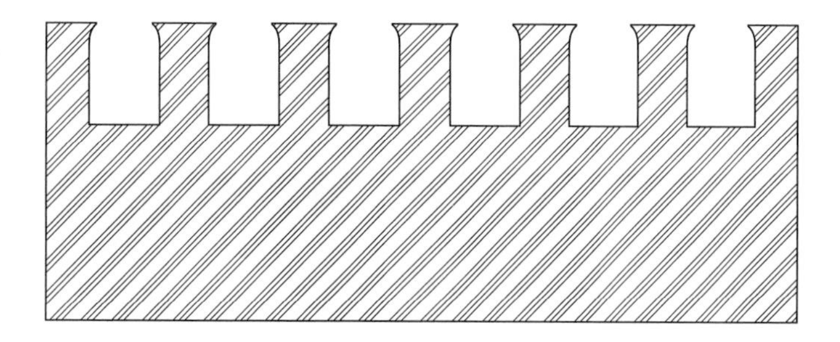

Figure 8-7. Grooved prepared surface for metallizing (magnified).

Note that once the object has been covered by a layer of sprayed material, the droplets for the next layers land on hot droplets so that the cohesion between droplets is fairly good; the bonding problem occurs mostly between the object and the first layer of droplets.

A second variable that is important to high density and good bonding is the velocity of the melted droplets as they hit the object. The slower the speed at which they hit, the less they are able to flow into pockets and displace the air that is there. Therefore, a high droplet velocity is desirable.

A third variable is the temperature of the flame. Ceramics (or other inorganic coatings) and binders for carbides require a much higher flame temperature than is available in the methods that use an oxy-fuel flame.

Metallizing Processes

Several metallizing processes are commonly used, including the following:

1. Oxy-fuel flame spraying
2. Electric-arc spraying
3. Plasma-arc spraying
4. Detonation spraying

Oxy-Fuel Flame Spraying

Flame-spray metallizing uses an oxy-fuel flame at 4000 to 5000°F (2200 to 2700°C) to melt the coating material and compressed air to atomize the melted coating material and propel it against the object being coated. Figure 8–8 shows a view of a typical flame-spray metallizing gun where the coating material is in the form of a wire or rod. The spray gun for powdered coating materials is similar; the difference is that the compressed air picks up the powdered material from the source and carries it to the flame where it is melted, and then to the object being coated. The principle is similar to that of a paint spray gun. Figure 8–9 is a photograph of a powder spray gun, and Figure 8–10 shows the gun in schematic.

The droplets in powder flame spraying have the lowest velocities of all of the metallizing methods: 200 to 400 feet/second (60 to 120 metres/second). The droplets in wire or rod flame spraying typically have velocities of 700 to 900 feet/second (200 to 280 metres/second), producing a more dense coating and a stronger bond. However, it is not practical to form coating material such as ceramics into wires or rods; therefore, powder flame spraying is the only option available for these materials.

Electric-Arc Spraying

If the coating material can be formed into wires, then the wires can be used as the electrodes for an arc. The arc melts the electrodes and a jet of compressed air blows the melted droplets against the object being

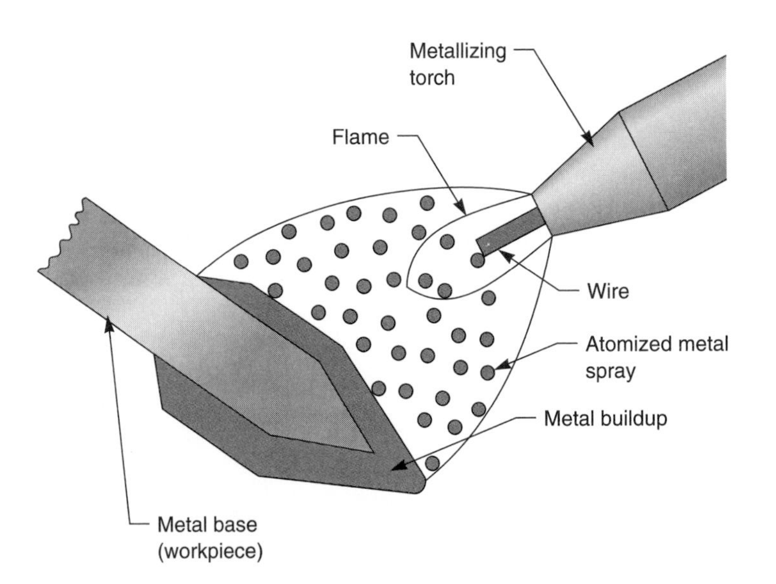

Figure 8–8. View of a wire flame-spray nozzle.

coated at speeds around 1000 feet/second (300 metres/second), creating a strong bond (Figure 8–11). Coating rates up to 60 pounds/hour (30 kilograms/hour) are possible with *electric-arc spraying.*

Although this metallizing method produces excellent results, its use is limited to coating materials that conduct electricity and can be formed into wires. Fortunately, many coating materials have these characteristics.

Plasma-Arc Spraying

In *plasma-arc spraying,* a chemically inert gas (such as helium or argon) flows into the arc where it is ionized by the high temperature (30,000°F or 17,000°C). A powdered coating material is introduced into the plasma. This material melts and is carried to the workpiece at speeds up to 1800 feet/second (550 metres/second) (Figure 8–12). This

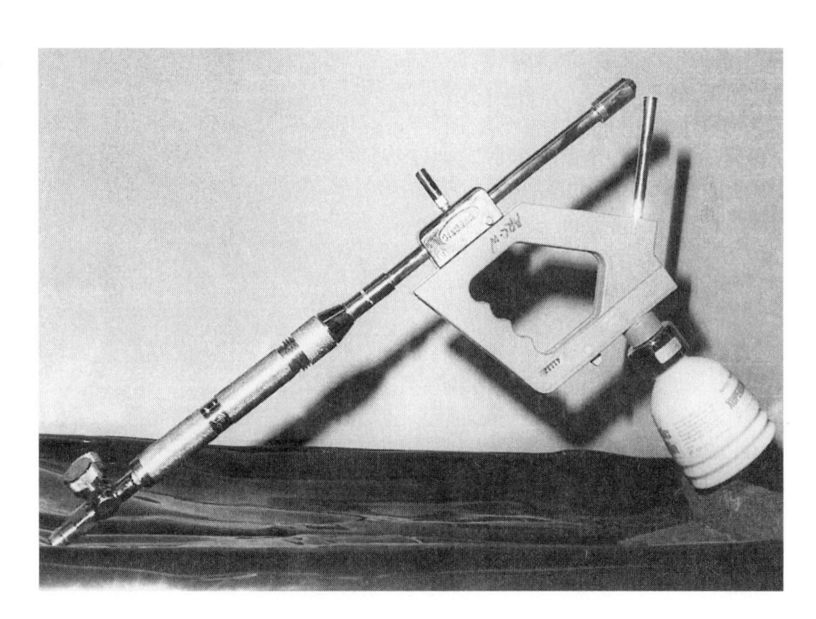

Figure 8–9. Powder spray torch.

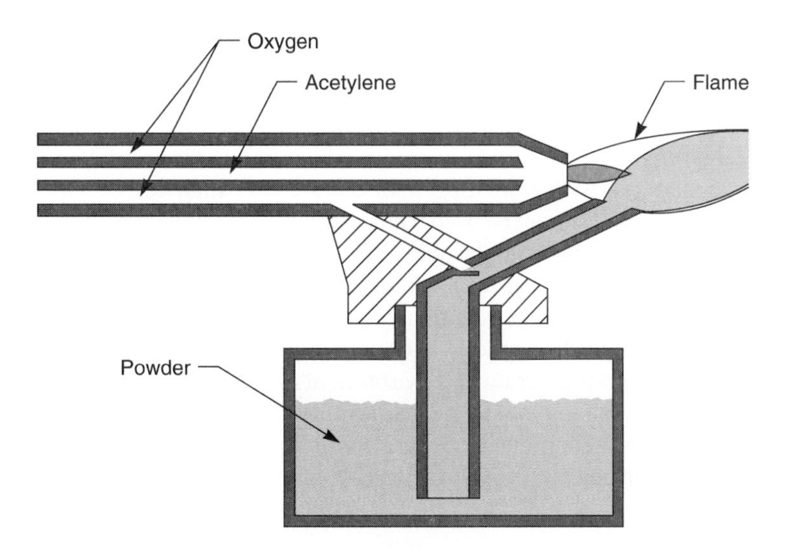

Figure 8–10. Schematic of powder spray torch.

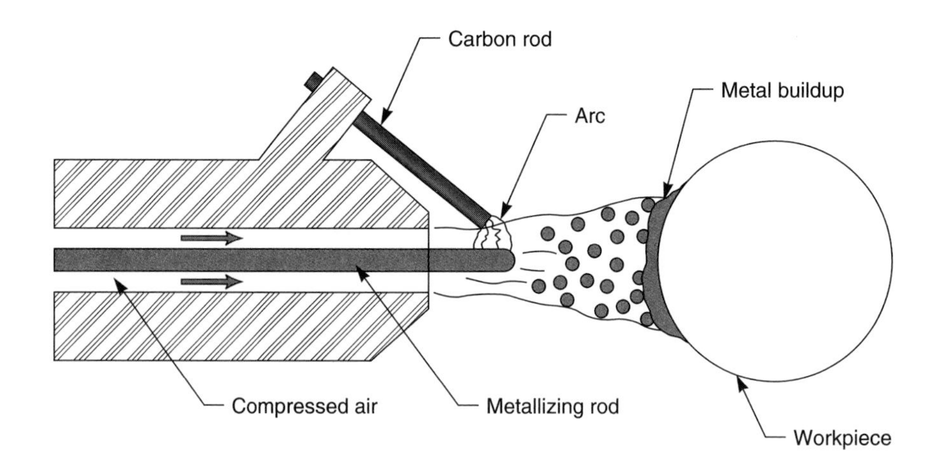

Figure 8-11. Electric-arc metallizing spray gun.

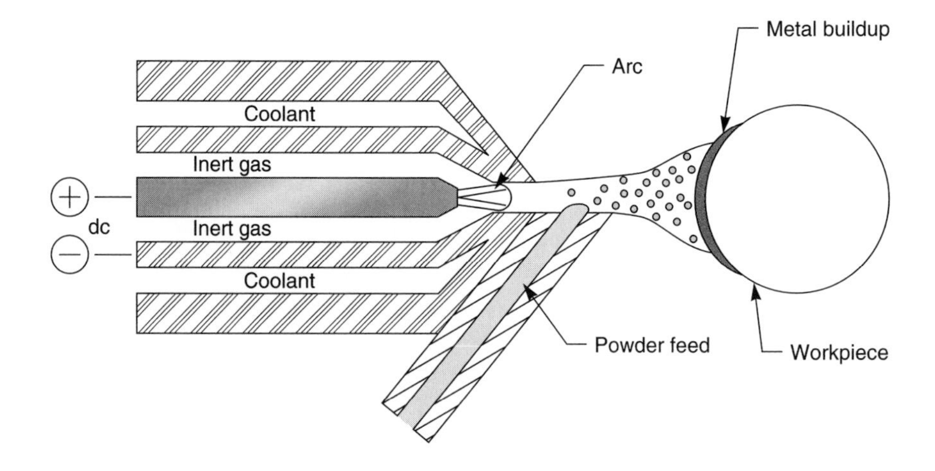

Figure 8-12. Plasma-arc spray gun.

produces a high-density coating with high bonding and structural strengths.

Any coating material with a melting temperature greater than 6000°F (3000°C) can be used with plasma-arc equipment. Typical materials would be high-temperature metals (such as tungsten), ceramics, cermets, carbides, etc. Coating materials that have high melting temperatures would be used to protect structural elements in high-temperature environments, such as rocket-engine exhaust nozzles. Coating rates up to 30 pounds/hour (14 kilograms/hour) are possible with plasma-arc metallizing.

Detonation Spraying

Another way of obtaining very high temperatures and high velocity sprays is through *detonation spraying.* In this method, oxygen and a fuel gas (acetylene, hydrogen, or other) are mixed in the chamber of a "cannon" and detonated by a spark several times per second. The powder is then sprayed on the surface by means of this very-high-temperature flame. Temperatures of 6000°F (3000°C) and droplet speeds up to 2500 feet/second (760 metres/second) are possible, producing very high bond strengths and densities (Figure 8–13).

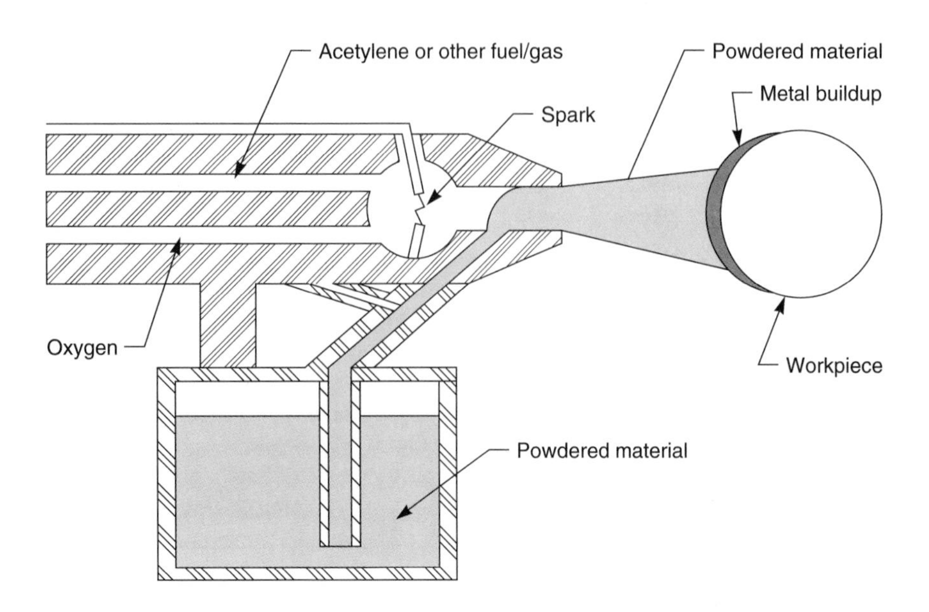

Figure 8–13. Detonation spraying.

VAPOR-DEPOSITED COATINGS

Vacuum Coating

Just as water-vapor molecules (from a tea kettle) will condense on any nearby cool surface, so also will atoms of a vaporized metal condense on any cool nearby surface. In practice, the atoms of the plating material are released by heating the sample (by passing an electric current through it). This process generally takes place in a high vacuum chamber to prevent air molecules from interfering with the process (Figure 8–14).

Virtually any metal can be plated onto virtually any surface by such a process. Commonly used metals include aluminum, chromium, gold, nickel, silver, and platinum. This process is a very slow one because of the setup and postprocessing tasks; the actual film deposition may require only a few seconds.

Because the vaporized atoms tend to travel in a straight line in the vacuum, it is necessary to rotate the object being plated if more than one surface is to be coated. Film thicknesses as thin as 1 microinch (0.025 micrometre) are possible, although thicknesses on the order of 20 microinches (0.5 micrometres) are more common. Common applications are low-reflectance coatings on lenses, highly reflective coatings on telescope mirrors, conductive coatings on plastics, precious-metal coating on jewelry, microelectric circuit fabrication, etc. More unusual applications include coatings on textiles and paper products.

Sputtering

In vacuum coating, the atoms of the plating material were vaporized by a heat source. By contrast, *sputtering* (or ion sputtering) refers to dislodging atoms (of the plating material) by bombarding the source electrode with ions of an inert gas at low pressure. These ions are often created by imposing a high voltage on the gas; when these ions are created by a radio-frequency source, the process is called radio-frequency sputtering. The bond between the sputtered atoms and the coated surface is considerably better than the bond formed by vacuum coating.

Figure 8-14. Vacuum coating.

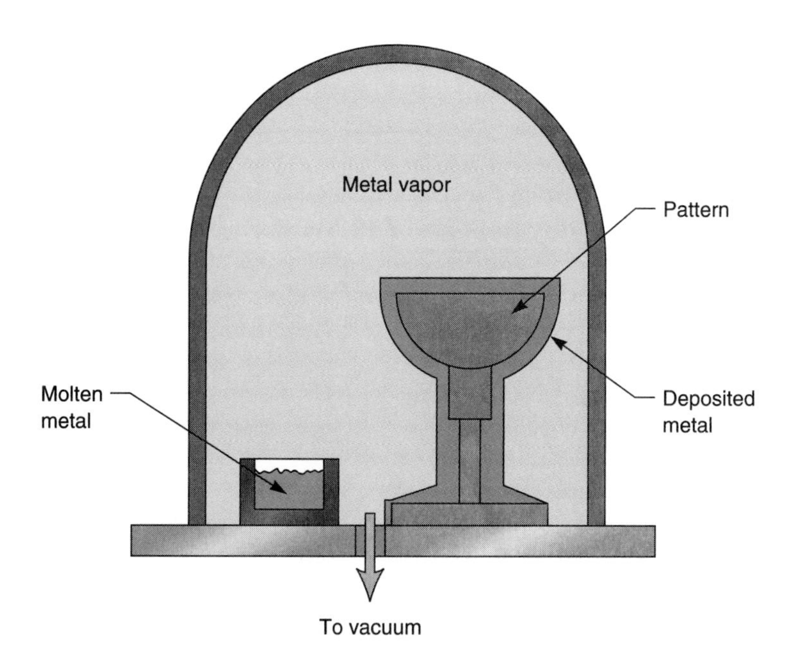

Figure 8-15. Sputtering.

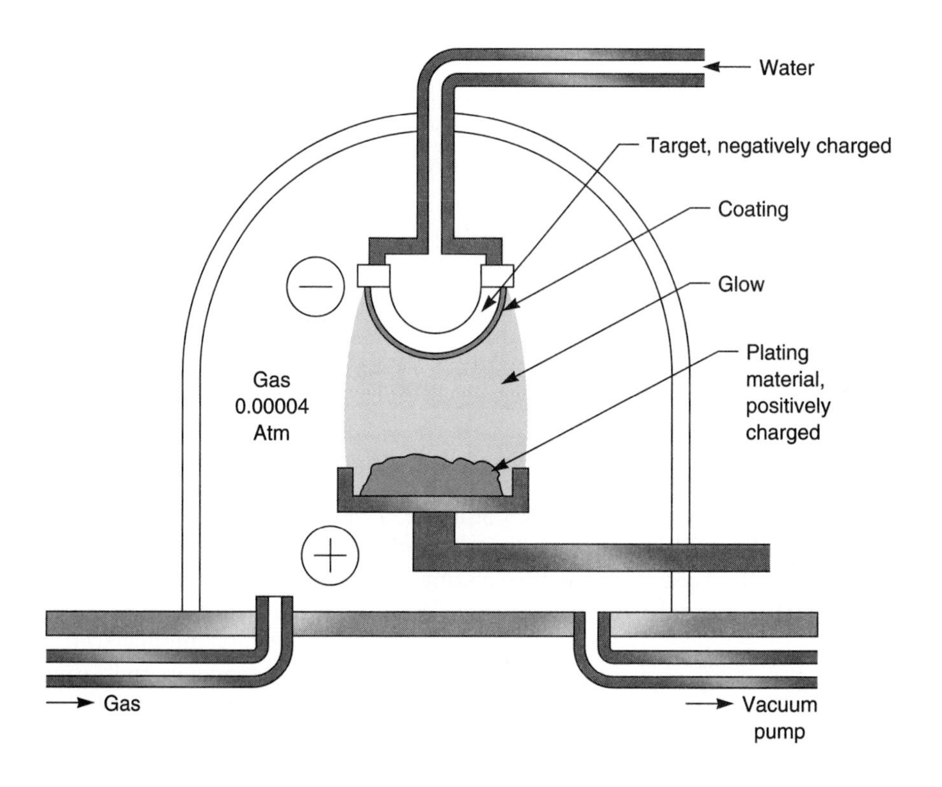

SIMILARITIES AND DIFFERENCES IN COATING METHODS

There are some striking similarities and differences among the sprayed-on material-addition processes. A summary of these similarities and differences follows.

The *similar* material addition processes, which could be lumped under "sprayed (or similarly applied) metallic or ceramic coatings" can be organized as follows:

Sprayed Metallic or Ceramic Surface Coatings

1. Forceful Processes ("Hardfacing")
   a. Fusion processes
      - Arc-welding variations
      - Torch-welding variations
   b. Nonfusion processes: thermal spraying variations ("metallizing")
      - Oxy-gas fuel
      - Electric arc
      - Plasma arc
      - Others
2. Vapor-Deposited Coatings
   a. Vacuum coating
   b. Sputtering

In fusion processes, both the object being coated and the coating material itself are melted so that they fuse; these are essentially welding processes.

In nonfusion processes, only the coating material is melted (and then sprayed onto the workpiece).

In vapor-deposited coatings, nothing is melted; the fragments of coating material are released to "find" the object to be coated and bond chemically or physically to it.

These surface-coating processes *differ* from the electrochemical or dipping processes in that (1) they are essentially thermal processes, or (2) the source for the droplets (or ions) of coating material is at some distance from the object being coated, or (3) both.

Uses

Some of the functions that the sprayed surface coating might serve are:

- *To provide corrosion protection.* Spray-coating is an alternative to hot-dipping for galvanizing iron and steel.
- *To apply a thin hardface.* This is useful when a thin coating will suffice.
- *To replace worn material.* For example, a crankshaft's bearing surfaces can be rebuilt to exceed their nominal diameters and then machined to achieve the original dimensions.
- *To apply expensive coating materials efficiently.* Thin coatings can be added uniformly; also, the coating can be applied to small areas (which would be difficult to do by dipping).
- *To change surface physical properties.* The applied coating may be harder, or more reflective, or a better conductor of heat or electricity, or just more attractive than the surface of the original object.

Materials

Commonly used coating materials are iron-based alloys, manganese steels (for toughness and resistance to impact), high-chromium steels (for abrasion resistance), carbon-chromium (for resistance to corrosion, heat, or oxidation), copper-based alloys (for corrosion and wear resistance), and tungsten carbide (for resistance to abrasion wear).

PROBLEM SET 8-2

1. List three reasons for hardfacing a workpiece.
2. What is the difference between metallizing and dipping?
3. List the variables that must be controlled in metallizing.
4. List five possible applications of metallizing.
5. What is the difference between metallizing and hardfacing?
6. Explain why a bonding problem occurs when metallizing a part with a material other than that of the workpiece. Explain why the bonding problem disappears as the added layer gets thicker.

7. List three applications for which vacuum coating would be a good method of material addition.

8. What materials are commonly used as the coating material in metallizing?

9. What is the purpose of grooving or roughening the surface of a part before metallizing it?

10. We want to add a thick layer of tungsten carbide to a lawn mower blade. Outline a process by which this could be done successfully.

11. We want to add a thin layer of an electrically conductive material to a block of ceramic. Outline a process by which this could be done successfully.

12. Could a metallized spray of ceramic oxide be applied to an aluminum part? Why or why not?

13. If you need to apply a refractory ceramic coating to the inside surface of a furnace, which of the processes could you *not* use? Why?

14. In photographs used to illustrate metallizing, the "object" frequently shown is a cylindrical, metal object mounted in a lathe. Give two reasons why such a shape would be an ideal one to add a coating to by metallizing.

Addition or building up of a material can be accomplished by several processes. Some processes are better suited for a given purpose than others. Electroplating, chemical plating, spraying, and vacuum depositions are all useful if the thickness of the material to be added is not more than a few thousandths of an inch. When thicker coatings are needed, electroforming, dipping, or metallizing should be used. Dipping and electroforming make possible the formation of parts that are irregular in shape and could not be made by mechanical processes. Electroplating and electroforming can be used only on metals, or parts coated with a metal by vacuum coating or dipping. Powder metallizing (the plasma type) can be used with metal oxides and other materials as well as metals.

SECTION

IV

Change of Form

INTRODUCTION

Previous chapters discussed methods whereby the volume of a piece of material could either be reduced or increased. In changing the form of a workpiece, only the *shape* of it is changed. This can be accomplished in several ways. The major methods of manufacturing discussed in this section involve:

- Melting the material
- Powdering the material
- Deforming the material

These processes include many methods of casting materials, powder metallurgy, forging, extrusion, rolling, bending, drawing, spin forming, and high-energy-forming methods.

Chapter 9

Thermal Methods of Change of Form

■ CASTING

Casting involves placing a liquid material into a mold where it hardens or solidifies into a final shape. It is one of the oldest methods of manufacturing. Today, many types of casting exist. The variations of casting discussed in this section are:

- Sand Casting
  - Green sand casting
  - Dry sand casting
- Investment Processes
  - Lost-wax casting
  - Shell mold casting
  - Centrifuged casting
- Permanent Mold Processes
  - Gravity feed casting
  - Low-pressure feed casting
  - Hot-chamber die casting
  - Cold-chamber die casting
- Centrifugal Casting
- Slush Casting
- Continuous Casting

Sand Casting

Sand casting is an ancient practice and is generally what most people think of when they hear the word "casting." Many of the terms used in sand casting apply to other forms of casting as well. It is appropriate, therefore, that these terms be discussed early in this section.

The basic concept in sand casting is that fine sand is packed around a *pattern* to form a *mold.* The pattern is then removed, leaving a cavity into which molten metal, a ceramic slurry, or other liquefied material is poured. Once the metal or other cast material is solidified, the mold is opened and the workpiece is removed. It's simple in concept, but there are a lot of complex variations on this simple theme.

Preparation of the Pattern

To generate a solid shape by casting, a pattern must first be made. For mass-produced sand castings, the pattern is generally made of hardwoods (Figure 9–1). Since most metals shrink upon cooling, the patterns must be made oversized by the amount they are expected to shrink. This is done by highly skilled

Figure 9-1. Casting pattern.

craftsmen, known as *pattern makers.* They make the patterns oversized by using *shrink rules.* A shrink rule is made for each different type of metal. If a metal is expected to shrink 2% upon cooling, then the shrink rule for that metal is 2% longer than a standard ruler. In other words, 10 inches on a 2% shrink rule is actually 10.2 inches long (Figure 9–2). Shrinkage of steel castings varies from about 0.5% to 2% depending on the alloying elements in the steel.

There are two significant restrictions on the shape of an object that can be sand cast. First, there must be a plane (associated with the shape of the workpiece) that will become the plane of separation for the two halves of the mold (Figure 9–3). This *parting plane* should be through the largest horizontal section of the pattern. For example, if we want to cast a shape like an incandescent light bulb, the plane would either be through the widest part of the bulb (if the axis of symmetry of the bulb is vertical) or the plane of symmetry (if the axis is horizontal). The parting plane of a more complicated object is shown in Figure 9–4.

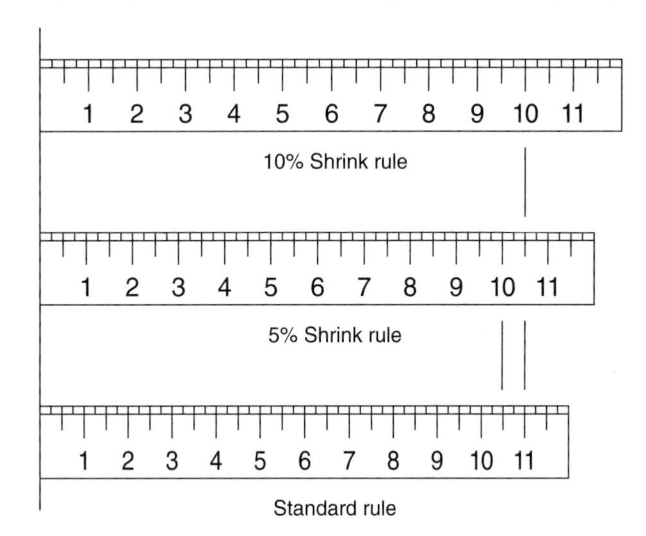

Figure 9-2. Shrink rules.

The second restriction is that the pattern should slip cleanly out of the sand without disturbing the sand. Further, the sides perpendicular to the parting plane need to be tapered slightly. The angle of the taper is called the *draft angle* and is typically about 2 degrees.

Admittedly, it is *possible* to make a pattern and mold in more than two parts, but that requires a lot more work and significantly increases the cost of the finished part.

Sand Casting Procedure

If you have ever made a sand castle on the beach, you know that dry sand will not stick together. Only damp sand will hold a shape. The sand used in casting must therefore be mixed carefully with the proper amount of water or oil before it can be used for the mold. This mixing process is called *mulling.* In large foundries, the mulling is done in large tubs above the foundry level and is sometimes done by robots. The mulled sand is fed through hoses to the casting room below.

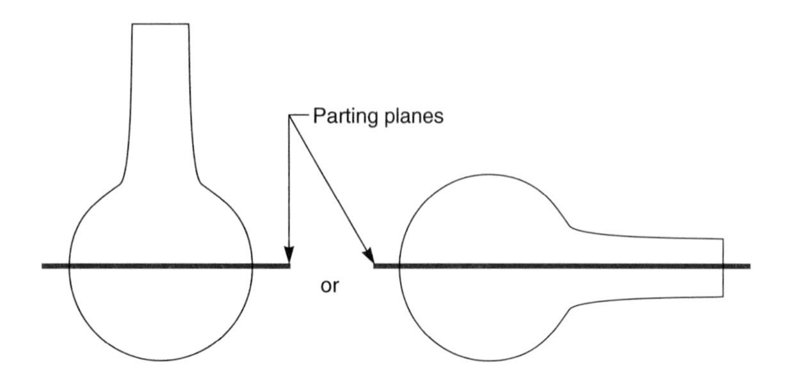

Figure 9–3. Parting planes.

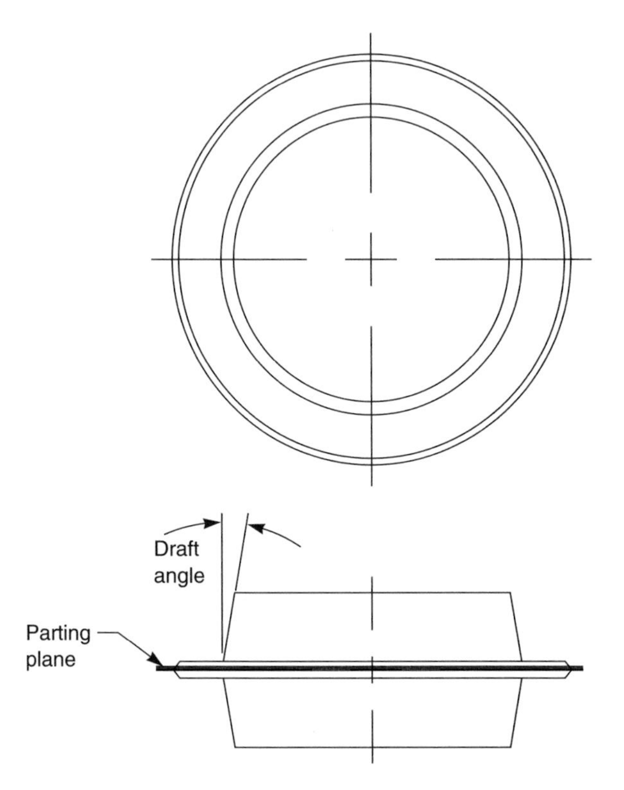

Figure 9–4. Parting plane for a complicated cast.

The container into which the sand is packed is called the *flask.* (Please refer to Figure 9–5 as we discuss these definitions.) The flask generally comes in two parts: the upper part is the *cope,* and the lower part is the *drag.* Depending on the shape of the part to be cast, the pattern may be entirely in the drag or split between the cope and the drag. A vertical hole called the *sprue* is cut through the sand in the cope beside the pattern. It is through the sprue that the liquid metal is poured. Horizontal *runners* are cut from the sprue to the pattern. Frequently, several workpieces are cast at a single pouring. In that case, the individual patterns are connected by runners from a common sprue.

In a typical sand casting, one surface of the pattern is flat. This pattern is turned over, the flat side placed on a smooth surface, and the empty drag placed around it (Figure 9–6). The pattern is then dusted with dry *parting sand* (a very finely ground white sand or talc) and the excess parting sand blown off by air. The mulled sand is then sifted or *riddled* into the drag and tamped or packed firmly in layers about the pattern. After the drag has been completely filled with sand and compacted, the bottom of the drag (presently on top) is *screeded* or scraped off to provide a smooth surface on which to rest. The packed drag is carefully lifted from the pattern and inverted.

To provide for internal cavities in the casting, *cores* must be prepared in advance. Cores are made of sand mixed with core oil. Cores made out of this specially prepared sand are formed in molds and baked or cured to a hardened state.* The core is placed in the drag so that it extends beyond the mold cavity and is held in place by the sand. *Core prints* are part of the pattern and leave holes in the sand into which the cores will be placed.

The drag is now ready to be mated with the cope. The cope is similarly packed (Figure 9–7). This time

* These cores can easily be shattered and removed from the internal parts of the casting.

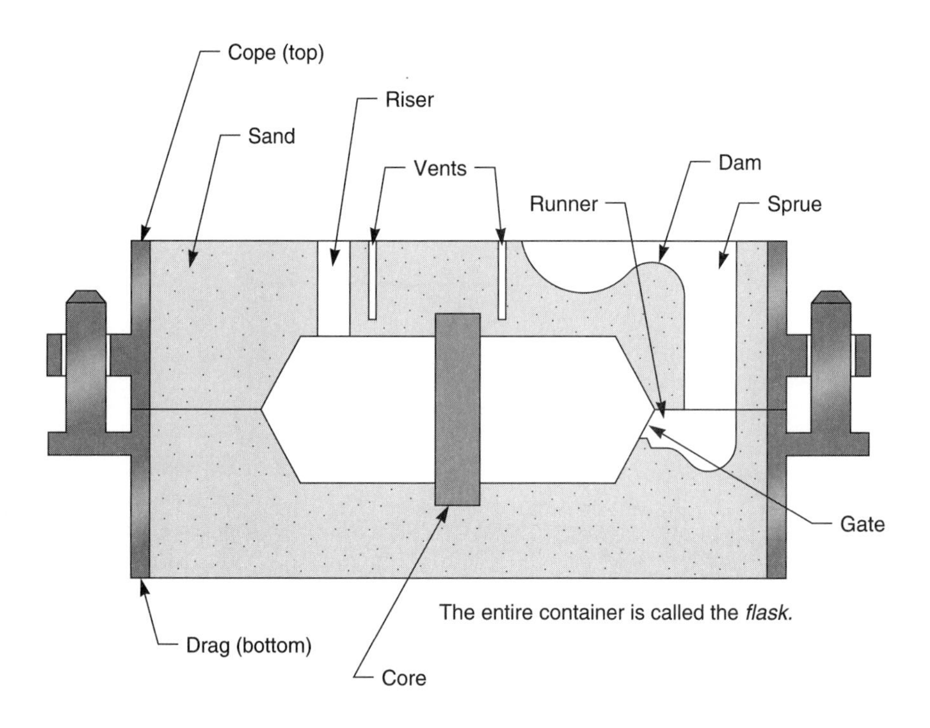

Figure 9-5. Diagram of sand casting process.

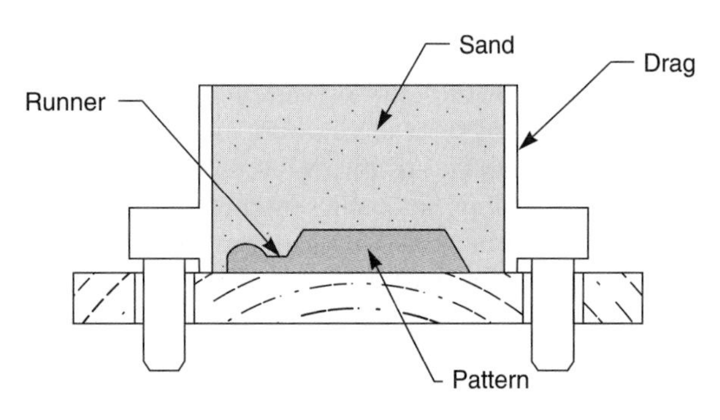

Figure 9-6. Packing the drag.

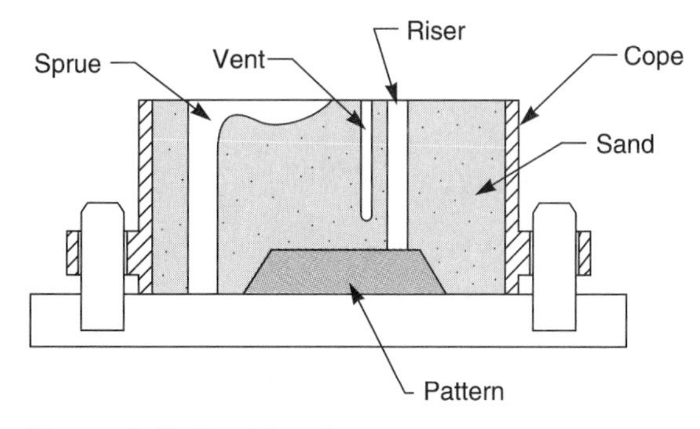

Figure 9-7. Packing the cope.

the surface to be mated with the drag is placed against the pattern. Cores or tubes are placed in position to form the sprue and risers. The pattern is dusted with parting sand and the mulled sand is riddled around the pattern and packed to fill the cope. Once filled, the sprue and riser tubes are removed. At the top of the sprue a depression is placed in the sand with a dam between the depression and the sprue. The molten metal will be poured into the depression. Since most hot metals oxidize rapidly, the top surface of the "puddle" of metal will be oxidized. The purpose of the dam is to allow the molten metal to flow from under the oxide into the sprue, thus preventing the oxidized metal from flowing into the mold.

Note that the sprue is never cut directly into the pattern but always to the side. The grain structure of the metal of the sprue and directly below it is not uniform. By placing a slight depression at the bottom of the sprue and allowing the metal to flow horizontally into the mold, the poor grain structure is averted. Furthermore, this depression at the bottom of the sprue forms a trap so that any loose sand carried down the sprue will have a place to settle out rather than enter the cavity and contaminate the casting.

In "green" or water-based sand molds, steam is formed where the hot metal comes into contact with the sand. Small holes or *vents* are often placed in the sand in the cope to allow the steam to escape.

The cope is then removed from the pattern and placed atop the drag. Large castings sometimes require weights to be added to cover the top of the cope to prevent the hydrostatic pressure of the liquid metal from pushing the sand out of the cope. The completed assembly is shown in Figure 9–5.

The mold is now ready for the pour. The metal is melted and poured onto a crucible for transfer to the mold. The metal is poured slowly and evenly into the sprue until it is seen in the risers. Then it is allowed to solidify. After the metal has solidified, the cope and drag are separated and the sand shaken from around the casting. The casting will have the sprues, runners, and risers still attached; these must be removed by sawing. The part is then ready to be finished by standard manufacturing practices discussed in other chapters of this text.

To ensure that the metal flows into all parts of the mold cavity, pressure is required. Further, most metals shrink on the surface as they solidify (Figure 9–8). Risers are cut from the mold cavity to the surface of the sand in the cope. Risers have several functions:

1. They allow the air to escape from the mold as the liquid metal is poured in.
2. As the liquid metal continues to flow up the risers, it creates pressure or a *hydrostatic head* on the metal to ensure that the cavity is completely filled.
3. They ensure that there will be no surface shrinkage in the mold.
4. When the metal flows to the top of the riser, the workers can see that the mold is full.

Note that pressure is a key consideration in both centrifuged and permanent mold casting, which are discussed later in this chapter. The higher the pressure, the smaller the detail that can be cast.

The rate at which the metal flows into the cavity is controlled by a *gate,* which is simply a restriction in the runner at the point where it meets the cavity. More delicate molds require the metal be fed into the cavity slowly so as not to damage the mold by the turbulence of the flow. There are other subtle but important points in the design of the runners and gates for a sand casting. Erosion will carry sand from the walls of the channels to the cavity. If the liquid metal has to flow around sharp corners in order to fill all of the mold cavity, holes might form in the casting on the downstream side of the corners. Also, the farther the liquid metal has to flow to fill the cavity, the cooler it becomes. There will be some solidification in the runners, which constricts the flow and prevents the cavity from being filled.

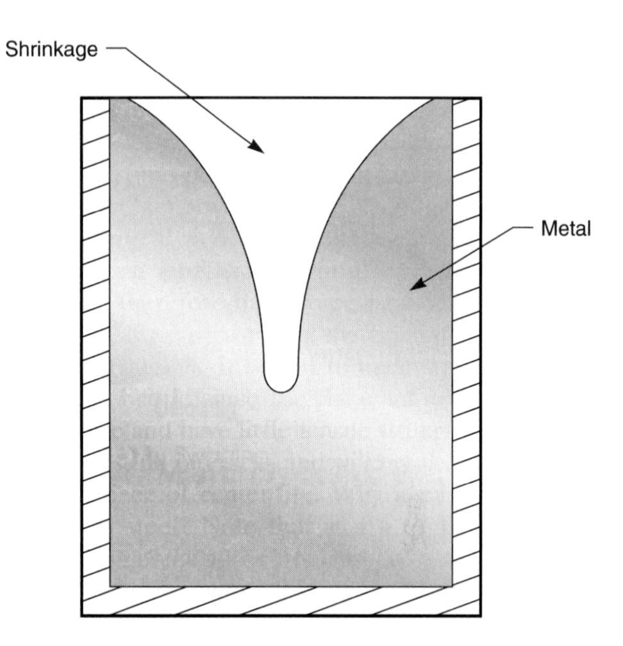

Figure 9-8. Shrinkage of metal surfaces.

If you get the sense that perhaps sand casting is as much an art as it is a science, you're absolutely correct! It is common practice to make one or more sample castings to determine the optimum choice of sprue and runner placements, gate sizes, and the size and placement of risers and vents. There are few rules to guide the foundry worker and there is no substitute for *experience* in sand casting!

Note that, once made, a sand mold has a short *shelf life.* Once the sand dries out, the mold is worthless and has to be discarded. The production of the mold needs to be coordinated with the melting of the metal.

The riddling, tamping, and other operations can be done by hand or by automated machinery. Air-driven tampers and riddles are used for mass production. For small-scale production jobs, much of the work is done by hand. Yet, in larger industries, the entire casting operation, from start to finish, is automated and done by robots. A representative assembly line production is shown in Figure 9–9. We'll discuss this in more detail in a later chapter.

Green Sand Casting

"Green lumber" is wood that is still wet with sap or other moisture. "Green sand" is sand to which water has been added to make it hold its shape between the removal of the pattern and the introduction of the molten metal. Casting that uses water-based sand is called *green sand casting.* The amount of water is critical. If the sand is too dry, the mold collapses before the metal can be poured. If it is too wet, the metal is chilled to the point that it will not properly fill the cavity before it solidifies. Further, if there is too much moisture in the sand, great amounts of steam may be generated and produce a back pressure such that the cavity does not fill completely or bubbles are produced in the casting. The sand should be just wet enough to stick together when it is compacted, and no more. The proper wetness of the sand can be learned quickly from experience. The water content of the sand will probably be between 5% and 6% by weight.

Green sand can be reused many times simply by adding a little water after each casting and remulling the sand.

Dry Sand Casting

In *dry sand casting,* the sand is "dry" only in the sense that water is not used for the bonding agent. Instead, the sand has been mulled with oil or some other petroleum derivatives to make it cohesive. These oil-based sands have a longer shelf life than the water-based sand, but the reuse of the sand is more limited. In many cases, the dry sand molds are baked in an oven to remove all volatile matter that might form gas pockets in the casting. The resultant mold is very dry and firm, but somewhat brittle. In the "old days"

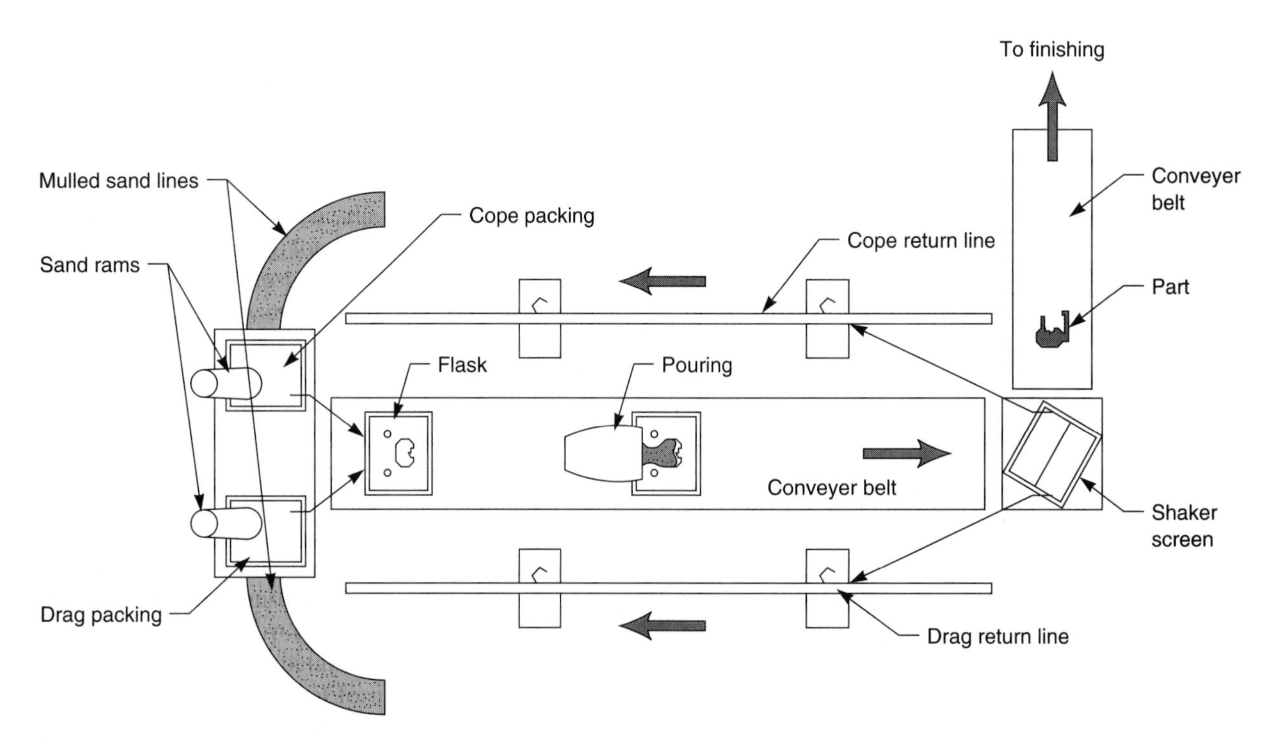

Figure 9-9. Casting assembly line.

before the petroleum-based bonding agents were developed, molasses was sometimes used to bond the sand. This worked very well except for the burnt sugar odor that arose when the molten metal was poured in the mold.

Green sand and dry sand castings each have their advantages and limitations. Green sands are cheaper and can be reused more readily than the oil-based sands. Sand can be made more cohesive with oil than with water. Therefore, larger castings can be made using the dry sand process. Castings as large as 3 tons can be made with the green sand process, whereas castings can be as large as 300 tons using the dry sand method. On smaller scales, molds from more delicate patterns can be made with the dry sand than with the green sand. While the green sand molds tend to generate steam when the hot metal is poured in, the oil-based sand molds tend to have a fire on the surface of the mold.

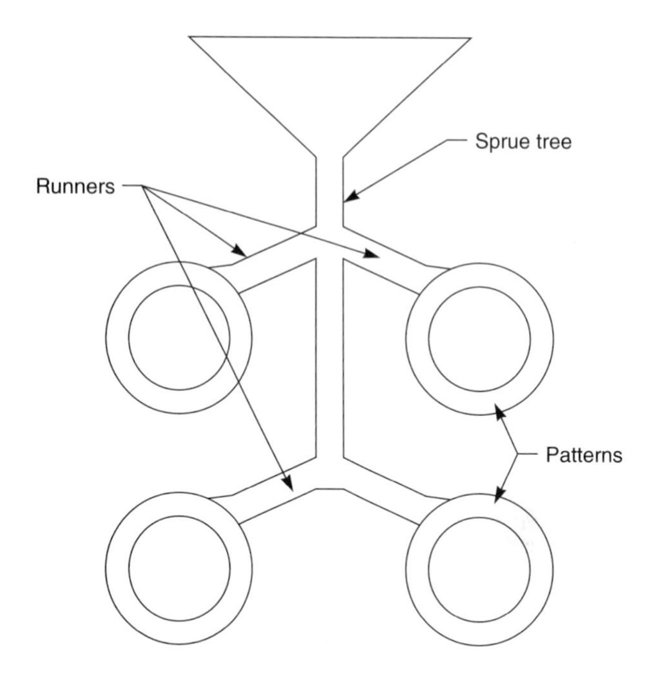

Figure 9-10. Sprue tree schematic.

Investment Casting Processes

The word "invest" comes from the Latin root meaning "to clothe or surround." In casting practice, the *"investment"* is a slurry of a refractory (heat-resistant) material that is used to coat the pattern, thus forming the mold. This process is called *investing the pattern.* Typically, the pattern is backed up with another coating of material to provide strength to the mold and support for it during the casting process. As with most general casting techniques, a wide range of methods is used in investment casting.

Lost-Wax Casting

The *lost-wax casting* technique is undoubtedly the oldest of the investment casting methods, dating back some 3000 to 5000 years (depending on which source you read). It should come as no surprise to find that it's a simple process. Its popularity comes from its ability to produce castings of incredibly complex shapes.

The pattern is made of wax, either carved and fabricated by hand or, in the case of high-volume production, cast in a mold. Typically the workpiece is small so that a number of copies of the pattern are attached to a central wax sprue *tree* by wax runners (Figures 9–10 and 9–11). The wax is dipped in a *debubblizer* to prevent air bubbles from adhering to the wax during the investment process. The assembly is then coated by the investment material and allowed to harden. This coating process is repeated until a sufficient thickness of investment is built up (rather like dipping a candle). In smaller parts, the wax is placed into a flask and the investment material is simply poured around it, filling up the flask (Figures 9–12 and 9–13).

Following a setting time for the investment, usually a few hours, the investment is placed in a furnace and heated to melt and evaporate the entire wax pattern, sprue, risers, and runners. With the wax gone, the investment now has a cavity to serve as the mold for the casting. Molten metal or other material can be poured into the mold. If the pour is made while the mold is still hot, the mold can be destroyed by plunging it into cold water. The part can be retrieved, the sprues and risers removed, and the part sent on to finishing processes.

Full-Mold Casting

Full-mold casting is also known as *lost-foam casting, expanded polystyrene casting,* or *evaporative pattern*

Figure 9-11. Sprue tree.

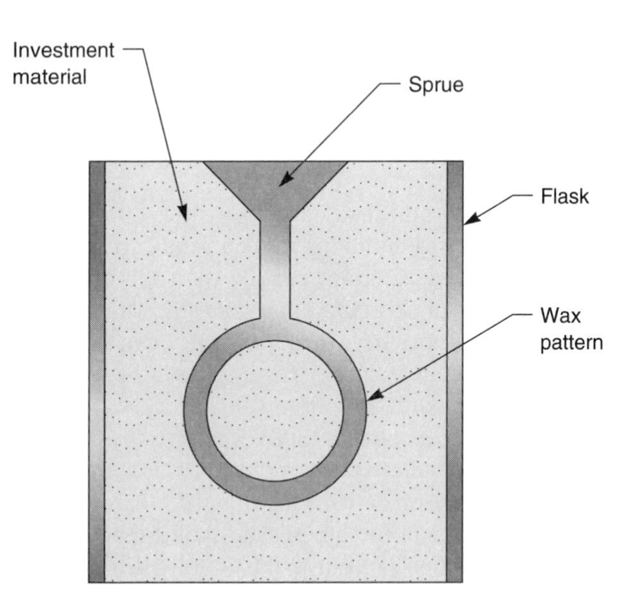

Figure 9-12. Investment flask.

casting. It is a method that has long been used by artists and sculptors for making one-of-a-kind objects. Now it is finding its way into industry since it is applicable to large, mass-produced parts. In this version of the basic investment casting process, the pattern is made of expanded polystyrene (trade name Styrofoam) rather than wax. As with lost-wax casting, the shape of the pattern can be very complex. However, the pattern can be much larger than those of the lost-wax process. Single workpieces can be made by this technique rather than having to cast them in several pieces that are later joined. Some automobile manufacturers are now using full-mold casting to produce engine blocks, heads, manifolds and other castings.

Once the pattern, sprues, risers, and runners are made from the expanded polystyrene, the pattern is invested. Once the investment has hardened, it is packed in dry sand for support.

In sharp contrast to the lost-wax process, the styrofoam is left in place. When the molten metal is poured into the sprue, the styrofoam is vaporized and the molten metal takes its place. The mold is never empty as in other casting techniques, hence the name *full-mold casting.*

Plaster Mold Casting

In *plaster mold casting* the pattern is typically made of metal or other material with a smooth finish, so that the surface finish of the casting is superior to that of a sand-cast piece. The investment material is plaster of paris (gypsum). The distinction of this method is that the investment is done in such a way that it can be separated into halves to allow the pattern to be removed. The halves are sealed together and the molten metal, plastic, ceramic, or other material is poured into the mold. The mold can be preheated, if necessary, to allow the entire cavity to be filled before the material solidifies.

Since the plaster mold material begins to decompose and fall apart at a temperature of about 1000°F, only plastics or low-melting-point nonferrous metals or alloys can be cast by this method.

Figure 9–13. Investment flask.

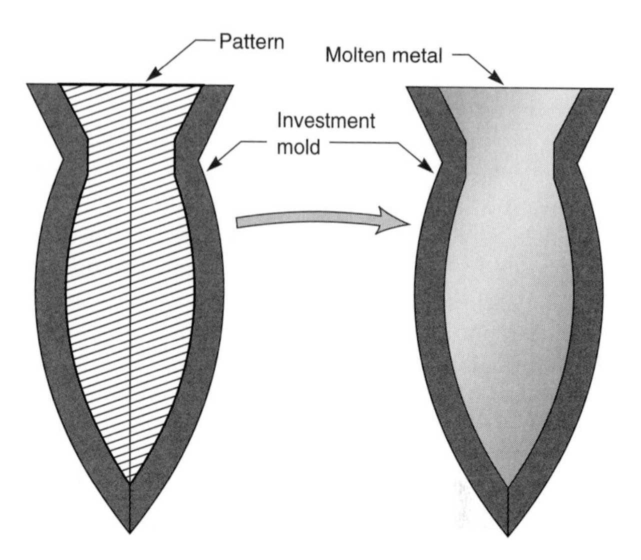

Figure 9–14. Shell mold casting.

Ceramic Mold Casting

Ceramic mold casting is a variation on the plaster mold process that uses a refractory slurry, instead of plaster of paris, as the investing material. These ceramic investments can withstand higher temperatures than the plaster. Firing the molds after investing forms a network of microscopic cracks, which allows the air to escape easily when the cavity is being filled with the liquid metal. This process is also known as the *Shaw process* after the two men who developed it.

Ceramic mold casting is especially valuable for the production of workpieces with materials or shapes that would make them difficult to machine. Since the shapes may be complex and the use of a metal pattern produces a smooth-walled mold cavity, resulting in a smooth-surface casting, little or no postcasting machining or finishing is required. Dies and molds for other castings are frequently made by this process.

Shell Mold Casting

Shell mold casting is an investment casting process, but the investment material is sand with a resin binder rather than a refractory slurry. The pattern is heated and then either dipped in the resin-treated sand, or the sand is sprayed or poured over it. The high temperature of the surface of the pattern melts the binder, and the coated sand grains form a shell around the pattern. Once the shell cools, it is split open, the pattern removed, the halves sealed together again, and the liquid metal poured into the mold to make the casting. After the casting solidifies, the shell is broken from it and the part finished (Figure 9–14).

As with plaster mold castings, parts cast by shell molding need to have only the sprue removed and require very little finishing. Unlike other investment castings, the shelf life of the mold is indefinite. Further, large castings can be made by this most useful process. High-temperature metals such as stainless steels and titanium alloys can be cast using shell molds. Turbine blades for jet aircraft are one example of parts that are made by shell mold castings.

Centrifuged Casting

Centrifuged casting is not to be confused with *centrifugal casting* to be discussed later. If the mold cavity is very complex, and especially if there are a number of small passageways in the pattern, the force of gravity may not be sufficient to make the liquid metal flow completely into all portions of the cavity. In that case, the flask can be rotated around an axis during the pouring and solidification steps. The centrifugal force will drive the liquid metal into the inner recesses of the mold cavity. Even in small centrifuged casting machines, 5-G forces can be attained. Most of the centrifuged casting machines rotate the mold about a vertical axis, in a horizontal plane as shown in Figures 9–15 and 9–16. Some centrifuged casting machines are vertical and they rotate the mold in a vertical plane. The action in these machines is similar to keeping water in a bucket while swinging it over your head.

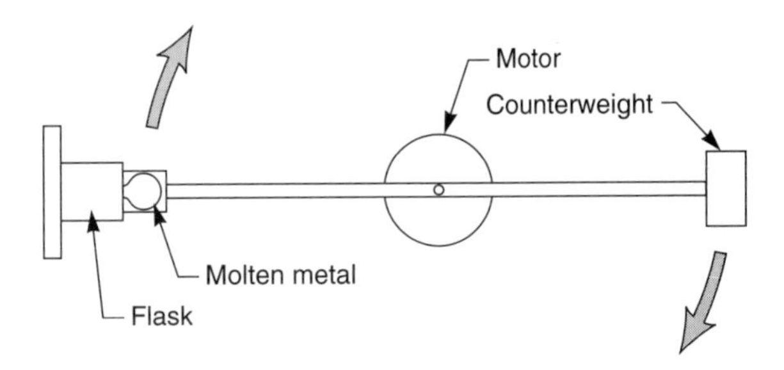

Figure 9–15. Centrifuged casting.

Centrifuged casting provides one of the most dimensionally stable castings available. It is used by dentists and dental technicians to produce inlays for teeth, dental crowns, and other false teeth. In industry the technique is used to cast small gears and other parts. Jewelers have long used this process to make settings for rings and other fine jewelry.

One significant drawback to centrifuged casting is that casting a mixture of metals having different densities often results in the separation of the different components prior to solidification. There is also a limitation as to the size of the machine. Generally, centrifuged castings are rather small and often weigh only a few ounces to a few pounds (kilograms).

Vacuum Casting

Instead of pushing the molten metal into the mold, *vacuum casting* draws the material into the mold by means of suction. This technique can be used as the final steps in the lost-wax process, with shell molds, or with other techniques. The plaster, sand, crystobalite, or other investment material is porous enough to allow air to flow through it. The flask is placed over a vacuum plate and a suction is drawn through the flask (Figure 9–17). The molten metal is then poured into the mold. The vacuum ensures that the molten metal fills all parts of the cavity. The vacuum also prevents air bubbles from forming in the metal.

One novel variation of this technique is sometimes used by jewelers. Instead of using a vacuum pump, a

Figure 9–16. Centrifuged casting machine.

sphere partially filled with water is used. The water is brought to a rolling boil with steam escaping from the vent. The vent is closed and the water allowed to cool. Once cooled, there is a vacuum above the water. As the molten metal is poured into the mold, the valve between the flask and bulb is opened, and the vacuum draws the metal into the mold cavity (Figure 9–18).

Permanent Mold Casting Processes

In all of the casting processes described to this point, the mold is expendable and is destroyed to remove the casting. However, a number of casting processes exist that use a reusable or *permanent* mold. Usually, the mold is made of forged or cast metal, but graphite and other materials are occasionally used. The cost of manufacturing and maintaining a permanent mold can be high compared with the cost of making a sand casting mold or an investment mold. For this reason, the permanent mold casting processes tend to be used for high-volume production runs. Sand casting and investment casting are generally more appropriate to one-of-a-kind and low-quantity jobs—but there are exceptions to any rule.

Permanent mold casting processes are often referred to as *near-net-shape* processes because little or no machining is required to convert the raw casting into a finished part. Frequently, the only machining needed is to pass the part through a shearing operation that removes the runner, sprue, and any flashing. With good design, only a brief abrasive tumbling, shot-peening, or other finishing operation is needed.

Since the mold is the major distinction between permanent mold casting processes and the casting processes discussed earlier, it is appropriate to examine a typical permanent mold, as shown in Figures 9–19 and 9–20.

Permanent Mold Design

Mold Halves. Typically, permanent molds have two parts similar to the cope and drag in sand casting. Occasionally, molds are made in several parts, but this can slow the production rate considerably and thereby increase the cost of production. Yet, it may be faster to use multiple-part molds and cast the part rather

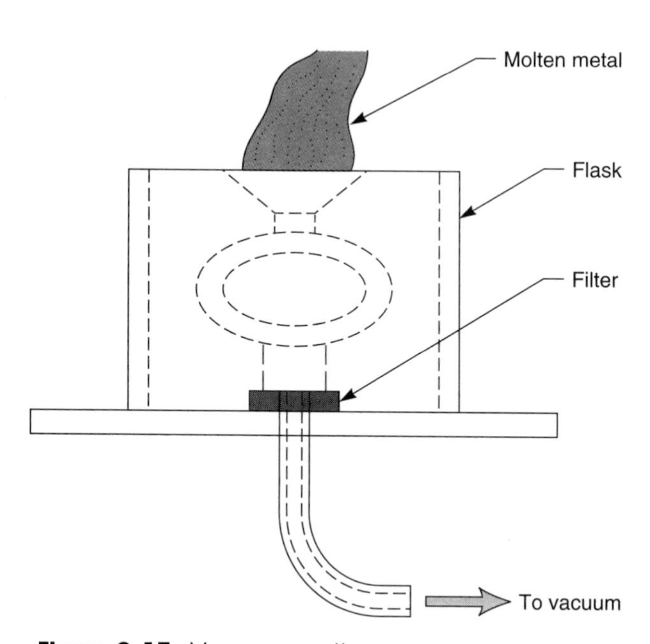

Figure 9–17. Vacuum casting.

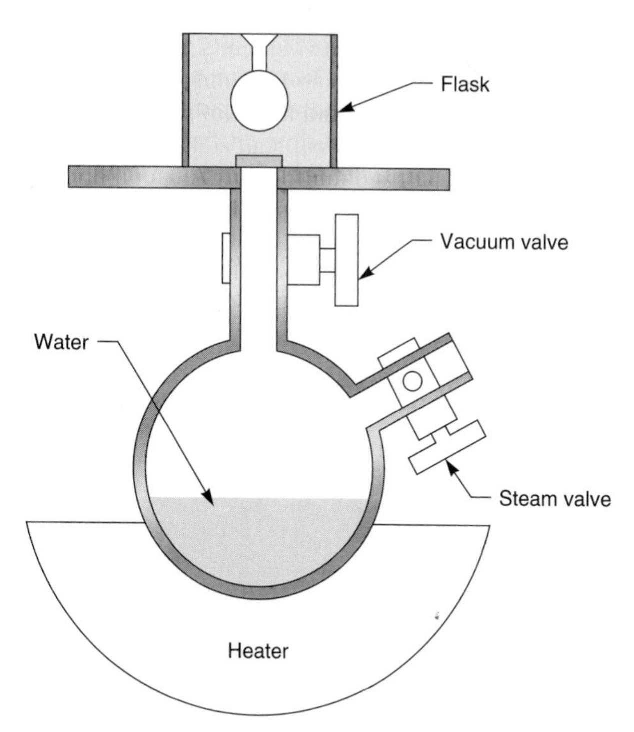

Figure 9–18. Vacuum bulb casting.

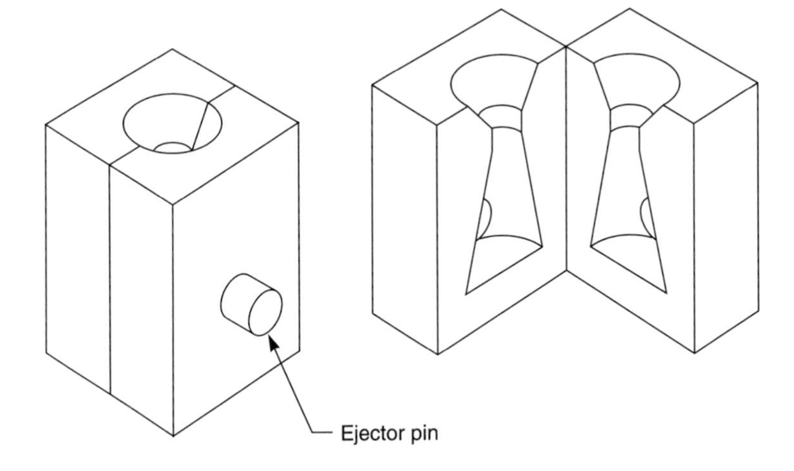

Figure 9–19. Reusable mold.

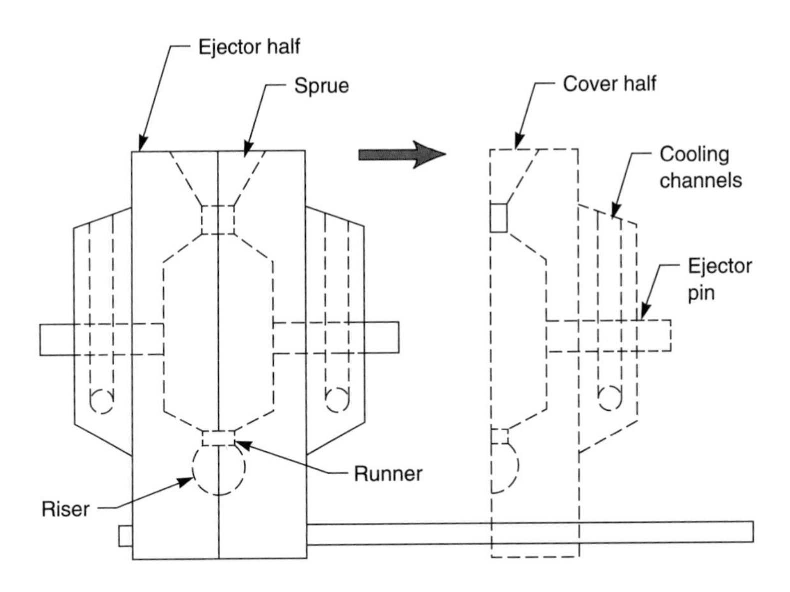

Figure 9–20. Reusable mold.

than machine it from a solid block of material. The slower rate is not automatically a valid reason to avoid using multiple-part molds.

The halves (or parts) of the mold are generally joined by some means. In Figure 9–20, the movable part of the mold (the *cover half*) slides along rods to release the completed casting. The cover half then slides back precisely into place for the next casting cycle. The "fixed" half of the mold is called the *ejector half.*

Another method of joining the molds is by using a hinge as shown in Figure 9–19. The movable part swings open to release the completed casting and swings closed again for the next casting cycle.

Sprues, Runners, and Risers. Sprues, runners, and risers are used in the same manner as in sand casting, except that in permanent molds, they are machined into the mold. Since permanent molds are used in casting techniques whereby the liquid metal, plastic, or other casting material is injected under pressure, the risers are often a closed chamber rather than being open to the atmosphere.

Cores. Cores can be placed into the mold by hand or by robot (as in sand casting), but this adds time to the cycle. In well-designed applications, the cores are a part of the movable part of the mold. After the mold cavity is filled and the cast material solidified, the moving part of the mold carries the cores with it. Closing the mold inserts the cores again, ready for the next cycle.

Inserts such as threaded studs, bushings, nuts, etc., can be accurately located in the molds and can be cast as an integral part of the workpiece.

Ejector Pins. Since the molds are not destroyed or broken from the casting, some method of removing the solid casting from the mold must be included in the mold. This is done by *ejector pins.* As the mold opens, the pins are forced against the casting to knock it out of the mold. If one examines some cast parts, especially plastic ones, small circular marks can be found. These are the marks left by the ends of the ejector pins.

Thermal Conditions. Permanent mold casting is ideal for many applications which use low-melting-point metals, alloys, or plastics. Since these materials tend to solidify rapidly, it may be necessary to keep the mold warm while the liquid is being poured in. After the material is in the cavity, the mold must be cooled to allow the casting to solidify. Heating of the molds can be done with electrical heating coils or with hollow tubes placed at the back of the mold through which hot water or other liquid can be run. Cooling can also be done by circulating a fluid through channels in the back of the mold, or by the use of cooling fins around the mold. Note that the higher the temperature of the molten casting material, the shorter the life of the mold.

Plastics that solidify upon cooling (thermoplastics) can be cast with standard casting procedures. Those that need heat to harden require special procedures. The casting of plastics and composites is discussed in greater detail in Chapters 18 and 19.

Mold Materials. The material from which the mold is made depends mainly on the material being cast. Other considerations that might also influence the selection of a mold material are the tolerances and quality of the product, the number of parts expected from the mold, and cost limitations. The material for a mold must have a high-melting-point temperature so that it will not be eroded by the molten material being cast. It must also have high enough strength to prevent it from distorting under pressure, a high resistance to thermal fatigue, and low adhesion to the casting material.

The majority of permanent mold cast parts are made either of plastic or of low-melting-point alloys of lead, tin, zinc, antimony, or bismuth. Molds for these materials can be made of bronze (copper-based alloys). Higher melting point metals such as aluminum, copper, brasses (copper-zinc alloys), and bronzes require a mold made of stainless steels, steels containing chromium, tungsten, or molybdenum, or nodular-grained cast iron. In the unusual cases where the parts are to be cast from cast iron or steel, the molds must be made of graphite or other refractory materials. Such molds are very expensive, resulting in a high cost per unit cast.

Although a mold set may cost from a few hundred to several thousands of dollars, the cost may be justified. It is possible to make up to 250,000 aluminum castings from one mold. For example, if a mold cost $2500 and could produce 250,000 castings, then the cost is only one cent per unit for the mold. This is very competitive with other casting techniques for high-production-rate items. For small production runs, expendable mold techniques such as centrifuged or shell mold casting might be more economical.

Coatings. Coating materials may be used on the inner surfaces of the molds for two reasons: (1) to protect the mold from heat damage by slowing the rate of heat transfer from the molten metal to the mold and (2) to keep the solidifying material from sticking or welding itself to the mold. Coatings made of a refractory powder, such as parting sand, will help insulate the mold from heat. Coatings of powdered graphite, silicones, or sodium silicate and water will make a satisfactory parting compound.

Multiple Cavities. When the cast part is small, has a simple shape, and the production quantity requirements are large, it is common practice to cast several identical parts at the same time by using a mold that has many identical cavities connected by runners from a central sprue. The process is similar to that used in the lost-wax technique.

The Pour Cycle. Permanent mold castings are often made by automated machines. A common pouring cycle for these machines would be as follows:

1. The mold is cleaned with a blast of air.
2. The cores and inserts are placed in the mold.
3. The mold is coated with mold release or refractory material.
4. The mold sections are locked or clamped together.
5. The mold is preheated if necessary.
6. The mold is filled with the liquid metal or other material.
7. The mold is cooled until the melt is solidified.
8. The sections of the mold are unclamped and the mold opened.
9. The casting is ejected.
10. Start over again at step 1.

It is clear from this listing of casting steps that the actual pouring step represents only a small portion of the time in the cycle. Therefore, many machines operate up to 12 molds sequentially so that melted metal is being poured nearly constantly.

Permanent mold machines can produce large numbers of parts in a short time. For instance, consider a machine that produces 8 parts per filling of the mold. If the machine has 12 molds and it takes one minute per complete cycle, then the production rate of the machine is:

$$8 \text{ parts/mold} \times 12 \text{ molds/cycle}$$
$$\times 1 \text{ cycle per minute} = 96 \text{ parts/minute}$$

That's 5760 parts per hour, 46,080 per 8-hour shift, or 230,400 per 40-hour week! If that seems like a lot, imagine that the parts are zipper teeth and it becomes clear that this production rate may be far smaller than the demand! Permanent mold casting methods are one solution to high-production-rate needs.

Types of Permanent Mold Casting Processes

There are four general types of permanent mold casting processes. They are distinguished by the amount of pressure used to force the melted material into the mold. They are, in order of increasing pressure:

1. Gravity feed
2. Low-pressure feed
3. Vacuum feed
4. Die casting

As the amount of pressure increases, the equipment becomes heavier, more complicated, and more expensive.

Gravity-Feed Casting. Gravity-feed permanent mold casting represents only a small fraction of all industrial permanent mold casting processes. Its most common use is for hand pouring of nonferrous metals. Its advantage is that the melted material enters the mold gently and without turbulence. This produces a casting with a fine grain structure and high strength. The size of a typical gravity-feed casting can be as small as a few ounces (a hundred grams or so) to 300 pounds (600 to 700 kilograms) with 20 to 30 pounds (44 to 66 kilograms) being a realistic upper range for high-volume production. One can expect 20 to 50 casting cycles per hour from a gravity-feed process, which is slower than die casting (discussed later). Figure 9–21 illustrates a gravity-feed permanent mold casting process.

Low-Pressure-Feed Casting. The basic concept of a low-pressure-feed casting is shown in Figure 9–22. The casting material is melted in a crucible, which is sealed off from the atmosphere and filled with an inert gas. The type of gas used depends on the material being melted but it can be nitrogen, helium, argon, or even carbon dioxide. The inert gas keeps the metal from oxidizing, which is a major cause of waste and flaws in open casting. Raising the gas pressure forces the molten metal through a feed tube to fill the mold cavity. Pressure is maintained on the mold until the casting becomes solid. Releasing the pressure allows the still-molten metal in the tube to flow back into the melting crucible while the casting is ejected and the mold made ready for the next cycle.

The mold can be filled without turbulence. This results in longer mold life and less trapped gas and porosity in the casting. Returning the metal from the sprue and feed tube reduces the amount of scrap that must be remelted.

Vacuum-Feed Casting. In some machines, the mold rather than the crucible is in a sealed chamber. Reducing the pressure in the chamber draws the molten metal into the cavity. Castings made by the vacuum technique have the advantage of even lower porosity than those of the low-pressure technique.

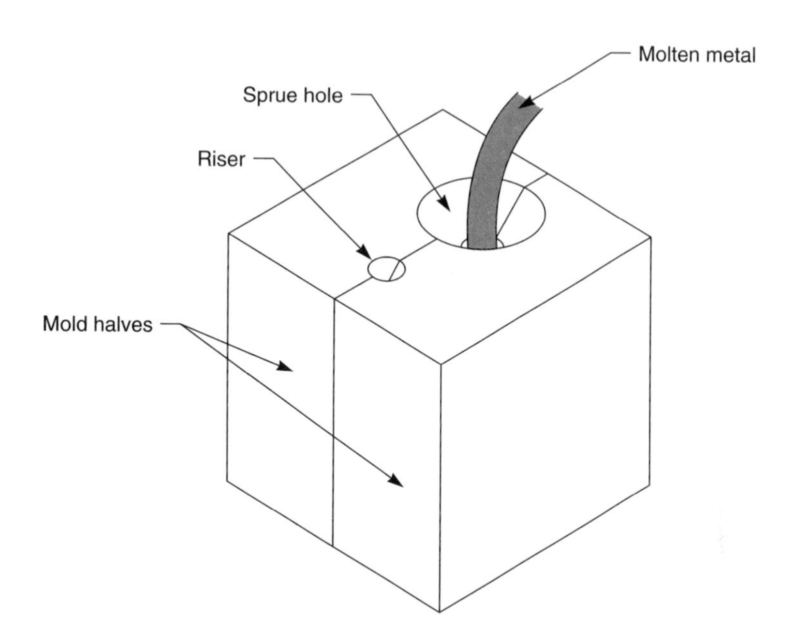

Figure 9-21. Gravity-feed permanent mold casting.

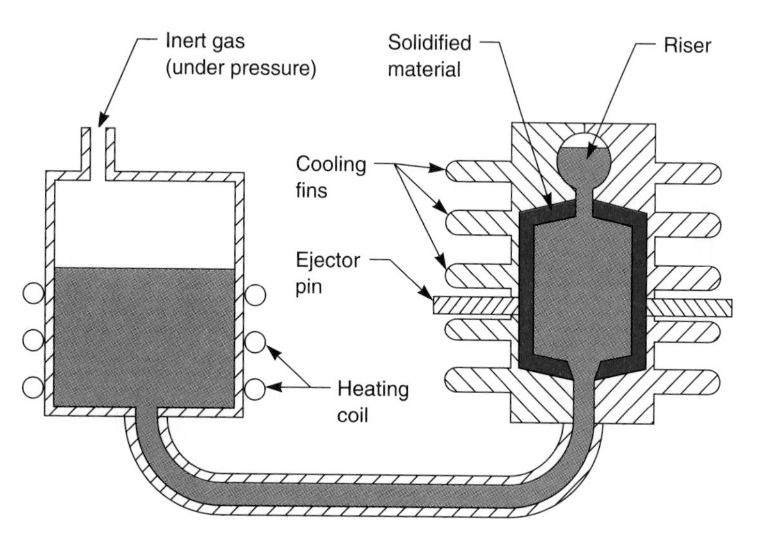

Figure 9-22. Low-pressure-feed casting.

Low-pressure-feed casting and vacuum-feed casting represent only a small fraction of all permanent mold casting. Their major use is in moderate-volume production of fairly large castings where the strength of the finished part is a key consideration.

Die Casting. To be absolutely correct, a *die* is "a (metal) shaping device with a cavity or hole machined into it," used especially for punching, forging, drawing, extrusion, etc. But the permanent molds used in casting also fit this definition well. Therefore, the term *die* is frequently used here and the process is called *die casting.* In this section the term *die* is used instead of *mold.*

Die casting is distinguished from other forms of permanent mold casting by the high pressure used to force molten metal into the die. Typical pressures range from 1000 to 30,000 pounds/inch^2 (7.5 to 225

MPa). Such high pressures are generally reached by filling a cylinder with the molten metal, then using a piston to ram the material into the die cavity. Two different arrangements of filling the die, the hot-chamber and the cold-chamber methods, are discussed in this chapter.

In the high-pressure method, there is such a force on the die that it must be clamped closed. In fact, the size of a die casting machine is generally given by how high a clamping pressure the machine can exert.

The high filling pressure has a number of advantages and limitations:

- Dies can be filled quickly. Production rates up to 100 cycles/minute are possible.
- The molten metal is under such pressure that it will find its way into any slight crevice in and between the die parts. Die-cast parts are more likely to require deburring and to need flashing removed than other casting techniques.
- The dies must be very strong and rigid.
- The die casting machines must be very rugged.

Die-cast parts may be made of zinc, lead, tin, aluminum, copper, magnesium, and alloys thereof. Zinc and aluminum are probably the most widely die-cast metal. Iron and steels are occasionally die cast, but they require special die materials due to the high temperatures involved.

In terms of weight limits, zinc parts as large as 75 pounds (34 kilograms), aluminum parts as large as 65 pounds (30 kilograms), and magnesium pieces weighing 44 pounds (20 kilograms) have been die cast.

Hot-chamber die casting is used for low-melting-temperature casting materials. The piston and mechanism for filling the cylinder chamber are actually immersed in the molten material. The piston can be withdrawn from the chamber allowing the liquid metal (or other material) to fill the cylinder automatically. Forcing the piston forward fills the empty die. This is a very fast process with up to 1000 cycles/hour possible (3.6 seconds between fillings). Pressures on the order of 2000 psi (15 MPa) are common. Figure 9–23 is a diagram of the hot-chamber process.

When higher melting temperature alloys or materials are being cast, the piston and cylinder assemblies can be damaged by constant immersion in the molten metal. For these high-melting-temperature materials, a piston/cylinder arrangement with an *outside* melting crucible is used. This is called *cold-chamber die casting*. The molten material is fed into the cold cylinder chamber from whence it is forced into the die cavity by the piston (Figure 9–24).

Pressures between 3000 and 10,000 psi (20 to 70 MPa) can be used. Cold-chamber casting is generally slower than hot chamber, but 1000 cycles/hour can be achieved.

Die casting is the main process for high-volume production of small, simply shaped parts. Production runs of 5000 to 100,000 parts are common. In addition to its obvious advantage of high production rates, die casting has the advantages of:

- Producing parts with excellent surface finish and close tolerances
- The nearly complete absence of any need for a finishing operation

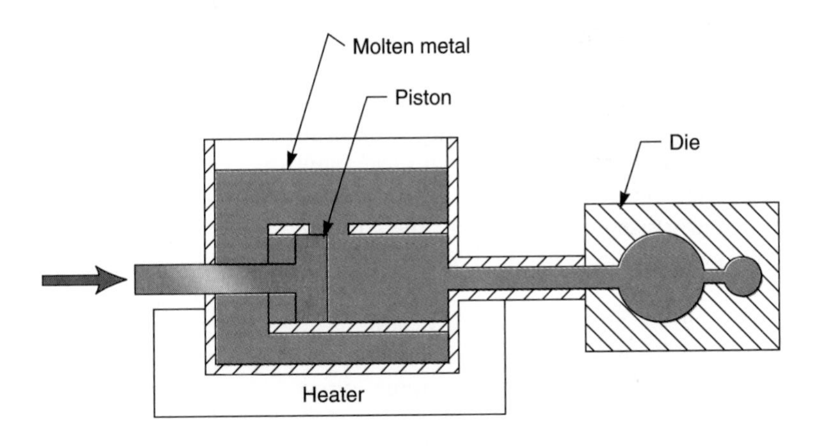

Figure 9–23. Hot-chamber die casting.

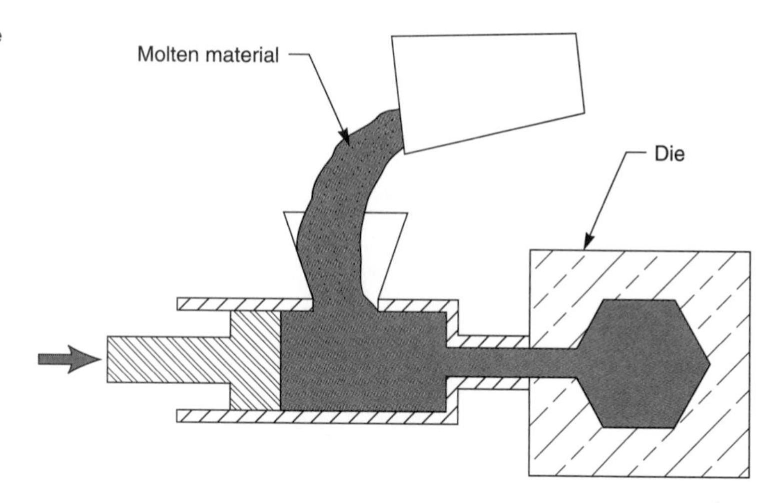

Figure 9-24. Cold-chamber die casting.

- A workpiece in which the surface is harder than the interior
- The ability to produce parts that have thin walls

The hard surface (yet soft interior) of die-cast parts makes them resist wear without being brittle. By means of die casting techniques, aluminum parts with walls as thin as 0.015 inch (0.33 millimetre) can be made.

Centrifugal Casting

Centrifugal casting must not be confused with *centrifuged casting* discussed earlier. In fact, three casting processes have nearly identical names: centrifuged casting, centrifugal casting, and semi-centrifugal casting. These names are often improperly interchanged in industry.

Centrifuged castings are made from a mold that typically has a center sprue with runners branching from it into the cavities. The mold is generally spun around a central *vertical* axis while the metal is forced into the cavity by centrifugal force as shown in Figures 9–15 and 9–16.

Semi-centrifugal casting also uses a vertical axis of rotation. The distinctive feature of a semi-centrifugal casting is that only a single, essentially ring-shaped piece is cast. The axis of revolution coincides with the centerline of the sprue and radial runners carry the metal from the sprue to the cavity. Centrifugal force is again used to force the liquid metal into small cavities and also to force any bubbles to leave the casting before it solidifies (Figure 9–25). Semi-centrifugal casting is rarely used today so it is not discussed further.

The molds for centrifugal castings have horizontal axes and use centrifugal force to keep the liquid metal against the mold until it solidifies. In the same way that a roller-coaster car is held against the rails even

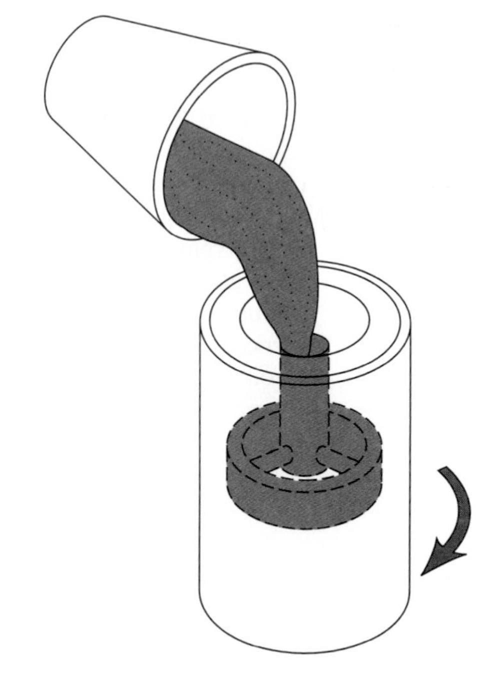

Figure 9-25. Semi-centrifugal casting.

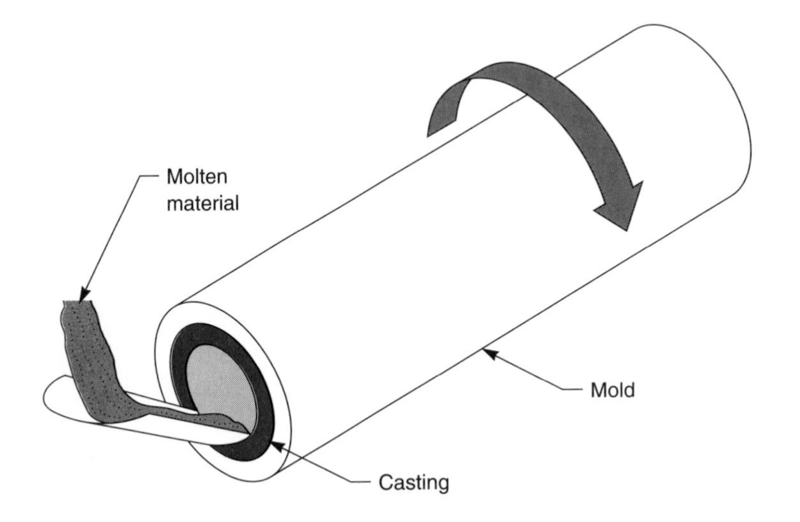

Figure 9–26. Centrifugal casting.

at the top of a loop, so the molten metal can be held against the top of a cylindrical mold if the mold is rotating fast enough. That's the secret to making a centrifugal casting (Figure 9–26).

The mold is generally a hollow cylinder (or more accurately, a surface of revolution) with some small changes at the ends and possibly some slight deviations from a cylinder on the inside. Molds can have diameters from 6 to 60 inches (15 to 150 centimetres) and mold length from 10 to 30 feet (roughly 3 to 10 metres). The cast part can have a hexagonal or octagonal outside surface with a cylindrical inside surface. The molds are rotated about a horizontal axis at speeds between 300 and 3000 rpm.

Cast iron, steel, nickel alloys, bronze and other copper alloys, and aluminum are the metals most commonly cast by this process. Pipes and other large-diameter parts are commonly made this way.

An advantage of centrifuged castings is that the lighter impurities, such as oxidized metal and slag, will end up on the inside of the cylindrical hole, requiring only a simple machining operation to remove them.

Slush Casting

In most casting procedures, the metal in the mold begins to cool and solidify where it comes in contact with the mold. The metal in the center of the mold remains liquid for a while. In *slush casting,* the mold is emptied of the molten metal after the outside has started to solidify. (Figure 9–27). This leaves a shell of solidified material. Besides metals, some plastics and porcelain clays can also be cast by this method. The workpieces made by slush casting have an attractive exterior but an uneven interior and wall thickness. They also generally have low strength.

Slush castings are used for making decorations, trophy statues, toys, ceramic castings, and other products. It is a relatively inexpensive process but does have limited applications.

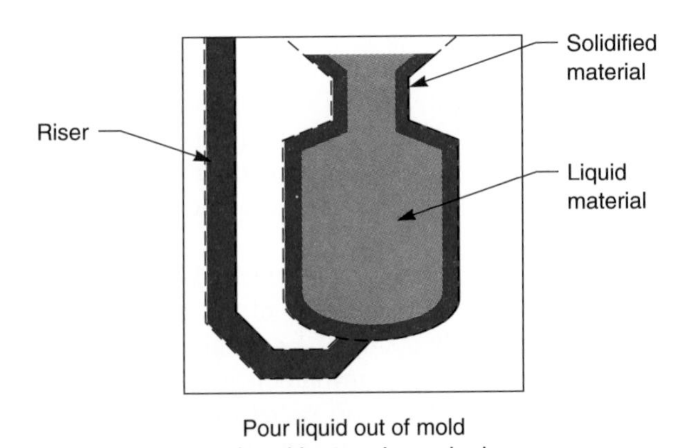

Figure 9–27. Slush casting.

Continuous-Pour Casting

A process used in the heavy metals industry is *continuous-pour casting.*

Even though individual steel ingots weighing several tons have been cast for more than a century, these ingots limited the maximum size to which beams and other shapes could be fabricated. The limit for continuous casting is the amount of liquid metal that can be melted at one time. Once the continuous-pour casting process was developed, ingot size was no longer a limitation.

The process starts with the metal being poured into a mold that has no bottom (Figure 9–28). Water is sprayed on the outside of the mold. If a correct balance is achieved between the temperature of the molten metal, the cooling rate of the mold, the length of the mold, the heat capacities and heats of fusion of the metals, and the pour rate, the metal will be solidified by the time it exits the bottom of the mold, and the casting is termed *continuous.* This takes place in a tower and the casting can be cut into sections as needed for fabrication.

Besides steel, continuous castings are made of copper, copper alloys, aluminum, and cast iron. After the castings are made, they may be formed into beams, rods, or other shapes by forging, rolling, and other techniques, which will be discussed later.

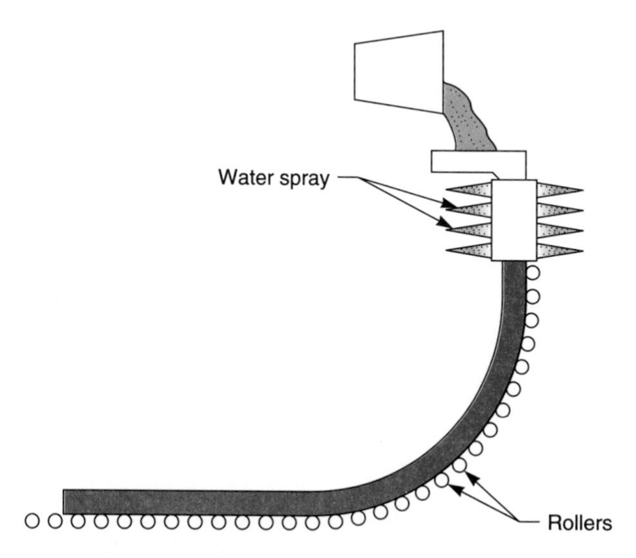

Figure 9–28. Continuous-pour casting.

SELECTION OF CASTING PROCESSES

With all of the methods available to create a workpiece, how can one possibly make the proper choice? Let us refer to the criteria as suggested in Chapter 1:

1. What are the physical properties of the material being used?
2. Does the process produce a part which meets all of the design specifications given in the plans?
3. Does the process meet the precision requirements of the product?
4. Does the process have the appropriate production rate?
5. Is the per-unit cost low enough?
6. Are the social or environmental consequences of the selected process justifiable?
7. Is the process available?
8. Is a trained operator available?

If these criteria suggest that the product should be made by casting, then we still need to decide *which* casting process should be used. Perhaps a few more questions particular to casting should be examined:

9. What equipment will be available in-house?
10. What are the shape-complexity and size of the finished part?
11. What physical properties must the finished part have?
12. What surface quality must the finished part have?
13. How many parts must be produced, and what production rate must be achieved?

No one process is the best process for all of these questions. The solution is to select the *optimum* method, not the ideal one. Further, there is no reason why the part has to be produced by a single manufacturing process. For example, the size and production quantity may suggest a process that produces too coarse a surface; but the processes that produce a smooth surface may not be appropriate to the desired production rate. The optimum choice here may have to be to use a finishing process (Chapter 17) after the casting process. Consider, however, that these multiple-step processes increase the unit cost of the finished product and that presents another problem.

One way out of the dilemma suggested here is to redesign the workpiece if possible to make it fit a particular casting process. In fact, an experienced casting craftsman or technician will generally work with the engineer or designer to suggest small changes in the shape of the part that will make the casting process more reliable, reduce waste, or make it possible to select a cheaper casting process. However, if the design of the part cannot be refined further, the problem is reduced to which casting process should be selected.

One caution: Don't take too literally the numbers or other information presented in Tables 9–1, 9–2, and 9–3. Casting processes are constantly changing as new ways are found to extend the range of existing processes. Some foundries specialize in a particular process and have developed techniques for extending the applicability of their favorite process. The entries in these tables are "typical" or representative, but they are not absolute limits.

Selection Based on Shape

The selection of the proper casting technique based on the shape of the cast object requires answers to two questions:

1. How complex is the shape generally in terms of internal cavities or internal detail, inserts to be cast in place, and the number and arrangement of the features in the workpiece?
2. How complex is the exterior surface? This determines the ease with which the casting can be removed from the mold.

Table 9–1 equates these criteria with the type of casting best suited for them. Of course, there are special types of shapes for which specific casting techniques are best suited. Some of these follow:

- Shapes with constant cross section (like an extrusion) are best suited to continuous-pour casting.
- If at least part of the shape has a constant cross section, then powder metallurgy is the best process (see Chapter 10).
- If the shape is essentially a solid of revolution with a central hole, it would best be produced by means of centrifugal casting.
- Nonsymmetrical shapes with thin walls are best produced by slush casting.

Selection Based on Weight Limits

The weight of the part to be cast is also a dictating factor in the selection of a casting process. Table 9–2 classifies the casting processes with the usable weights. Again, the numbers are approximate.

Table 9–1. Casting Processes Based on Shape

Complexity		Suitable Casting Process
Internal	**Surface**	
Very intricate	Very detailed	Investment molding including lost wax Full mold Centrifuged Plaster mold Shell mold
Very intricate	Moderately detailed	Green sand Dry sand
Moderately intricate	Very detailed	Die casting
Moderately intricate	Moderately detailed	Ceramic mold (all castable materials) Permanent mold (nonferrous metals)
Low intricacy	Low detail	Permanent mold (ferrous metals)

Table 9–2. Casting Processes Based on Weights

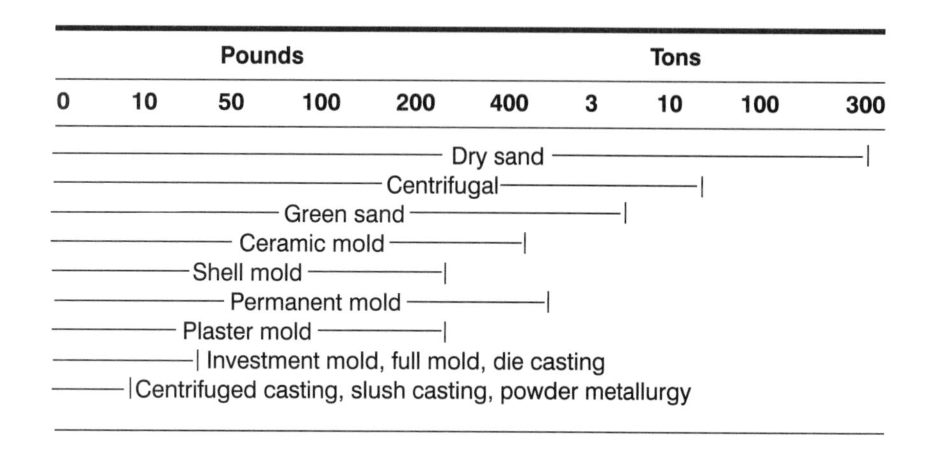

Pounds						Tons			
0	10	50	100	200	400	3	10	100	300

Dry sand
Centrifugal
Green sand
Ceramic mold
Shell mold
Permanent mold
Plaster mold
Investment mold, full mold, die casting
Centrifuged casting, slush casting, powder metallurgy

Selection Based on the Material to Be Cast

The choice of material for the workpiece or part is based on what the part is required to do. Such factors as strength, hardness, appearance, cost, toughness, elasticity, electrical conductivity, corrosion resistance, and many other factors must be considered. Few casting techniques will work equally well with all materials. Table 9–3 examines the restrictions of the casting processes with respect to the materials cast. From Table 9–3, we can see that aluminum, magnesium, copper, bronzes, and copper-based alloys can be cast by any of the techniques. The other materials are best suited to specific types of casting.

Tables 9–1, 9–2, and 9–3 do not cover all of the criteria applicable to the selection of a casting technique. Such items as the minimum quantity required to make the process profitable, cost per item produced, production rate, lead time required to produce the part, tolerances required, shrinkage limitations, and many other factors must be considered.

For example, if a single prototype of a part needs to be produced, die casting would not be a wise choice. The cost of producing the *die* is so high that the individual part could have been made quicker and at lower cost. Since casting is still somewhat of an art, experience in the field of casting is the most reliable asset when selecting a method of casting.

Table 9–3. Casting Process Based on Material Cast

High-Melting-Point Iron Alloys (Steels)	Cast Irons	Aluminum, Magnesium	Copper Bronzes	Low-Melting-Point Alloys of Tin, Lead, Zinc,etc.

Die casting
Slush casting
Plaster mold
Shell molds
Permanent molds
Powder metallurgy
Dry sand, green sand, ceramic mold, centrifuged
Centrifugal, investment, full mold

PROBLEM SET 9-1

1. Distinguish between:
   a. A pattern and a mold
   b. Green sand and dry sand casting
   c. A cope and a drag
   d. Mulling and riddling
   e. A sprue and a runner
   f. Centrifugal and centrifuged casting methods
2. Explain why the sprue in sand casting is never cut directly into the pattern.
3. List and explain three independent reasons for having risers in a sand casting.
4. Why does a green sand mold have a short shelf life but a shell mold can be kept for a considerable time?
5. What is the difference between *investment casting* and *die casting?*
6. What casting process (if any) should be recommended for making an automatic-transmission valve body, given the following information:

   Production rate:
   10,000 parts per week, 500,000 total
   Very intricate interior detail,
   simple external shape
   Material: aluminum
   Size: approximately 2 × 4 × 6 inches
   (50 × 100 × 150 millimetres)

   Explain the reasoning for your choice of process.
7. What casting process (if any) would you recommend for making a 400-lb brass sculpture? Explain your reasoning.
8. What casting process (if any) would you recommend for producing a steel transformer case that has a roughly cylindrical shape, 1 m long, with a 250-mm internal diameter and a wall thickness of 15 mm? The rate of production is to be 20 per week for the next five years. Explain your reasoning.
9. What casting process (if any) would you select for a crank used on automobiles to "crank" open windows? The crank is to have a splined hole on one end to fit on the shaft of the opening mechanism, and a steel insert for supporting the handle at the other end. The material for the casting is aluminum, magnesium, or zinc. The production rate is to be 10,000 per month for five years. Explain your reasoning.

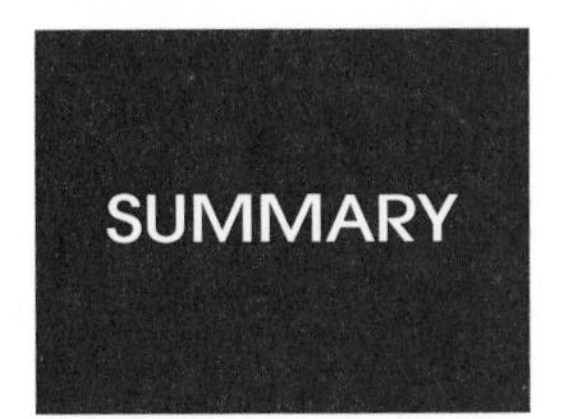

Change of form refers to those manufacturing processes that change the shape of the material without either adding to or removing any material. The essential idea is that the material is made to flow by some means to assume a new shape.

Casting refers to those change-of-form processes in which the material is made to "flow" by melting it. An enormous number of casting techniques have been developed over the centuries. The advantages and disadvantages of each of the techniques force the engineer or designer to make complicated choices as to which method to specify for a given part. The choices are based on the desired characteristics of the finished part as well as the type of material being cast and the limitations and purposes of the casting method.

Chapter 10

Powder Methods of Change of Form

INTRODUCTION TO POWDER METALLURGY

Powder metallurgy is similar to casting in that a shape can be created by the use of a mold and the powdered solid materials can be made to "flow" in much the same manner as a liquid. The material in this chapter could have been included in the chapter on casting. However, because the versatility of powder metallurgy is so great, and because it is such a widely used technique in industry, this separate chapter is devoted to the powder forming techniques.

POWDER METALLURGY

Powder metallurgy, sometimes called *powder forming,* involves the compaction of finely powdered solids into a mold and the fusing of the particles. The processes involved in powder metallurgy can be grouped into five divisions:

1. Preparation of the powdered material
2. Blending of the powdered materials
3. Filling the die with the powder, and compaction
4. Sintering (the process in which the particles are bonded together)
5. Postsintering treatment to produce the final dimensions and properties

Preparation and Blending of the Powder

Many types of metals and other materials can be used in powder form. The preparation of the powder depends on the physical properties of the material being used and the type of final product being made. Some brittle materials, such as ceramics, glasses, and brittle metals including bismuth, titanium, antimony, and others can be powdered by crushing them in a ball mill (Figure 10–1). For other materials, the powder is made by:

- Grinding
- Plating it onto a temporary surface then scraping it off
- Melting and atomizing the metal
- Chemical reduction from an oxide
- Chemical precipitation from solution
- Solidification from the vapor phase

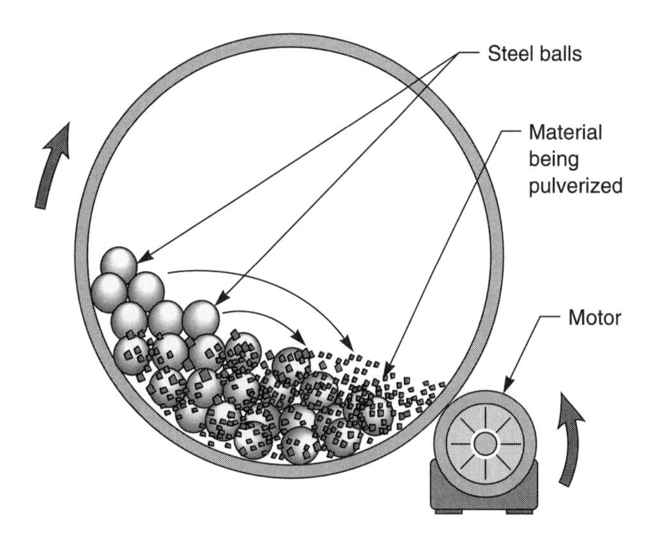

Figure 10–1. Ball mill.

You may notice that the floor around any well-used grinding wheel soon becomes covered with powder. Most of this powder is the metallic oxide of the metal that was removed from the workpieces. The sparks from the wheel are the oxidation or burning of the metallic particles resulting from the heating of the metal during the grinding process. Metal can be ground under a water spray or in an inert gas atmosphere so that it does not oxidize. The resulting powder can be used in powder metallurgy.

If metal is plated from solution by using too much voltage and current, it will pile up on the cathode surface and not adhere to the surface. The resultant, spongy material can be dried and pulverized for use in powder metallurgy.

Metal can be melted, then the liquid sprayed into an inert gas atmosphere in small droplets. The action is similar to that of a nasal spray. The small droplets will solidify into a powder suitable for powder forming.

Some metal powders are produced by chemical precipitation. One example of such a powder is rust. The oxidized iron does not stick well to the parent piece of iron or steel but can be scraped off in a powder form. If a piece of copper is immersed in a vat of silver nitrate or other soluble silver salt, metallic silver will rapidly build up about the copper. The silver can be removed and dried to form a powder. Many other metals can be prepared for powder forming by this method.

Powder particles that have been made by being pulverized in a mill or atomized may need to be annealed so that they will have the proper physical properties for compaction. Whatever the method used to form the powder, the resulting particles must be screened to provide an even size and then packaged and stored until needed.

Often the powders are blended with a dry lubricant or an antioxidant for uniform compaction. Typical lubricants are stearic acid, lithium stearate, or graphite. Some powdered metals, such as magnesium, aluminum, and zinc, are very flammable and must be mixed wet to reduce the hazard of explosion.

Immediately prior to use, the materials for the "alloy" are combined. The alloy could be a combination of various metals, a combination of ceramics and metals (to produce *cermets*), or a combination of abrasives and metal powder—whatever serves the needs of the final product. The most commonly used materials for powder metallurgy are iron-based or copper-based powders. However, aluminum, carbides of various metals, chromium, cobalt, graphite, lead, molybdenum, nickel, silicon, stainless steel, tantalum, tin, titanium, tungsten, zinc, and various metal oxides are used extensively.

Compaction

A die in the shape of the final product is filled with the prepared powder. Once the die cavity is filled, it is compressed from both the top and the bottom (Figure 10–2). This is necessary because the powder does not flow well and would tend to be compacted mostly at the point of pressure rather than throughout the cavity. This process is known as *briquetting.* The pressure used depends on the porosity desired in the final product, the type of material being compacted, and the size of the final product. For some parts, presses that exert forces up to 300 tons have been used. The part thus formed is called a *briquette* or *green compact.* At this stage the briquette is rather fragile.

Compaction can be done by other methods as well. The powder can be compacted between two rollers to produce sheet or plate stock. The powder can be packed into a mild steel tube, then forced through a die in a process similar to extrusion. Workpieces that have shapes that cannot be removed easily from a die

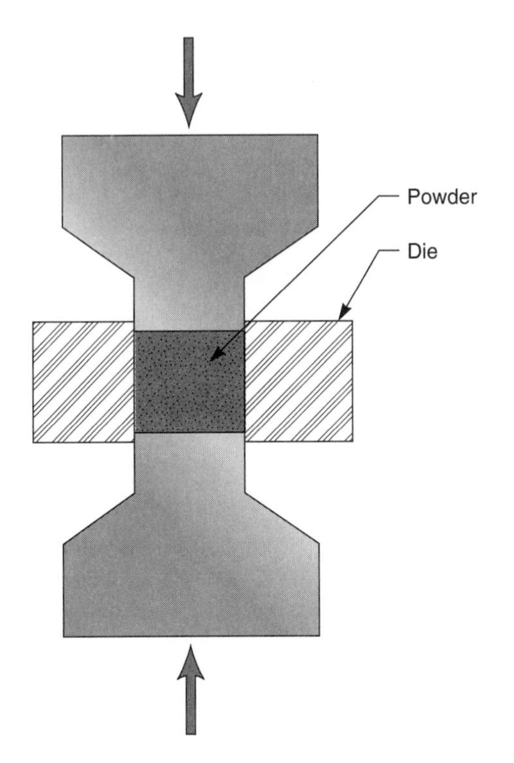

Figure 10–2. Briquetting.

can be compressed against the interior cavity of a separable die by an explosion (*explosive compacting*). Figures 10–3, 10–4, and 10–5 show examples of different types of compacting.

Parts up to 4 inches (100 millimetres) in diameter and 6 inches (150 millimetres) long are commonly mass produced by powder metallurgy techniques. Parts weighing up to 200 pounds (90 kilograms) are feasible. An ideal shape for parts made by powder metallurgy is uniform in cross section with a length-to-smallest width ratio of less than 3:1.

The speed of compaction and ejection of the parts is important for economic reasons. The minimum production quantity of a powder metallurgy product needs to be around 10,000 parts per die to offset the cost of producing the dies. Small parts can be produced at rates of 500 to 1000 per hour so the production of 10,000 parts is not an unreasonable requirement.

Sintering

In the process of *sintering,* the particles are not melted, so powder forming can be used for materials such as ceramics and cermets that cannot be melted and cast by other methods. The sintering process causes the faces of the individual particles to fuse by allowing atoms from the surfaces of each particle to diffuse into the surface of the adjacent particle, thus forming a true weld (Figure 10–6).

Figure 10–3. Roller compacting.

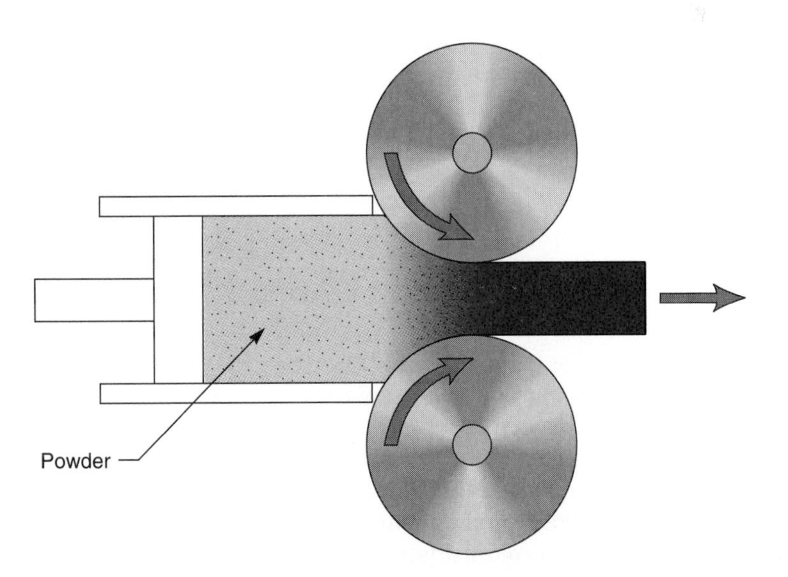

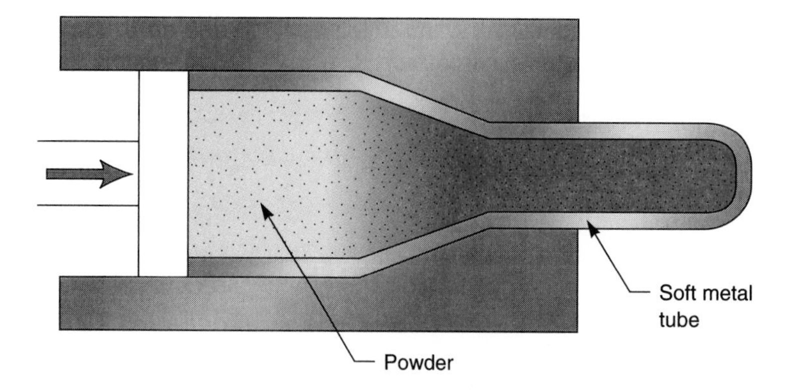

Figure 10-4. Extrusion compacting.

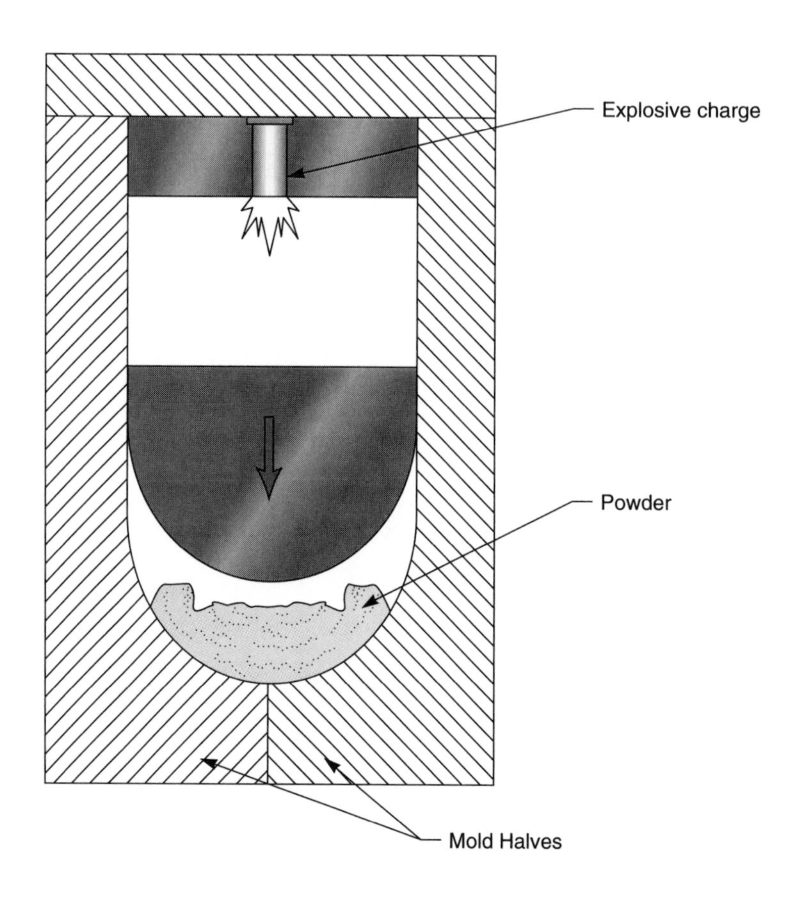

Figure 10-5. Explosion compacting.

To produce diffusion bonding between the particles of the briquette, the part is heated to 60% to 80% of the melting temperature of the chief component and held at that temperature for 20 minutes to an hour. A reducing atmosphere of hydrogen, ammonia, or an inert gas is used to prevent rapid oxidation, which would normally occur at the elevated temperatures. An atmosphere of carbon dioxide can be used if the part is to be carburized. Lubricants that were added to aid in the compaction are burned off at the beginning of the heating process. The sintered product is free of this contaminant.

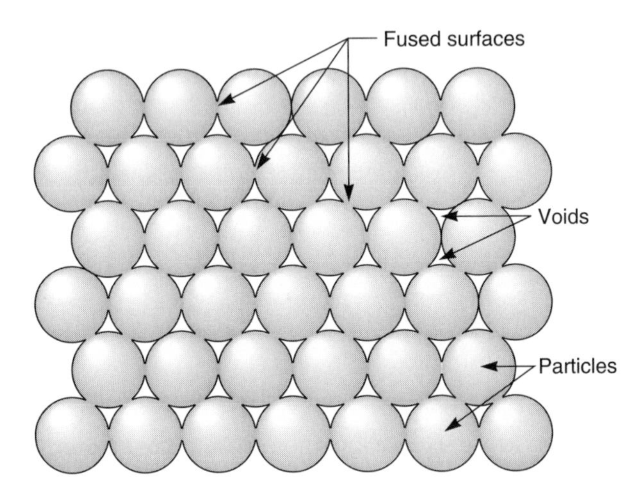

Figure 10–6. Sintering.

After the part has been heated sufficiently, it is removed from the oven and cooled. The cooling rate is controlled to produce the desired grain structure and hardness of the part.

Finishing of Sintered Parts

Many parts can be used as they emerge from the sintering oven and require no further processing. Others require a variety of postsintering operations. Three of the most common ones are infiltration, impregnation, and sizing.

Infiltration. Because powder-formed parts can be very porous, other materials can be placed in the voids to enhance the properties of the product. Frequently, iron-based powder-formed parts can have their strength increased 70% to 100% by allowing a blank of bronze or copper to melt on the workpiece. The copper-based alloy is then absorbed by capillary action into the pores of the workpiece. This also increases the density and the hardness of the part.

Impregnation. One of the most used products of powder metallurgy is that of "permanently lubricated" bearings for small machinery. These bearings are formed by impregnating the sintered sleeve with oil or other lubricant. Simply soaking the sintered bearings in the oil will allow the oil to fill the voids in the bearing. In many cases, the bearing will wear out before it needs further lubrication. Gears, cams, and pump parts are also impregnated with lubricants.

Sizing. Powder-formed and sintered parts can be finished by machining and other methods. Many times the sintered part needs only a minute amount of material removed to meet tolerances. One method of finishing these parts is to force the part through a finish die or to force a mandrel through the central cavity in the workpiece. In either case, no material is actually removed; rather, the surface is compacted to meet the final requirements.

Applications

Uses for powder-formed parts are increasing daily. Permanently lubricated bearings are used in kitchen appliances, motion picture projectors, automobile speedometers, and video cameras and recorders. If you examine the brushes in the alternator of an automobile engine, the chances are very high that they will be sintered copper and graphite. Powerful magnets, metals with imbedded abrasives, cutting tools and dies, and thread taps are regularly formed by powder metallurgy technology.

Powder metallurgy using coarse-grained materials is also used to make filters. Some more exotic uses of powder techniques are in the fabrication of fuel elements for nuclear reactors. In these, elements of the nuclear fuel (plutonium) are dispersed in aluminum or steel. Recently, superconductor technology has depended on powder forming to fabricate copper oxide–yttrium oxide–barium oxide and other materials into superconducting materials. It is a growing field of manufacturing. Figure 10–7 shows a magnified view of powder-formed alternator brushes.

New methods of powder metallurgy and modifications to existing techniques are being reported every month in the technical society publications. New methods such as *high-velocity oxygen-fuel (HVOF)* thermal spray coatings are being used to replace hard chrome plating methods in some applications.* The

* Because HVOF uses hexagonal chromium, which has been found to be cancer producing by the Occupational Safety and Health Administration, strict regulations are being put in place to limit the exposure of workers to this material.

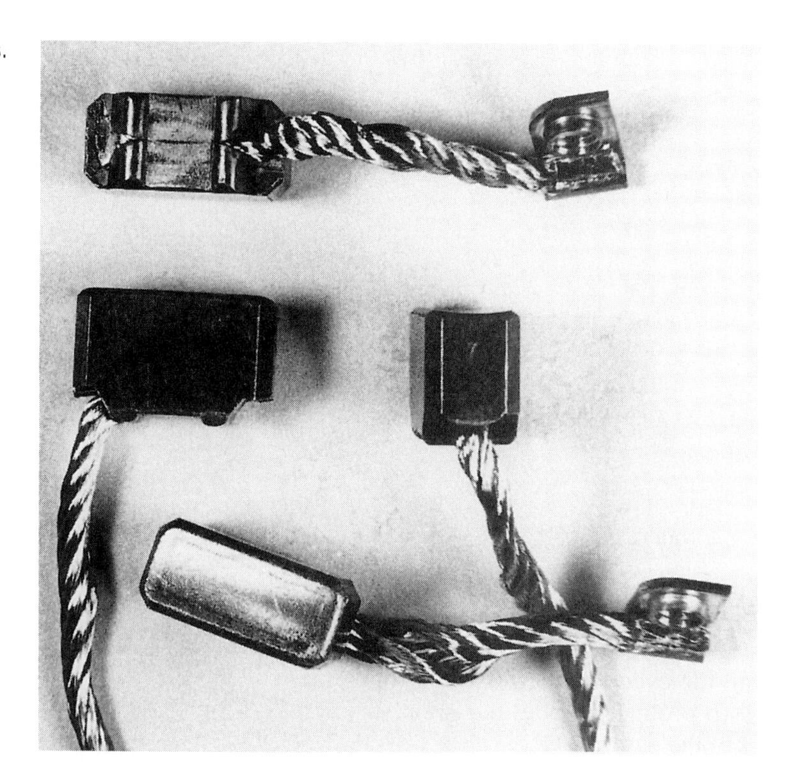

Figure 10-7. Alternator brushes.

HVOF thermal spray process uses an internal combustion jet to generate supersonic gas velocities of about 67 miles per hour. Powders of the coating materials are injected directly into the combustion region of the spray gun. This forces the molten powder onto the surface of the workpiece with considerable impact, producing a very hard surface.

Spray forming is another technique in which molten metal is sprayed onto a shape to build up entire billets and other shapes. The process is applicable to many types of materials from aluminum to copper to superalloys of very high strength.

Powder metallurgy is also being used in the medical field. Lightweight, high-strength prosthetic devices such as artificial arms and legs have been made by powder-forming techniques. Sintered titanium surfaces on hip, knee, and other joints are made using powder-forming and metallizing techniques.

The applications of powder metallurgy are limited only by the imagination of the designers. All technicians and engineers should join one or more technical societies and attend their seminars just to keep up with the new techniques. As more engineers and designers become aware of and trained in the use of powder techniques, the uses of powder-forming methods are certain to expand rapidly.

PROBLEM SET 10-1

1. Make a list of objects you have seen that were made by powder techniques.
2. List three parts of some machine that could have been made by powder forming and state why these parts might function better if they had been powder formed.
3. Explain, in your own words, what *sintering* is.
4. Make a list of the equipment needed to make a simple sintered powder-formed "washer."

5. Examine the following list of parts and decide which ones would *not* be best made by powder forming. List the reasons for your decision.

a. Wood screw
b. Bearing for a sewing machine motor
c. Screwdriver
d. Coil spring
e. Fuel filter for a lawn mower
f. Ball peen hammer
g. Electrical contact in a switch
h. Knife blade
i. Housing for a small electric motor
j. Gear housing for a jeweler's lathe
k. Gear for a hand drill

6. We want to make a ceramic mold for die casting. Could this be done by powder forming? Explain your reasoning.

7. What is the difference between sintering and briquetting?

8. What production technique could be used to produce a very hard lathe cutting tool, formed of fused ceramic and metal? The shape is a rectangular solid roughly 5 × 40 × 50 mm with embossed numbers on one face. The production rate is to be 5000 per week for a total of 50,000 pieces. Explain your reasoning.

Powder metallurgy refers to those change-of-form processes in which the material is made to "flow" by powdering it. Powder metallurgy allows parts to be made from materials that have very high strengths, high melting temperatures, high hardness, or are very brittle. Powder forming can also be used to make permanently lubricated bearings, gears, cams, and other parts. The voids in powder-formed parts can be filled with other metals to enhance the hardness, strength, and other properties of the part.

Chapter 11

Mechanical and Other Methods of Change of Form

FORMING

Another way to get metal to flow into a desired shape is to exert enough force on the workpiece that plastic deformation will occur. The processes that do this are collectively called *forming.* Forming includes the following processes:

1. Forging
2. Extrusion
3. Rolling
4. Bending
5. Drawing
6. Spin forming
7. High-energy-rate forming
   a. Explosive forming
   b. Magneforming

These are not gentle processes—far from it. Forming processes are generally described as "bulk deformation processes," which generally involve a lot of force. Forming can be done on either hot or cold metals.

A few new terms are used specifically with forming processes. The piece of material being formed is called a *billet.* The moving part that delivers the force or energy to the billet is known as the *ram* or *punch.* The tool with the cavity into which, or through which, the billet is forced is the *die.*

Forging

The word *forging* generally conjures up an image of a muscular blacksmith, sweating profusely as he pounds on a red hot bar of iron taken from the forge. Although the basic concept of "plastic deformation by compressive forces" hasn't changed, the details certainly *have* since the first recorded use of forging in roughly 5000 B.C. Several methods of forging are commonly used.

Hand Forging

As shown in Figure 11–1, hand forging is exactly what the blacksmiths did. Although there are few blacksmiths still working their trade, those who are still around generally work on one-of-a-kind tasks in areas far removed from large-volume manufacturing. Horseshoes are still hand forged and fitted to the hooves of the horse. However, today most forging is done by machines.

Drop Forging

In its simplest form, a drop forge raises a massive weight and lets it fall. It is essentially a mechanized form of the blacksmith's hammer. In practice, however, the process is a lot more complicated than that.

Figure 11-1. Blacksmith.

The two basic types of forging machines are *presses* and *hammers.* Presses exert enormous forces, which are applied slowly enough that the metal has time to "flow." They are generally the machine of choice for workpieces made of aluminum, magnesium, beryllium alloys, bronzes, and brass. Presses can be operated by hydraulic pressure exerting forces up to 75,000 tons or 670 meganewtons. Some presses convert energy stored in a flywheel in "screwpresses" to exert forces up to 31,500 tons (280 meganewtons). Lighter presses use mechanical linkages to exert forces up to 12,000 tons (110 meganewtons). Figures 11–2(a) and (b) show the action of presses.

The hammer machines are designed to raise a massive weight and let it drop. *Power hammers* add to gravity with pneumatic or hydraulic assistance. Besides the vertical (drop) type hammers, there are horizontal power hammers. These machines actually use two opposed hammers called *counterblow hammers* that work together (Figure 11–3). These machines produce much less vibration than the single-hammer types. Hammer machines have been made that are capable of transferring up to 900,000 foot-pounds (1200 kilojoules) of energy to a workpiece. Hammers are the machine of choice for workpieces made of copper alloys, steel, titanium, or refractory alloys.

Hydraulic presses can be operated at repetition rates of a few strokes (blows) per minute, whereas power hammers can be operated for small parts at rates up to 300 strokes per minute (that is, 5 strokes per second).

Presses or hammers can be used in a variety of ways. Several of the common ways follow:

- Open forging
- Cogging
- Closed forging
- Coining
- Heading
- Swaging

Open Forging

Open forging simply *presses* the billet between two flat plates to reduce its thickness. This action can be repeated (just as a blacksmith beats on a bar to flatten it) in order to produce a rough shape for further forging processes. In all of the forging processes, the billet may be worked when red hot or when cold. Cold-worked metals may have to be annealed before further forging or work on them can be performed. Open forging is illustrated in Figure 11–4.

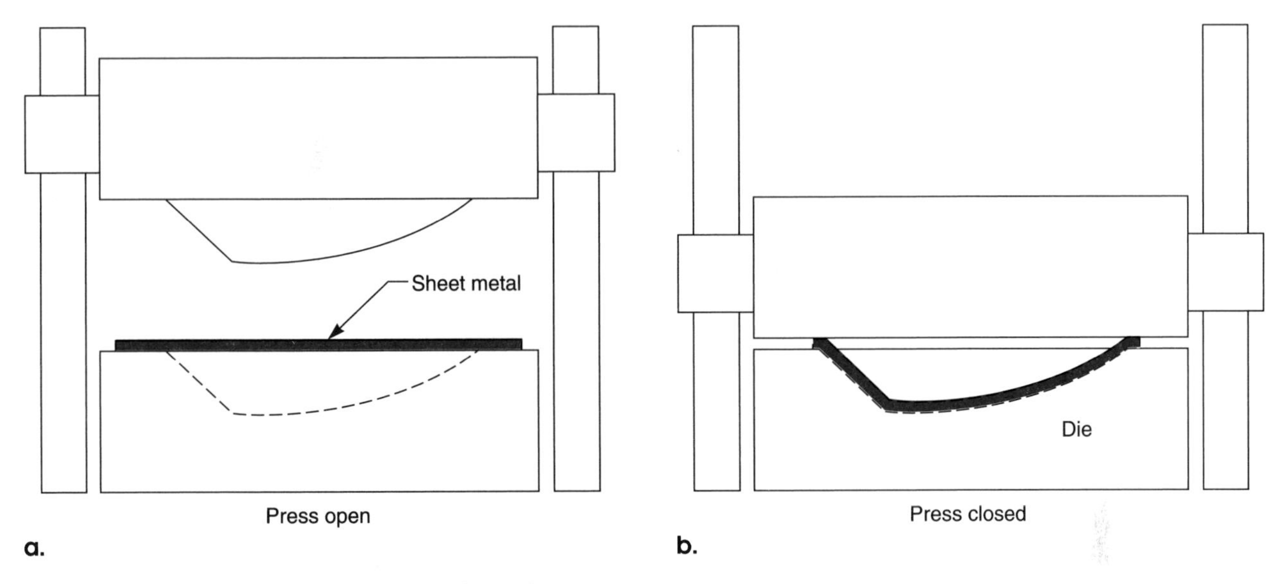

Figure 11–2. Press action: **a.** Open. **b.** Closed.

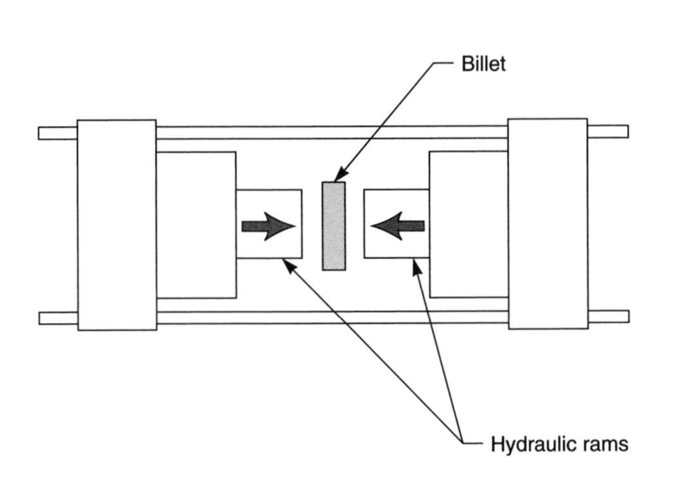

Figure 11–3. Counterblow forging.

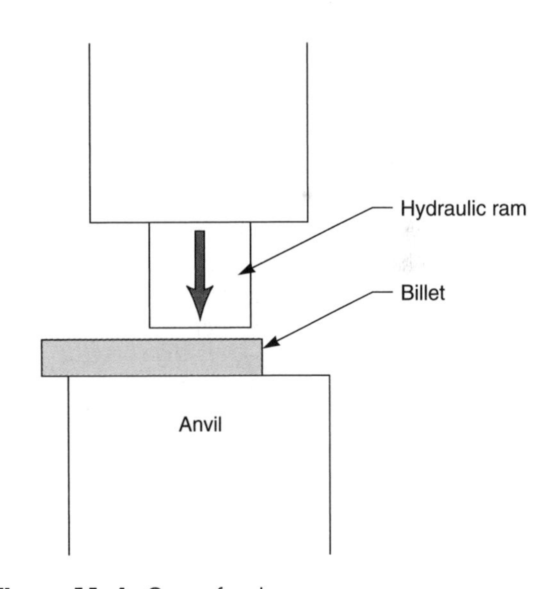

Figure 11–4. Open forging.

Cogging

Cogging is a process that reduces the thickness of a billet by small increments by means of a hydraulic, pneumatic, or mechanically driven hammer. It is similar to open forging in that no dies are used (Figure 11–5).

Closed Forging

Closed forging involves the use of dies. The billet is forced into the cavities of one or more dies. To fill a die completely with a solid material, some of the material must become *flashing.* Flashing is the thin sheet of material that leaks into the space between the

Figure 11-5. Cogging.

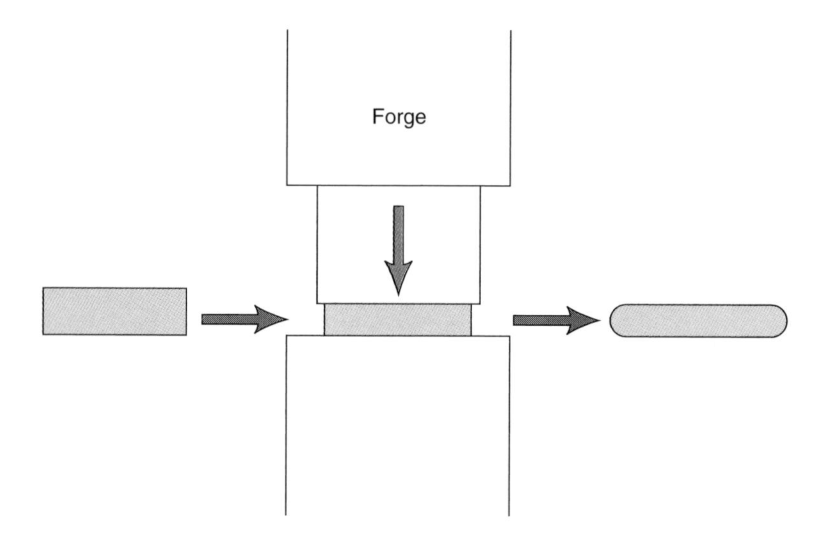

die halves in a casting or forging. In cast parts, it is considered a nuisance because it must be removed before the part can be used, but it serves a positive function in forging.

Visualize a solid billet of material in the die cavity as the press ram descends in slow motion [Figure 11–6(a)]. As the billet comes into contact with more and more of the die surface, it doesn't flow well into the corners. It does flow into the gap between the die halves and thereby *seals* the cavity [Figure 11–6(b)]. As the ram continues to descend, the material is forced into the rest of the dies until they are completely filled [Figure 11–6(c)]. Only then can the pressure in the material rise high enough to force the last bit of metal into the flashing. Figure 11–7 shows a classic example of flashing left in forming a plastic. Metal flashings look about the same and must be removed after forming.

Coining

Coining is, not surprisingly, the process used to form faces of coins on coin blanks. The pattern produced has a depth measured in thousandths of an inch (hundredths of a millimetre). Depending on the metal being coined, it may take considerable force to cause the metal to flow from one part of the blank (billet) to another.

Heading

Heading is the process of "upsetting" metal to form heads on nails or screws. The shaft (or shank) of the nail or screw is clamped in such a way that some of the material sticks above the clamp. It is then rammed into a die cavity that has the dimensions of the head to be formed (Figure 11–8). Nail-making machines do both the cutting and heading operations very quickly. The clamp on the machine draws out the proper length of wire for the nail from a roll of wire, cuts it off so that there is a point left on the roll of wire, moves the nail to the die, heads the nail, and drops it into a chute where it is sent to packaging. Depending on the size of the nail, the whole operation takes between a fraction of a second and a second. Figure 11–9 illustrates the nail making operation.

Swaging

Swaging is sometimes called *radial forging.* Swaging is the process by which tubing is hammered against a mandrel alternately by two to four dies so that its diameter is reduced (Figures 11–10 and 11–11). The tubing will fit tightly around the mandrel. A cable clamp can be swaged around a wire cable with such force that the cable will break before the clamp is pulled off. (More about this in the chapter on material joining.) Internal shapes can be formed in tubular billets by swaging. The tubular billet is swaged about specially shaped mandrels to form internal splines, or the rifling in rifle barrels. Swaging is also used to *clad* workpieces. To clad a workpiece, a tube of cladding metal is placed around the part and the tube swaged firmly around it.

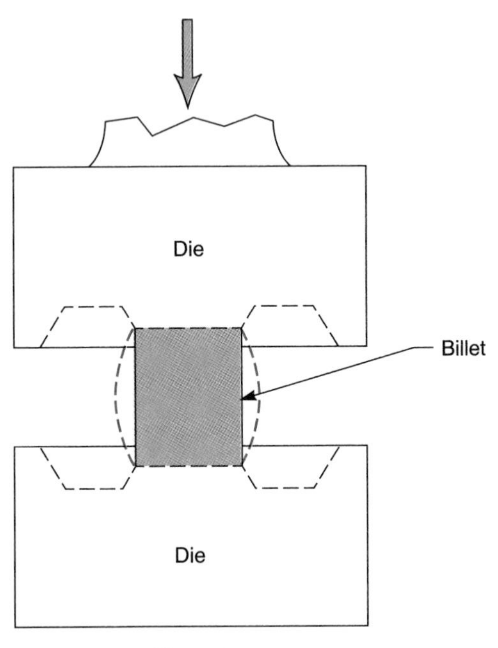

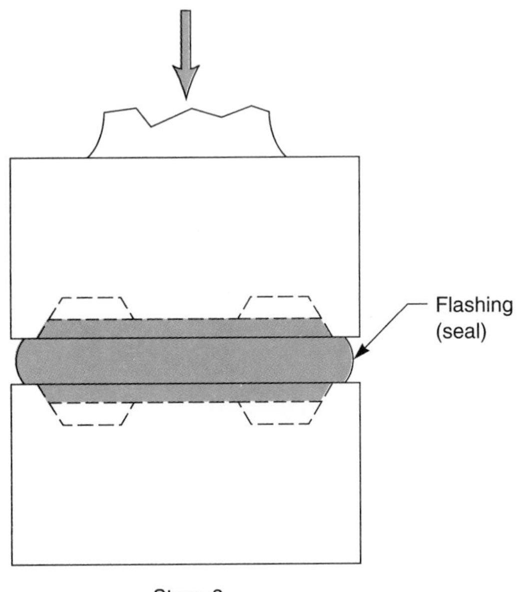

Figure 11-6. Closed forging: **a.** Stage 1. **b.** Stage 2. **c.** Stage 3.

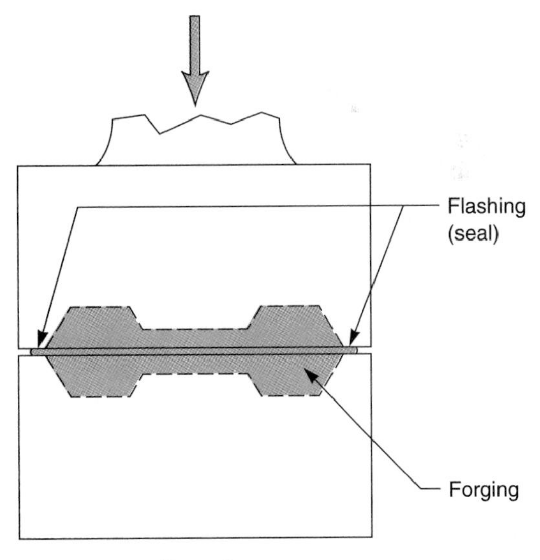

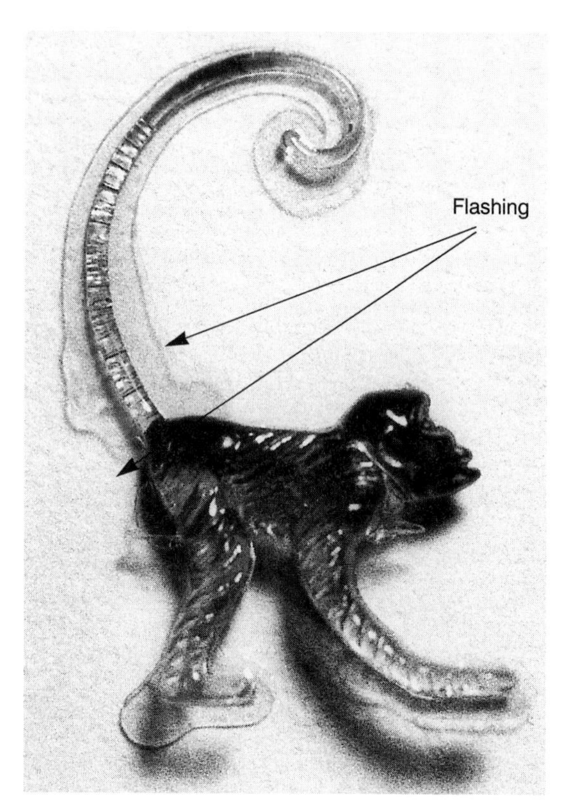

Figure 11-7. Flashing left on plastic.

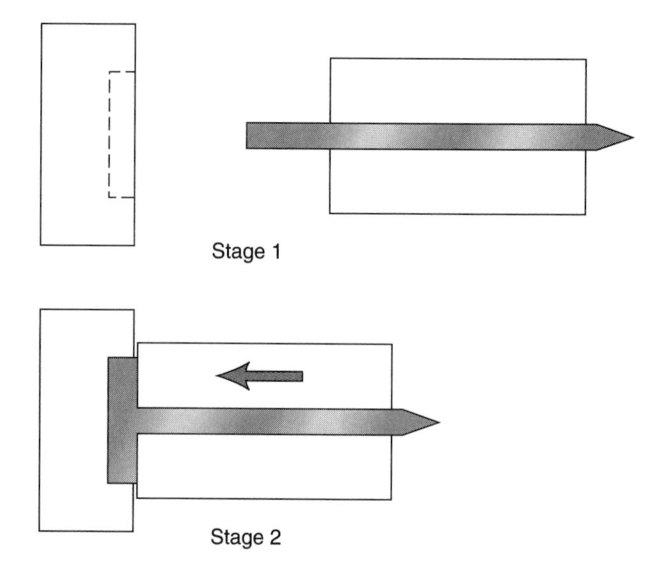

Figure 11-8. Heading.

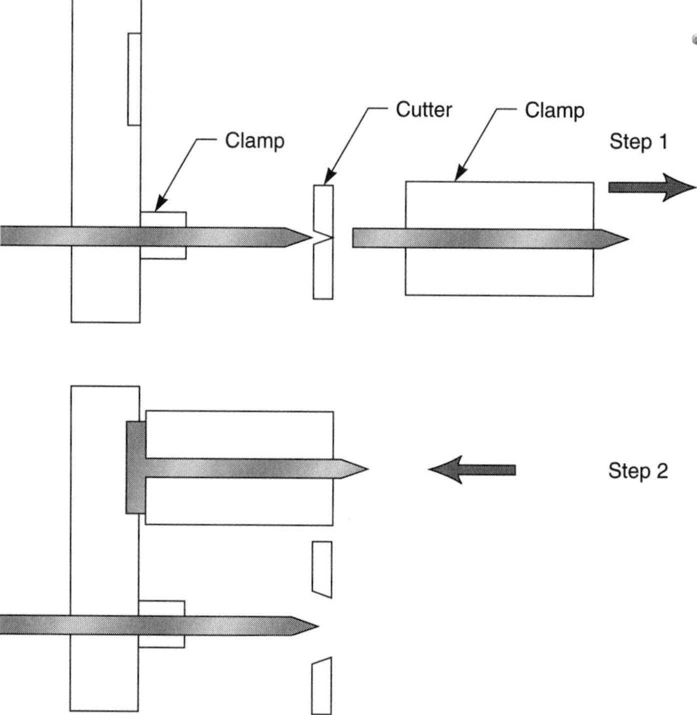

Figure 11-9. Nail making.

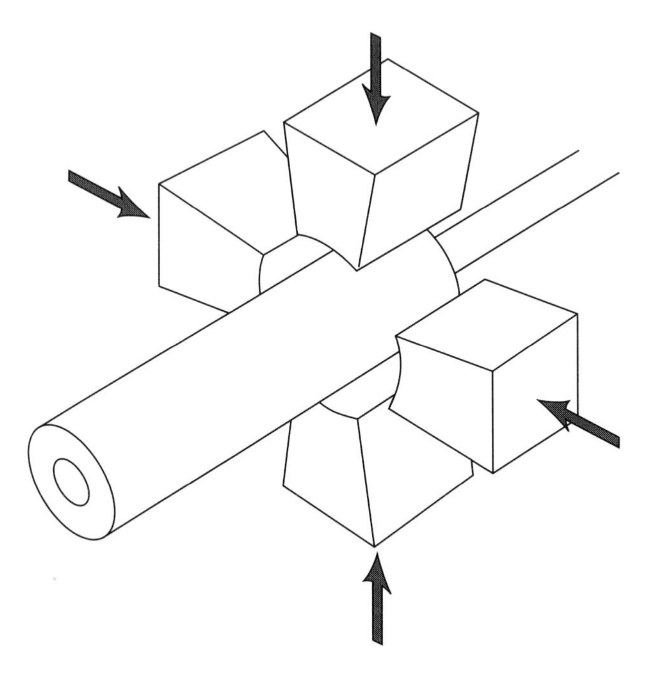

Figure 11-10. Swaging.

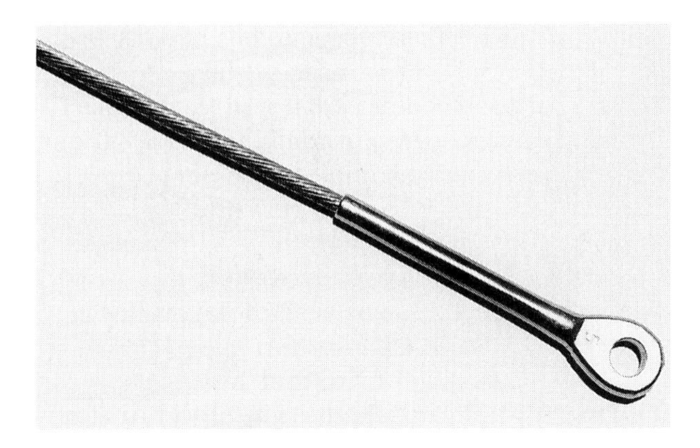

Figure 11-11. Swaging.

Swaged parts have a work-hardened exterior surface and can have good dimensional accuracy and precision.

Lubricants for Forging

Lubricants are a vital part of the forging process. Most billets used in forging are coated with a lubricant before forging begins. The billets may require further coating for subsequent forgings. The purposes of this coating are to improve the flow of the material into the dies, to reduce die wear, to control the cooling rate, and to serve as a parting agent, which allows the formed part to slip easily out of the die or off of the mandrel.

For hot forging operations, the appropriate lubricants would be graphite, molybdenum disulfide, or powdered glass. The lubricants generally used in cold forgings are mineral oil, soap, and silicones. Some low-friction lubricants such as polytetrafluoroethene (Teflon is one trade name) are also used.

Pressures Involved in Forging

Forge presses are made in all sizes and exert from a few pounds of force to several tons. The force needed to forge a part depends on (1) the compressive strength of the metal, (2) the area, including flashings, of the metal being forged, (3) the temperature at which the forging is being done, and (4) the amount of deformation each compressive stroke of the ram or hammer performs. High-carbon steels require more force than low-carbon steels and lower strength metals such as bronze, copper, zinc, etc. More force is also required at the bottom of the ram stroke than at the beginning.

The forging force is calculated in various ways. One recommendation is to multiply the total area of the forging by approximately one third the *cold* compressive strength of the material being forged. Other sources recommend calculating the force required using the *hot* compressive strength. These calculations may use from 3 to 10 times the hot compressive strength of the material, multiplied by the area being forged. As with all engineering design, the design of the forging press for a specific project is a compromise. The greater the force required, the larger, heavier, and more expensive the equipment and operating costs. However, to do the forging job properly, certain minimum forces are needed. It is prudent to design the tools to adequately perform their function, but not to overdesign them beyond the requirements of the product.

Hydroforming

A variation of forging is *hydroforming.* Hydroforming can be used to form thin metal sheets into an irregularly

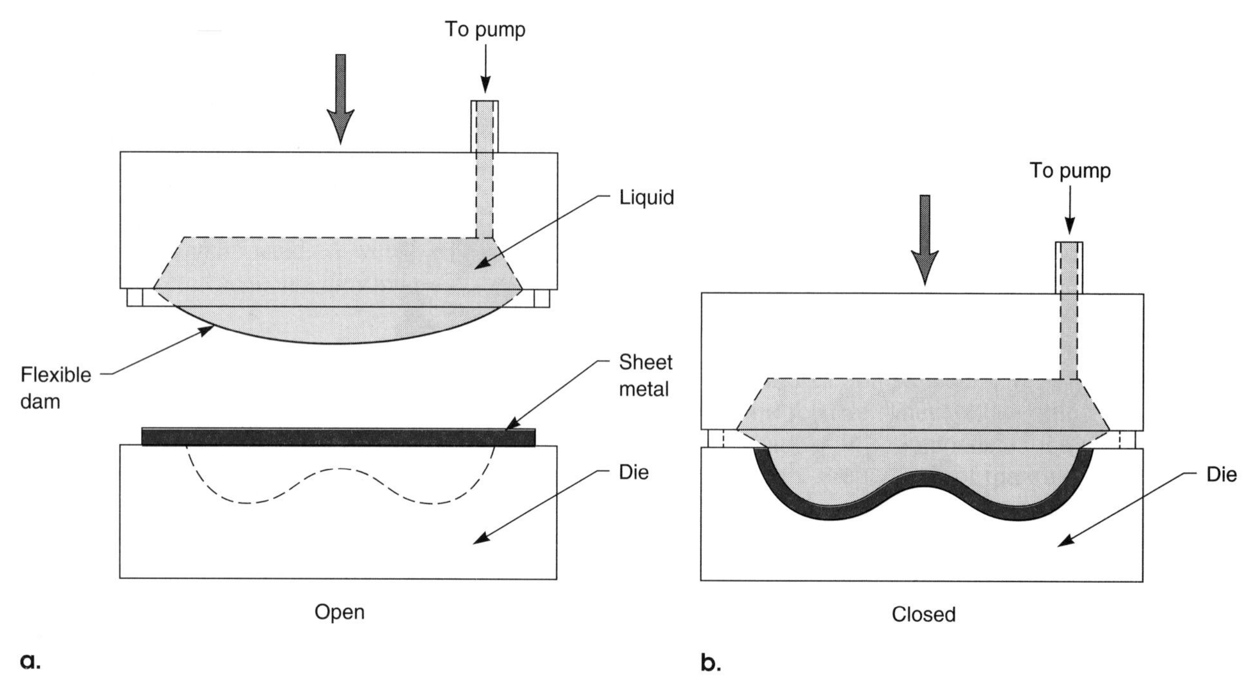

Figure 11–12. Hydroforming: **a.** Open. **b.** Closed.

shaped die. One of the dies in a permanent mold is replaced by a cavity into which oil or water can be pumped. This die is covered by a *dam* made from a sheet of rubber. When the die is closed, the fluid is pumped into the cavity, forcing the metal into the other die with an even pressure. Pressures up to 15,000 psi (100 megapascals) can be used. The pressure is then released, the dies separated, and the workpiece ejected [Figures 11–12(a) and (b)].

Hydroforming has the advantage of low tooling costs, flexibility and ease of operation, low die wear, and the ability to form complex shapes. After all, only one die needs to be made, not two. Laminated parts can easily be made by clamping the appropriate sheets in place and pressing them together using a hydroforming press. Because there are no sharp corners on the rubber dam, sharp corners will not be formed on that side of the sheet metal. This can be an advantage in that stress risers are not formed, but hydroforming should not be used for parts that require a sharp corner or a mating surface.

Extrusion

Extrusion was developed in the late 18th century. This process is frequently compared to squeezing toothpaste from a tube or the process by which decorative icing is forced through a die to make cake decorations. Extrusion is the process in which the material is forced through a die to produce a very long workpiece of constant cross section. A more accurate model would be that of forcing a caulking compound from a tube with a caulking gun. This task involves the use of a ram or piston as well as the cylinder and extrusion die. Plastics, ceramics, and other materials can also be extruded but in this section our discussion focuses on the extrusion of metals.

Extrusion can be done "cold" (at room temperature) or "hot" so that the material is softened slightly. The forces on the die are very high for metals. Therefore, the die must be supported well and designed to reduce friction if optimum production is to be achieved. Lubricants must be used if the material

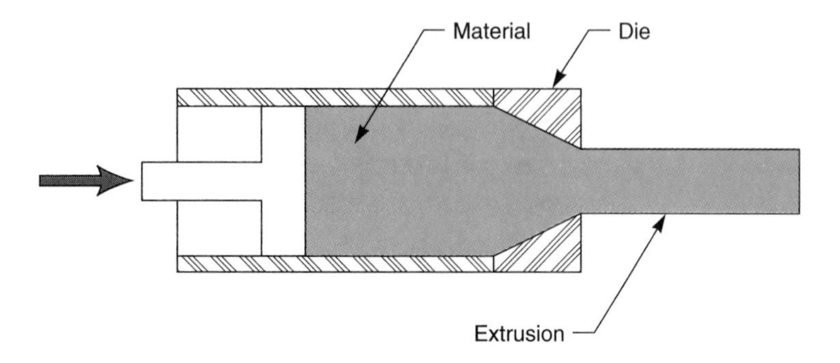

Figure 11-13. Direct extrusion.

being extruded has a tendency to stick to the die. Glass is sometimes used as a lubricant in hot extrusion. Furthermore, the material being extruded may be given a conversion coating or a "jacket" of soft metal before extrusion.

Extrusion can be divided into five major types:

- Direct
- Indirect
- Hydrostatic
- Impact
- Hollow

Direct Extrusion

Direct extrusion is the principle used in the caulking gun model mentioned earlier (Figure 11–13). In this process, the metal is simply forced through a die by a ram. Cold extrusion is generally done vertically and therefore bears some resemblance to continuous casting. However, there is an important distinction between casting and extrusion. In casting the material being formed is a liquid and is poured into a mold. In extrusion the material is solid. This means that the die in continuous casting must withstand high temperatures but little force, whereas the die in extrusion must withstand large forces but little heat.

Indirect Extrusion

In *indirect extrusion,* the ram is also the die. The billet is placed in a closed cylinder and the ram pressed against it. The billet is then forced back into the cavity or die built into the ram (Figure 11–14). Indirect extrusion is essentially a small-batch process and is limited to relatively small parts.

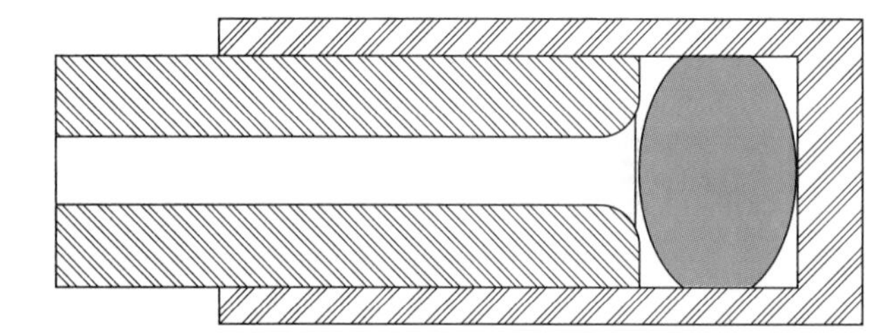

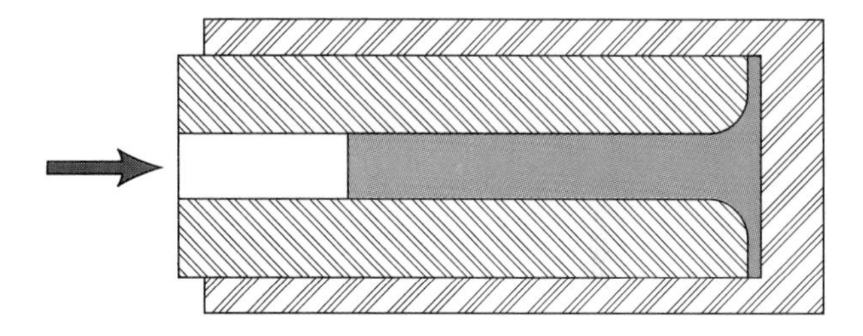

Figure 11-14. Indirect extrusion.

Hydrostatic Extrusion

Hydrostatic extrusion is similar to direct extrusion with some important variations. In hydrostatic extrusion a fluid is placed between the ram and the metal being extruded. The fluid forces itself between the metal and the walls of the cylinder. This produces two advantages: (1) The fluid presses radially inward on the billet, which helps guide it into the opening in the die, and (2) the fluid lubricates the walls of the cylinder, which reduces the friction forces in the extrusion process. Clearly this method is only applicable to cold extrusion, not to hot extrusion. Figure 11–15 illustrates hydrostatic extrusion.

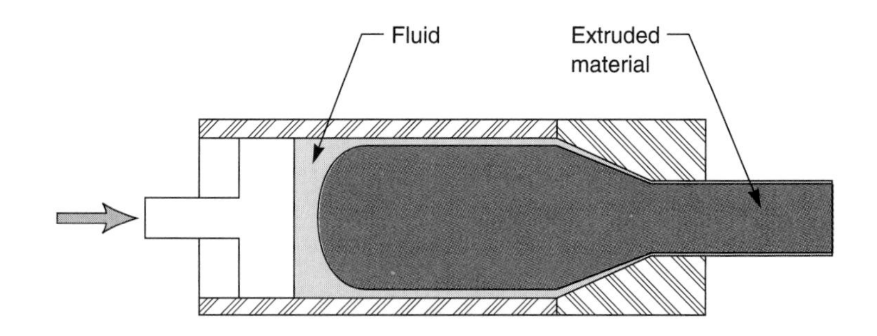

Figure 11-15. Hydrostatic extrusion.

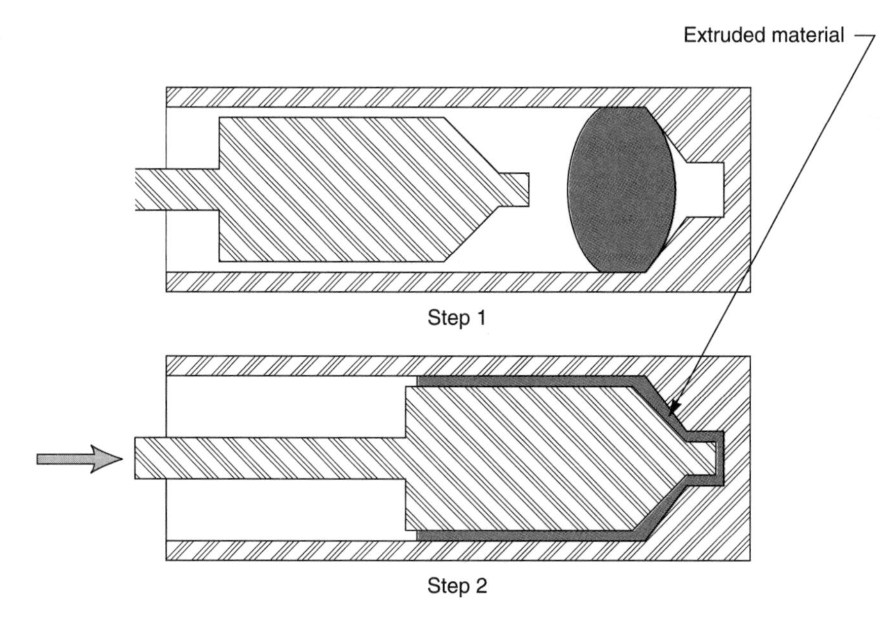

Figure 11-16. Impact extrusion.

Impact Extrusion

Impact extrusion is a process for making hollow containers from ductile materials. Containers such as roll-up tubes or deep cylinders are made by this technique. In impact extrusion, the *die opening* is the space between the ram and the cylinder wall (Figure 11–16). The billet of material is placed in the die and the ram driven suddenly against it. The metal then flows around the ram and into the opening in the die face. If the product is a roll-up tube, such as those used for toothpaste or medicinal ointments, the die will form the cap end of the tube at the same time the walls are being formed. The ram is then withdrawn and the tube removed from it. This is an efficient process in that the entire tube is made in one operation.

Hollow Extrusion

Hollow pieces such as pipes and tubing can be made by extrusion if some "obstacle" is part of the die design (Figure 11–17). The obstacle must be supported in place by some structure and the extruding material must flow around it. In order for the design to work, the extruding material must reform downstream of the support structure. Aluminum is one of the metals that can be extruded into hollow tubes and piping by this method. This process is used extensively to make tubes from a variety of plastics.

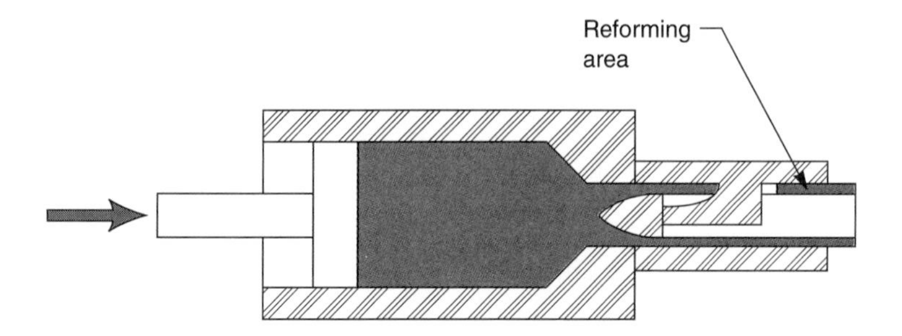

Figure 11-17. Hollow extrusion.

Figure 11-18. Rollers.

Rolling

Rolling of metals was developed in the late 16th century. Even though several sophisticated modifications have been introduced, much of the roll-forming equipment used today is not substantially different from that used 400 years ago. The power plants that drive the rollers are larger, the rollers more massive, and the materials used in the machines are different, but the principles are the same (Figure 11–18).

It is little wonder that rolling is one of the most used processes in today's industry. It is so versatile that it can be used for the following processes:

- Bend rods or sheets into curved surfaces
- Change the grain structure of cast bars or sheets
- Form billets into structural shapes such as flanges, channels, or railroad rails
- Produce tapers or threads on rods
- Straighten bent sheets, rods, or tubing

The basic concept is simple. If the rollers are pressed together with enough force, whatever passes between them must take the shape of the space between the rollers. If the rollers are perfectly cylindrical, the billet will be flat and plate or sheet material

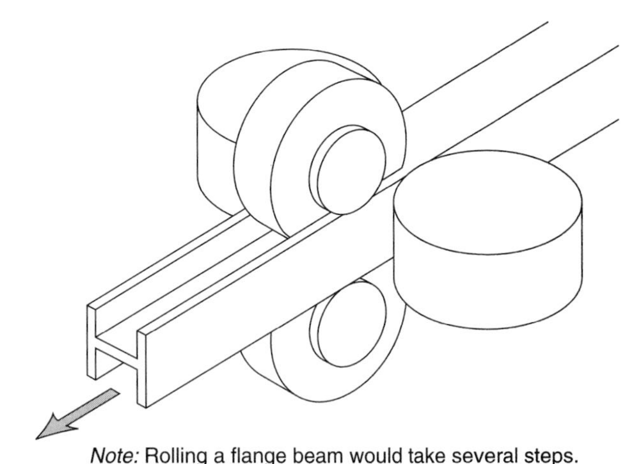

Figure 11-19. Rolling shapes.

is produced. If the rollers are curved, then the billets can be made into rods, channels, or any variety of shapes (Figure 11–19).

Even designs on material can be produced by rolling. Pastry chefs have used rolling pins with designs in the rollers to produce cookies. Metals can be similarly embossed by rollers. If the rolls are parallel to each other, then the material being passed between them will be flat. However, the rolls can be tilted slightly to produce beveled stock. Knife blades with a tapered thickness can be produced cheaply this way.

Bending by Rolling

If a set of three rollers is used, the stock can be curved by rolling (Figures 11–20 and 11–21). One roller depresses the sheet or plate between the other two. The

Figure 11-20. Bending by rolling.

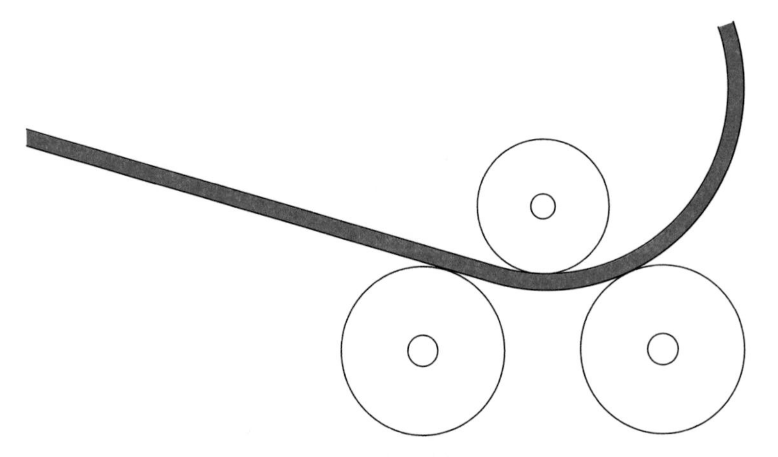

Figure 11-21. Bending by rolling.

radius of curvature is set by the amount of distance between the two bottom rollers and the distance between the top roller and the plane of the bottom rollers. The rolling machines are adjustable for different radii of curvature. Large steel culverts and other structural steel shapes are fashioned by these types of rollers.

Metal can also be *crimped* by rolling. For example, sheet metal workers often run a piece of sheet stock through crimping rollers to reduce the length of one side of the sheet. This allows the metal to be rolled into a cylinder or pipe in which one end will fit inside another cylinder of the same basic diameter. This process is shown in Figure 11–22.

Strip stock can be rolled to form tubing for welded pipe by the use of several sets of rollers. The strip stock is sent through several sets of rollers until the cylinder is formed. The pipe is then seam welded to complete the operation. Figure 11–23 is a schematic of the process, and the finished product is shown in Figure 11–24.

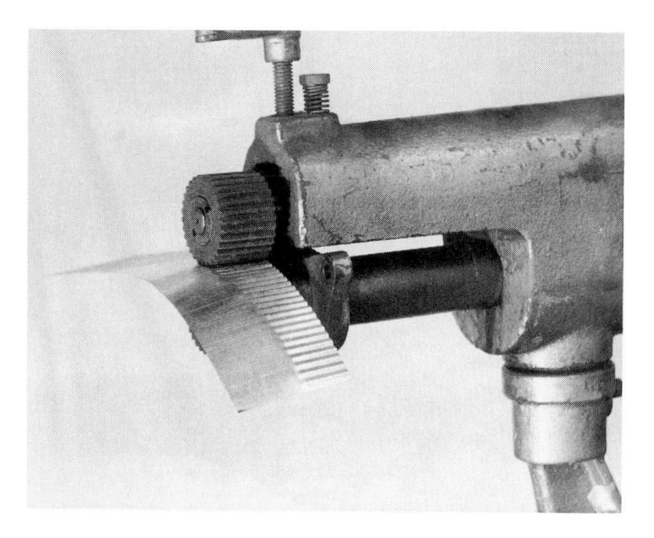

Figure 11-22. Crimping by rolling.

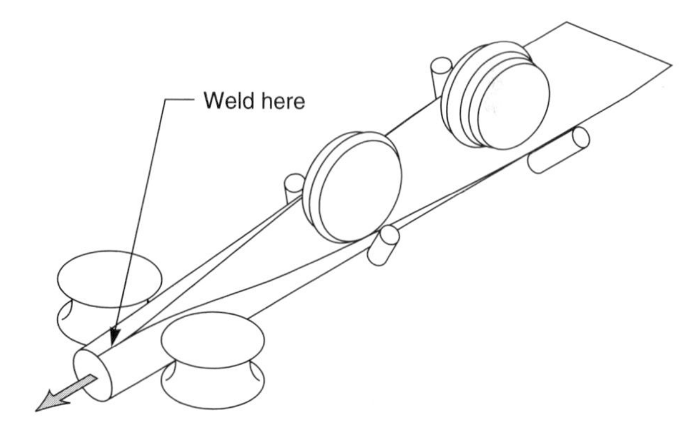

Figure 11-23. Rolling of pipe.

Figure 11-24. Welded pipe.

Threaded parts can be made by rolling the threads onto the shank of the part (Figure 11–25). Note that part of the thread is formed by raising or "upsetting" the metal. In cut threads the major (outside) diameter is the diameter of the shaft (unless the shaft is reduced in size later). In rolled threads the major diameter is larger than the shaft.

Machine screws can be made in the following way: A washer is placed on the shaft, then the thread rolled on so that the washer cannot slip from the screw. Not only is the rolling process faster than machining the threads, it leaves a harder grain structure. Modern screw machines can form one threaded part per second if the part is large and up to eight parts per second or approximately 30,000 per hour if the part is small.

Spur gears, helical gears, ball bearings, and roller bearings can also be made by rolling. Large gears are usually cut on a machine called a *gear hob,* but small, nonferrous gears can also be rolled (Figure 11–26). Ball bearings can be formed by rolling as shown in

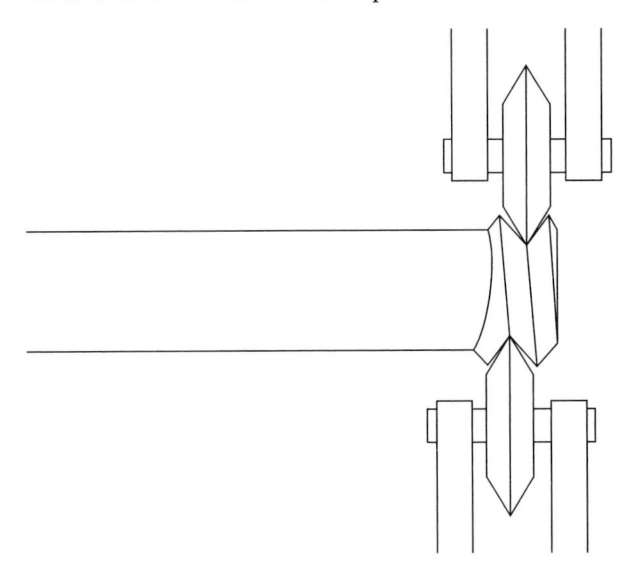

Figure 11-25. Rolling threads.

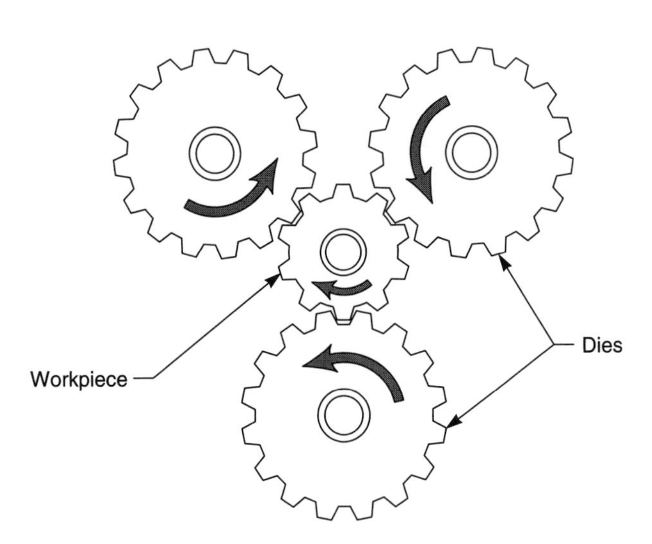

Figure 11-26. Rolling gears.

Figure 11-27. Rolling ball bearings.

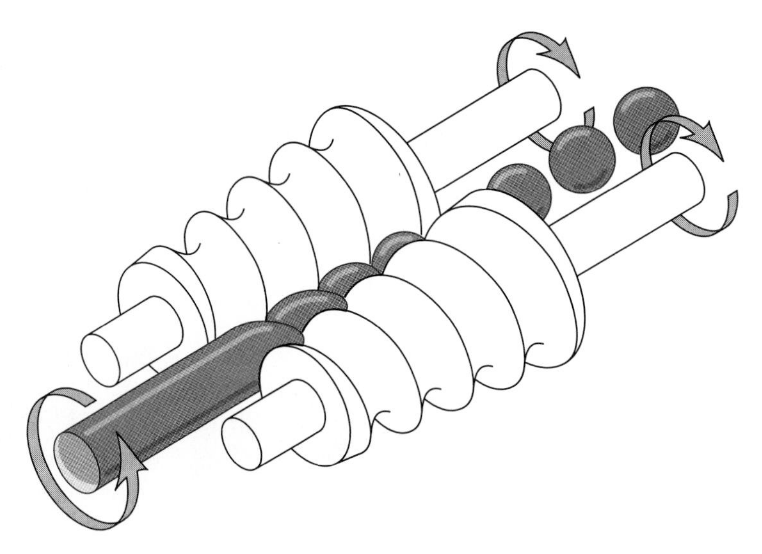

Figure 11–27. Here a shaft is fed into the rollers at an angle to the axis of the rollers. Note that in rolling ball bearings, the rollers are tapered. The farther the shaft goes into the rollers, the more it is compacted and shaped until it is finally cut off by the rollers and ejected at the end of the roll as a completely formed ball bearing.

This may sound like a paradox, but sheet stock, bar stock, rod, and tubing can be straightened by bending. Just as wire can be straightened by bending it backward to its original shape, some metals can be straightened by rollers. The stock is passed through a series of rollers that bend it in alternating directions by a smaller and smaller amount, until finally the bends are removed and it is not bent at all (Figure 11–28). Of course, straightening by bending does somewhat work harden the material. Subsequent annealing may be required of the straightened material. Beware, however, because annealing can also cause the metal to bend again.

Rolling Shapes

Sheet and plate are made by rolling. In industry, *plate* is defined as stock that is thicker than 0.25 inch (6 millimetres), whereas *sheet* runs from 0.25 inch down to about 0.0003 inch (0.008 millimetre). Metal less than 0.0003 inch thick is considered to be *foil.*

Large flange beams (I-beams), channels, and even wire are made by rolling. In rolling, the billet can be deformed only so much per "pass." Therefore, many stages or passes are required to make the completed shape. A standard flange beam (I-beam) starts out as a rectangular billet from the steel mill. Several passes are needed to reduce the billet to the width and depth required. Then shaped rollers start to reduce the middle section to the web. Each time the billet passes the rollers, the cross-section size is reduced, the shape is changed, and the billet gets longer.

Steel wire also starts out as a cylindrical billet. Each time the billet is fed into the rolls, it goes through smaller openings in the rolls, until the final diameter is reached (Figure 11–29). A 10-foot-long billet of steel 4 inches in diameter could be rolled into an 8-gauge wire (diameter = 0.16 inch) over one mile long.

Hot Versus Cold Rolling

Some materials are rolled while the metal is still red hot, whereas others are rolled at room temperature. Which is better? As with almost everything, there are advantages and disadvantages associated with each. Billets heated to the red hot range rapidly form an oxide coating or *scale.* Scale affects both the working of the billet and the wear on the roller. In hot rolling, the rollers must be made of a heat-resistant material or be replaced quite often.

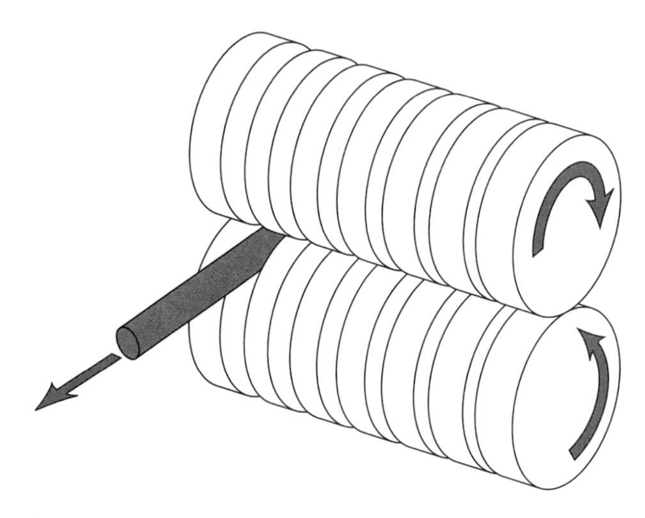

Figure 11-29. Wire rolls.

Figure 11-28. Straightening by bending.

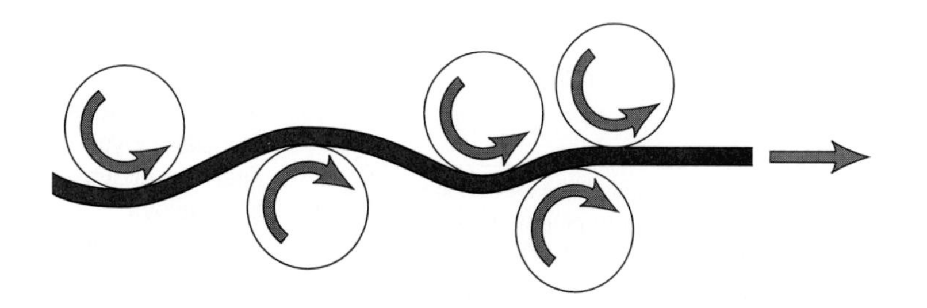

Cold rolling, on the other hand, requires more force and, hence, more massive equipment. High-stiffness materials, which have a high modulus of elasticity, are generally rolled hot to reduce the forces required. Steels of various types are usually hot rolled. Softer materials such as aluminum and copper are cold rolled.

A consideration of the material being rolled is the amount of deformation it can take without "tearing." This is especially important in cold rolling. There is a limit to which cold-rolled materials can be deformed without having to be reheated and annealed.

When a workpiece or billet is rolled, the grain structure is changed. Billets that were cast and allowed to cool slowly have large, even grains. (This is discussed at greater length in the section on heat treatments.) As the billet is run through the rollers, the grains are mechanically broken and elongated (Figure 11–30). When a metal billet is rolled hot, the large grains are broken by the rolling, but the grains will regrow to large grains after the rolling occurs. In cold-rolled steels, the grains don't regrow. Cold rolling produces a *wrought** grain structure, which is characterized by fine grain size. The finer the grain, the harder and less malleable the metal becomes. Therefore, if a hard steel is desired, it is cold rolled. When a softer, more malleable steel is desired, it is hot rolled.

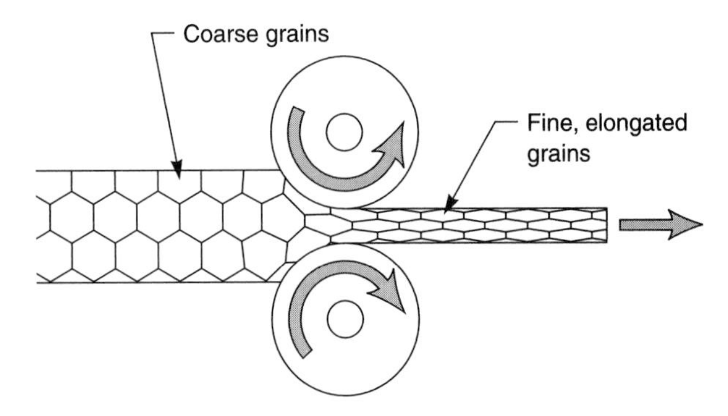

Figure 11-30. Grain changes in rolling.

* The word *wrought* is a verb meaning "to work." Ancient ironmongers hammered the cold iron to pound out many of the impurities and produce the fine-grained structure; hence, the term *wrought iron.*

Shapes that require many forming steps are usually hot rolled. Otherwise the billets would have to be annealed prior to each successive operation. Reheating of billets after they have been cooled would not be economically efficient. The combination of continuous casting followed immediately by hot rolling has been developed as a highly efficient combination to produce good quality steels in many accurately dimensioned shapes.

The temperatures at which various metals are hot rolled are shown in Table 11–1.

Factors Affecting Rolling

Many factors must be considered when rolling metals or other materials. Among these factors are the following:

- The material being rolled
- The material of the rollers
- The shape being rolled
- The size of the stock being rolled
- The size of the rollers
- Power requirements

Besides reducing the thickness of the stock, rollers must also pull the stock through the rollers. The friction between the rollers and the stock must be high enough to pull the stock between the rollers. If the stock hits the rollers too high up on the roller, the rollers will not be able to drag it between [Figures 11–31(a) and (b)]. Billets that must be drastically reduced in thickness require either large-diameter rollers or reduction by gradually putting the billet through many passes with small-diameter rollers. The use of small rollers results in more time being needed

Table 11-1. Hot Rolling Temperatures

Material	Temperature Range	
	Fahrenheit	Celsius
Aluminum alloys	750–850	400–450
Copper alloys	1150–1750	625–950
Titanium alloys	1400–1800	750–975
Steel alloys	1700–2300	925–1250
Refractory alloys	1800–3000	975–1650

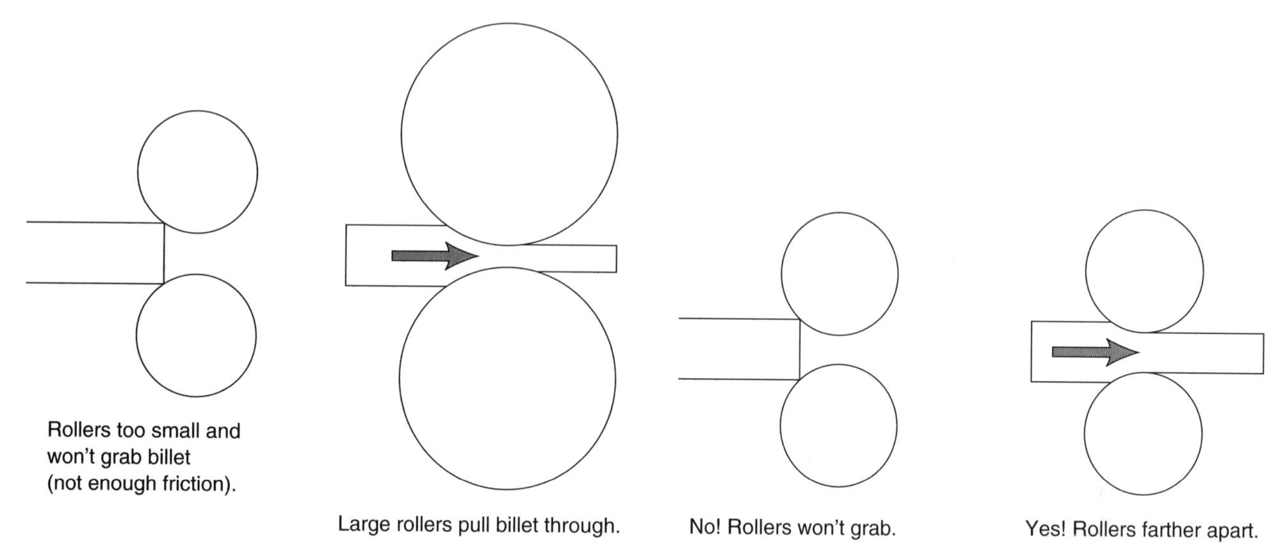

Figure 11–31. Size restrictions in rolling.

to reduce the size of the stock. Large rollers can reduce the size of the billet more on each pass but the trade-off is that they require more power to run and are more expensive.

An analysis of the forces involved in roll forming turns up the fact that the smaller the diameter of the roller, the smaller the force required to press the metal through the opening. Large rollers are very heavy and have large leverages for pulling the metal through the rolls. For cold rolling, the forces involved would require a large-diameter roller for strength and stiffness. A small-diameter roller can be used, however, if properly supported. The easiest way to do this is with "backup" rollers (Figure 11–32). It is not unusual to find three layers of rollers in a cluster so that the forces can be provided by large outside-layer rollers, while the small size of the active roller reduces the forces required to do the forming. In the case where large rollers are needed to provide the friction to draw the billet between the rolls, several sets of small rollers can be used in sequence, each reducing the stock slightly as depicted in Figure 11–33. Because the volume of the billet remains constant, its length and width increase as the thickness decreases. This means that the downstream rollers must turn faster than the rollers preceding them in order to keep the billet from jamming behind the rollers.

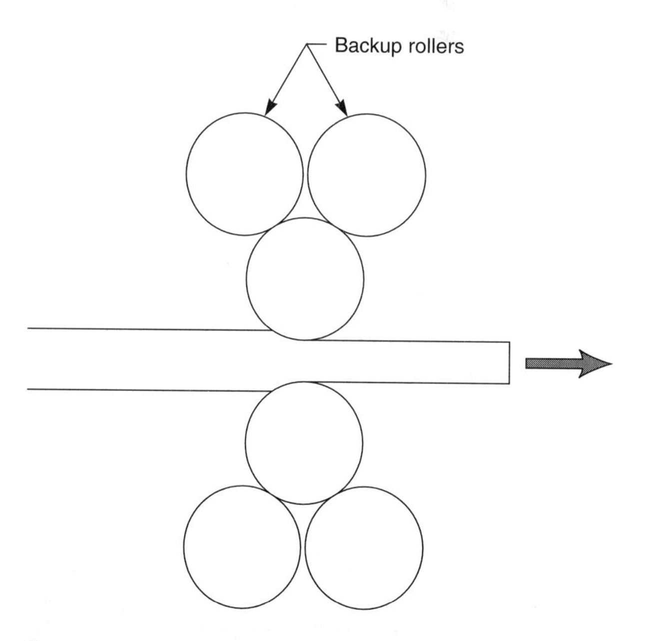

Figure 11–32. Backup rollers.

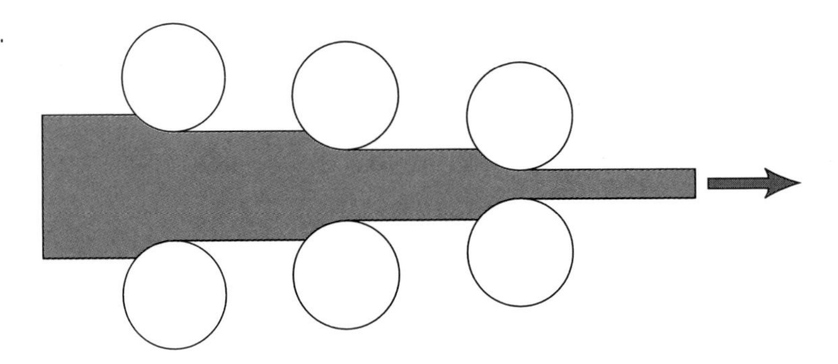

Figure 11–33. Sequential rolling.

Bending

Bending is one of the most basic processes available. Many methods are available for bending materials, including the following:

- Continuous roll-bending (covered earlier in this chapter)
- Punch and dies
- Wiping (sliding) die braking
- Hinged braking
- Press braking
- Air bending
- Beading
- Flanging
- Dimpling
- Hemming and seaming
- Tube flaring
- Tube bending
- Roll forming

Bending processes are used mostly on flat sheet, plate, extrusions, rods, and tubing stocks. Some of these have already been discussed.

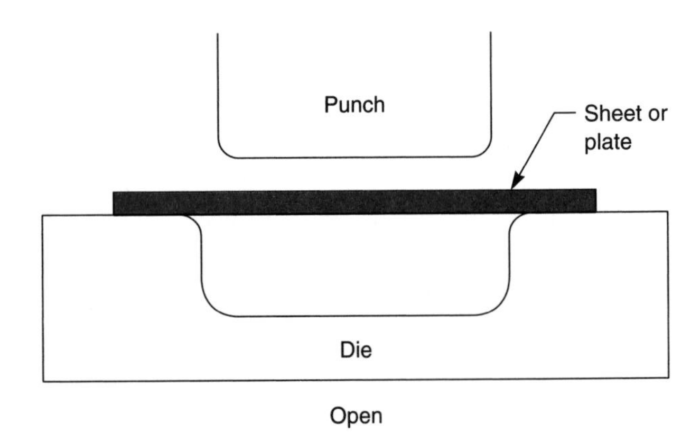

Figure 11–34. Punch and die, open.

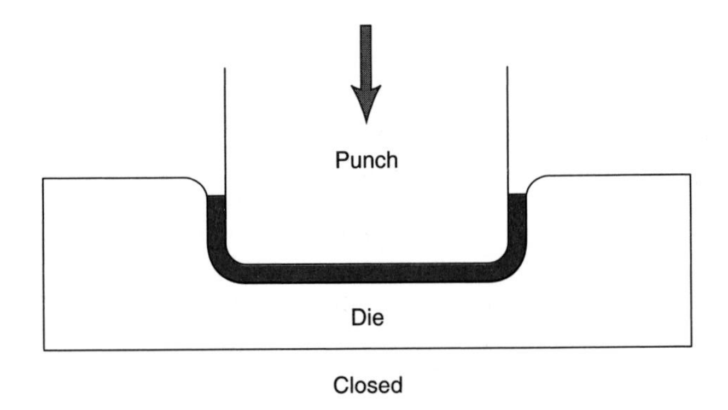

Figure 11–35. Punch and die, closed.

Punch and Dies

Sheet metal and flat bar stock can easily be shaped by punching the material into a die (Figures 11–34 and 11–35). For some very complicated bend patterns, a *series* of punch-and-die sets is used to form the successive bends in the workpiece. Zinc cups used in electric dry cells are formed by punching them into dies. Strap hangers for automobile muffler pipes and metal beam hangers for buildings are but a few examples of parts made by the punch-and-die process.

Wiping Die

Sliding a movable die across an extended piece of metal will bend the piece around a mandrel. The process requires simple tooling and it is easily adjustable to changes in material thickness. The bend angles are generally 90° or less but can be more if the material being bent can withstand the deformation. Wiping is often used to make large radius bends in sheet metal workpieces (Figure 11–36).

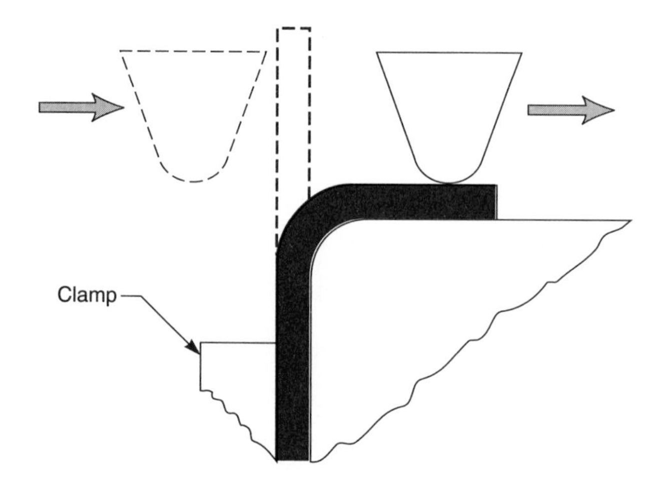

Figure 11-36. Wiping.

Hinged Braking

A *brake* is a general-use device for bending sheet metal. The machine shown in Figure 11–37 and illustrated in Figure 11–38 can be used for one-of-a-kind production as well as for mass production. The "fingers" hold the metal in place at the point of the bend. The hinged bed is then raised to bend the metal. The fingers can be adjusted so that only part of the extended metal is bent. The *finger brake,* as it is sometimes called, can be used to bend sheet metal from 0 to 180° by means of multiple bends. Whereas wiping is used to bend metal around a large radius, brakes can make bends with radii as small as the thickness of the sheet or plate being bent.

Press Braking

The *press brake* is an extension of the punch-and-die set extended along one dimension to make complex bends in a long piece of sheet stock (Figures 11–39 and 11–40). The workpiece can be bent in successive passes with a series of different die sets to create a complex shape.

Air Bending

Air bending doesn't use air in any way. The name merely refers to the fact that there is no defined cavity or die. The operator uses the press to force the metal into the die opening "just far enough" for the correct

Figure 11-37. Finger brake.

Figure 11-38. Finger brake.

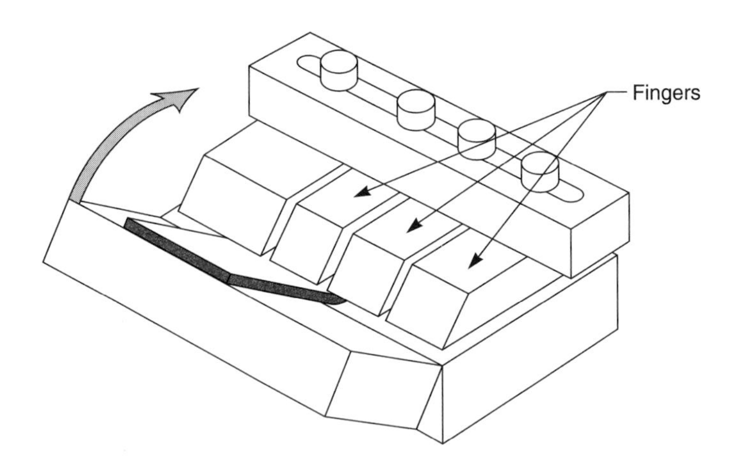

Figure 11-39. Press brake.

Figure 11-40. Press brake.

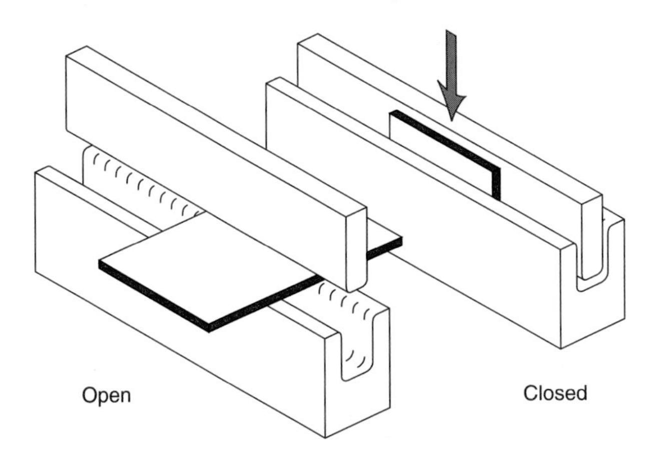

bend (Figure 11–41). Basically it is a "freehand" bending operation. Bicycle handlebars are one example of a product made by this process. This is a very versatile operation—one for which one machine can perform many functions, but one that requires a skilled operator.

Beading

Beading is the forming of the rolled edge on sheet metal (Figure 11–42). It gives strength and stiffness to the edge as well as turning under the sharp edge. Beading can be done with a finger brake, by rolling, or by making two or more passes on a press brake.

Flanging

A *flange* is a piece of an object that protrudes at 90 degrees to the main body of the object. High-pressure pipes are joined by flanges, perhaps with a gasket between two flanges. Sheet metal parts can have flanges in them for stiffening as well as to join two parts (Figure 11–43). Sheet metal flanges can be made with brakes and other machines.

Dimpling

Dimpling refers to the formation of a cylindrical flange at a location other than at the edge of a sheet metal workpiece (Figure 11–44). Often a circular hole is first punched into the metal at the desired location, then a punch-and-die operation is used to bend the entire perimeter of the hole into a cylindrical form. Dimples are used for stiffening the sheet or for a hole to be threaded by a self-tapping screw. Dimpling can also be used to provide spaces between sheets.

Hemming and Seaming

Hemming refers to the process of folding over the edge of a piece of sheet metal and pressing it flat. This stiffens the sheet and creates a safer, non-jagged edge.

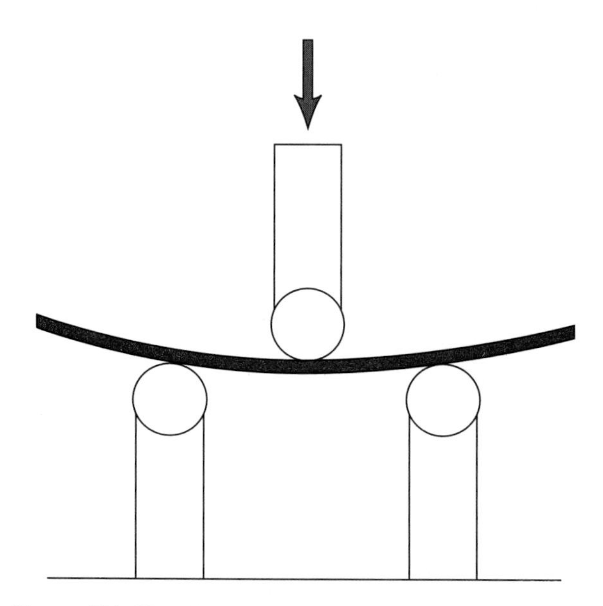

Figure 11–41. Air bending.

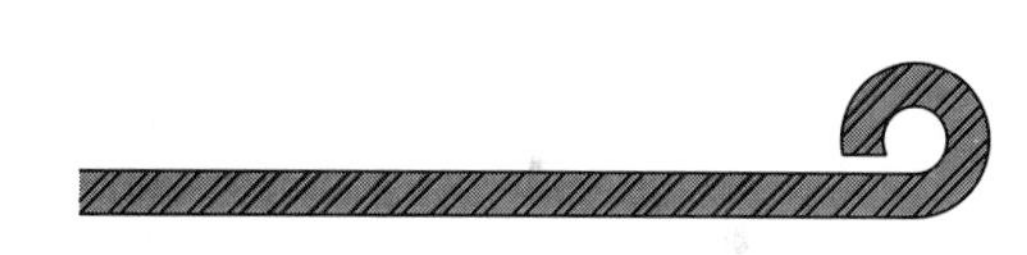

Figure 11–42. Beading.

Figure 11–43. Sheet metal flanges.

Figure 11–44. Dimpling.

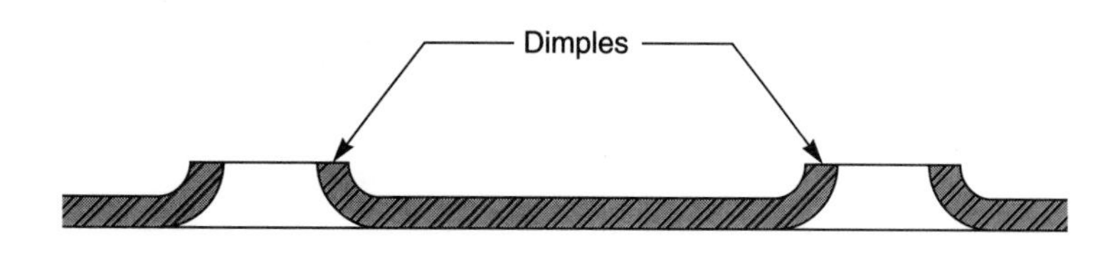

Hemming can be done by means of a hinged brake or press, then rolled or forged closed (Figure 11–45). Hemming and beading are similar operations.

Seaming is the joining of two edges as in the joints of a steel can (Figure 11–46). Seaming involves forming the bent shape of two open hems, sliding the two edges together, and then pressing them flat. The hem can be further coated with plastics or molten metal to form seams that are watertight and airtight.

Tube Flaring

Flaring differs slightly from flanging. In flaring, the metal is bent less than 90 degrees. Flaring is generally used on tubing to produce a tapered section rather than a flange. Some thinning of the tubing occurs as the metal is deformed. The greater the angle of the flare, the greater the thinning. If need be, the tubing can first be forged with a thicker end so that flaring merely thins the tubing to the thickness of the rest of the wall (Figure 11–47).

Tube Bending

Straight tubing is bent into curved sections for a variety of products as diverse as French horns and electrical conduit. But *tube bending* has an inherent problem. The tube tends to collapse at a single point if not prevented from doing so (Figure 11–48). This collapsing can be prevented in two ways. An inner core can be placed in the tube that holds the walls of the tube apart and therefore prevents the collapse; or an outer constraint can be used to keep the sides of the tubing from spreading apart. The first method exerts force from the inside to keep the opposite walls from coming together and the other method exerts force from the outside to keep the tube from flattening (the first stage of collapsing).

Both methods of preventing collapse are in common use. The conventional tool used to bend electrical conduit, copper pipe, or muffler pipe uses the outside constraint (Figure 11–49). This is simply a tool having a bend radius many times larger than the tube or pipe. This tool

Figure 11–45. Hemming.

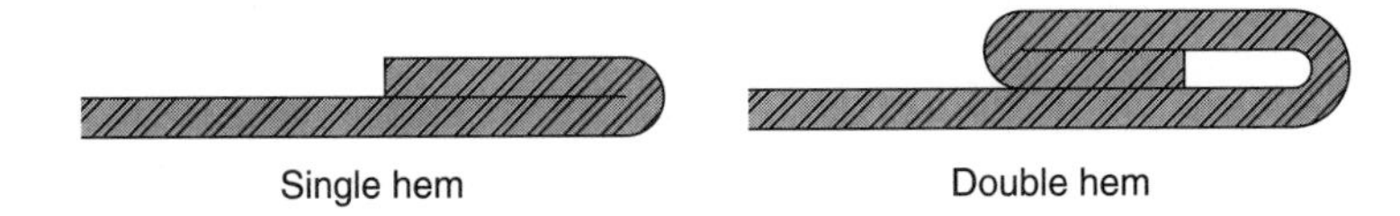

Figure 11–46. Seaming.

Figure 11–47. Flaring.

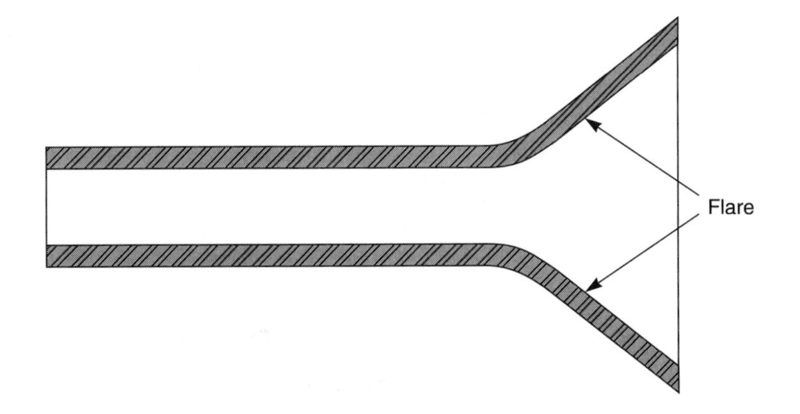

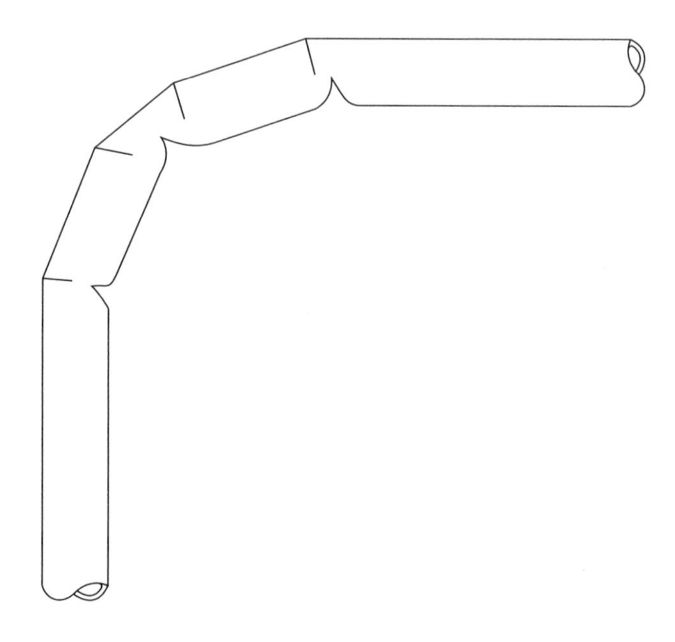

Figure 11-48. Tube collapse.

grips the outside of the tubing to prevent the crimping of the tube. This type of bend can result in some flattening of the inside radius of the tube. For small, lightweight tubing, a coil of stiff spring wire can be slipped over the tube and the whole apparatus bent. The spring prevents the tube from crimping and ensures a large radius bend. The spring is slipped off the end of the tube after the bend is made (Figure 11–50).

Musical instruments and thin-walled tubes in which the internal cross section must be kept constant are often made by filling the tube with packed sand, pitch, or small steel balls prior to the bend. One technique used by some musical instrument manufacturers is to fill the tube with water, freeze the tube and water, and then bend the tube with the ice inside. After the bend is made, the internal material is removed. Many modifications of this process are used (Figure 11–51).

Roll Forming

Roll forming differs from roll bending, which was shown in Figure 11–20. Roll forming uses rolls with matching contours so that the sheet material is bent to the desired shape as it passes through the rollers. This process is used to form seamless rain gutters and other long pieces. Custom-made parts can be made on location by this process (Figure 11–52).

Figure 11-49. Pipe bending.

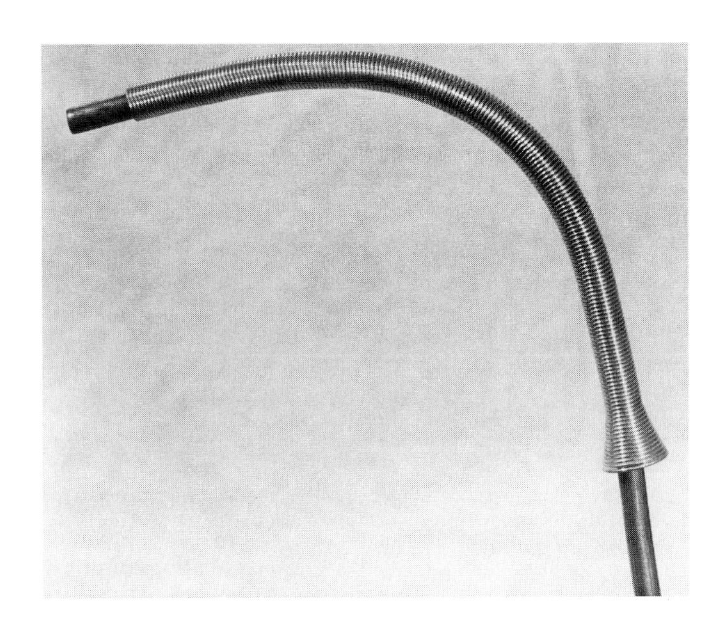

Figure 11-50. Bending with a coil.

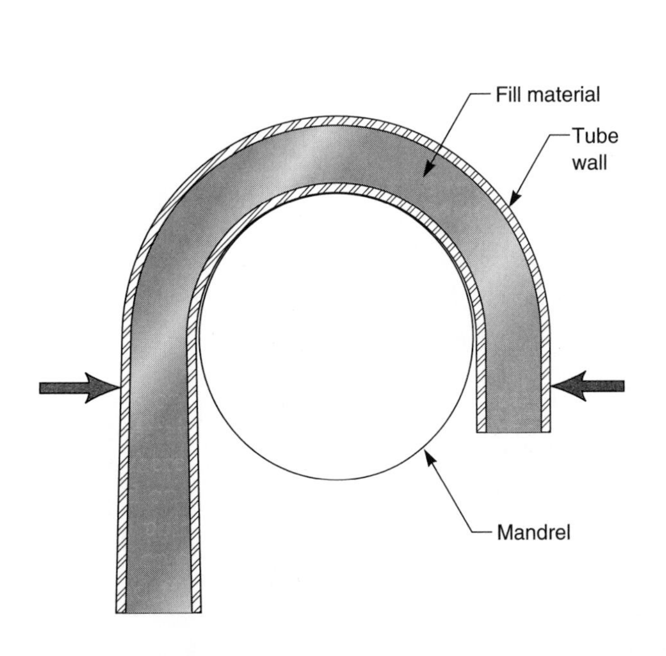

Figure 11-51. Bending with internal filling.

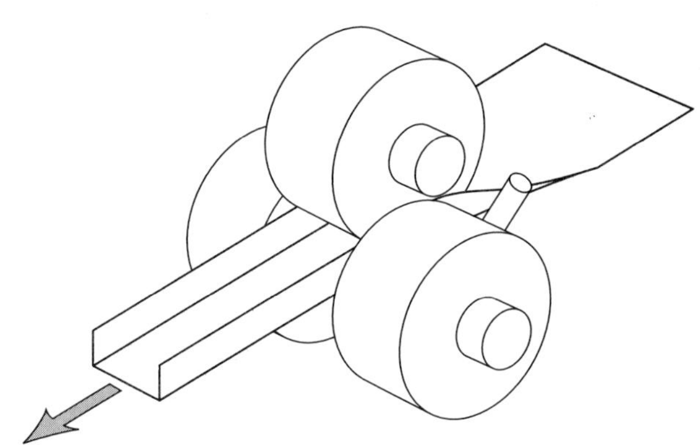

Figure 11-52. Roll forming.

Drawing

Drawing has been the primary process used to form wire and metallic thread since the 11th century. The principles involved have not changed. Both drawing and extrusion processes force a billet through a die in order to shape or reduce the size of the billet. In extrusion, the billet is *pushed* through the die, whereas in drawing it is *pulled* through the die. Extrusion works by using pressure in the input side of the die, whereas drawing works by placing tension on the output side.

In drawing, the stock is given a point by means of a swaging process (as discussed earlier), then threaded through the die opening. A clamp is then attached to the end of the stock and it is drawn through the die (Figure 11–53). If there is to be a large reduction in diameter from the original billet, it may be necessary to repeat the drawing operation several times, reducing the diameter a small amount at a time. Since the stock is work-hardened as it is drawn through the die, it may require annealing between drawing operations.

To reduce the power requirements for drawing, the billet is lubricated. Many polymer lubricants, such as silicones and polytetrafluoroethenes (Teflons), are currently available. However, one of the most commonly used lubricants is still soap. Soap works well with most drawing operations, is cheap, washes off easily, and does not require disposal at a toxic fill site.

In drawing titanium or steels, a soft metal sleeve is placed over the billet and the sleeve and metal are both drawn through the die. The sleeve provides the lubrication for the die. One method of providing the sleeve is to dip the harder metal into the molten softer metal and let it solidify around the hard metal.

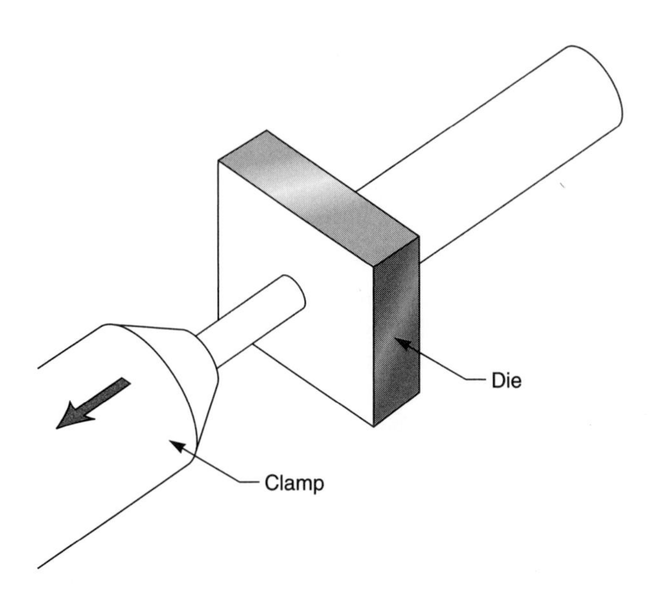

Figure 11-53. Drawing.

Spin Forming

Spin forming is a process that deforms a disk of material around a mandrel whose shape could be turned on a lathe. The disk is mounted against the mandrel and clamped in position. The mandrel and disk are rotated at several hundred rpms and a smooth tool is forced against the disk until it "flows" by plastic deformation onto the mandrel. The metal thereby takes the shape of the mandrel. Cymbals used in the percussion section of bands and orchestras are made by this technique. The advantages of spin forming are that the tooling is inexpensive and versatile. The disadvantages of spin forming are that it does require considerable skill on the part of the operator, and the production rate is limited (Figure 11–54).

Another method of spin forming is a variation of the process that potters use to form clay pots (Figure 11–55). A tube of metal, or even a hot solid piece of metal, can be anchored to a spinning platen or table. Rollers or smooth tools can be forced against the metal. As the tools are forced into the metal, its diameter is reduced and it becomes elongated in much the same manner that clay takes shape on the potter's wheel. The tube of metal can be forced against a mandrel and the mandrel removed after the spin forming is completed.

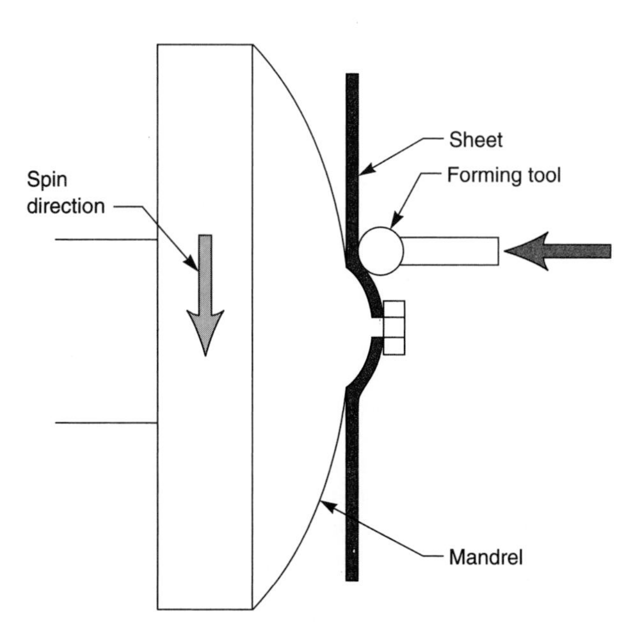

Figure 11-54. Spin forming.

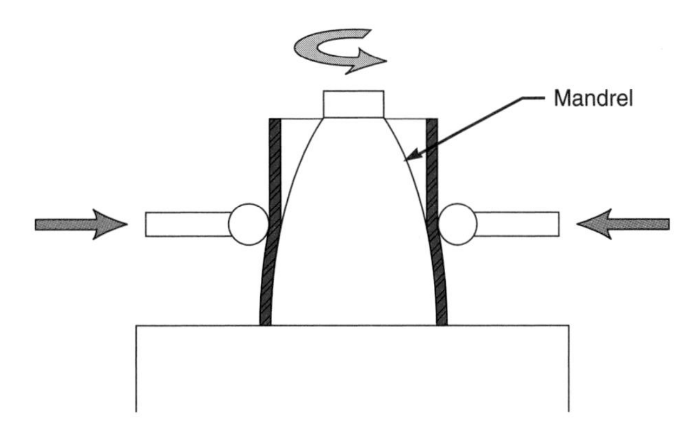

Figure 11-55. Another spin forming method.

There are some limitations as to the shapes that can be produced by spin forming. For instance, if the shape is such that it can't slip easily off a mandrel, such as a sugar bowl, then the mandrel must be constructed so that it can be disassembled into several pieces and removed piece by piece. This is a slow production technique. Although it is possible to make complex parts by spin forming, it isn't generally practical or commercially profitable to spin form recurved shapes.

HIGH-ENERGY-RATE METHODS

Several innovative methods of forming parts involving the sudden application of pressure have been developed in the past few decades. Among these methods are explosion forming and magneforming.

Explosion Forming

In *explosion forming,* the workpiece, usually sheet metal or plate, is placed over a die cavity and the volume in back of the plate is evacuated. An explosion of some form is detonated over the workpiece. The sudden pressure or shock wave presses the metal firmly and evenly against the surface of the cavity, producing the desired shape. The size of the parts that can be made by explosion forming varies from a few ounces to hundreds of pounds.

Explosion forming of small parts can be accomplished as shown in Figure 11–56. The workpiece is placed over the cavity and sealed in a container. The explosion increases the air pressure above the stock, driving it into the cavity. The amount of explosive must vary with the thickness of the stock and the size of the equipment. For very small parts, a .38-caliber blank pistol cartridge may be sufficient. Larger powder cartridges and specialized explosive charges are used on larger parts.

Larger parts, sometimes weighing hundreds of pounds, are explosion formed using a different technique. The die, with the cavity of the desired shape, is lowered into a tank of water. The metal plate is placed over the cavity and the water removed from the cavity. An explosive charge is detonated in the water above the plate. The pressure is not only applied by the force of the explosive, but by the hydrostatic pressure of the water and the force of the water as it rushes back after being blown away by the explosion. For these larger pieces, sticks of dynamite are sometimes used. Figure 11–57 illustrates this technique.

Surprisingly, the dies for explosion forming need not have extremely high strength. Dies may be made of aluminum alloys, steel, zinc alloys, reinforced concrete, wood, or even plastics. Composites are also gaining favor for constructing these dies.

Magneforming

A special metal-forming process is based on the principle that "like magnetic poles repel each other"; hence the name *magneforming* or *magnetic pulse forming.* Two systems are in general use.

One system is used to form flat sheets or plates into a die cavity. The sheet of metal is placed over the die cavity under a coil of wire as shown in Figure 11–58. A sudden pulse of electric energy is discharged through the coil from a very large electrolytic capacitor, which has been charged to a high voltage (Figure 11–59). When the great pulse of current flows through the coil, a magnetic field suddenly builds around the coil (Figure 11–60). As this magnetic field penetrates the sheet metal, it induces an electric current in the metal in the same manner that current in the primary winding induces a current in the secondary winding of a transformer (Figure 11–61). This induced current produces

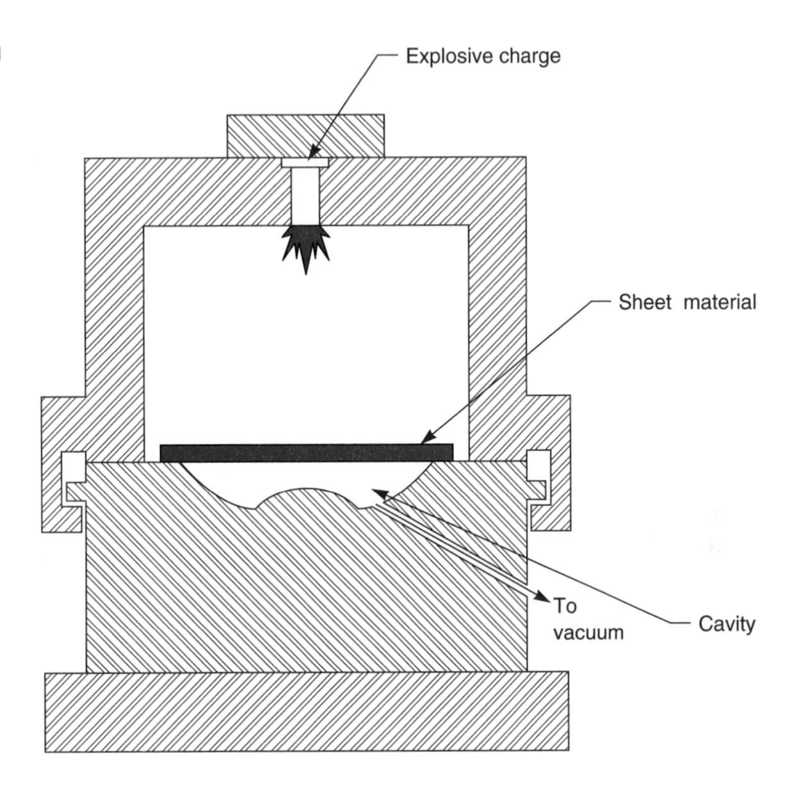

Figure 11-56. Explosion forming of small parts.

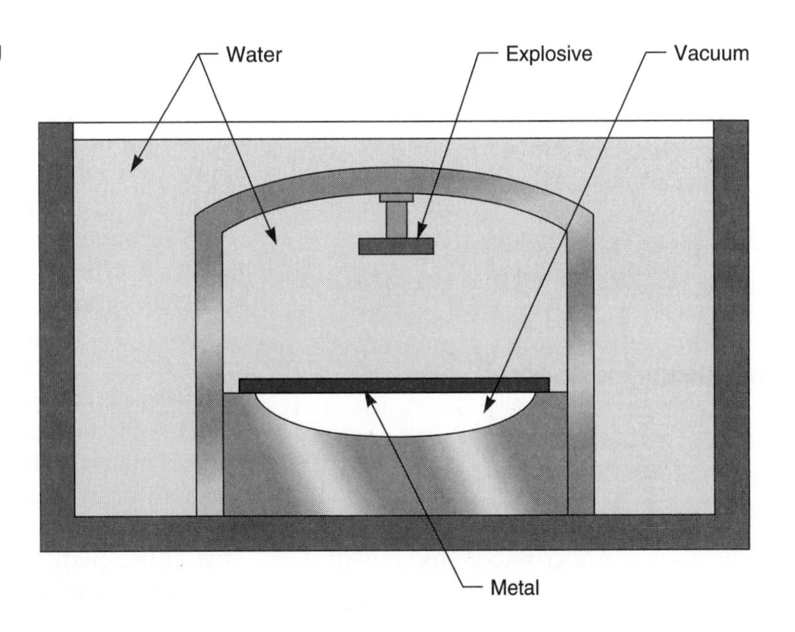

Figure 11-57. Explosion forming of large parts.

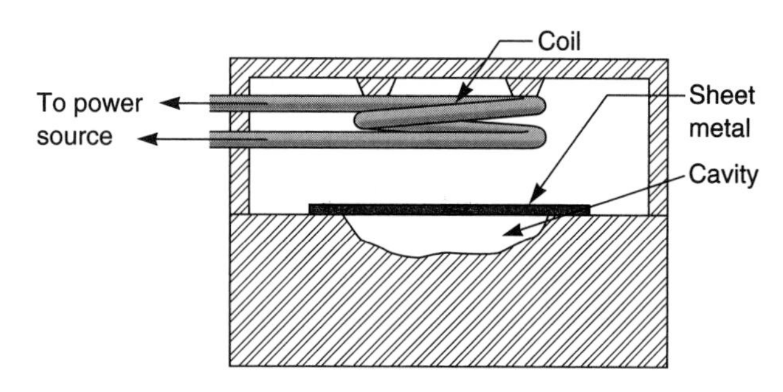

Figure 11-58. Magneforming.

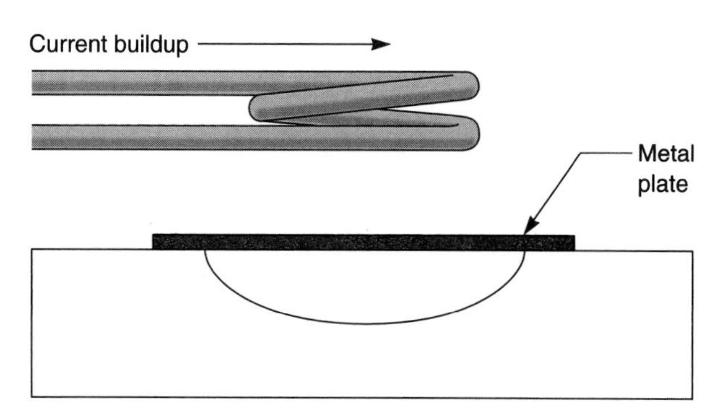

Figure 11-59. Magneforming: step 1—current buildup.

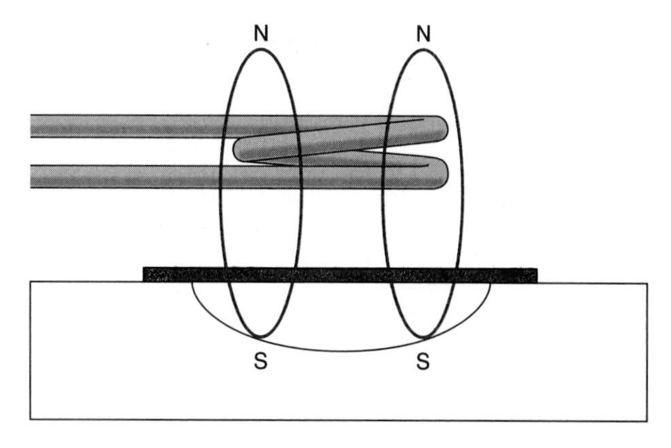

Figure 11-60. Magneforming: step 2—magnetic field buildup.

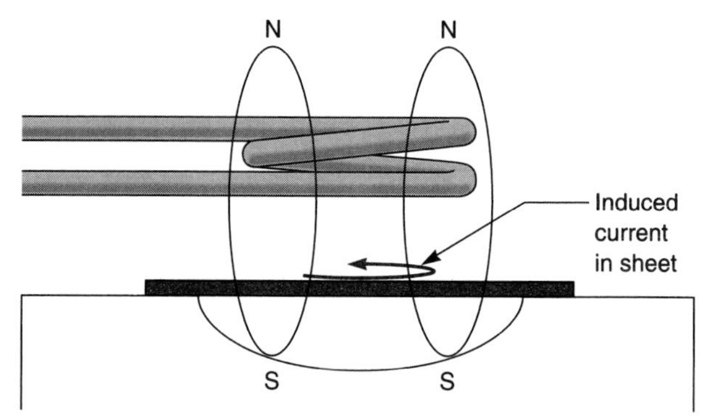

Figure 11-61. Magneforming: step 3—current induced.

another magnetic field about the metal plate (Figure 11–62). The magnetic field around the coil and the one around the plate have the same polarity. Therefore, these magnetic fields repel each other. Since the coil is rigidly fixed to the frame of the apparatus, the only object that can be moved by the repulsion is the plate. The plate is repulsed into the cavity and takes its shape (Figure 11–63). The entire process is completed almost instantaneously. A second "shot" is sometimes needed to complete the forming, but time must be allowed for the capacitor to recharge.

An interesting aside to this process is that as the magnetic field collapses about the coil (after the current is shut off), an electric current is regenerated in the coil that can be fed back into the capacitor to help recharge it. Of course, the capacitor will not be fully recharged by this regenerated current, so some time must pass before the capacitor is recharged completely by an external power source.

A second variation of magneforming can be used to swage cylindrical tubing onto a mandrel, wire, or other cylindrical stock, as shown in Figure 11–64. Here the coil is placed around the metal tubing. When the switch is closed to fire the current through the coil, the tubing is collapsed about the wire or mandrel.

In both magneforming and explosion forming, the metals are deformed so rapidly that little grain deformation occurs. Since the pressures, either pneumatic, hydrostatic, or magnetic, are evenly distributed over the entire plate or tube, the tendency to tear the formed material is reduced.

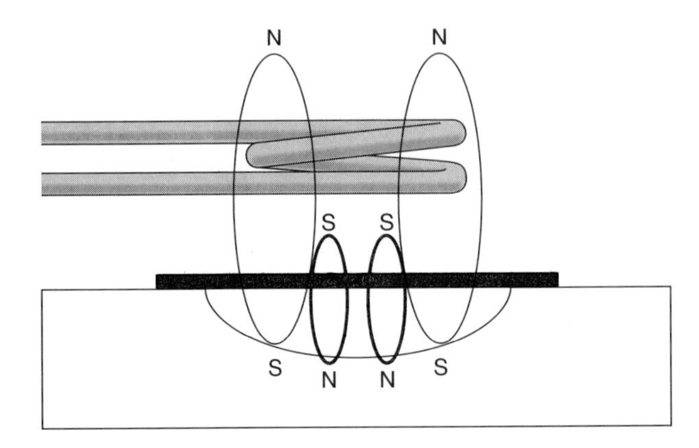

Figure 11-62. Magneforming: step 4—opposing magnetic field created.

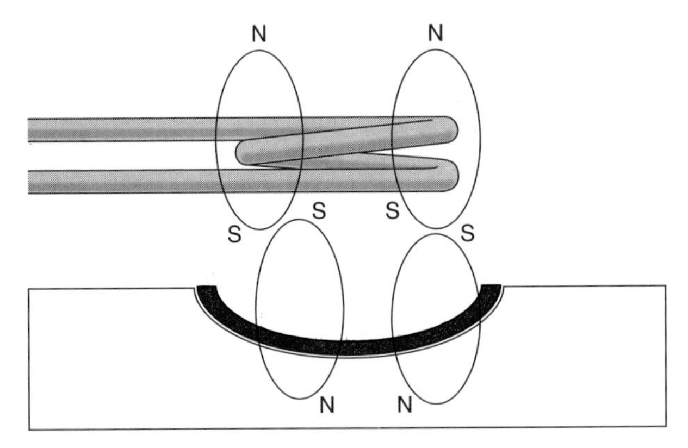

Figure 11-63. Magneforming: step 5—metal deformed into cavity.

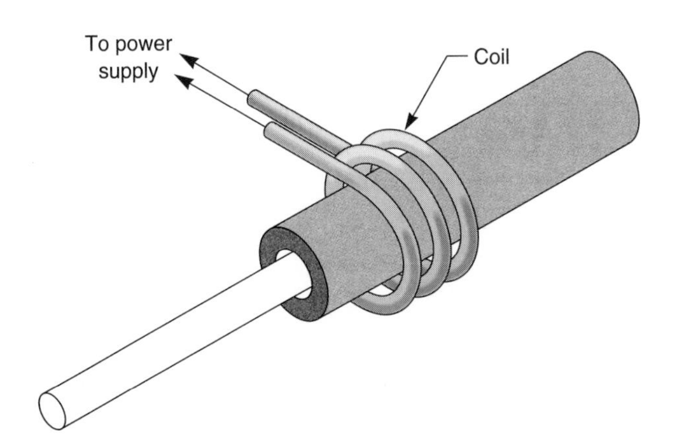

Figure 11-64. Swaging by magneforming.

Some metals can even be painted before magneforming. Many modern paints and coatings can withstand the forces in magneforming without being damaged. Magneforming is used in making many types of products, from toys to aircraft.

PROBLEM SET 11-1

1. Could magneforming be used to shape a sheet of Lucite or Plexiglas? Why or why not?
2. An adjustable crescent wrench has three working parts: the handle and one jaw, the movable jaw, and a worm gear. Outline processes whereby all three of these parts could be made by methods discussed in this chapter.
3. A 5/8-inch end wrench is to be made in large quantities.
   a. Outline a process whereby this wrench could be made as cheaply as possible.
   b. Outline a process that would produce a wrench with high strength and high precision.

   Use only methods discussed in this chapter.
4. List three examples of products made by extrusion.
5. List three examples of products made by magneforming.
6. A copper U-tube has an inside diameter of 0.5 in., a wall thickness of 0.05 in., and an inside radius of the bend of 6 in. Outline two processes whereby this tube could be made without any kinks in the tube.
7. Outline two processes whereby a conical nozzle for a 0.75-in. hose could be manufactured at a minimum rate of 50 per day.
8. A "domed" shape cover for an electric motor is to be made from 20-gauge mild sheet steel. The radius of the dome is to be 5 in. List three processes whereby this piece could be made.
9. What is the difference between a "finger brake" and a "press brake"?
10. Explain in your own words how magneforming works.
11. Look at any ductwork, such as an air-conditioning system, and list examples of flanging, seaming, hemming, or beading.

12. List the advantages and limitations of magneforming.
13. List the advantages and limitations of explosion forming.
14. List three materials that cannot be used in explosion forming and the reasons why they can't be used.

Forming refers to those change-of-form processes whereby the material is made to "flow" by exerting enough force to cause plastic flow to occur in a solid material. The major forming methods are forging, extrusion, rolling, and bending. Each of these has several subcategories. The forces are applied either by presses or hammers.

Explosion forming and magneforming are techniques in which the forces are applied very suddenly. These methods of forming allow parts to be made without upsetting the grain structures and other properties of the metals being formed.

SECTION

V

Change of Condition

Chapter 12
Change of Condition

■ INTRODUCTION

The cold working of a metal workpiece (as a result of forging, forming, rolling, punching, or other methods of change of form) often leaves it in such a hard, brittle condition that it is unsuitable for further manufacturing processes or for use as a final product. Fortunately, we have processes whereby the internal structure of metals can be altered, thereby changing their properties.

Change of condition refers to the changing of the internal crystal and/or grain structures of a metal to alter its physical properties.

Chapter

12

Change of Condition

In the design of many products, it is not enough to specify the material of construction. Sometimes we also need to specify the surface hardness, toughness, or other property of that material. In many materials, the hardness, toughness, resilience, and other properties are a function of the elements and cannot be changed. Fortunately, however, the surface and interior properties of some materials can be changed radically. This chapter discusses the methods by which properties of a material can be controlled and changed. Basically, the surface conditions of a metal can be changed by four methods:

1. Mechanical processes
2. Thermal processes
3. Chemical processes
4. Tempering processes

Mechanical techniques for changing the internal properties of a material include the cold-working processes as discussed in Chapter 11 under the topics of rolling, forging, bending, etc.

Thermal methods can be used to soften the metal by full annealing, box annealing, bright annealing, and normalizing. They can also be used to harden metals by means of quenching with brine, water, oil, or air, or they can be hardened by means of the surface hardening techniques of flame hardening and induction hardening.

Chemical methods for changing the internal condition of a metal include case hardening by such techniques as carburizing, nitriding, or carbonitriding.

Tempering or drawing processes for steels modify the internal conditions to relieve internal stresses. This can be done by processes such as tempering and spheroidizing.

Few new processes exist for the heat treatment of steels. Those available include martempering, austempering, and patenting.

Nonferrous metals can also be heat treated using such techniques as annealing, precipitation (age) hardening, and solution heat treatments.

■ MECHANICAL TECHNIQUES

The properties of hardness, brittleness, toughness, ductility, and others depend somewhat on the size of the grain. We discussed how grains are nucleated and grow in Chapter 2. Slip always occurs between planes of the atoms. Ductile materials will allow the planes of atoms to slip easily over each other (Figure 12–1).

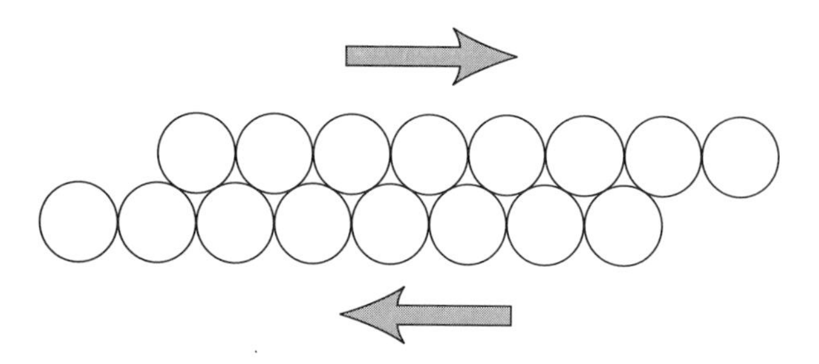

Figure 12-1. Slip planes.

Planes that can move over each other are called *slip planes*. It is logical, therefore, that the larger the planes of atoms, the more ductile the material. This translates to the larger the grain size, the more ductile the material. Small grains have small slip planes and therefore are more brittle.

Cold Working

Large grains in metals can be broken into smaller grains mechanically by rolling, forging, beating, bending, or crushing. This is called *cold working* the metal. Cold working makes the metal harder but more brittle.

A simple experiment will demonstrate the effect of cold working. Simply draw a 6- to 8-inch length of heavy-gauge copper wire (8- to 12-gauge solid wire will do) through a gas burner flame. Allow the wire to become red hot before moving it to the next section. Let the wire cool slowly to room temperature. Now, the first time this wire is bent it will require very little force. Try, however, to bend it again and notice two things. First, a greater force is required to bend it back and, second, the bend will usually move over to a softer unbent portion of the wire. The bending action has cold worked the wire, broken up the large grains formed by the heating, and hardened the copper.

Metal beams, slabs, rods, and other shapes are formed after the molten metals have been cast into ingots. The rolling, forging, stamping, or other forming process can break up the large grains. If the metal is worked while it is still very hot, usually red to orange in temperature (1400 to 1800°F), the small grains will again form large grains and revert to their softer, more ductile condition. This process is known as *hot working,* hot rolling, or hot forging. Hot-worked metals are easier to shape than cold-worked ones. Hot working requires less mechanical force, and the parts can be formed faster than by cold working. Cold-worked metals, however, will hold tolerances for machining better than their hot-worked counterparts, but are usually more difficult to cut and machine. Further, the cold-worked metals damage cutting tools more than the hot-worked ones.

Cold working can deform the grains to the point that cracks start to form. Once a crack is initiated, the material is destroyed for all practical purposes. In many manufacturing processes, very severe deforming by forging, rolling, punching, or some other process is required. In these cases, it is essential that the material be either hot worked or forged, rolled, stamped, or punched, then heated to just below the melting point to allow grain regrowth before further deforming is done. In many cases, this cold working and reheating process must be done several times to complete the forming process.

Materials can be peened or cold worked only on the surface to provide a hardened surface. Surface hardening is discussed in detail later in this chapter.

■ THERMAL PROCESSES

Theory

Why can we change the properties of a metal by heating and cooling it? What happens to the metal undergoing heat treatments? To understand the principles of heat treatments, it is necessary to have some knowledge of the phases and phase structure of materials.

Steel is composed of iron and carbon. Other elements may also be added to give the steel specific properties. When two or more *components* are mixed, two possibilities occur. The components can form a *mixture* such as oil and water, water and ice cubes, or sugar and coffee; or they can chemically react to form a new *compound.* In mixtures, the components can be physically separated by filtering, distilling, skimming, or other mechanical technique. When the components form a compound, however, a new product is formed and the components can no longer be separated by mechanical means. If sodium metal is mixed with gaseous chlorine, the product sodium chloride is formed, which is common table salt. The chlorine and sodium cannot be separated and the product has characteristics of its own that are different from those of the sodium and chlorine individually.

A *phase* is a macroscopically distinct portion of matter. Three qualities are necessary to define a phase: (1) Separate phases will always have a sharp boundary line between them. (2) Phases exist on a large enough scale to be seen optically. However, some phases such as those in steel may require a microscope to make them visible. (3) Phases can be separated by physical (as opposed to chemical) means. (Admittedly, the separation of the phases in steel would be difficult, but theoretically it could be done.) The individual *states* of matter (solid, liquid, gas, and plasma or ionized gas) may form single phases by themselves. But it is possible for two phases to exist together in equilibrium in the same system. A common example of a two-phase system is oil and water. When equilibrium is reached, the two liquids will form two layers or phases. Note that there is a sharp boundary between the oil and the water phases. The oil and water could be separated simply by drawing the water from the bottom.

We also have solid-liquid two-phase systems such as ice cubes and water. Again there is a sharp boundary line between the solid and liquid water phases and the cubes are easily removed from the water by the use of a filter or tea strainer. Visible smoke and high-altitude clouds and contrails left by aircraft are examples of solid particles suspended in air making up solid-gas two-phase systems. Different solid phases can be seen by examining polished and etched two (or more) component metals such as solder, bronze, brass, or steel with a microscope.

A *phase diagram* is a graph showing the parameters in which the phases of a system exist. The parameters could be temperature, percent composition, or pressure of the system. If the system has just one component, the phase diagram might show the temperature and the pressure at which the solid, liquid, and gaseous phases of that component exist. Figure 12–2 shows the phase diagram for a single component system such as water. Note that temperature is plotted as the horizontal axis and pressure is plotted on the vertical axis.

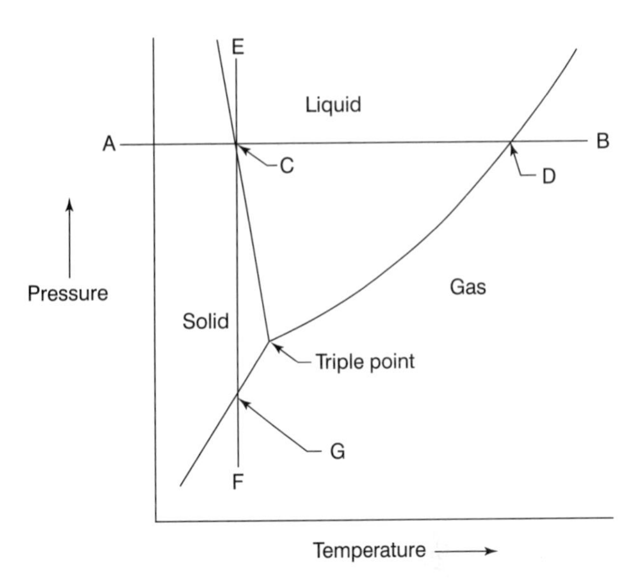

Figure 12-2. Single-component phase diagram for water.

At a given pressure, shown by line A–B on the diagram, there is one temperature (shown at letter C) at which both a liquid and solid can exist in equilibrium with each other, and one temperature (D) at which both the liquid and gas can coexist. If a liquid is cooled along line A–B, at temperature C it will freeze and become a solid. If the pressure is then lowered on the solid, keeping the temperature constant, shown by line E–F, at point G the solid will go directly to a gas without passing through the liquid state. This process is known as *sublimation.* This is the principle by which "freeze-dried" coffee is made.

There is one temperature and pressure at which the liquid, solid, and gas phases can exist in equilibrium. This is known as the *triple point.* For water the triple point is at about 0°C and 0.0886 pounds per square inch pressure. The graph shown in Figure 12–2 is commonly shown for water, but many compounds exhibit similar single-component phase diagrams.

If two components are mixed, then the parameters of the phase diagram are usually percent composition (horizontal axis) and temperature (vertical axis). The two components could be anything—two different chemical compounds or two different metals. When two different metals are mixed, several things might happen. One of the metals might dissolve completely in the other, the first metal going into the other metal either interstitially or as substitutional atoms; the two metals might react chemically with each other to form alloys; or the two metals might form separate phases and just coexist in equilibrium with each other. Very often, when two metals are mixed, the melting point of the mixture is lower than the melting point of either of the components. There is then one composition at which the two-component system has the lowest melting point. This point is known as the *eutectic* composition and temperature. Figure 12–3 shows a typical two-component phase diagram. Note that on either side of the eutectic temperature there is a region between the liquid and solid regions. This is a two-phase region in which both liquid and solids exist together. The regions marked alpha (α) and beta (β) on the phase diagrams are the areas in which one of the metals will dissolve completely into the other metal without separating into two different phases.

The tin-lead two-component system is typical of many mixtures of metals (Figure 12–4). The eutectic point of the tin-lead system is at a composition of 61.9% tin, 38.1% lead. Tin melts at 450°F and lead melts at 621°F, yet the eutectic temperature of this system is only 182°F. Electricians like to use this composition as a solder for joining wires because it has a very low melting point and solidifies from a liquid to a solid without going through a "slushy" two-phase system. On the other hand, plumbers like to use a solder that is not at the eutectic composition to join cast iron pipe since it will go through a two-phase region on cooling. This gives them a little time to "wipe" the joint to produce a smooth surface. Because lead has been found to be detrimental to human health and has been prohibited for use in water systems in recent years, the lead in plumber's solder has been replaced by indium. The tin-indium phase diagram is very similar to the tin-lead diagram, producing similar characteristics in the solder.

The Iron-Carbon Phase Diagram

Steel is defined as an iron-carbon system having less than 2% carbon. It is fundamentally a two-component system, although other alloying elements may be added. The iron and carbon in steel form both mixtures and compounds, depending on the percent of carbon and the temperatures involved. Because the phase diagram for steel (the iron-carbon diagram shown in Figure 12–5) is used so much, many of the points and areas have been given specific names. The eutectic temperature of the iron-carbon system is 2066°F at a composition of 4.3% carbon. This is about

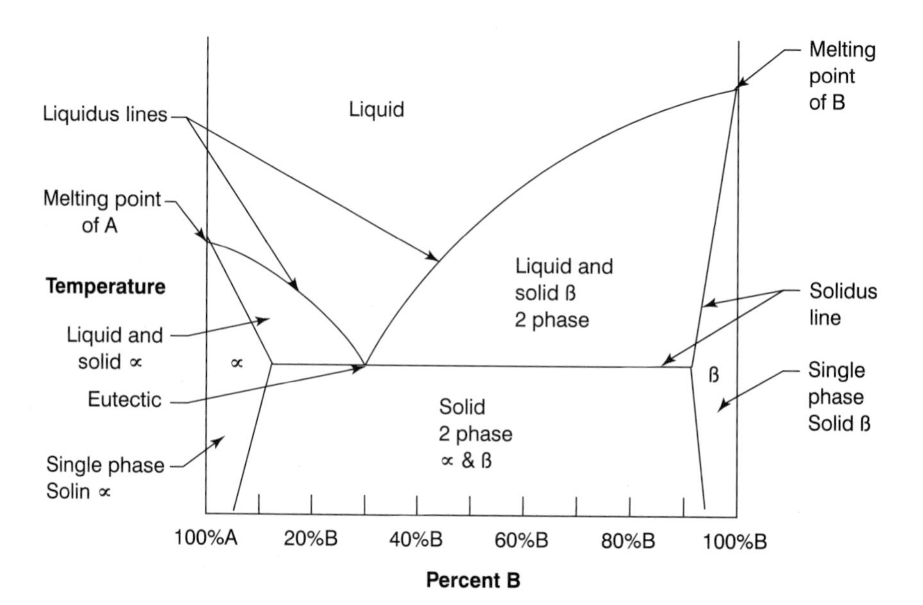

Figure 12–3. Two-component phase diagram.

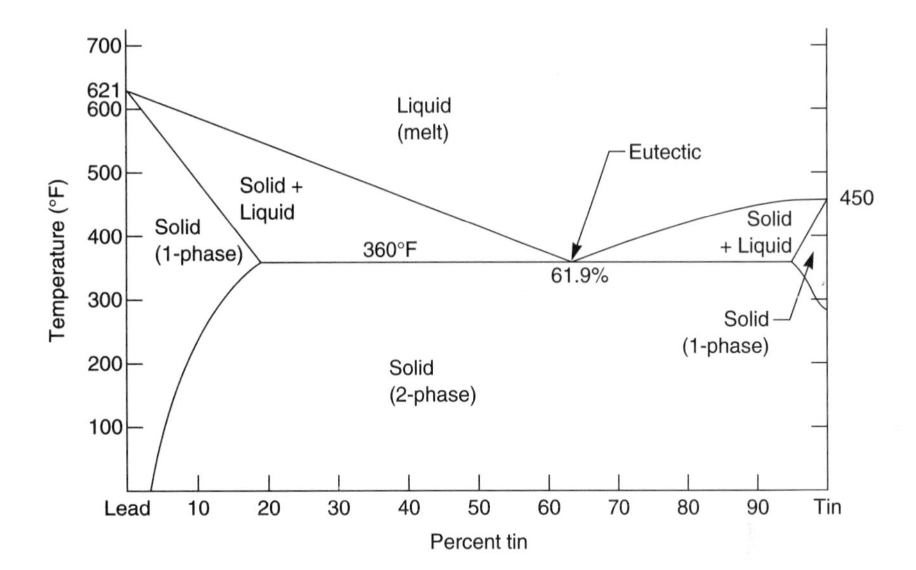

Figure 12-4. Lead-tin phase diagram.

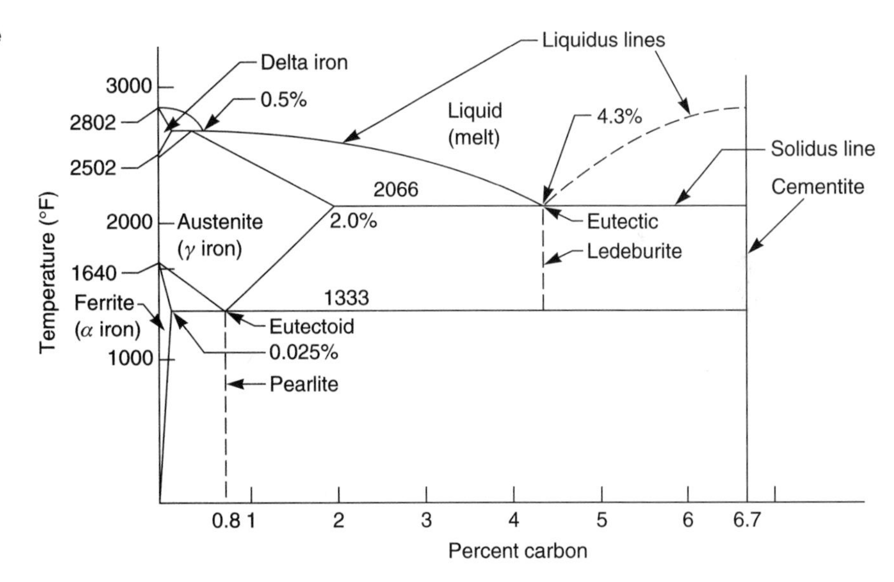

Figure 12-5. Iron-carbon phase diagram.

the composition of cast iron. The eutectic composition is known as *ledeburite.*

The "diamond-shaped" area on the iron-carbon phase diagram, between the temperatures of 1333 and 2800°F with less than 2% carbon, is known as *austenite.* Austenite, also known as *gamma (γ) iron,* has a face-centered cubic structure (Figure 12–6). The lowest temperature and composition at which austenite can exist is 0.8% carbon and 1333°F.

The small triangular region on the left side of the phase diagram having less than 0.025% carbon at temperatures below 1640°F is known as *ferrite* or alpha (α) iron. For all practical purposes, ferrite is pure iron and is body-centered cubic in structure (Figure 12–7). Note also that the tiny triangular region on the 0% carbon line between the temperatures of 2802 and 2502°F is known as delta (δ) iron. At one time it was thought that this region had a different structure than

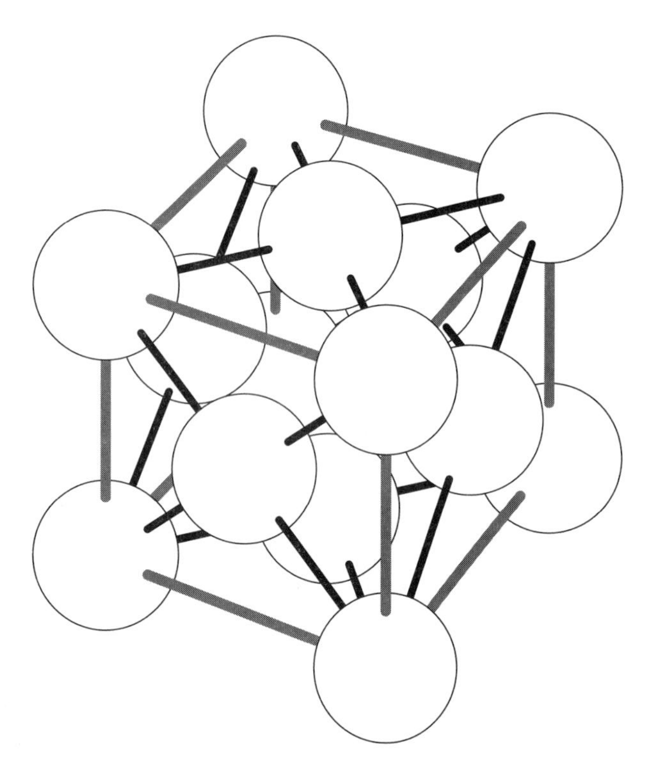

Figure 12–6. Face-centered cubic crystal.

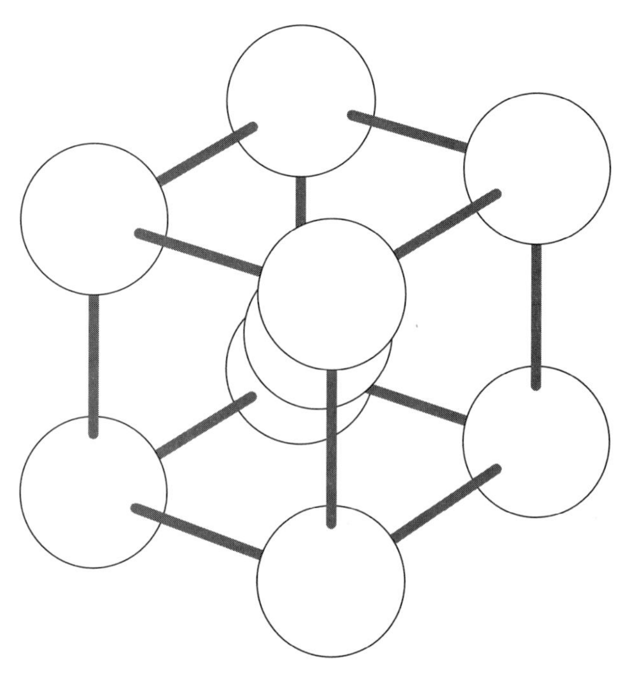

Figure 12–7. Face-centered cubic crystal.

the other phases. In recent studies, delta iron has been found to have the same body-centered cubic structure as ferrite, only at a higher temperature.

At a composition of 6.7% carbon, an intermetallic compound is formed. It has the chemical composition of Fe_3C and is known as *cementite.*

The lowest temperature at which austenite can exist is 1333°F at a composition of 0.8% carbon. This point is known as the *eutectoid.* The general definition of the eutectoid is that temperature and composition at which a single-phase solid goes directly, on cooling, to a two-phase solid. The eutectoid composition of the iron-carbon system is called *pearlite.* Pearlite has a body-centered cubic structure. When pearlite is heated above the eutectoid temperature, it reverts to the face-centered cubic, austenite.

Earlier in this chapter, steel was defined as an iron-carbon alloy having less than 2% carbon. Using the phase diagram, steel could also be defined as an iron-carbon alloy that can be completely converted to austenite when heated.

Although almost all metals including aluminum, magnesium, copper, bronzes, and others can be cold worked to increase slightly their hardness, yield strength, and tensile strength, a few metals can have their hardness, ductility, and toughness *radically* changed by heating and cooling. Steel is one of these materials. The difference between steel and these other metals is that **steel changes its crystal structure on heating.**

In converting the steel from body-centered cubic to face-centered cubic or from face-centered cubic to body-centered cubic, the atoms must break and reform bonds with their neighboring atoms. This requires time. If sufficient time is allowed for cooling of the austenite, it will revert completely to pearlite. However, if the steel is cooled quickly from the austenite, an intermediate *body-centered tetragonal crystal* called *martensite* is formed (Figure 12–8). A martensite crystal has a lot of internal energy due to the stresses formed during rapid cooling and is the hardest structure of steel. Martensitic steel has a Rockwell C hardness of about 66. Pearlite is relatively soft in comparison. The hardness of pearlite changes

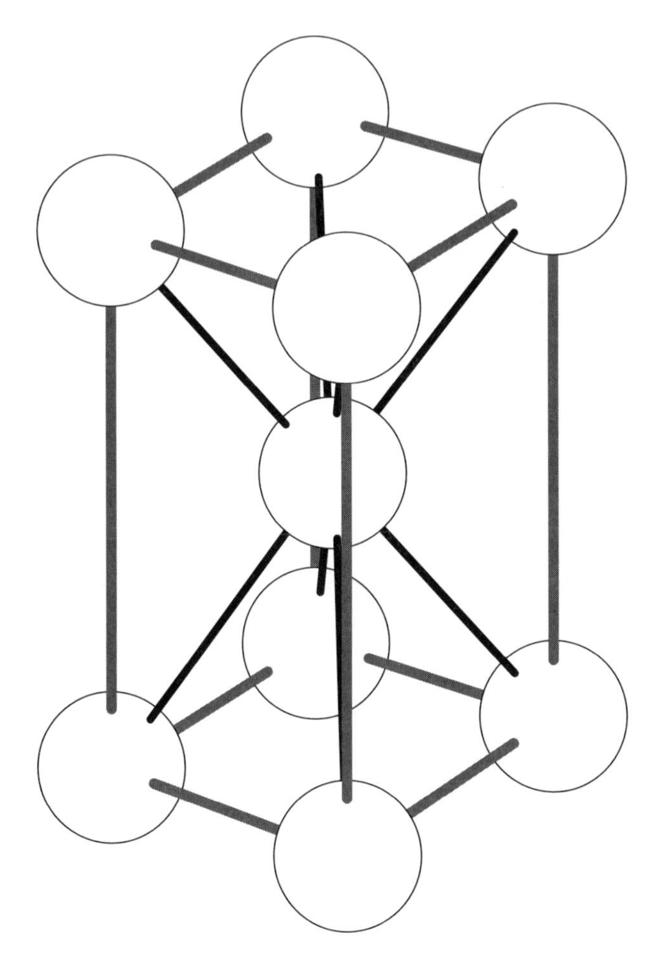

Figure 12–8. Body-centered tetragonal crystal.

with its grain size. Of course, the rapidly cooled martensite is always very fine grained.

The phase diagram also shows a region between austenite and liquid. This is a two-phase region made up of solid austenite and liquid. In this region, solid particles of austenite are suspended in a liquid, forming a "mush" similar to shaved ice in a glass of water. Other two-phase regions occur between the cementite and liquid, austenite and ferrite, austenite and cementite, and ferrite and cementite.

Because pearlite falls between ferrite and cementite, it too is a two-phase system. If a piece of pearlitic steel is polished, etched with a solution of 2% nitric acid in alcohol, then observed with a microscope, alternate layers of ferrite and cementite can be seen. This is an example of a solid two-phase system (Figure 12–9).

Pearlite must have 0.8% carbon. Steels having less than 0.8% carbon, known as *hypo eutectoid steels,* will form some pearlite but also have some ferrite (pure iron) left over. Figure 12–10 shows a low-carbon steel.

Most cast irons have around 4% carbon. Cast irons will therefore have some pearlite and excess cementite. In gray cast iron, the cementite occurs in flakes or platelets. It is easy to understand why cast iron is so brittle since the flakes of cementite are very brittle and have little tensile strength. Basically, cast iron is full of cracks and will break at its largest flaw or piece of cementite. Why is cast iron used instead of steel? Note that at 4% to 4.5% carbon, the melting point of the iron carbon system is nearly 500°F lower than the temperatures required to melt steels. This lower melting point makes cast iron much cheaper to produce. In casting applications where high tensile strengths and ductility are not required, the more inexpensive cast iron is usually sufficient. Cast iron can be made into a much stronger product by alloying and heat treating. In nodular or ductile cast irons, the cementite is made to form little spheres rather than flat flakes. This reduces the small "cracks" to small "holes" and greatly improves the ductility, strength, and dependability of the cast iron. Ductile cast iron is more expensive than gray cast iron.

Time-Temperature-Transformation Curve

The change of austenite to pearlite or to martensite is a time-dependent function. A graph showing the temperature of the steel versus the time it takes to cool it from the austenitic temperature is called the *time-temperature-transformation curve* (Figure 12–11). It is also known as the:

- TTT curve
- Bain S curve
- C curve
- Isothermal transformation curve
- I-T curve

Figure 12-9. Pearlite.

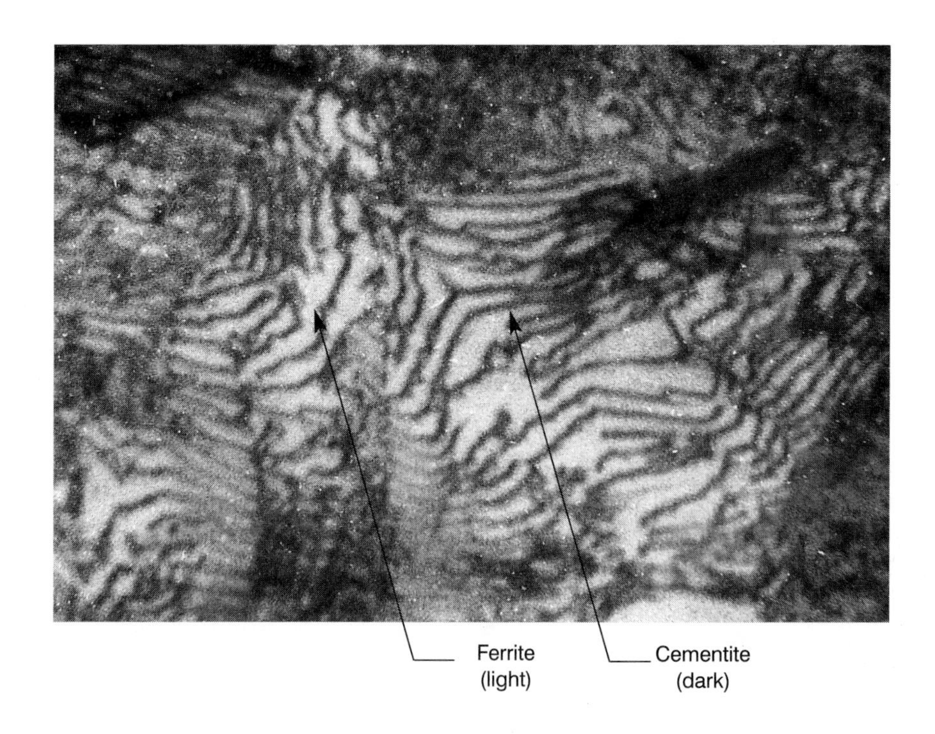

Figure 12-10. Low-carbon steel.
Photograph courtesy of USS Technical Center, Monroeville, Pennsylvania.

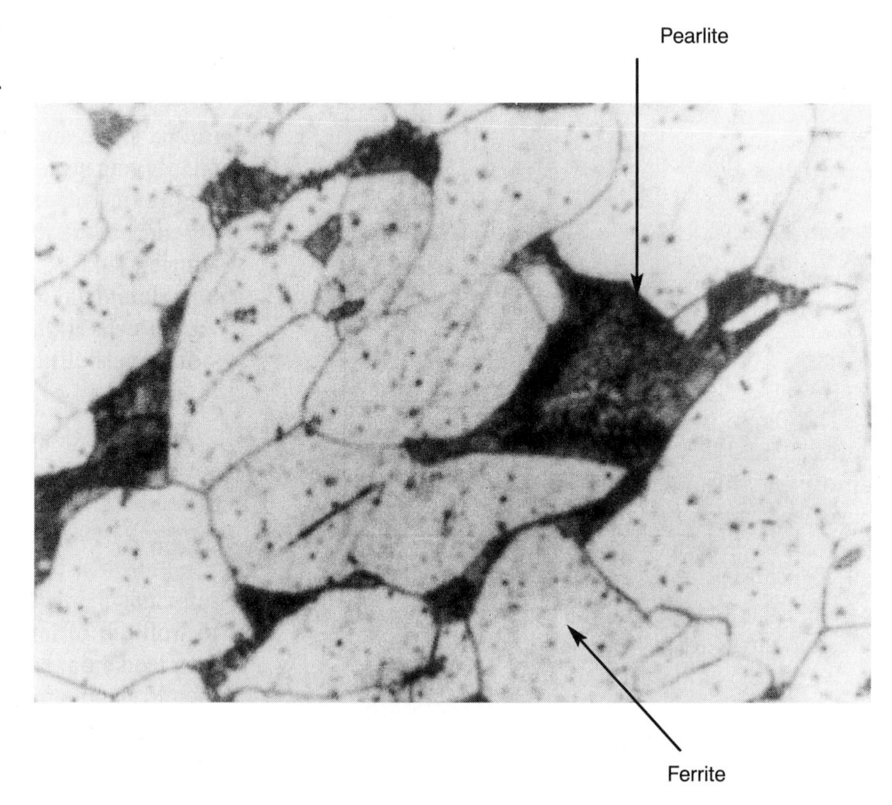

The outer curved line on the TTT curve is called the pearlite start line or Ps line. The inner curved line is the pearlite finish or Pf line. The top horizontal line is the martensite start (Ms) line, and the lower horizontal line is the Mf or martensite finish line.

Once the steel is heated into the austenitic range, it will remain as austenite until it crosses one of the lines on the TTT curve. A grain of austenite can go either to pearlite or martensite, but not both. If a specimen is cooled rapidly enough from the austenitic temperature to miss the Ps line, no soft pearlite grains will be formed and it will start to form martensite when the temperature is at the Ms line. When the temperature is lowered to the Mf line, all of the grains of austenite will have been changed to the hard martensite with no pearlite being formed. This is shown as path A in Figure 12–11. On the other hand, if the austenitic steel is cooled along path C in the TTT curve, the austenite will start transforming to pearlite at the Ps line and be completely transformed to pearlite at the Pf line. This cooling rate can form no martensite since all of the grains of austenite have already changed to pearlite. Once the grains have passed the Pf line and are still at relatively high temperatures, the grains will become larger, providing larger slip planes, which improves the ductility and softens the steel a little further.

A cooling rate along path B in Figure 12–11 crosses the Ps line but not the Pf line. Therefore, some of the grains of austenite will go into pearlite but those remaining as austenite as the temperature crosses the Ms line will be converted to martensite. This will result in a fine-grained steel with some martensite and some pearlite. Its hardness will be between that of fine-grained pearlite and martensite.

Alloying other elements besides iron and carbon with steel modifies the steel in several ways. One effect of alloying chromium, vanadium, nickel, molybdenum, or other similar elements with steel is to shift the TTT curve to the right. This allows more time to cool the steel from the austenitic temperature and still allow only martensite to form. In a plain carbon steel such as a 1060, the austenite must be cooled below 600°F in less than 0.5 second to produce 100% martensite. In some *alloyed* steels, as much as 100 seconds can be allowed when cooling the steel from the austenitic stage and pure martensite will still form. Figure 12–12 shows a TTT graph with curves for both a plain carbon and an alloyed steel.

Figure 12-11. Typical TTT curve.

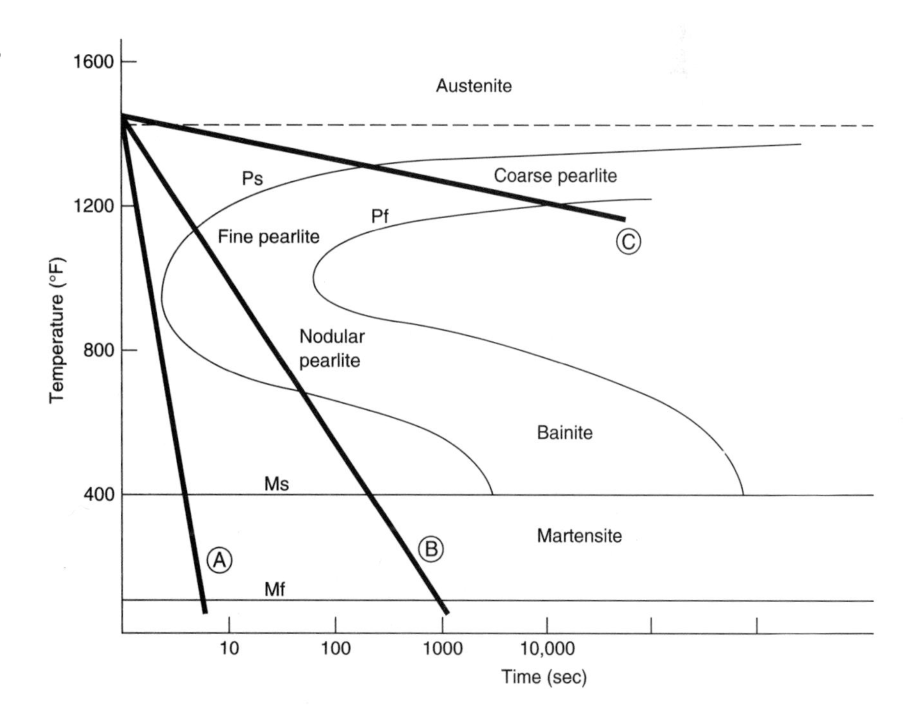

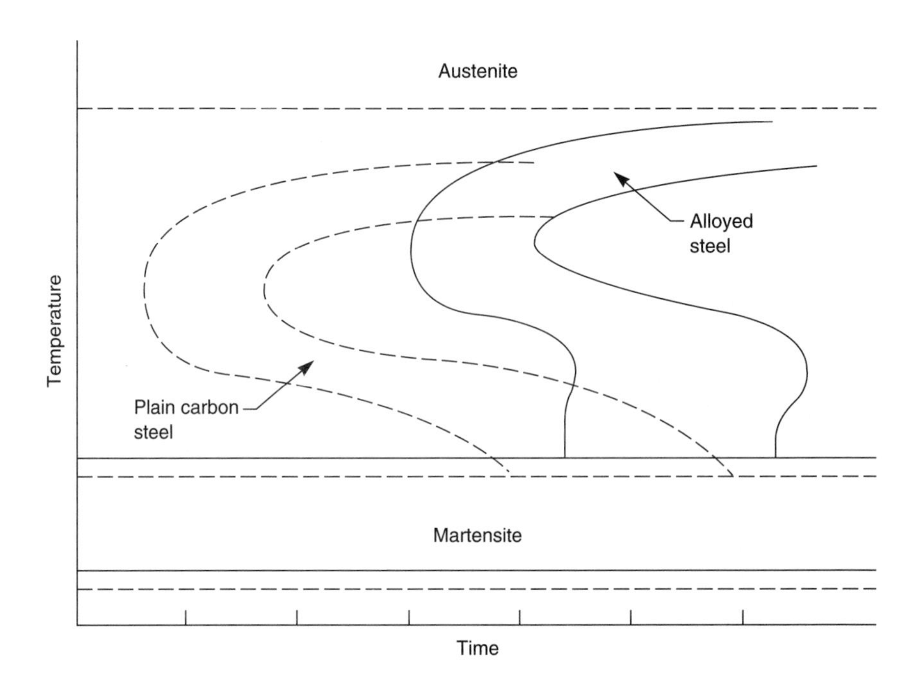

Figure 12-12. TTT curves for plain and alloyed steels.

By using the information gained from the iron-carbon phase diagram and the TTT curve for an individual steel, several different standard heat treatments have been developed. These can be divided into three major categories of heat treatments:

1. Methods of softening steels
2. Methods of hardening steels
3. Methods of modifying the properties of steels

The methods of modifying the properties of steels include Tempering, Spheroidizing, Martempering, Austempering, and Patenting which will be discussed later in this chapter.

Methods of Softening Steels

Full Annealing

Full annealing softens a steel to its softest possible condition. To full anneal a steel it must be heated 50°F into the austenitic region. The heating rate should be slow enough to prevent warping or distortion of the steel. High-carbon steels should never be placed directly into the hot furnace. They should either be preheated in a separate furnace, or placed in a cool furnace and the temperature of the steel raised as the oven is heated. The parts are then held at the austenitic temperature for one-half to one hour per inch of thickness to allow the part to stabilize throughout as austenite. The steel is then very, very, very (put a few more very's in if you wish) slowly cooled. Full annealing a steel often requires from three days to a week to complete. Note from the iron-carbon phase diagram (Figure 12–5) that low-carbon steels must be heated to a higher temperature than high-carbon or eutectoid steels. The annealing temperature of a 1020 steel would be around 1650°F, whereas a 1080 steel would need a temperature of about 1450°F.

Besides creating all pearlite, full annealing also gives the steel an extremely large grain size. However, the equipment for full annealing is expensive and often not available. Controls on the ovens that lower the temperatures a few degrees per hour are required. It is a slow and, therefore, expensive process. Steel mills usually are equipped for full annealing, but many smaller heat treatment shops use another method. Figure 12–13 is a picture of a relatively large heat treatment furnace. This is a soaking-pit type installation in which the furnace is mounted on a movable

Figure 12-13. Heat treatment furnace.

track and blows the flame and heat downward into the hole containing the steel to be annealed.

Box Annealing

In *box annealing,* the steel is placed in a box furnace, which is a well-insulated oven. Large box furnaces can be made at relatively low cost. The steel is again heated 50°F into the austenitic region, held at that temperature for one hour per inch of thickness, then the oven is simply turned off. Generally, about 24 hours are required for the oven to cool to room temperature, at which time the oven doors can be opened and the steel removed. Box annealing does not produce quite as soft a steel nor grains as large as a full anneal, but it is much easier and less expensive to do. Box annealing is usually sufficient for most manufacturing requirements.

Bright Annealing

One of the drawbacks to annealing a steel is that the surface of the steel becomes oxidized and scaly. This is especially true for plain carbon steels or if the steel is held at the high temperature too long in the oven. A method has been developed to prevent this. *Bright annealing* is a form of box annealing that is done in a muffle furnace. A muffle furnace can be sealed after the steel is placed in it and the air pumped out and replaced with an inert gas. Helium or nitrogen is often used. The steel is then annealed using the box annealing procedure just outlined. The steel emerges from the oven as shiny as when it started. Using nitrogen in the muffle furnace can produce a blue color on the surface of the steel since the nitrogen reacts with the iron and carbon to form a ferric ferrocyanide coating on the surface. Ferric ferrocyanide is known as Prussian blue. This finish is discussed in the chapter on finishings as a method of protecting the surface of a steel against further corrosion. An old method of bluing guns used this technique.

Normalizing

Normalizing is a technique used prior to other heat treatments or to give the steel good machinability. Steel is normalized by heating it 100°F into the austenitic region, holding it there for one hour per inch of thickness, then removing it from the oven and allowing it to air cool at room temperatures. Normalizing does not produce an extremely soft steel, but it does give the steel an even grain size throughout the part. This even grain size gives the steel its workability. Steels with uneven grain sizes will have hard spots in them and will not form extremely smooth surfaces when machined.

Note that hot-rolled steels are self-normalizing. Steels are in the austenitic range while they are red hot. Although the grains are broken up while being hot rolled, the normalizing process allows regrowth of the grains. Cold-rolled steels will have uneven grain sizes and should be normalized prior to other heat treatments or machining.

Methods of Hardening Steels

To obtain the maximum hardness in a steel, martensite must be produced. This involves heating the steel about 100°F into the austenitic or red hot region and rapidly cooling it. For a medium- to high-carbon steel the temperature should be between 1450 and 1550°F. To get the hardest steel, the cooling rate must be rapid enough to miss the "knee" of the pearlite start line on the TTT curve (Figure 12–11). The cooling can be accomplished by several methods.

Quenching

The faster a steel can be cooled from the austenitic temperature, the more martensite will be formed and the harder the steel will get. To remove the heat quickly, the steel must be immersed in a liquid that has a high heat capacity. One of the best is brine or saltwater. However, plain tap water works almost as well and is cheaper and easier to obtain. A water quench, however, may be too severe for delicate parts and could crack them. This is especially true for parts that have thin areas that cool and shrink faster than adjacent thick areas. In this case an oil quench may be required. Oil quenches do not produce quite as hard a steel as a water quench of the same steel, but then the parts aren't cracked in the rapid cooling either (Figure 12–14). It is a trade-off. The part is quenched as rapidly as possible without causing breakage.

If the part is simply placed in the cooling bath and held there motionless, a steam jacket will form about it. The steam jacket becomes an insulator and hinders the cooling of the part. Therefore, the part should be rapidly agitated in the cooling medium or the coolant should be sprayed or pumped rapidly past the steel. This prevents the buildup of the steam jacket and quenches the steel to its maximum hardness.

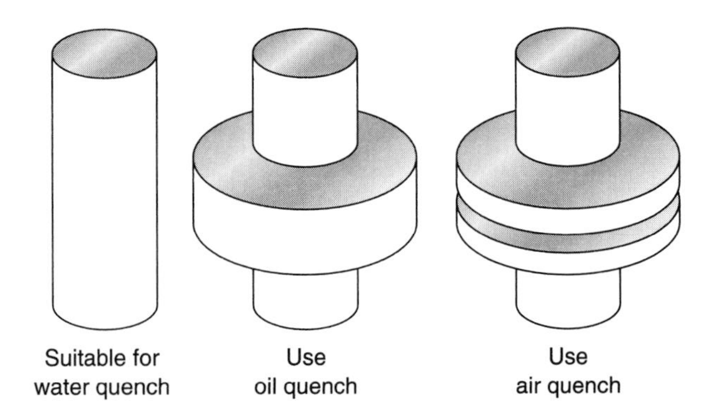

Figure 12-14. Types of quench problems.

In some very delicate parts even an oil quench is sufficiently severe to break it. Here about all that can be done is an air quench. In the air quench, the part is removed from the oven and placed in front of a fan that can deliver high-velocity air against the part.

Surface Hardening

If a steel is hardened all the way through the part, it will be brittle. In parts that have wearing surfaces such as gear teeth, shafts, lathe beds, and cams, only the surface of the part should be hardened so as to leave the inside soft and ductile. In this condition, the surface of the steel would withstand the wear while the interior of the part could absorb any impact without breaking. Several methods are available for surface hardening.

Flame Hardening. In *flame hardening,* the surface of the part is quickly heated into the austenitic range by a torch or other locally applied flame. The surface must be heated quickly enough so that the interior of the part does not reach the red heat, signifying that it is in the austenitic range. The part is then immediately quenched by immersing it into water or oil, or a jet of water or oil is sprayed on the surface of the part. The depth of the hardened layer will vary from one-sixteenth to one-quarter inch. Remember, however, that only medium- and high-carbon steels (0.35% carbon or more) will significantly harden. Therefore, flame hardening is only applicable to medium- or high-carbon steel parts. Flame hardening is illustrated in Figure 12–15.

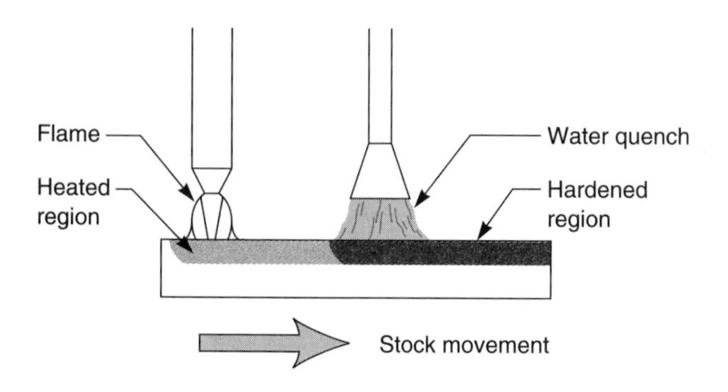

Figure 12-15. Flame hardening.

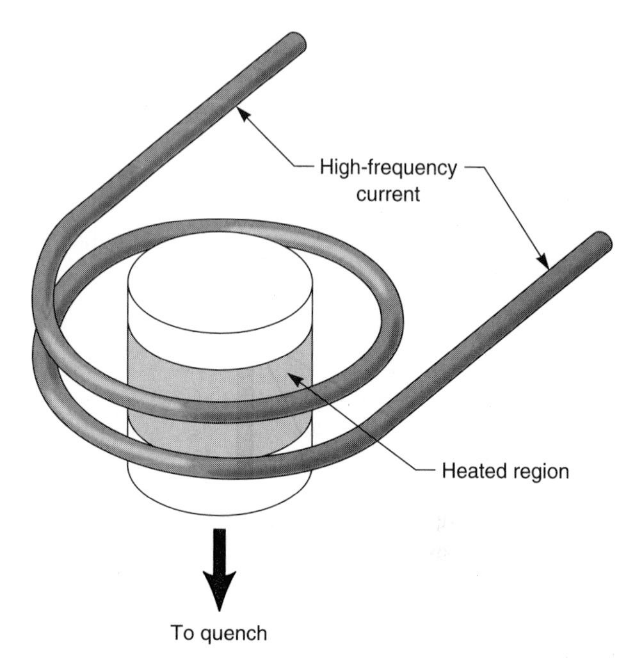

Figure 12-16. Induction hardening.

Induction Hardening. A variation of flame hardening is *induction hardening.* The part to be hardened is placed in a coil of wire through which a high-frequency electric current is run. The alternating current induces a current of electricity to flow in the part in the same way that a current is generated in a transformer. The induced current heats the surface of the part into the austenitic range from which it can be hardened by quenching. As with flame hardening, induction-hardened parts must be made of medium- or high-carbon steels. Figure 12–16 depicts induction hardening.

CHEMICAL PROCESSES

Case Hardening

If a low-carbon steel is needed to give toughness to the workpiece, its surface cannot be significantly hardened. There are techniques whereby carbon or nitrogen can be added to the surface of the part to allow the surface to be hardened. The method, known as *case hardening,* includes *carburizing* (also called *carbonizing*), *nitriding,* and *carbonitriding* or *cyaniding.*

Carburizing

If a low-carbon steel is placed in a high-carbon atmosphere, the carbon will diffuse into the outer layers of the steel. The longer the steel is left in the carbon, the deeper the penetration of the carbon. Several methods can be used to add the carbon steel. One popular technique is to pack the steel part in a mixture of barium carbonate and graphite or charcoal. This technique is known as *pack carburizing.* The entire pack is then heated to 1700°F and left for several hours. A rule of thumb is that the carbon will penetrate the steel at a rate of about 0.005 inch per hour. To obtain a one-sixteenth- (0.0625)-inch-thick case, an immersion time of about 12.5 hours would be required. This figure varies with the type of steel and the temperatures used. The surface of the steel will have enough carbon to allow it to be hardened by quenching, but the low-carbon core will remain ductile. Please note that adding carbon to the surface does not automatically harden the steel. The steel must also be heated into austenite then quenched to produce the hard surface.

One word of caution about carburizing. It should not be done in an electric furnace where the electric coils are exposed. The carbon used to form the case can be absorbed by the nichrome wire heating coils. This will embrittle the coils, causing them to burn out and shatter. Gas furnaces with neutral or oxidizing flames are preferred.

There is an interesting background to the case hardening of steels. Long before the principles of case hardening were known, blacksmiths would heat knives, swords, and other cutting tools to a red hot condition then let them burn their way into a wood log or bury them in sawdust to increase the hardness of the blades. The carbon in the wood, in effect, case hardened the steel.

Gas Carburizing

One disadvantage of pack carburizing is that it is a batch process. As such, it is time consuming and expensive. A second method of adding carbon to the surface of a steel is *gas carburizing.* In this technique the part is placed in a furnace containing carbon monoxide, methane, ethane, propane, or a mixture of these gases. The part is then heated. The rate of absorption of carbon is much higher with this technique. A 0.04- to 0.05-inch case can be obtained in about 4 hours with this technique. The case thickness is more even and the thickness is easier to control than with the pack method.

Nitriding

Both carbon and nitrogen can be used to increase the hardness and hardenability of a steel. Ammonia gas (NH_3) can be used to add nitrogen to the surface of the steel using the same technique as gas carburizing. In *nitriding,* the steel is heated only to 950 to 1000°F but requires 50 to 100 hours to produce a case. The advantages of nitriding lie in the fact that it is done at temperatures that do not distort the steel and it *requires no quenching* to produce the hard surface. Nitrided steels will also retain their hardness and wear resistance at higher temperatures than carburized steels. The disadvantages of nitriding are the time required, the resultant cost, and the specialized training needed for the people doing the nitriding.

Carbonitriding (Cyaniding)

If only a thin case of high-carbon steel is required, *carbonitriding* or *cyaniding* is the preferred technique. The part is immersed in a bath of liquid sodium cyanide (NaCN). The temperature of the bath is held between 1500 and 1650°F. The diffusion rate of the carbon and nitrogen is very rapid. Case depths of 0.01 inch/hour of immersion can be obtained. Quenching should be done directly from the cyanide bath. For case depths of more than 0.03 inch, gas or pack carburizing is more practical. This is an inexpensive method for mass production. Figure 12–17 shows a section of a case-hardened wrist pin from an internal combustion engine.

Of course, there is one problem with cyaniding—hydrogen cyanide fumes are extremely poisonous. As long as the sodium cyanide is kept chemically basic, that is, with a pH greater than 7.0, no hydrogen cyanide will be emitted. It is only when the cyanide becomes acidic that problems arise. (Acids have a pH range from 0 to 7. In bases the pH ranges from 7 to 14. A pH of 7.0 is neutral. Pure water has a pH of 7.0.) It is customary to add sodium carbonate to the solution to slow down the decomposition of the sodium cyanide. Further, the baths must be well ventilated and fume hoods are essential. Instrumentation to indicate the pH level should be installed.

The main difference between flame hardening and case hardening is that *in flame hardening a high- or medium-carbon steel is used* with only the outside being hardened. *In case hardening a low-carbon steel is used*

Figure 12–17. Case-hardened wrist pin.

with carbon or nitrogen being added to the outside. In either case only the surface of the steel is hardened.

METHODS OF MODIFYING THE PROPERTIES OF STEELS

Tempering

No steel should be left in its "as-hardened" condition. Quenched steels have internal stresses that make the steel brittle. They must be tempered to relieve these stresses. Steels are tempered by heating the quenched part back to a temperature between 200 and 1200°F, leaving the part at that temperature for one hour per inch of thickness, then cooling it in air at room temperature. The temperature of tempering depends on the type of steel and the amount that the hardness is to be drawn back.

Table 12–1 is a table of the approximate hardness that can be obtained from various carbon contents and tempering temperatures for plain carbon steels. Alloying of various elements can significantly change the hardness of the steels.

Spheroidizing

One of the problems encountered with high-carbon, hardened steels is that they are generally brittle. Much of the brittleness is caused by the grains having sharp corners on them (Figure 12–18). Studies have shown that the grains will be changed into small spheres if the steel is heated to a temperature just under 1300°F and held at that temperature for at least an hour per

Table 12–1. Rockwell C Hardness of Tempered Steels

Carbon Content	As Quenched	Tempering Temperature (°F)			
(%)		200	600	1000	1200
0.35	55	52	42	27	20
0.40	60	56	47	32	23
0.60	62	60	52	34	25
0.70	64	63	53	37	26
0.80	66	65	55	39	30

Figure 12–18. Nonspheroidized steel.

Photograph courtesy of USS Technical Center, Monroeville, Pennsylvania.

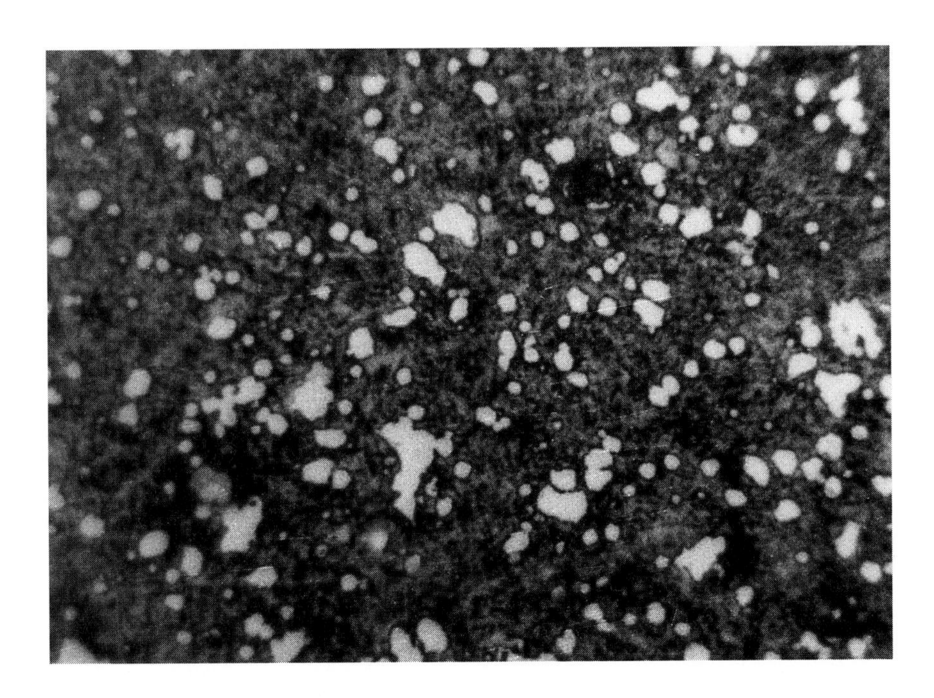

Figure 12–19. Spheroidized steel.

inch of thickness. Higher carbon steels require longer times at that temperature to complete the spheroidization. *Spheroidizing* improves the machinability and the toughness of the steel. Figure 12–19 shows a spheroidized steel.

Martempering and Austempering

One of the problems encountered in quenching a steel is that the outside of the part cools at a faster rate than the inside. This produces a nonuniform grain structure at the time that martensite is to start forming. The nonuniform grain size will weaken the part. Figure 12–20 shows the TTT curve for a normally quenched and tempered steel. Two techniques have been developed to avoid this problem: martempering and austempering.

In *martempering,* a steel is heated into the austenitic range just as in a normal quenching operation. The part is then quenched to a temperature just above the point at which martensite will start to form. This is usually about 500 to 600°F—just above the Ms line. The steel is held at that temperature for a few seconds to allow the inside and outside to stabilize at a uniform temperature. This is done by removing the steel from the austenizing furnace and quickly placing it in a furnace at a temperature of about 500 to 600°F. From here it is quenched to its final temperature and tempered to remove residual stresses. The part should not be held at the intermediate temperature long enough to allow it to pass through the Ps line. If the Ps line is crossed at this temperature, bainite will start to form, which prevents the maximum formation of martensite. Refer to Figure 12–21 for the TTT curve of the martempering process.

Austempering is a similar technique. The steel is heated into the austenitic range and quenched to about 700°F then held at that temperature from 20 minutes to 3 hours depending on the type of steel. After this time the part is quenched. In this technique the grain structure of the steel will be entirely bainitic. See Figure 12–22 for the TTT curve of an austempered steel. Both martempered and austempered steels are stronger than normally quenched and tempered steels.

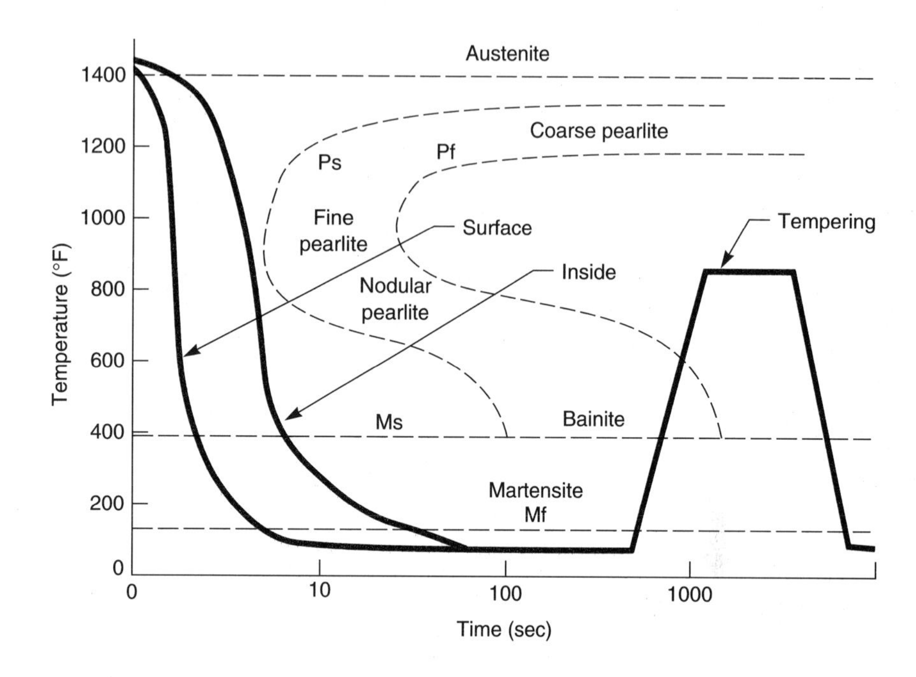

Figure 12–20. Normal quenched steel cooling rates.

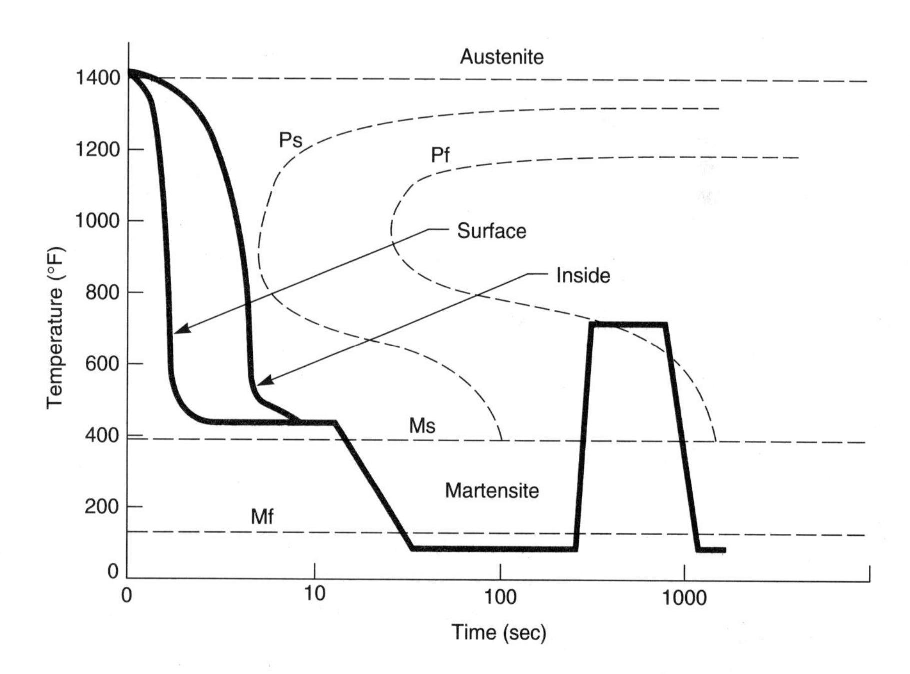

Figure 12–21. Martempered steel cooling rates.

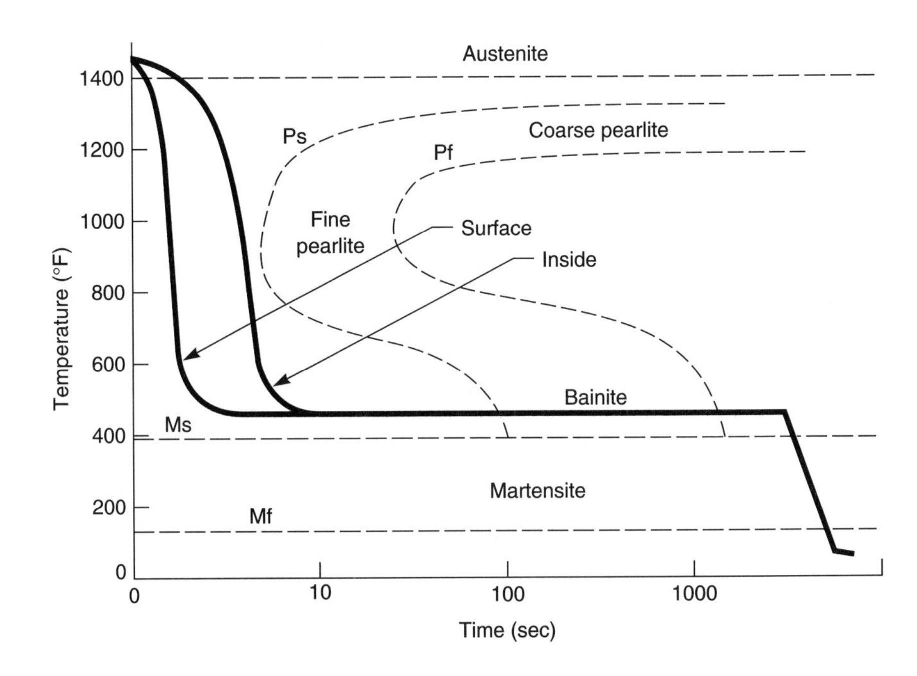

Figure 12–22. Austempered steel cooling rates.

Patenting

A process very similar to austempering is *patenting.* In patenting the austenized steel is removed from the oven and immediately quenched in molten lead. Since lead melts at 621°F, the effect is much the same as that of austempering. Patenting is used prior to drawing the steel out into wire because it gives the metal good ductility.

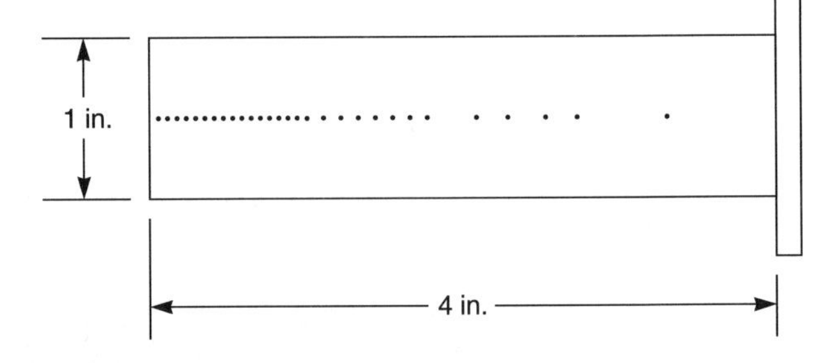

Figure 12–23. Jominy bar.

■ JOMINY TEST

A test to determine the hardenability of a steel is the *Jominy test.* Do not confuse the terms *hardness* and *hardenability.* They are two different concepts. Hardness is a resistance to surface deformation, whereas hardenability is the depth to which a steel can be hardened. In conducting the Jominy test, a standard Jominy bar (Figure 12–23), 1 inch in diameter and 4 inches long, is used. It has a steel ring attached to one end. The bar is heated to 1650°F in an oven, then placed in the Jominy bucket as seen in Figures 12–24(a) and (b). A stream of water is played against one end until the entire bar is cool (usually about 20 minutes). The bar is removed from the bucket and a flat area about one-quarter inch wide is prepared along the length of the side by using a water-cooled sander. Rockwell C hardness tests are taken every sixteenth of an inch for the first inch, every eighth of an inch for the second inch, then every quarter of an inch for the last two inches. A graph of the Rockwell hardness readings versus the distance from the quenched end is then prepared as in Figure 12–25. The Jominy bar represents a cross section of any steel piece that has been quenched.

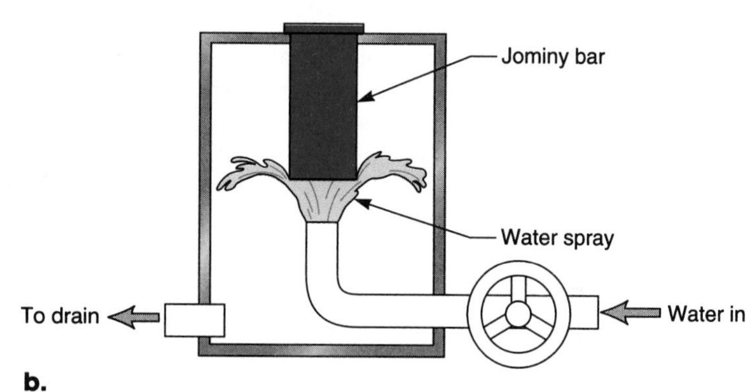

a. b.

Figure 12–24. **a.** Jominy bucket. **b.** Jominy test.

The Jominy (hardenability) depth is defined as the distance from the quenched end at which 50% martensite will be formed. In a plain carbon steel, this depth may be as low as an eighth of an inch. The reason for this is obtained from the TTT curve of Figure 12–11. The rapidly quenched end has a cooling rate that misses the "knee" of the Ps line on the TTT curve. A point a few sixteenths up the bar from the quenched end would have a slower cooling rate, one that would allow the formation of some pearlite. A point an inch or two from the quenched end is cooled at a rate slow enough to allow all of the austenite to be changed into pearlite and no martensite is formed. The alloying of various elements such as chrome, nickel, vanadium, or molybdenum with steel shifts its TTT curve to the right, allowing more time for the

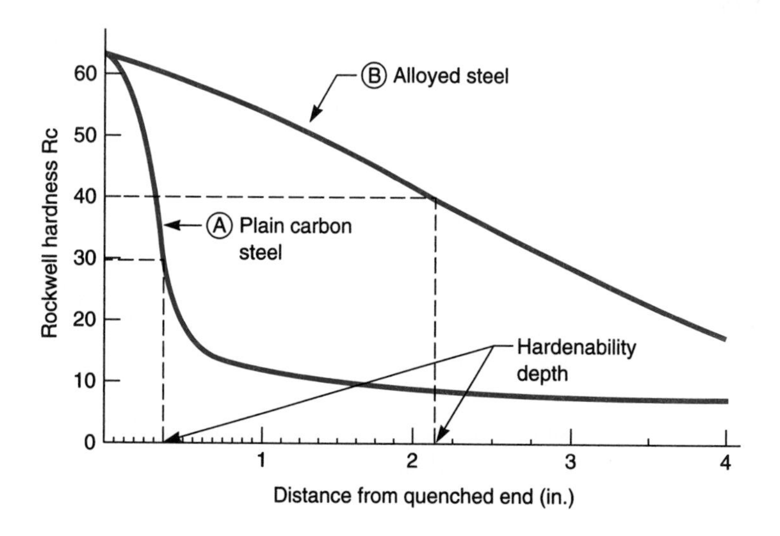

Figure 12–25. Jominy hardness graph.

formation of martensite. Line A in Figure 12–11 would be for a plain carbon steel, whereas line B in the graph represents an alloyed steel.

The Jominy graph must be considered in the selection of a steel to be used where wear resistance is important. If a plain carbon steel was used on a bulldozer blade, earth grader blade, or plow blade, for instance, it could only be hardened to a depth of an eighth of an inch or two. Once the outer hardened region was worn off, the soft, inner steel would be more rapidly worn away and the useful life of the blade shortened. On the other hand, an alloyed steel could be hardened much deeper and a longer life guaranteed for the blade.

SELECTION OF HEAT TREATMENTS

The procedure for selecting a steel that must be hardened is as follows:

Step 1. Decide on the properties needed for the core and for the surface of the part.

Step 2. Select a carbon content and alloying elements that will give the desired properties for the core of the part.

The carbon content of the surface should be sufficient to produce a Rockwell C hardness from 3 to 5 points harder than the final product requirements.

Jominy graphs should be consulted to help determine the alloying elements needed. Corrosion resistance, machinability, and other properties should also be considered in selecting these alloying elements.

Step 3. Anneal, normalize, or soften the part to permit fabrication of the part (if necessary).

Step 4. Fabricate the part to rough but oversized tolerances.

Step 5. Select the method of hardening. Case hardening may be necessary for low-carbon steels.

Step 6. Fully harden the steel part.

Step 7. Temper the part by selecting a temperature that will draw the hardness back to the desired final product.

Step 8. Finish the part by reducing the dimensions to the design limits, and applying the desired finish.

HEAT TREATMENT OF NONFERROUS MATERIALS

Annealing

Although many nonferrous metals do not change their crystal structure, the grain sizes of these metals can be altered by heat treatments. Large grains produce soft metals; fine grains result in harder ones. Cold-worked nonferrous metals such as copper, magnesium, aluminum, and zinc are hard and brittle. Often in the forming of parts such as battery cases, cups, automotive body parts, and the like the pieces of metal can only be deformed or bent so far before they will fatigue, tear, or break. In these instances the metal must be deformed part way, placed in a furnace, and annealed. In the annealing process, new grains form and grow, replacing the old distorted small ones. The partially formed product can then be deformed further and the process repeated. This is especially true for forged parts. The fabrication of many parts requires several steps of forging and annealing before the final product is attained.

Precipitation Hardening

If an alloying element is added to a metal, it may form a second phase in the solid. This two-phase system is similar to an oil-water mixture: The oil and water phases just don't mix. If the oil-water mixture is severely shaken, the phases will be broken into minute particles so fine that the entire mix will seem like a single cloudy substance. It takes time for the oil and water phases to separate again.

Consider a metallic two-component material such as an aluminum-copper alloy. At low temperatures, the aluminum and copper exist as two phases, much like the oil and water analogy. If the alloy is heated to around 1000°F and rapidly cooled, the two phases are not immediately precipitated. Over a period of a few months, the copper will begin to emerge as tiny bits of impurities within each grain. These impurities interrupt the slip planes of the once ductile aluminum, thereby hardening it. This process is known as *precipitation hardening* (Figure 12–26).

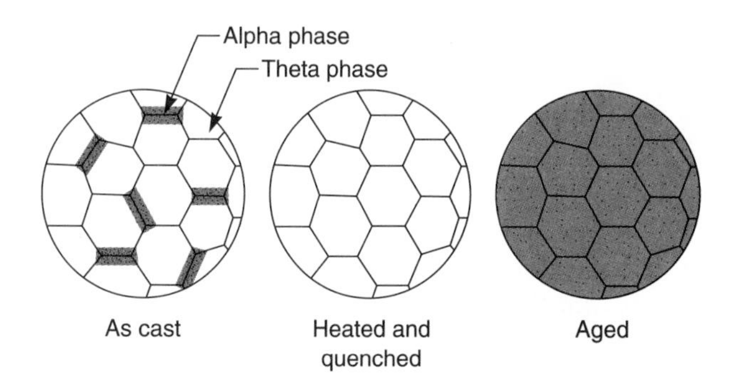

Figure 12-26. Precipitation hardening.

Precipitation hardening is also known as *age hardening* because it will happen at room temperature given a time span of several months. The higher the temperature, the faster age hardening occurs. Conversely, the process can be slowed by keeping the material cool. Aluminum rivets and other parts that must be fabricated while ductile are often kept refrigerated until it is time for their use.

Solution Heat Treating

It is possible to speed up this age hardening process to get to the more stable, harder state of the metal quickly. This can be done by heating the quenched alloy back up to about 200 to 400°F (for aluminum alloys) and holding it there for an hour or so, depending on the thickness. The metal is then quenched for maximum hardness. This method of accelerating the age hardening process is known as *solution heat treating.*

Some stainless steels and other alloyed steels can also be hardened by solution treating. Parts can be machined while in the soft state. The finished part can then be hardened by solution treating.

PROBLEM SET 12-1

1. Why can a piece of high-carbon steel be hardened significantly, while a similar part made of aluminum, magnesium, or copper cannot?
2. You have an annealed part made of high-carbon steel. What heat treatment(s) could be used to harden its surface?
3. What is the purpose of the Jominy test?
4. What information can be obtained from a TTT curve for a given steel?
5. Outline the steps by which a file could be made into a hunting knife blade.
6. You have a piece of low-carbon steel. Outline a procedure whereby the surface could be hardened.
7. Aluminum rivets may become age hardened to the point where they are not usable in as little as six months. What steps should be taken in a factory to keep usable aluminum rivets on hand?
8. Why must hardened parts always be tempered before use?
9. Compile a list of six items that require hardened parts for their successful operation.
10. Lathe beds and milling machine beds are usually flame hardened. Why is this necessary?
11. List the reasons why flame hardening a table for a band saw would be better than using a high-carbon steel and quenching it.
12. What is "work hardening" of a metal and what can cause it?
13. List three applications in which work hardening a steel would be beneficial. List three applications in which it is detrimental.
14. How is work hardening removed from a metal?
15. Refer to the iron-carbon phase diagram (Figure 12–5) to answer the following questions:
    a. What is the eutectic composition of iron and carbon?
    b. What is the eutectoid composition of iron and carbon?
    c. What is the composition of cementite?
    d. What is the melting temperature of steel with 0.5% carbon?
16. What is the designation of a nickel steel having 0.60% carbon? Is this a high-, medium-, or low-carbon steel?

In this chapter several methods of changing the internal conditions of a material have been discussed:

- The theory underlying the cold working and thermal methods of changing the hardness of metals. Included in the theory are the concepts of the iron-carbon phase diagram and the time-temperature-transformation curve.
- Mechanical methods of hardening metals such as cold working.
- Methods of softening steels including full annealing, box annealing, bright annealing, and normalizing.
- Methods of hardening steels including brine, water, oil, and air quenches.
- Flame hardening and induction hardening.
- Case hardening methods including carbonizing, nitriding, and carbonitriding (also known as cyaniding).
- Tempering of steels.
- Austempering and martempering.
- The Jominy test and its applications.
- Heat treatments for nonferrous metals including annealing, cold working, precipitation hardening (age hardening), and solution heat treatments.

SECTION

VI

Material Joining

INTRODUCTION

We want to join two parts together. These parts may have to be joined permanently or only temporarily. The joint may have to allow one part to move over the other, or the parts may need to be rigidly fixed together. In this section, various methods of joining two workpieces to meet the required specifications are discussed.

The topics in this section could be diagrammed as shown in the chart on this page.

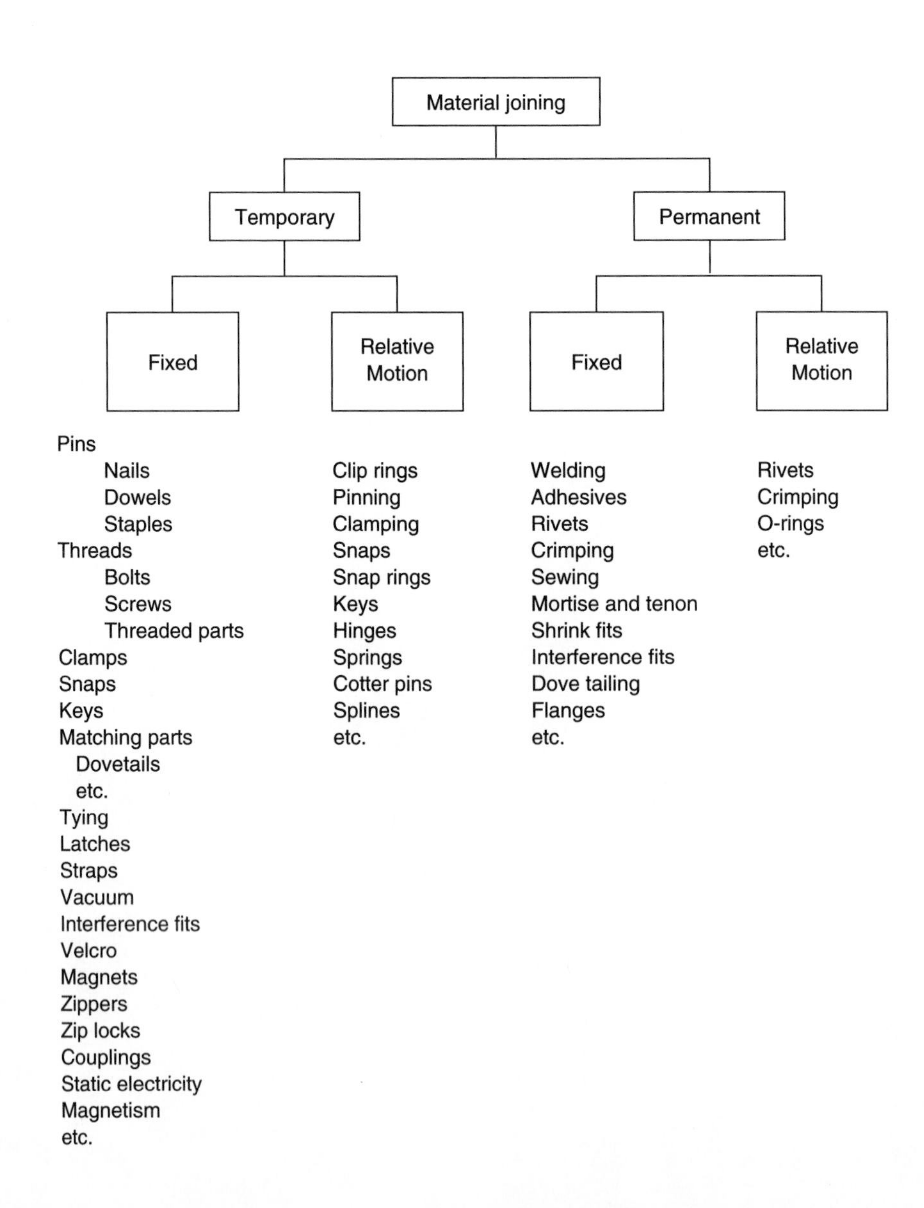

Note that in this chart some of the methods of joining appear in more than one category. For instance, we could use pins to hold two pieces in a fixed position or, used in a different way, to allow relative motion between the two parts. Figure VI–1 illustrates this concept.

Of course, the list from the chart is not complete. Let your imagination run free. With a little thought, you could add many items to this list. Only a few of the methods used in industry are discussed in this section.

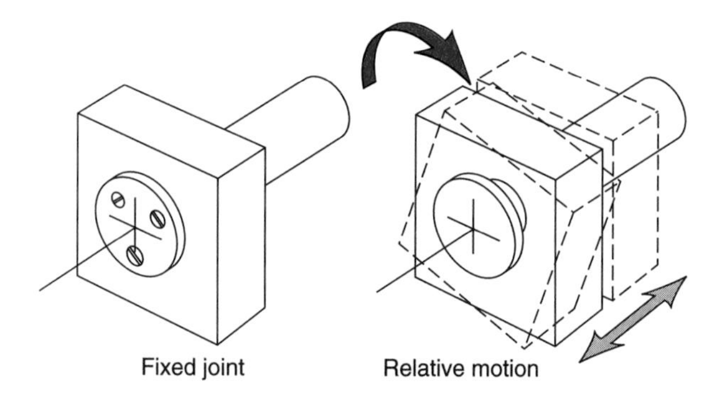

Figure VI–1. Relative motion and fixed joint.

Chapter 13

Adhesives

Adhesive joining can be defined as a method of joining whereby the two materials are held together by a chemical, thermal, or mechanical bonding agent. These vary from library paste to super glues and epoxies and include solders and brazes. In adhesives, the parts are held together mechanically by the glue. The individual parts do not become one piece as in welding. Only the ability of the adhesives to "grab" and hold onto the individual pieces keeps them together. Some adhesives, however, have greater tensile or shear strength than do the parts they hold together. This is especially true of some wood glues. Many times, glued wood parts fail through the wood rather than through the adhesive.

PERMANENT ADHESIVES

Although many glues, solders, and even brazes can be removed by chemicals or heating, they bond the surfaces of the materials together with such force that their removal usually leaves the surfaces damaged. These types of adhesives would be considered "permanent" and should not be used where disassembly of the parts may be required. Many "permanent" adhesives are available at any hardware store and include the epoxy glues, caseine glues, super glues, solders, and brazes.

TEMPORARY ADHESIVES

Parts often need be held together during assembly and then released. Some adhesives are designed to do just that. The adhesives used on drafting tape and some note pads are designed to detach easily from paper. Rubber cement and contact cement can be considered temporary adhesives since the glue can be dissolved and easily removed.

SOLDERING AND BRAZING

Soldering and *brazing* are adhesive processes. They do not fuse the surfaces of the parts but merely hold the surfaces securely in a metal "glue." Although solder can be removed by melting and removing the solder, these are considered permanent adhesives. Neither brazing nor soldering results in very strong bonds. However, some brazing material does have shear and tensile strengths sufficient for many moderate-strength applications.

Solder has traditionally been a mixture of tin and lead. Eutectic or electrician's solder has a composition of 60% tin and 40% lead. It melts at 182°C (360°F). Plumber's solder is a 50% tin, 50% lead mixture. It does not have a single melting point but melts between 182 and 210°C (360 and 410°F). The melting range allows plumbers to smooth or wipe the partially molten solder to ensure a good joint between pipes.

Lead-based solders have been declared hazardous to health and are being phased out for use in plumbing and other health-related applications. Indium is now used instead of lead, but it is very expensive. New nonlead metallic solders are becoming available to the public. Higher melting point silver solders are used in applications requiring greater strength and corrosion resistance.

Brazing metals are often copper-based alloys. Whereas solder melts at a few hundred degrees Celsius, brazes have melting points around 830°C (1500°F). Brazes are often used to join metals that are difficult to weld. Cast iron tends to crack or warp unless skillfully welded. Cast iron can be easily brazed. Brazes are also used to bond two dissimilar metals. Welding steel to aluminum is almost an impossibility, but these can be brazed fairly successfully.

SELECTION OF ADHESIVES

All too often, glued parts fail because of the improper selection of the adhesive to hold them. Using library paste to attach the wings on a jet fighter would be folly. Similarly the use of *masking* tape to temporarily hold a piece of paper on a drafting board would be a poor choice because this type of tape usually tears paper when it is removed. The glue on *drafting* tape, however, is designed not to tear paper. A white or caseine glue will not hold glass or other nonporous material together for a long period of time. Examples of poor choices of adhesives for joining materials are endless.

This may sound startling, but often a properly glued joint is stronger than one that is bolted or riveted. Riveting and bolting require holes to be drilled through the two parts being joined. These holes weaken the parts. If holes are drilled too close together through a part, it is severely weakened and can fail in much the same manner as paper being torn on a perforated line. Further, riveted joints place the entire stress on the area of the rivet. Glued joints require no holes and use the entire bonded surface to support the stress (Figure 13–1).

In selecting an adhesive, several factors must be considered. The first item to consider is the **materials being glued together.** A glue that works well on wood may not work at all on glass. Glues designed for metals may be unusable on plastics.

The second consideration is the **type of stress** that will be applied to the joint (Figure 13–2). Contact cement and rubber cement have tensile strengths much higher than their shear strengths. This allows two pieces of material such as a table top and Formica to be glued together, then the Formica can be slid over the top to make a proper fit. But the Formica cannot be pulled straight up without breaking it. A glue used on drafting tape must be weak in peel, but fairly strong in shear.

A third consideration is the **atmosphere** in which the adhesive must be used. The difference between interior and exterior plywood is the glue used to hold the various layers of wood together. Exterior plywood is made with a waterproof glue, whereas interior plywood is not. Rubber cement must not be used around chemical solvents such as alcohols, acetone, or trichloroethylene. Such adhesives quickly disintegrate in contact with chemical fumes. Similarly, some adhesives cannot be used at extremes of temperature.

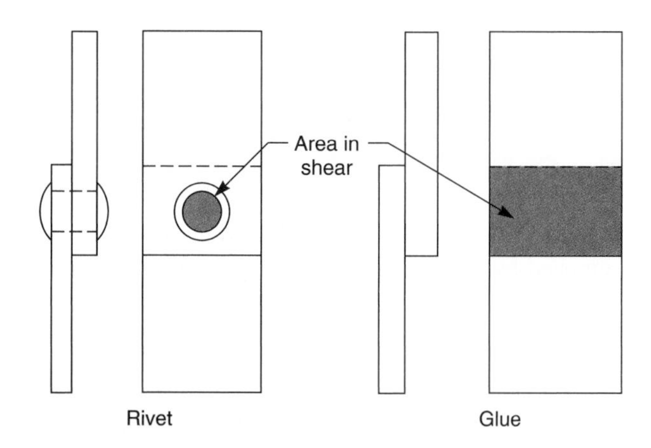

Figure 13–1. Glued versus riveted joints.

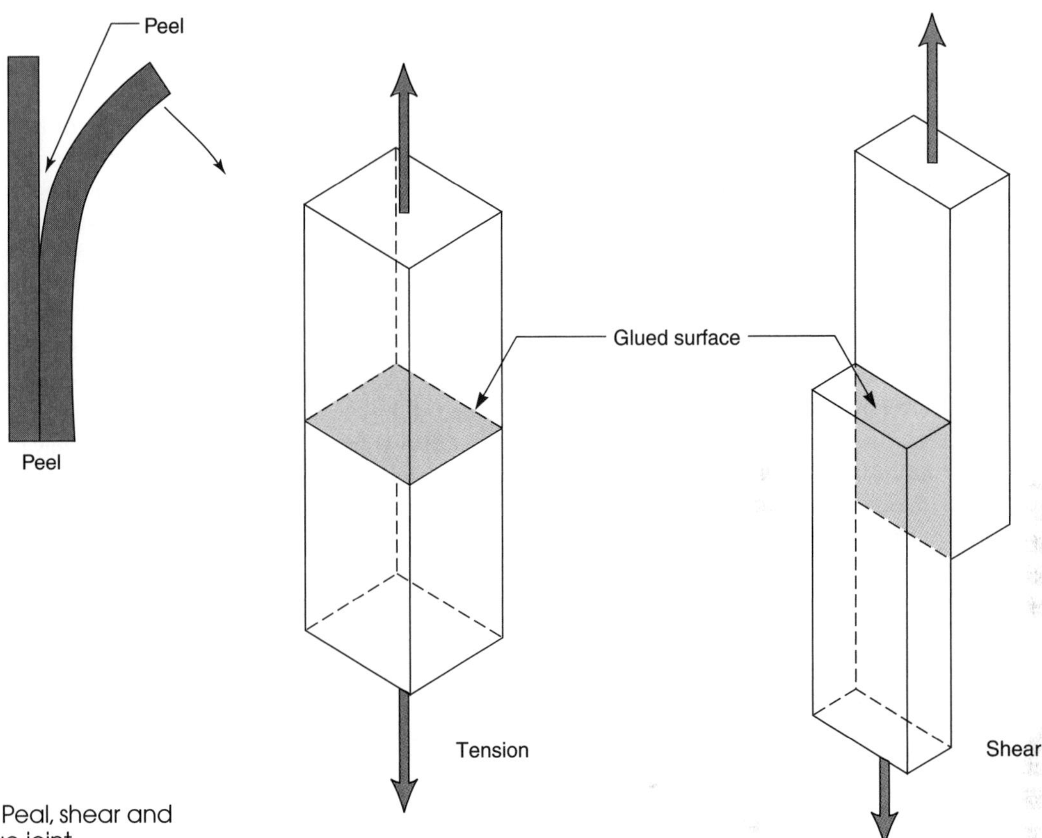

Figure 13-2. Peal, shear and tension on glue joint.

Manufacturing considerations must also be taken into account in the selection of a suitable adhesive. Glues will often polymerize and harden even though they are still in the unopened container. The length of time a glue will remain usable in an unopened container is termed its *shelf life.*

Epoxy, caseine, and certain other glues must be mixed with hardeners, water, or other components just prior to use. Once the components are mixed they must be applied and the parts joined in the short time before they harden. The length of time a glue will remain workable after mixing is its *pot life.* If glue must be applied to a large area, requiring considerable time for application, a glue with a long pot life must be selected. However, this slows production. Joints that have a small surface can be made with short pot life glues. The pot life for glues such as some epoxies may be as short as 3 to 5 minutes, while other adhesives might have pot lives of more than an hour.

Another factor that affects production rate is the *cure time* of the adhesive. Cure time can be divided into the time required for an initial set and the time for total hardening. The parts must be clamped or otherwise held together at least for the initial set time and if possible until the total hardening time is reached. Stresses must not be applied until total hardening has occurred.

Any adhesive requires that the surfaces to be joined be properly prepared. Since glues and adhesives bond the surfaces by chemically or mechanically "grasping" the surfaces, as opposed to fusing with the surfaces as in welds, the surfaces should be clean and slightly roughened. The surfaces may have to be sanded or etched prior to the application of the glue. Dirt, grease, dust, and even oils from the operator's hands can prevent the glues from adhering properly to the surfaces. The surfaces may require cleaning with air, water, alcohol, or other solvent to remove all of the contaminants. Further, the surfaces must fit together properly. Gaps between the surfaces can produce air bubbles and weak spots in the joint. Once the surfaces are prepared, the adhesives must be applied to each surface and worked into them. The parts are then clamped together and allowed to cure.

Adhesives are now available for almost any application. It is difficult to believe that much of the skin on our modern aircraft is glued onto the frame. If you looked at an old DC-3, C-47, or World War II bomber or fighter, you would see thousands of rivets placed in several rows very close together. Now many of the rivets are replaced by an adhesive between the plates. Rivets are used nowadays merely to provide the resistance to peel and are much fewer in number, yet the joined surfaces are stronger than the older counterparts. It is amazing to note that glues are now being used by surgeons to replace stitching for closing cuts in the skin after surgery.

Perhaps the best suggestion for the selection of a glue is just "read the label on the container" and "use as directed." Sales representatives and vendors will gladly work with engineers and technicians and explain the proper applications, use, and procedures for their products. Remember, however, that a glued joint is only as good as the skill with which it is applied.

One caution about glues. Glues do not dry on curing. They react chemically to form long-chain polymers that grab onto rough spots on the surface to hold them together. The chemical reaction in making these polymers is relatively slow, often requiring several hours for completion. For this reason, never place a stress on the joints before the recommended curing time. The longer the clamps can be left on the joint, the better. Early removal of the clamps can allow *delamination* to occur, letting air pockets between the glued surfaces and resulting in a weakened bond.

■ MANUFACTURING WITH ADHESIVES

The quality of a glued, soldered, or brazed connection depends on the skill with which the joint is made. Not only must the proper adhesive be chosen for the designed application, but the surfaces to be bonded must be *carefully prepared,* the adhesives must be *thoroughly mixed* (if two-part adhesives), the glues must be *properly applied,* and the joint must be *properly clamped* until the adhesive has set.

Adhesives mechanically bind surfaces together. The glues work their way into each surface to be bonded. Few (if any) glues will stick to a highly polished surface. Therefore, the surfaces of any glued joint must be *clean* and *slightly roughened.* In solder joints, this can be accomplished by the use of a flux, which causes the solder to bind to the metal. A common flux for copper pipe is hydrochloric acid. An old blacksmith's trick was to use borax soap as a flux for brazing. Wood surfaces should be sanded and wiped clean before applying the glue. Also make certain that there are no "gaps" between the surfaces to be bonded. Both surfaces must fit together very closely for a good bond.

Two-part glues always carry instructions for their proper mixing on the package. Make sure the glues are mixed according to the instructions, and *completely* mixed before applying them to the surfaces to be bonded.

Since adhesives must be worked into the surfaces of the materials to be joined, the glues must be applied, individually, to **both** surfaces. Do not apply the glue to just one surface and then lay the mating part on it. Any air bubbles should be worked out of the glue before the parts are put together.

Care must also be taken when putting the parts together. For soldered and brazed joints, the parts are clamped together before the solder or brazing material is applied. If "contact" type cements are used, they cannot be pulled apart to reposition the joints. (In some cases, the parts can be "slid" over each other to get their locations properly aligned.) These cements demand the proper alignment of the parts on the first try.

Once the glued surfaces are put together, they must be held together while the adhesive cures. This can be done using clamps, or the parts can be held together by nails, screws, bolts, springs, or even tape. Use caution to ensure that the parts do not "slip" during the clamping or curing stage. Remember, in most cases the joined parts cannot be easily taken apart after the adhesive cures.

Once the glued joint has cured, remove the clamping devices, inspect the glued joint, and finish the joint as necessary by removing any excess bonding agent. A well-made glue joint should be barely noticeable. Be sure soldered joints have a "shiny" surface. Dull or crystalline looking joints are referred to as "cold joints" and will often fail under mild stresses.

Soldering, brazing, and gluing are simple operations, but do take a little practice to produce good results.

PROBLEM SET 13-1

1. Visit a hardware store and make a list of the different types of adhesives available. Also make a list of the recommended applications for each adhesive.
2. What is the difference between "welding" and "brazing"?
3. What is the difference between "soldering" and "brazing"?
4. List five situations in which a "temporary" adhesive might be used.
5. Check with a lumberyard to find out what the difference is between an "interior" and an "exterior" plywood.
6. List five applications for which rubber cement would be a good choice of adhesive.
7. If a glue had a shear strength of 120 lb/in.^{2}, how many square inches of surface must be glued together to hold 800 lb in shear? (Assume that the glue is weaker than the material it holds together.)
8. Make a list of the items in your automobile that are joined by adhesives.
9. What precautions must be taken in order to obtain a good glue joint?
10. Masking tape and drafting tape appear to be the same. What is the difference between them?

The emphasis in this chapter has been on the need to analyze the type of application so as to select the best fastener. A primary consideration is whether or not the fastening is going to be permanent. Then, should relative motion be allowed? What are the materials to be joined? and so forth.

Adhesives were considered, with emphasis on the strength of the bond, consideration of the materials to be joined, and the use of adhesives in manufacturing.

PHOTO ESSAY 13-1

Gluing

STEP 1 Prepare the workpieces. The surfaces to be glued together should fit snugly with no gaps or bumps. The glued surfaces should be roughened slightly to allow the glue to form a mechanical bond with the surfaces.

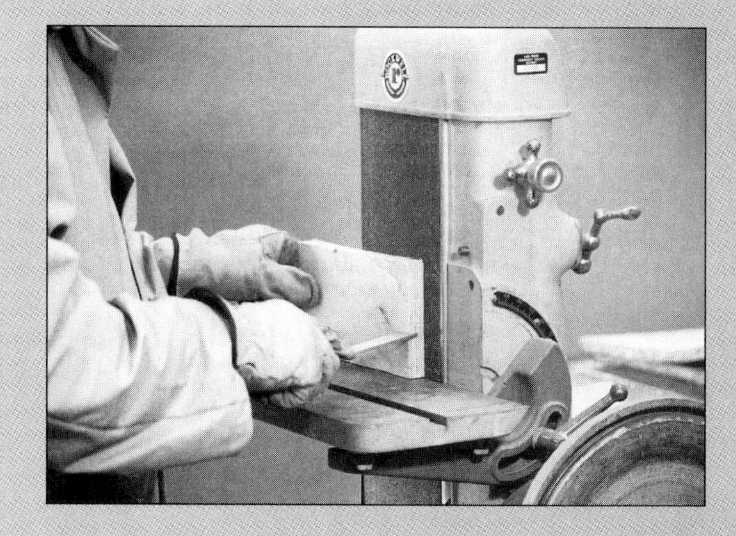

STEP 2 Mix the glue if necessary (some glues require no mixing). Epoxies, caseine glues, and resin glues must be mixed either with water or with some other catalyst prior to use. Make sure the components are thoroughly mixed before application.

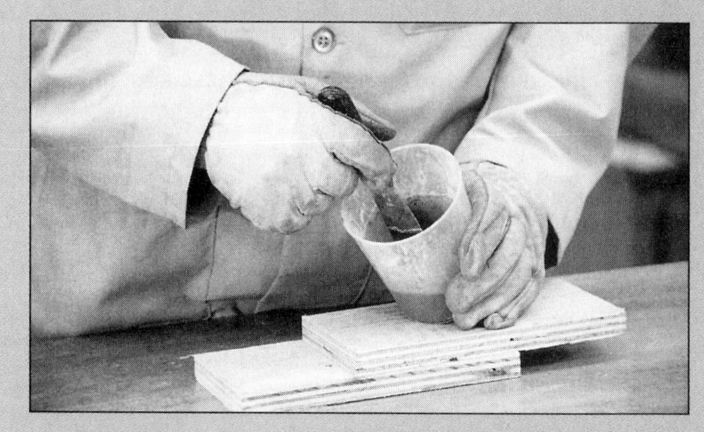

STEP 3 Apply the glue to *both* surfaces of the workpieces. Work the glue into the surfaces.

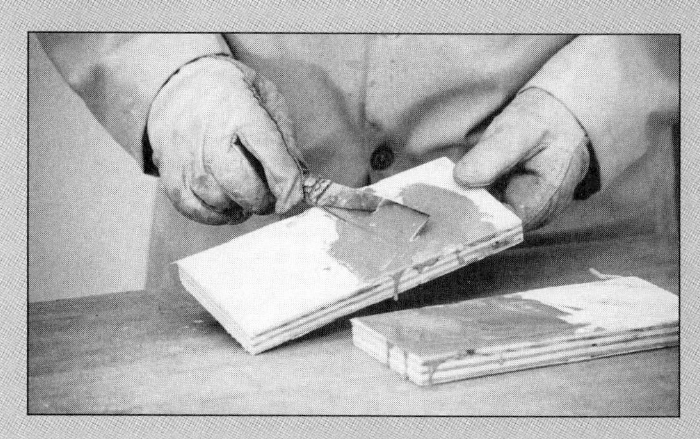

STEP 4 Put the pieces together and apply pressure with a clamp, vise, or other fixture.

STEP 5 Remove any excess glue from the workpiece. The workpiece should remain in the clamps until the glue is thoroughly set.

Chapter

14

Welding

Soldering and brazing are *adhesive* bonds, whereas welding is a *cohesive* bond. In welding, two surfaces are literally melted together, allowing the material of each part to diffuse into the other. Welds basically make one part out of two and are always considered a permanent method of joining two pieces. To separate welds, you must saw or cut them apart. In soldering and brazing, a sharp boundary can be seen between the parts and the solder or brazing material. In welded surfaces, there may be a change of grain size, but there are no sharp boundaries at all. Figure 14–1 illustrates the difference between a weld and a brazed joint.

A good weld will not break at the weld when subjected to stress or loading. However, a weld is only as good as the skill of the welder. Critical structural welds should always be tested by magnefluxing, x raying, sonic imaging, or holographic or other nondestructive techniques.

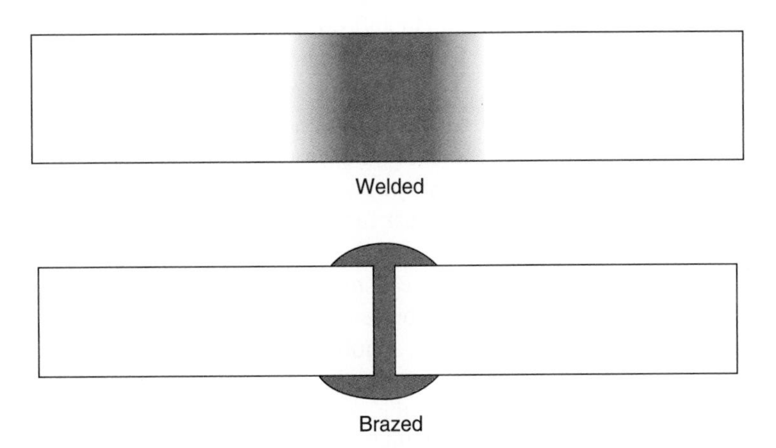

Figure 14–1. Welded and brazed joints.

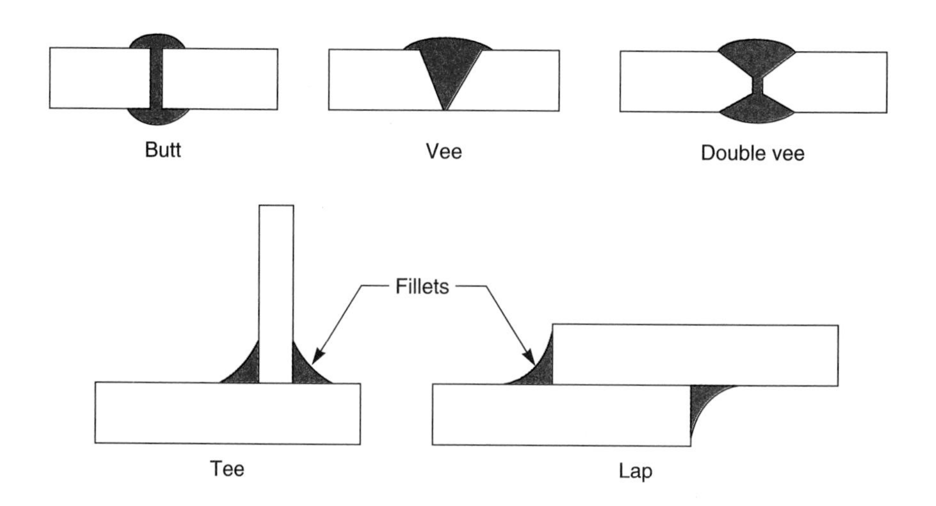

Figure 14–2. Types of joints.

■ JOINT PREPARATION

Several types of joints are used in welding. These include butt joints, vee joints, double-vee joints, tee joints, which require a fillet weld, and lap joints. Figure 14–2 shows examples of these.

Butt joints are used on metal that has a thickness of one-quarter inch (6 millimetres) or less. Metal of one-eighth inch can be butt welded from one side. Thicker butt-welded plates would require welding from both sides to obtain complete penetration through the plate.

Metals thicker than one-quarter inch need either a *vee* or *double-vee* joint, which must be welded in several layers or *passes.* Plates or pipes have the angles either ground or milled to bevel the edges. Once the plates are beveled to the appropriate angle and set to the gap specified by the engineer, they are tacked together by welding just a few points along the weld. A root pass or weld bead is applied to the complete length of the prepared groove; then successive layers of weld metal or filler rod are added until the vee or groove is completely filled. Each pass must fuse the new weld metal into the face or surface of the vee shape and into the previously laid weld metal. Between each pass, the slag or flux must be removed and the surface cleaned. Figure 14–3 shows a multiple-pass vee weld.

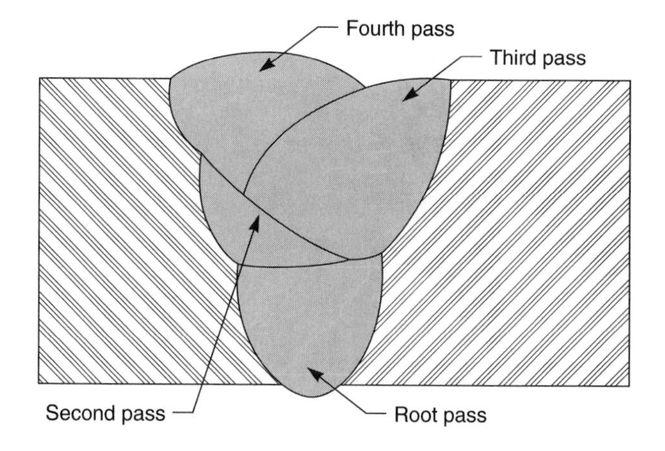

Figure 14–3. Multiple-pass weld.

The joined plates of *tee welds* can have the ends of the plates flat or beveled depending on the thickness of the metal to be welded. After welding a tee joint, the weld metal will be in a concave shape called a *fillet* (Figure 14–4). *Lap welds* also require a fillet type weld.

Preparation for Weld Joints

As with adhesives, welded joints must also be properly prepared. The surfaces to be joined must be ground to the weld specification. Any slag, corrosion, or other foreign material must be removed from the surface being welded and the nearby surfaces. Because these surfaces will be melted during the welding process, the surface roughness is usually not a factor in the quality

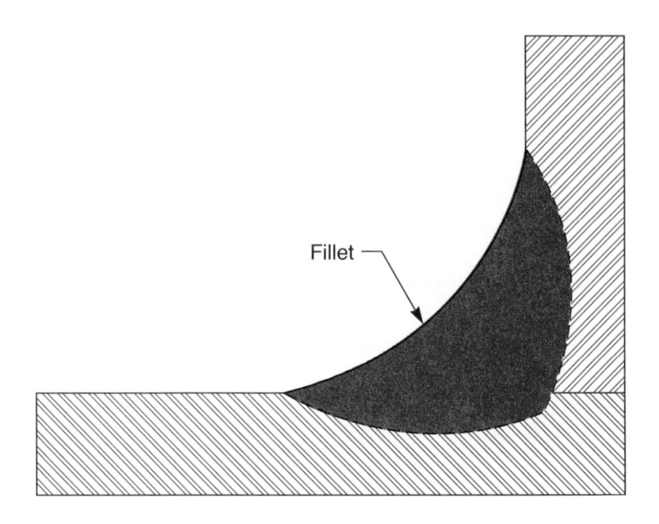

Figure 14–4. Fillet weld.

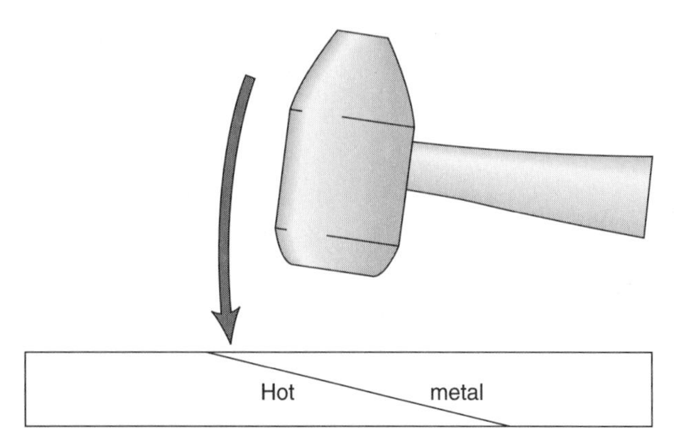

Figure 14–5. Forge weld.

of the weld. (Electron-beam, laser, and friction welds can be an exception here.) Care must be taken to securely anchor the parts being welded. The added heat from the torch or other welding devices can cause clamps to "relax" and let the parts slip. This not only can cause a bad weld, but can be hazardous to the person doing the welding.

■ FORGE WELDING

For thousands of years, blacksmiths have used *forge welding,* which makes it the oldest type of welding process known. In this modern age of high technology, it may seem strange, but forge welding is still used. Surprisingly, even some parts on the landing gears of large aircraft are attached by a modernized version of forge welding.

In making a forge weld, the surfaces of both parts are carefully prepared to mate very closely. Flat surfaces are either lap welded or beveled as shown in Figure 14–5. Forge welds can be made on butt joints also. The surfaces to be welded are heated well into the red to yellow heat range, between 1600 and 2000°F. This heating was done in a blacksmith's forge in the old days but modern techniques use gas flames. A flux is applied to remove the scale on the surfaces and the two parts are pounded together. The atoms of the mated surfaces will diffuse into the other surface to form as strong a bond as there is between the atoms of the parent metals. Properly done, these welds are at least as strong as the parent metals.

In forge welding both surfaces must be exactly the same temperature before they are pounded together. The old blacksmiths often guarded the secret of their flux. However, one flux commonly used is nothing more than *borax soap.*

In a modern version of the forge weld, parts are often heated with a gas flame ring. At the proper temperature, the two ends are very forcibly "rammed" together to create the weld. Aircraft landing gears and other parts are made in this manner by some companies. It is a very reliable method of welding.

■ GAS WELDING

With the discovery of acetylene gas (the correct chemical name is ethyne), very high temperatures became possible. Acetylene (C_2H_2) can be made by mixing water with calcium carbide (CaC_2), often called *miner's lump.* The gas coming from the reaction is acetylene. It is captured, bottled, and sold for use in welding and other heating and chemical applications. Acetylene can also be produced from ethane, a petroleum and natural gas derivative.

Oxygen and acetylene can be bought from any welding supply company in various size tanks. Standard

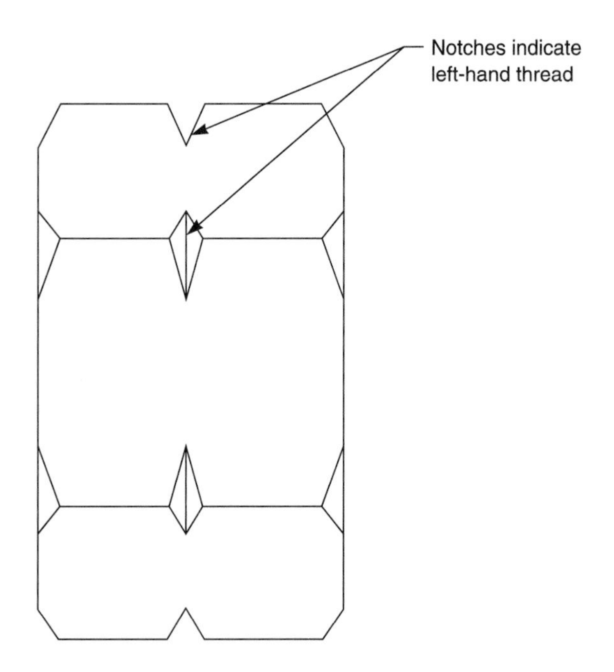

Figure 14–6. Left-hand nut.

color codes dictate that the oxygen tank be painted green and the acetylene tank red, or black with a red top. The pressure valves are attached directly to the tanks. The pressure gauges and hoses cannot be interchanged. The oxygen pressure valve has a right-hand internal thread, whereas the acetylene pressure valve has an external left-hand thread. The left-hand threaded nut can be easily spotted because it has a little "vee" notch on the corners of the hexagonal nut (Figure 14–6). Oxygen hoses are green and acetylene hoses are red. The connections to the torch are again right- and left-hand threads to prevent incorrect connections to the torch. Figure 14–7 is a photograph of a typical oxy-acetylene apparatus.

Oxygen-Acetylene Welding

An oxygen-acetylene flame is very hot, approaching 3500°F. It can melt cast iron (melting point: 2300°F) and steel (melting point: 2800°F). All that is required for an *oxy-acetylene weld* is to place the two pieces against each other and melt their surfaces together (Figure 14–8). This is known as a *fusion weld.* If carefully done, no filler rod is needed for this type of weld.

Figure 14–7. Oxy-acetylene welding apparatus.

Gas welding heats a large section of the metal, disturbing the grain structure for a considerable range about the weld. Further, if the flame is not adjusted correctly, carbon can be added or removed from the steel by the flame, so proper adjustment of the torch is essential.

In lighting the torch, the acetylene is adjusted to the proper pressure for the tip being used. The pressure is regulated by the valves on the tanks. For small tips, acetylene pressures ranging from 2 to 10 psi are required with oxygen pressure being set at about twice the acetylene pressure. The acetylene is ignited and the flow of acetylene is adjusted at the torch until a slight *feather* and slight *roar* of the flame is reached. The oxygen is then adjusted with the valve on the

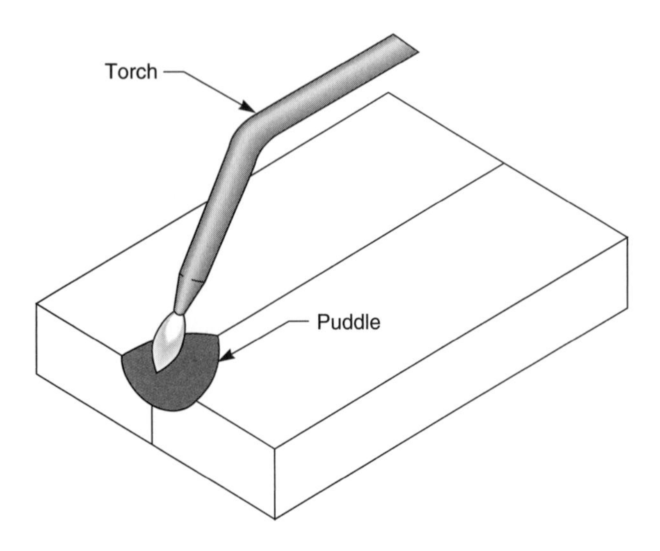

Figure 14–8. Acetylene fusion weld.

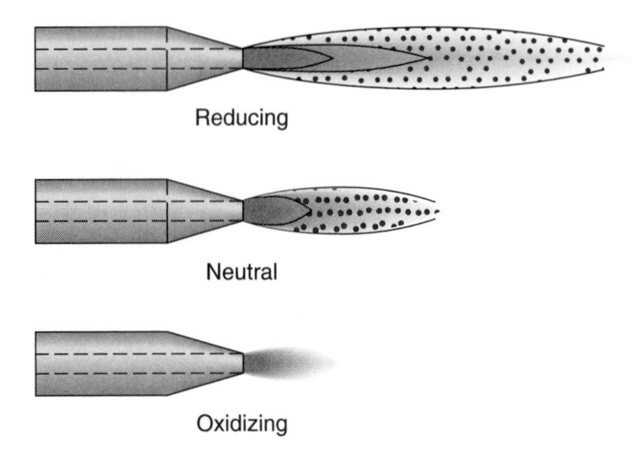

Figure 14–9. Types of oxy-acetylene flames.

torch. The pure acetylene flame is bright orange with a considerable amount of black soot being produced. As the oxygen is first turned on, the flame will turn from orange to a light blue. The flame has three different sections: The end of the flame will be a transparent blue and there will be two successively brighter cones. With two cones apparent, one longer than the other, the flame is a *reducing flame* (Figure 14–9). A reducing flame is used to melt low-melting-point metals and alloys because it does not oxidize or corrode the metals. It is not an extremely hot flame. By adding more oxygen, the longer outer cone can be brought back to the inner, bright blue flame. The point at which the two cones of flame just touch provides a *neutral flame.* The neutral flame is the hottest one possible and is the proper adjustment for welding. The addition of more oxygen past the neutral flame provides an *oxidizing flame* that can cause corrosion in the metal. It is only used for cutting flames or burning pieces of metal from a piece of stock.

When turning off an oxy-acetylene flame, the oxygen is always turned off first, then the acetylene. Turning off the acetylene first will cause a loud pop and can damage the tip of the torch.

A good oxy-acetylene weld requires that the filler rod be thoroughly melted into and mixed with the molten surfaces of the workpieces. To do this, the welder must first melt a small part of the workpieces. This process is called *puddling* in the trade. Once the puddle is started, the filler rod is melted into the puddle and worked into the surfaces of the workpieces. The puddle must go slightly deeper than halfway through the stock. If this is not done, there is said to be a *lack of penetration* and a gap will be left in the middle of the weld, which can cause it to fail (Figure 14–10).

The advantages of an oxy-acetylene weld are that it is inexpensive and requires very little specialized equipment. Its disadvantages are that since there is carbon in

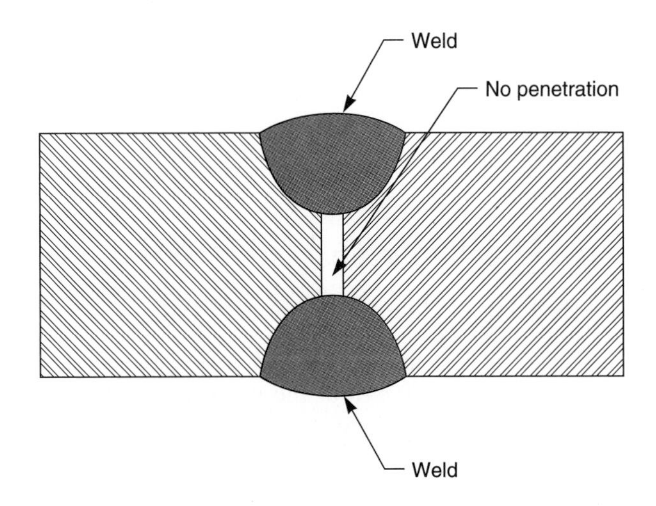

Figure 14–10. Illustration of lack of penetration.

the acetylene gas, any incomplete combustion can leave traces of carbon in the weld, thereby weakening it. It is not considered a very clean weld. A *clean weld* refers to the fact that there are no impurities or imperfections in the weld. In some welding techniques, pieces of flux, metal oxides (slag), and even air bubbles are left in the weld. These imperfections often embrittle and weaken the weld. A weld containing these impurities is referred to as a *dirty weld.*

Oxygen-Hydrogen Welding

Besides acetylene, other gases can be used to melt the materials. An *oxygen-hydrogen* mixture is often used. The oxygen-hydrogen torch can reach temperatures much higher than the oxy-acetylene torch. The procedure for oxy-hydrogen welding is the same as for oxy-acetylene. The oxy-hydrogen weld is often quicker and is a cleaner weld than the oxy-acetylene weld.

The problems with oxy-hydrogen welding are that it is more expensive than oxy-acetylene welding and does involve the flammability risk that accompanies the use of hydrogen.

Plasma Welding

A *plasma* is an ionized gas. If hydrogen is run through a tube that has a positive electric charge, the electrons will be stripped from the hydrogen atoms, creating hydrogen ions or plasma. Hydrogen plasma burns even hotter than hydrogen gas. This permits welding of extremely high-melting-point metals and even some ceramics. Plasma welding is also a very clean procedure that results in very little slag or foreign matter in the weld.

■ ELECTRICAL WELDING

Resistance Welds

As an electric current is run through a wire, it generates heat due to the *resistance* of the metal to the flow of electrons. The higher the resistance, the more heat generated. If a high current is run across the point of contact between two pieces of metal, sufficient heat will be generated to melt the metals at the point of contact. After the joint is melted, the current is switched off and the melt allowed to solidify, which creates a weld. This entire process can take a fraction of a second.

Two major methods of resistance welding are *spot welding* and *ribbon welding* (sometimes called seam welding). In spot welding the two pieces to be welded are clamped between the electrodes while the electric current is applied. In ribbon welding, the electrodes are rollers. The parts to be welded are drawn between these rollers while electricity is applied. Figures 14–11 and 14–12 illustrate spot welds and Figure 14–13 is a schematic of a ribbon weld.

Spot and ribbon welds are very fast, inexpensive, clean, and reliable. They weld a "blind" spot between metals that would be impossible for a torch or welding rod to reach. Preparation of the weld points is important. There must be no paint or corrosion at the point of weld, and the joints must be well mated. Often the weld points are punched or raised above the surrounding surface to ensure proper contact. This is sometimes referred to as *upset welding* (Figure 14–14).

Resistance welds are used extensively in the automotive industry and are readily adapted for robotics

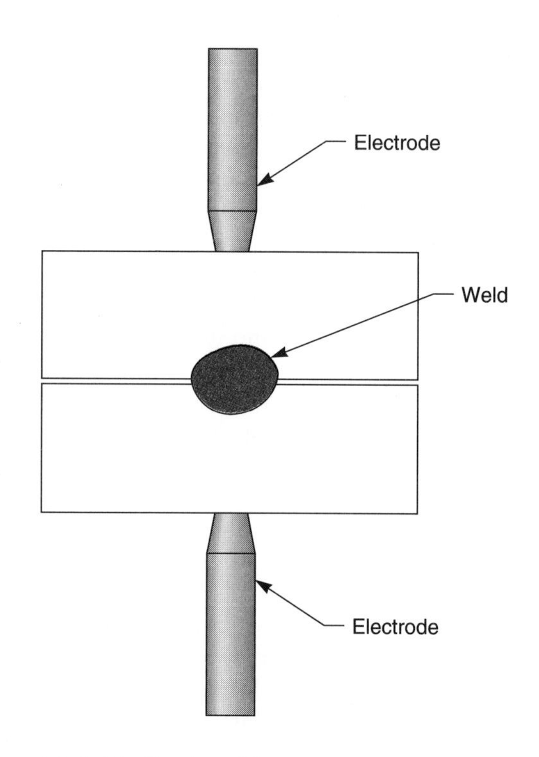

Figure 14–11. Spot weld.

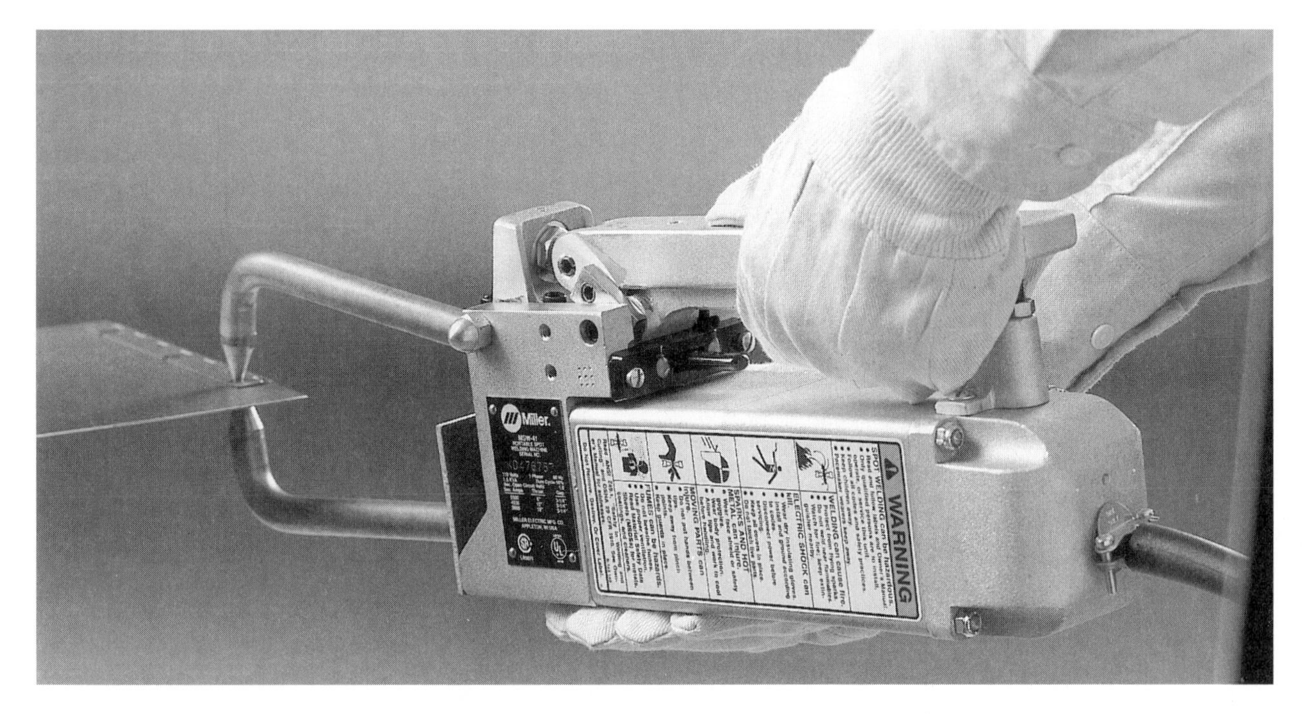

Figure 14-12. Spot welding apparatus.
Photograph courtesy of Miller Electric Manufacturing Company, Appleton, Wisconsin.

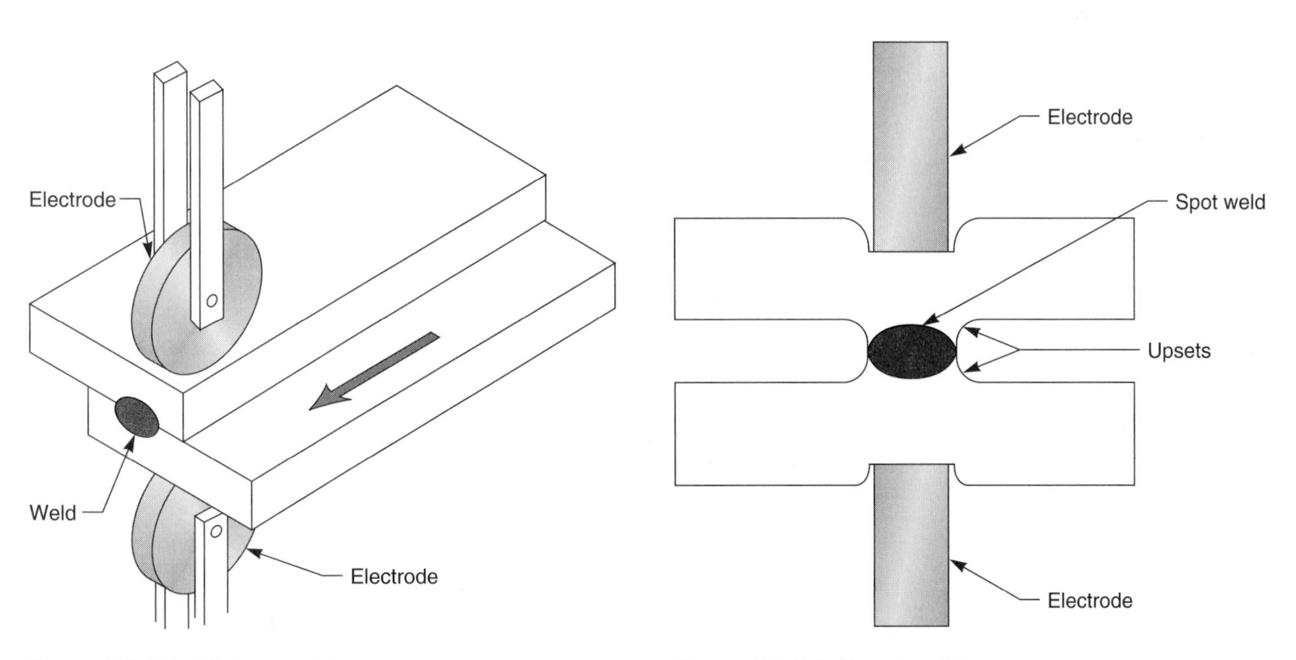

Figure 14-13. Ribbon weld.

Figure 14-14. Upset weld.

applications. What we think of as the "hand" of the robot is replaced with a spot welder.

Arc Welds

Everyone has seen a spark. A spark is created by electrons under high voltage that are forced from one conductor through the air to another conductor. The passage of the electron through the air causes the atoms in the air to glow. This is the principle on which lightning is based. These electrons, being driven by a voltage, carry a lot of energy. Upon hitting another metal, they can transfer that energy to the metal, causing its temperature to rise and the metal to eventually melt. The melted metal can then flow together to form a weld. This is the principle behind all types of arc welding.

Carbon Arc Welds

Arcs can be used in several different ways in welding. The simplest is to use an arc from a carbon rod to melt the metal, and the filler rod is fed into the base of the arc and puddle formed by the arc (Figure 14–15). In *carbon arc welding,* either alternating current (ac) or direct current (dc) is used. If direct current is used, the welded material becomes the positive electrode and the carbon rod is the negative electrode. This method is seldom used anymore. In another variation of carbon arc welding, the arc is struck between two carbon electrodes and the heat from the arc melts the metal and the filler rod (Figure 14–16). This method is also rarely seen except for a few backyard applications.

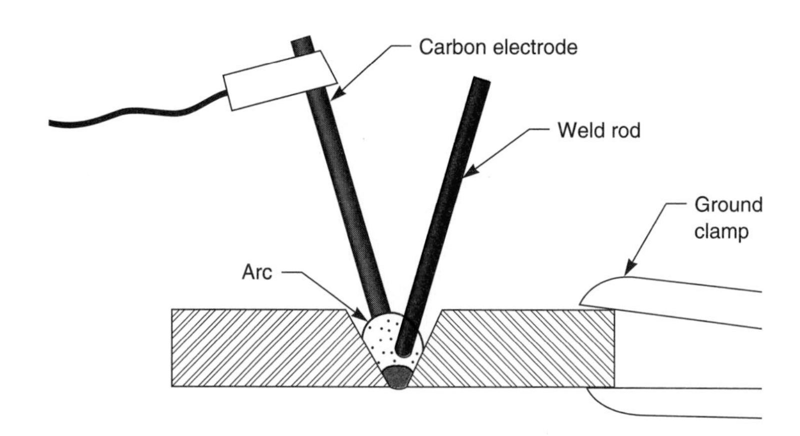

Figure 14-15. Carbon arc welding.

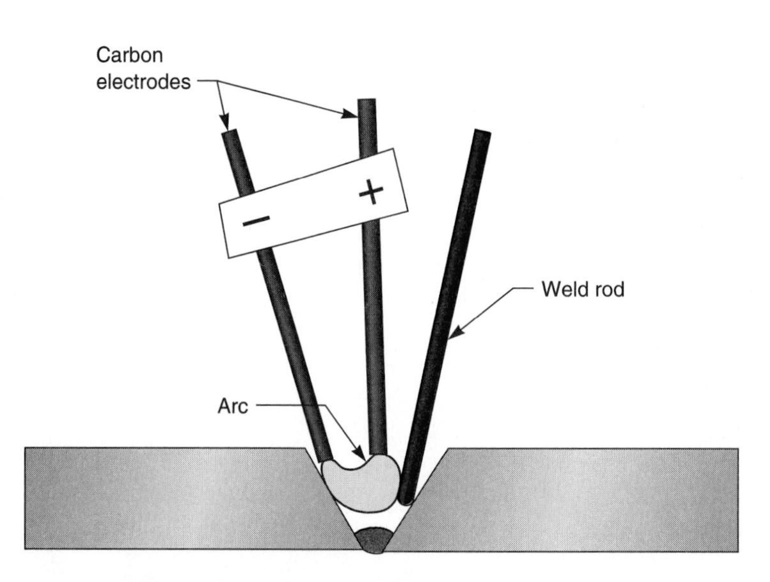

Figure 14-16. Dual carbon arc welding.

Shielded Metal Arc Welding

A modification of the carbon arc weld is to use the filler rod as one electrode of the arc. Here both the metal being welded and the tip of the filler rod are melted. In *shielded metal arc welding (SMAW),* the flux is coated around the weld rod (Figure 14–17). Traditionally, welders refer to this method as *stick welding.* Figure 14–18 is a photograph of a stick welder at work.

Selection of Welding Rods. The selection of the electrode or rod in arc welding is critical to the success of the weld. Several factors must be considered. First, consider the metal being welded. Since a good weld must not fail, a weld filler rod should be selected that has a tensile strength greater than the metal to be joined.

The filler rod must also be compatible with the welded metal so that it will fuse evenly with it. Other factors to consider are the welding positions required, the welding current (ac or dc), the joint design (groove, butt, fillet, etc.), the thickness and shape of the base metal, the service conditions and specifications, and the production efficiency and job conditions.

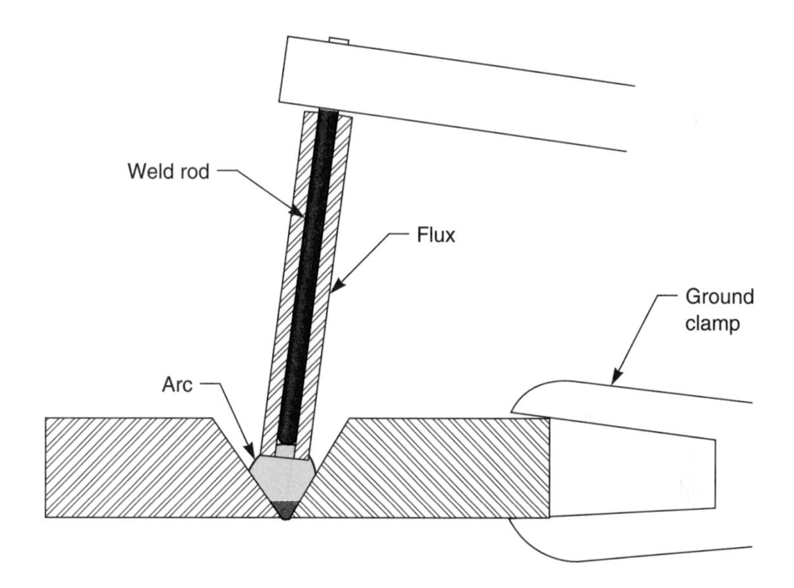

Figure 14–17. Shielded metal arc welding.

Figure 14–18. SMAW welder.

The types of SMAW rods are designated by either a four- or a five-digit system designed by the American Welding Society. Examples of these are E-6010, E-7018, E-7025, and E-11020. The *E-* stands for electrode. The first two or three numbers indicate the tensile strength of the rod in *thousands* of pounds per square inch. The next-to-last number gives the welding positions for which the rod may be used. The number indicating position will range from one to six. The welding positions are as follows:

1. All positions
2. Flat and horizontal
3. Flat position only
4. Vertical
5. Flat, vertical, overhead (pipe)
6. Inclined

Either fillet or groove welds can be made in all of these positions. Thus a 4G weld is a vertical-groove weld, but a 3F weld is a flat fillet weld. Plates are welded in the 1, 2, 3, 4, and rarely in the 5 or overhead positions. Pipes can be welded in the 1, 2, 5, and 6 positions depending on the situation. Figures 14–19(a) and (b) show the welding positions. Figure 14–20 shows a welder doing a vertical weld (position 4). It is

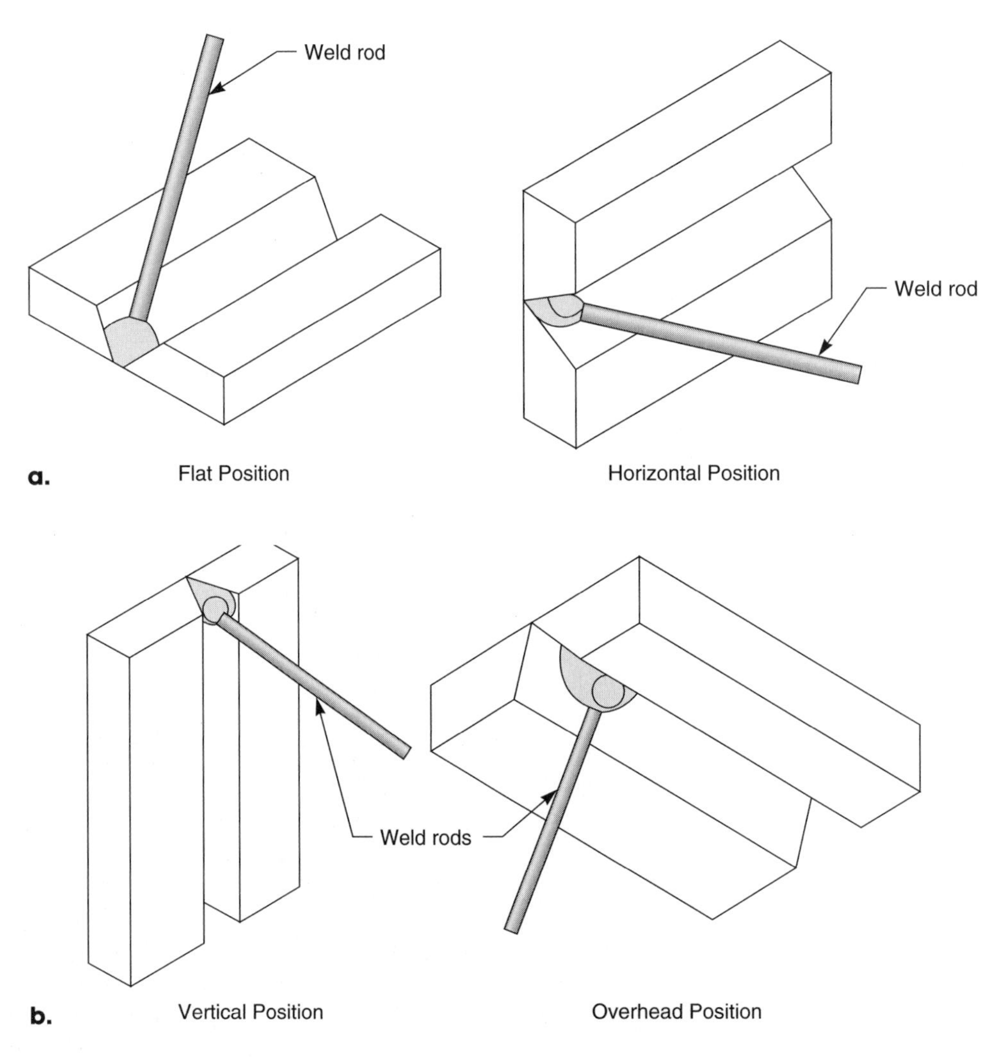

Figure 14–19. Welding positions.

Figure 14–20. Welding in vertical position.

easy to see from Figure 14–21 that pipe welding involves several positions.

The last digit of the weld rod number indicates the type of current for which the rod may be used (ac, dc straight, dc reverse), the penetration, and the type of flux around the rod. *DC straight* and *dc reverse* refer to the polarity of the weld rod. If the weld rod or electrode is connected to the negative terminal of the power source, it is said to be in the dc straight system. If the weld rod is connected to the positive terminal, it is considered to have reverse polarity. Remember that electrons, because they are negatively charged, are attracted to positive charges. In the dc straight configuration the electrons flow from the rod to the base

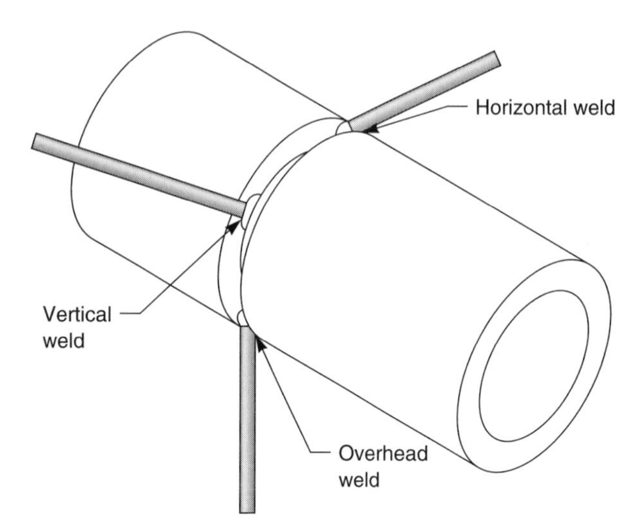

Figure 14–21. Pipe welding.

metal. The bombardment by these electrons heats the metal to the melting point and the heat then melts the rod. In the dc reverse configuration, the reverse is true. Table 14–1 is a list showing the variables indicated by the last number of the SMAW welding rod.

All low-hydrogen rods must be preheated to about 200°F to dry out the moisture in the rods. Failure to do this will cause pitting in the weld.

From Table 14–1, we can see that an E-6010 rod would have a tensile strength of 60,000 psi, could be used in all positions, has a cellulose-sodium flux, could give deep penetration, and must be used with dc reverse current.

SMAW welding is a relatively low-voltage, high-current weld. Voltages in the range of 20 to 35 volts with currents of 75 to 130 amperes are used. The voltages and currents depend on the type and diameter of the rod, the flux, the polarity, and even the skill of the welder.

Considerable practice is needed to become proficient at stick welding. One has to strike the arc by touching the weld rod to the metal, then immediately back it off to get the arc started. Further, the tip of the rod must be kept at a constant distance from the weld to keep the arc going, the current constant, and provide an even depth and width of weld.

Table 14–1. Last Digit Indications of Weld Rods

Rod	Current	Penetration	Flux
E-xxx0	dc reverse	Deep	Cellulose-sodium
E-xxx1	ac and dc reverse	Deep	Cellulose-potassium
E-xxx2	ac and dc straight	Medium	Titanium oxide-sodium
E-xxx3	ac and dc straight and reversed	Shallow	Titanium oxide-potassium
E-xxx4	ac and dc straight and reversed	Shallow	Titanium oxide-iron
E-xxx5	dc reverse	Medium	Low hydrogen-sodium
E-xxx6	ac or dc reverse	Medium	Low hydrogen-potassium
E-xxx7	ac and dc straight and reverse	Medium	Iron-iron oxide
E-xxx8	ac and dc straight and reverse	Medium	Low hydrogen-iron, iron oxide

If the rod is not removed quickly enough when striking the arc, it melts and fuses itself to the metal. This is very often the case with beginning welders. They can be seen frantically trying to break the sticking rod away from the weld, providing much amusement to more experienced welders. Some rods are easier to use than others and all too often the rod, if selected by the welder, is based on the ease of use rather than the engineering considerations. Almost anyone can do a fairly good weld with an E-7018 rod, but a great deal more skill is needed to make a good weld using an E-6010 rod.

Properly done, SMAW or stick welding provides a stronger, more reliable weld than a gas weld. It is faster and usually cleaner than the gas type welds.

Submerged Arc Welds

Submerged arc welds use the flux to prevent the oxygen of the air from forming corrosion in the weld. The flux, in the form of a powder, is laid out along the path of the weld and the weld rod electrode is run through the flux. It is a bit of a blind weld, but it is used very often in manually controlled welding machines as well as computer and robotically controlled welding machines. Figure 14–22 illustrates submerged arc welding.

Inert Gas Welding

Oxygen must be kept away from the hot metal during welding to prevent corrosion both on the surface and within the weld metal. One method of doing this is to keep the weld surrounded by an *inert gas* such as helium, argon, krypton, nitrogen, or carbon dioxide, which is chemically incapable of corroding the hot metal. There are several variations of this technique.

Gas Metal Arc Welding. One of the problems incurred in welding is that oxygen from the air gets into the weld metal. The oxygen corrodes the weld both externally and within the weld metal itself. The corrosion significantly weakens the weld. To prevent this, several techniques involving shielding the weld with an inert gas have been developed. *Gas metal arc welding* (GMAW) is but one of these methods. The old name for GMAW was metal inert gas or MIG welding.

In GMAW, a wire is continuously fed from a spool through the weld nozzle. An inert gas of argon, argon-carbon dioxide mixture, or even helium is blown around the weld wire. Once the arc is struck, the welder need only to control the speed with which the weld is laid down. The speed or rate of travel of the rod controls the depth of penetration and the width of the bead.

The variables in GMAW that must be controlled include the following:

- Size and type of filler wire
- Type of shielding gas
- Joint preparation
- Gas flow rate
- Voltage
- Arc length
- Polarity

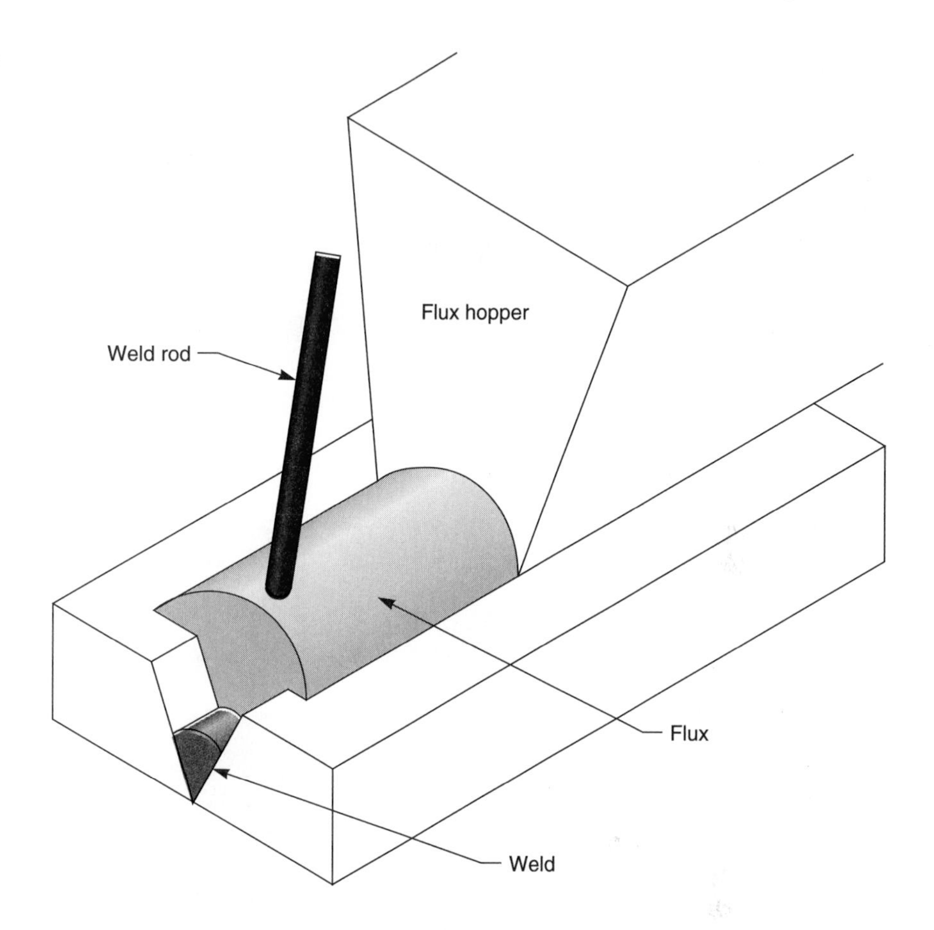

Figure 14-22. Submerged arc welding.

- Wire stick-out
- Nozzle angle
- Wire-feed speed

These variables are generally specified by the engineering drawings. In making the specifications for the weld the engineer consults the welding codes. However, engineers usually discuss these variables with their expert welders in order to arrive at their specifications.

The size and type of filler wire are determined by the requirements of the weld. The wires have number designations very similar to the SMAW designations. The diameters of the wire vary from a very fine 0.030 inch (0.8 millimetre) to 1/8 of an inch (3 millimetres). The current used depends on the arc length, but may vary from 50 to 140 amperes for the 0.030-inch wire to close to 600 amperes for the 1/8-inch wire. The selection of the inert gas depends on the type of metal being welded, filler wire, and the welding position. Argon-carbon dioxide mixtures can add carbon to the surface of the weld to provide a harder surface. The flow rates of the inert gas vary from 25 to 100 cubic feet per hour.

If the wire is fed too slowly, the arc will not be continuous. A wire feed rate that is too fast, however, will cause the unmelted wire to pile up on the weld. Wire stick-out refers to the length of wire protruding from the nozzle before entering the arc. The current used in the weld depends on the wire stick-out. Generally the stick-out is about 3/8 of an inch. The nozzle should be held at an angle that will allow the gas to cover the weld and provide a good weld bead. Figure 14–23(a) shows how a welder would get in position to start a GMAW weld, and (b) shows the weld in progress. Figure 14–24 is a photograph of a GMAW torch, which is illustrated in Figure 14–25.

Figure 14-23. Gas metal arc welder.

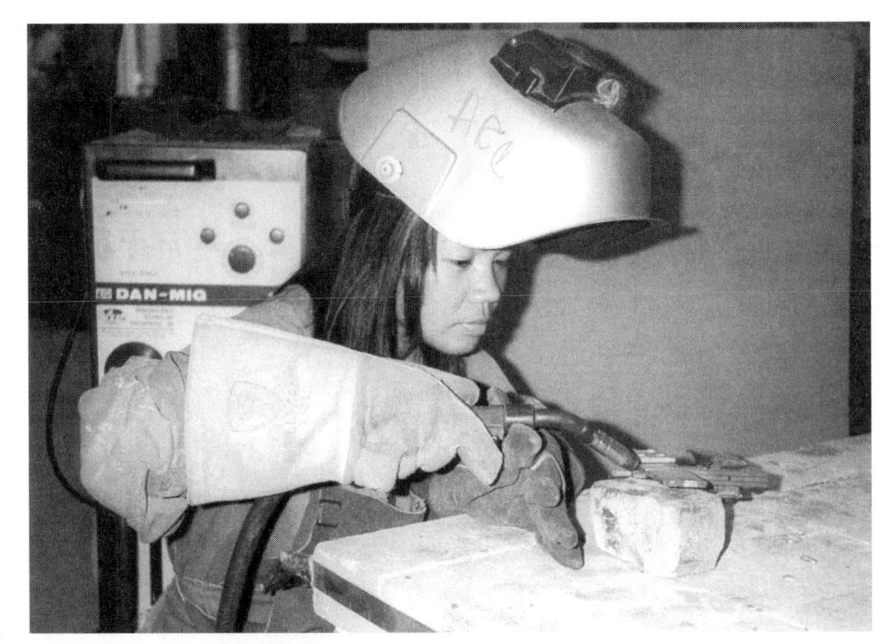

a.

b.

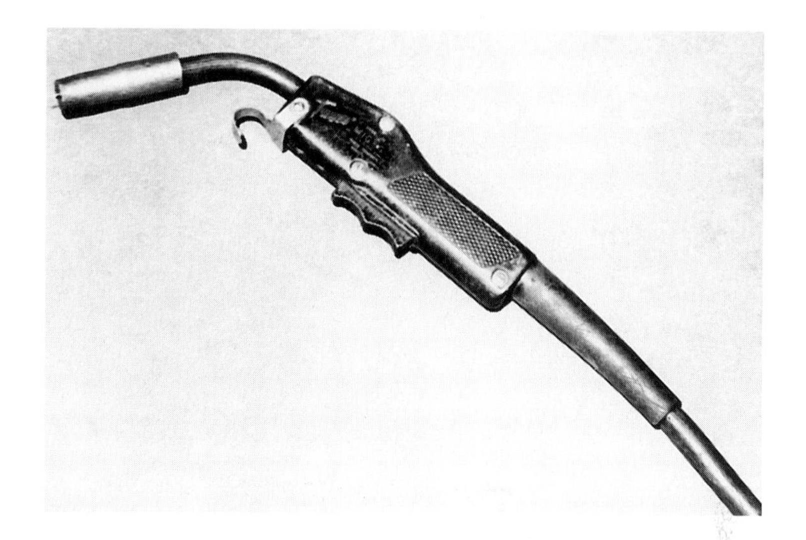

Figure 14–24. GMAW torch.

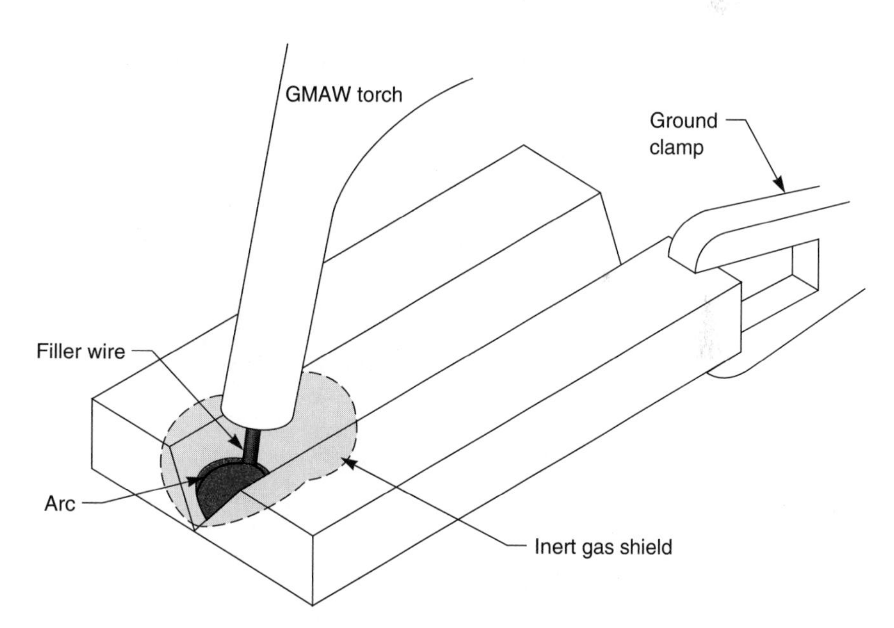

Figure 14–25. GMAW schematic.

GMAW can be done by hand but is easily converted for automatic welding machines, computer-controlled welding machines, and robotics control. Many present-day automotive parts are manufactured by robotics GMAW welding machines.

Flux Core Arc Welding. *Flux core arc welding* (FCAW) uses a spool of filler wire fed through the handpiece similar to gas metal arc welding except there is no inert gas. Instead, a core of flux is inside the wire. Voltages are lower but currents are much higher than in GMAW. Voltages from 15 to 35 volts with currents of 200 to 400 amperes are the norm. Since there is no large pipe or gas cup around the electrode, it is easier for the welder to see the weld being made. Figure 14–26 shows a FCAW machine. Figure 14–27 is a photograph of a welding torch, and Figure 14–28 is a schematic of the flux core weld.

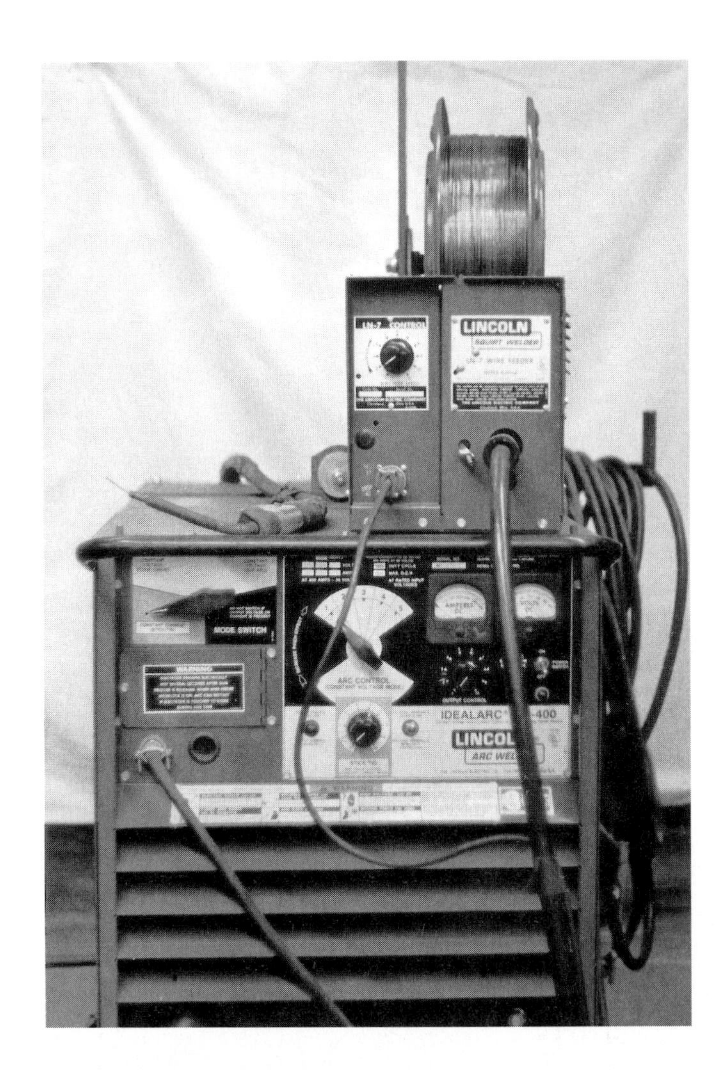

Figure 14-26. Flux core arc welding machine.

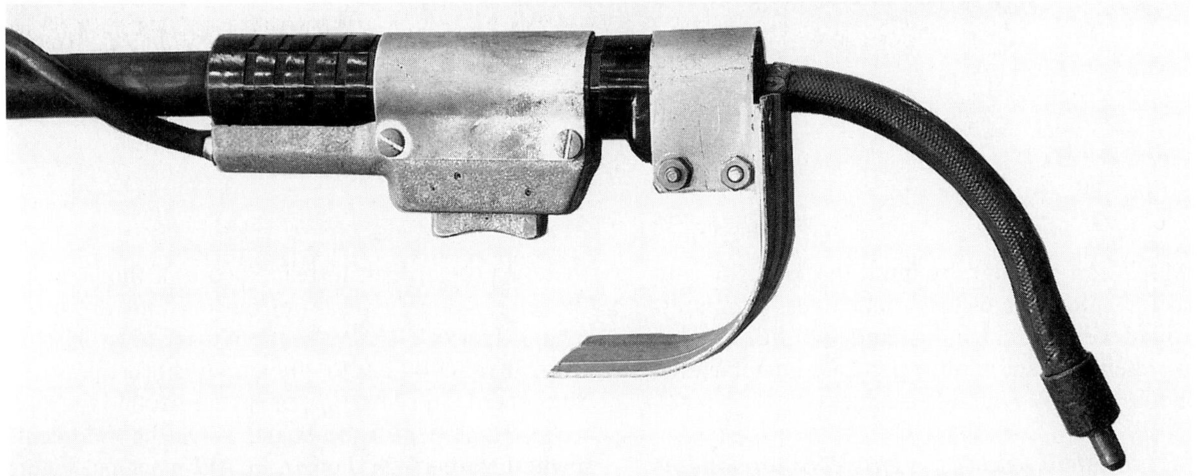

Figure 14-27. FCAW torch.
Courtesy of The Lincoln Electric Company, Cleveland, Ohio.

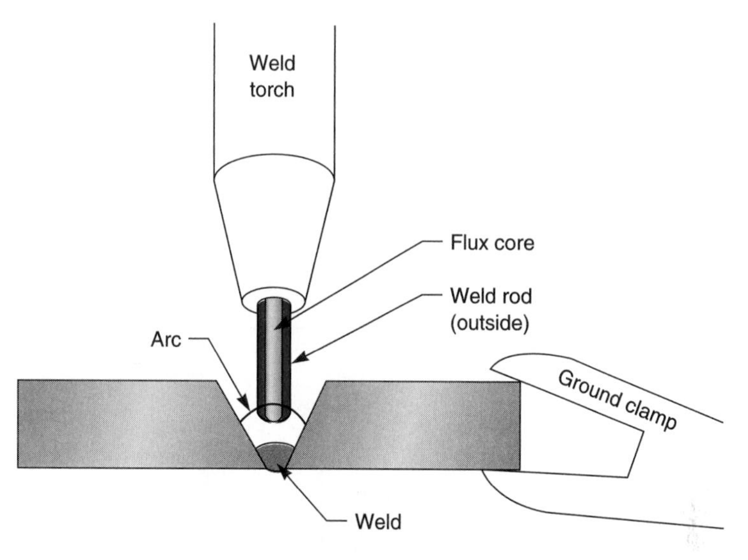

Figure 14-28. FCAW schematic.

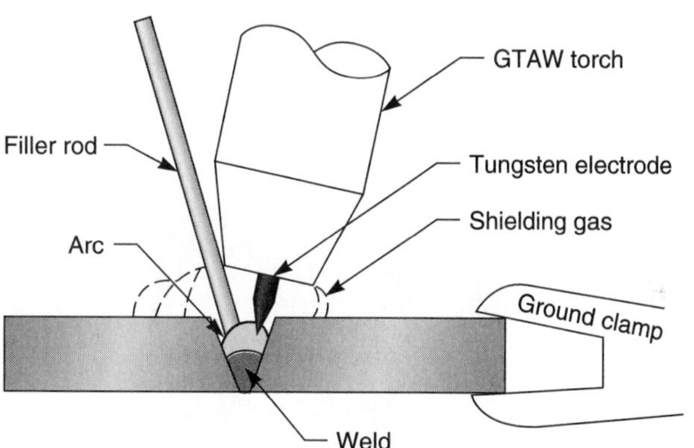

Figure 14-29. Gas tungsten arc welding schematic.

Gas Tungsten Arc Welding. *Gas tungsten arc welding* (GTAW) was formerly known as tungsten inert gas or TIG welding. In this method, a tungsten electrode, shielded by an inert gas, is used to strike an arc with the base metal. The arc melts, or puddles, the base metal and the filler rod is added to the puddle. In the manually controlled units, the current is adjusted by a foot pedal control. The inert gas can be argon, argon-carbon dioxide, or helium. In fact, the earliest units used helium and the process was called *heliarc* welding. Figure 14–29 depicts a GTAW welding torch in schematic. Figure 14–30 is a photograph of the GTAW torch and Figure 14–31 shows a GTAW (or TIG) welder at work.

Gas tungsten arc welding does not produce as deep a penetration as stick or other types of welding and requires more passes to fill a vee joint. As a result, GTAW is a slow method of welding, which results in an expensive product. It does, however, produce one of the cleanest arc welds possible. GTAW can also be used to weld aluminum, magnesium, titanium, and stainless steels. Since aluminum and magnesium oxidize and burn very easily and at relatively low temperatures, an inert gas weld is one of the few types of welding that can be used for these metals. Many aircraft and spacecraft parts are GTAW welded because of the reliability, cleanliness, and strength of the weld.

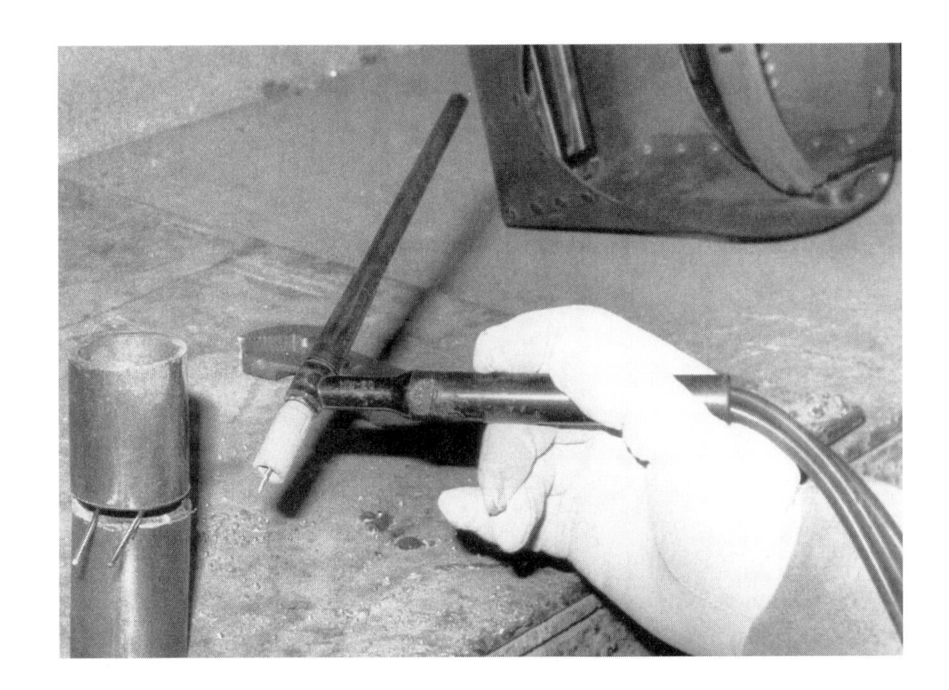

Figure 14-30. GTAW torch.

Figure 14-31. GTAW welding.

OTHER WELDING TECHNIQUES

Several exotic methods of welding have been developed, including the following:

- Electron-beam welding
- Laser welding
- Friction welding
- Thermite welding
- Chemical welding
- Robotics welding

Electron-Beam Welding

Impurities in the weld have always been a concern to engineers and welders. Impurities are created by the flux, air, overheating of the metals, moisture in the flux, and surface corrosion on the base metal. In every case, impurities weaken the weld, make the weld more brittle, increase its tendency to fatigue under repeated stresses, and lower the reliability of the weld. One solution to all of these problems is the *electron-beam weld.*

To understand the process of the electron-beam weld, a few principles of electronics are in order. While trying to invent the incandescent light bulb, Thomas Edison discovered that when metals are heated into the red- or white-hot temperatures, electrons are rapidly emitted from the metal. If a positive charge is placed nearby, the negatively charged electrons will be drawn toward it. The higher the voltage of the positive charge, the faster the electrons will be accelerated. When the electrons collide with another metal, they heat it by the energy of the collision and there is often enough energy to melt the metal. In electron-beam welding, electrons are generated from a hot wire, accelerated toward the metal joint, and focused carefully on the joint to fuse it together. Figure 14–32 illustrates this principle. Figure 14–33 is a photograph of a completed electron-beam weld. The depth of penetration shown in the photograph is about 1.5 inches.

Electrons are very small particles, roughly 1/1839th the mass of a hydrogen atom. If they collide with an atom of oxygen or nitrogen in the air, they will be absorbed. Therefore, electron-beam welding must be done in a vacuum. Since there is no filler rod used in

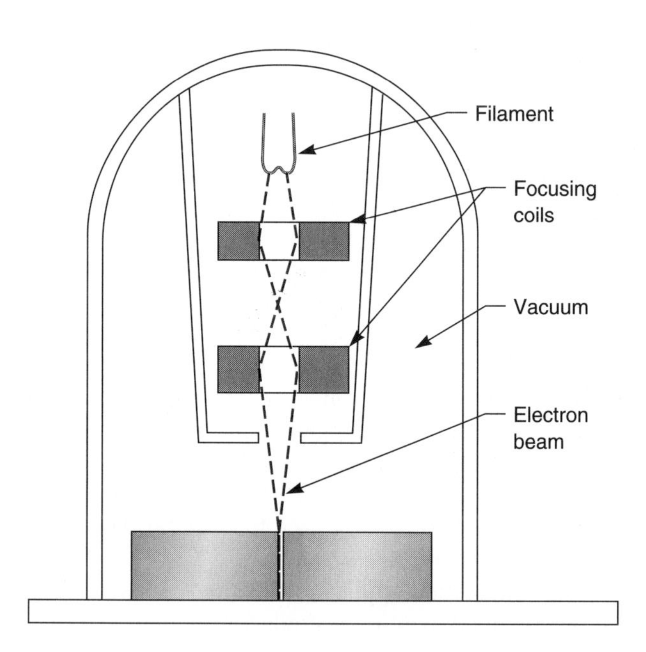

Figure 14–32. Electron-beam welding schematic.

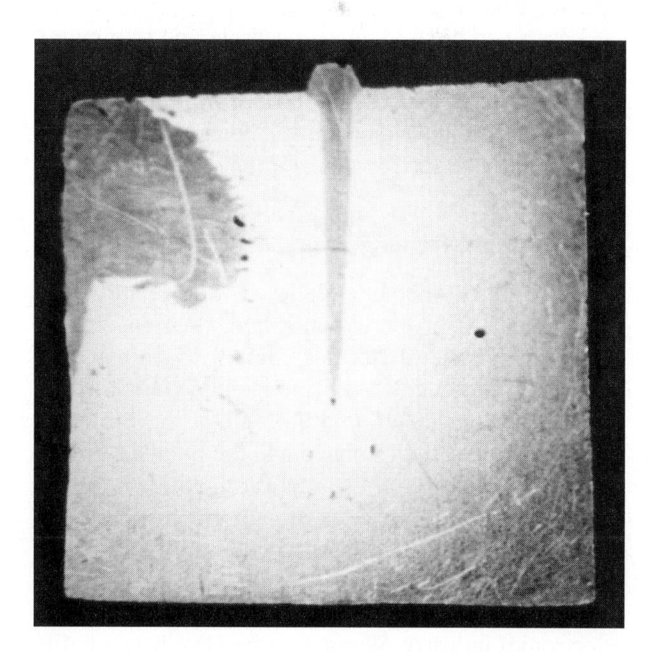

Figure 14–33. Electron-beam weld.

electron-beam welding, the surfaces to be welded are mated very closely and butted together. Penetration can be very deep and welds several inches thick can be made in a single pass. Electron-beam welding is recognized as one of the cleanest welds currently possible.

Laser Welding

Light is a form of energy that can be used to provide the heat source with which two pieces of material can be fused. A magnifying lens can be used to focus the rays of the sun to a point that will burn a hole in paper. The problem with using sunlight as a heat source is that the light from the sun has many different wavelengths. Different wavelengths are bent at different angles by a lens. Therefore, it is difficult to focus these different wavelengths to a single point. A *laser,* which stands for *l*ight *a*mplification by *s*timulated *e*lectromagnetic *r*adiation, produces a very intense and powerful light of a single wavelength. This single wavelength can be focused very precisely and generate a very high temperature at that point. This heat will melt—even vaporize—metals. It is an ideal method of welding because no external gas, except for the atmosphere, is needed.

Lasers can be used on very delicate and thin metals as well as larger sizes. The major problems with the current lasers lie in the cost and bulk of the power source. Lasers for use in welding are still being developed and no doubt will be used to a great extent in the future.

Friction Welding

Rub two sticks together and they get hot. Friction generates heat. If properly controlled, the heat will be localized sufficiently to melt metals. Machines have been built that will vibrate one metal against the surface of another metal until both metals have melted. The vibration is then stopped and the puddle of metal solidifies, welding the two metals together. It is simple, clean, quick, inexpensive, and effective. *Friction welds* have thus far been used mainly for very small applications.

One major application of friction welding is in the electronics industry. Wires must be cleanly welded to transistors, diodes, and integrated circuits. However, heat destroys semiconductors. Friction welding is often the answer. The very fine, hair-thin wires are friction welded to contact points on the integrated circuits. Figure 14–34 is a diagram of the friction welding process.

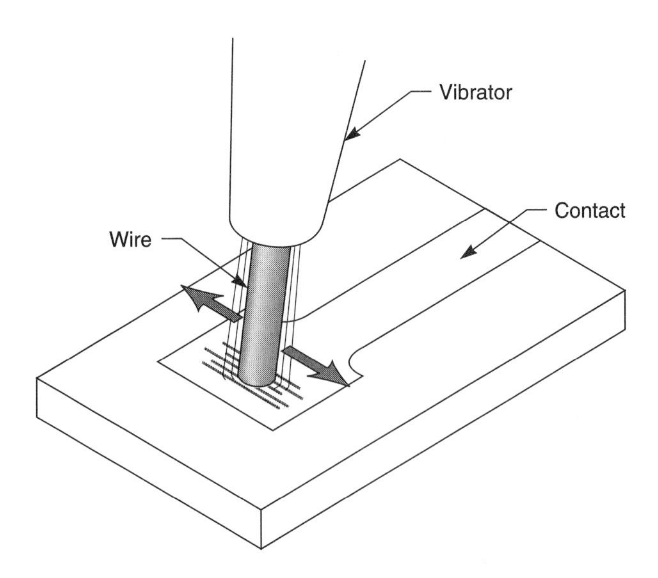

Figure 14-34. Friction weld.

Thermite Welding

A well-known chemical reaction is the reduction of iron oxide with aluminum, which produces iron, aluminum oxide, and a tremendous amount of heat. Chemists write the equation as follows:

$$Fe_2O_3 + 2Al \rightarrow 2Fe + Al_2O_3 + \mathit{HEAT}$$

If powdered iron oxide (rust) is gently mixed with aluminum powder and the mixture ignited, the reaction will start and heat will be generated. In fact, this mixture is used in incendiary bombs by the military. This heat can be used to create a weld. Further, the pure iron generated will serve as the filler metal. This technique is known as *thermite welding.*

One interesting application of thermite welding is in the joining of continuous rails for the railroads. Traditionally, 20-foot sections of rails were bolted

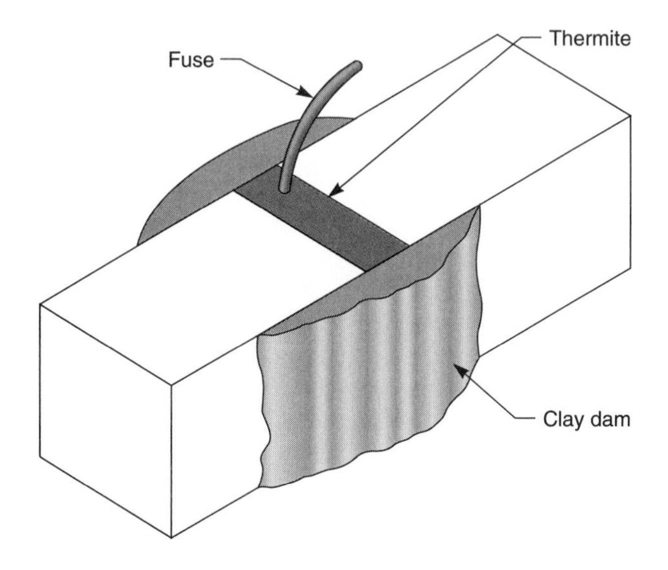

Figure 14–35. Thermite weld.

together with a gap between each section. The wheels running over these gaps made the "clickity-clack" sound heard by passengers. Now the rails are welded together. Since the rails are quite thick, it would be difficult to weld by arc, acetylene, or other common techniques. The rail joints are encased in a clay *dam,* thermite is packed between the butted ends, and the mixture is ignited by touching it with a torch. The heat welds the ends of the rails and all that is left to do is to remove the dam and grind the surface smooth. High speeds for trains are now possible with these continuous rails and there isn't a "clickity-clack" to be heard. A sketch of this technique is shown in Figure 14–35.

Chemical Welding

Welding can be done to materials other than metals. Plastics are often welded. Sheets of Lucite, Plexiglas, or acrylic (all are trade names for polymethyl methacrylate) can be fused by acetone or methyl ethyl ketone (MEK). The chemical simply dissolves the surfaces of the plastic. When the solvent evaporates, the surfaces repolymerize to form a true weld. This method is called *chemical welding.*

Robotics Welding

In today's industry, robots are increasingly being used to perform many functions, including welding. Just about any method of welding can be done by robots, but some methods have been used extensively. Resistance welding (spot welding) and the various forms of ARC welding are especially compatible with robots. Automobile assembly lines use robotic welders for complete fabrication of many of the parts of an automobile.

■ ANNEALING

Welding deforms the grain structure of the metals. Some of the grains formed when the molten metal solidifies will be larger than others. This difference in grain sizes creates internal stresses, which cause the metal to be brittle and easily broken or fatigued. To prevent the weld from failing, the internal stresses throughout the welded joint must be relieved and the grains must be of an even size. Welds should always be allowed to cool slowly. It is often necessary to heat the entire joint and surrounding metal into the red heat range, then cool it very slowly. This process is called *annealing* the weld. Whether or not a weld needs to be annealed depends on the type of metal welded, the filler rod used, and the type of weld. Welds that are quickly applied and retain the heat locally around the weld site generally require annealing.

■ SPECIFYING A WELD

The engineer designing a part is responsible for specifying the weld. Working prints do this by use of standard symbols. While it is not the purpose of this text to delve deeply into welding design, the student of manufacturing should understand at least the basics and be able to understand a welding drawing. Refer to Figure 14–36 as we explain the different weld symbols.

The leader (arrow) always points to the position of the weld. At the bend of the leader, a solid dot, or "flag" would indicate a field weld. A circle around the joint means to weld all around the object. Any

Figure 14-36. Weld symbol notation.

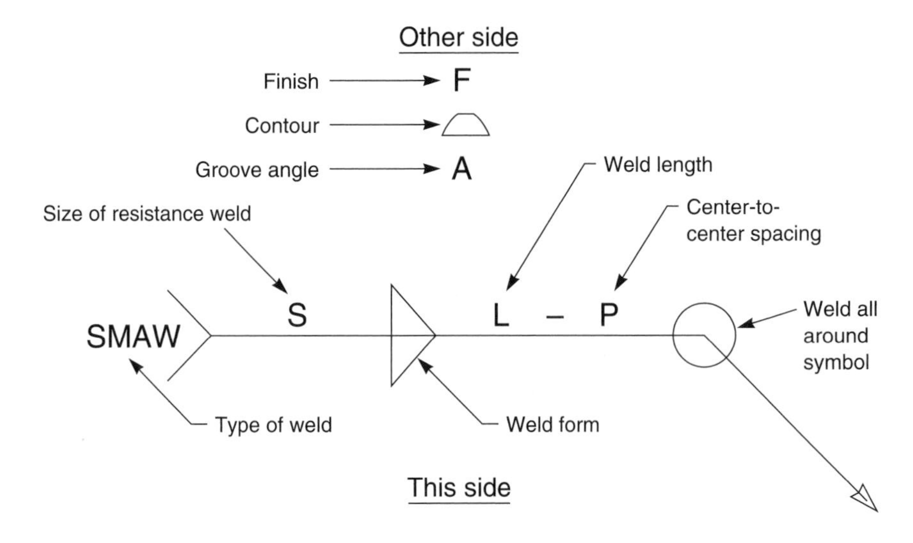

Figure 14-37. Examples of weld symbols.

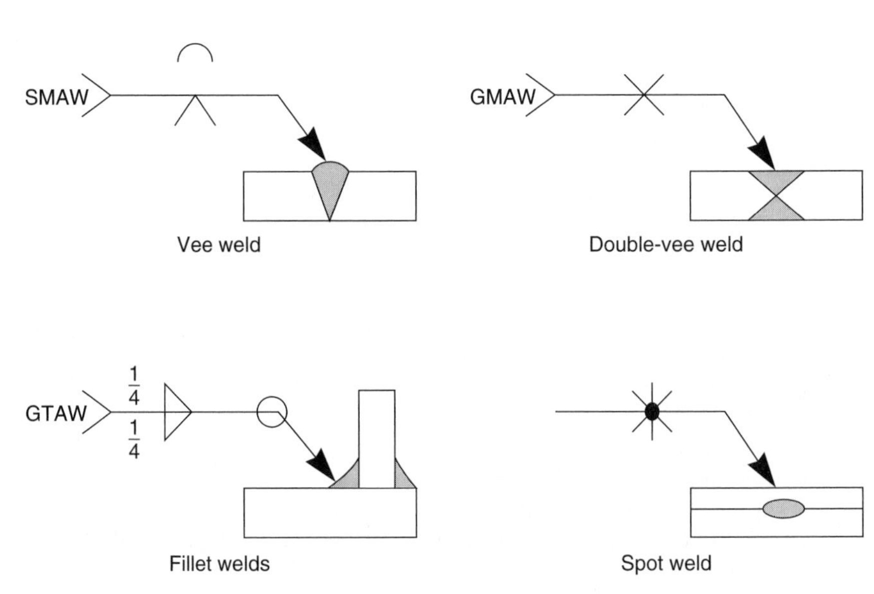

markings on the top of the horizontal part of the leader pertain to the weld on the side opposite the arrow, whereas the markings on the bottom are for the weld on the side of the material to which the arrow is pointing. Circles, triangles, U's, or other symbols on the arrow designate the type of weld, (i.e., bead, fillet, groove, bevel, spot, upset, etc.). The length and width of the weld are also designated. The pitch or center-to-center spacing of the welds can be specified as well as the number of spot or upset welds. A tail may also be added to the arrow to indicate the process or type of weld (i.e., SMAW, GMAW, etc.)

A few examples of these weld symbols are shown and explained in Figure 14–37. A more detailed explanation of these weld symbols can be found in any engineering handbook or welding handbook. One word of caution. Welding symbols, drafting symbols, and standards change from time to time. Further, different industries often have their own standards and symbols. Older drawings may use one symbol, whereas newer plans may show different ones.

IMPERFECTIONS IN WELDING

Welds make a continuous piece of material out of the parts welded together. Flaws anywhere in this metal will weaken the part. Welding often leaves imperfections in the metal. Oxy-acetylene welds often have air bubbles in them, caused by too much oxygen in the flame. Slag is sometimes found in forge welds. Inclusions of slag can come from the flux around arc welding rods. Improper cooling of a weld can cause microscopic cracks in the weld.

In many applications, the weld is sufficiently strong even with these imperfections. But, all too often, people's lives depend on a weld being without flaws. The loss of the *U.S.S. Thrasher,* a Navy submarine, was traced to a faulty weld. The failure of several bridges and other structures have been attributed to poor welds. Machines that produce vibration have been known to fail because of faulty welds. Large cargo ships have broken apart because the welds were too weak to withstand the force of ocean waves. Therefore, all welds in critical areas of stress should be inspected.

WELD TESTS

As stated earlier, a good weld should not fail in the weld. To receive their certification for each specific type of welding, welders are required to demonstrate that they can produce welds that will not fail. After a specified weld has been completed, the welds are tested by several methods.

Visual Tests

A preliminary inspection of a weld will often detect flaws. After the slag has been removed with a chipping hammer, the welds should be of even thickness, have no pits or pinholes, show complete fusion with the base metal, and have no cracks. Figure 14–38 shows a good weld and Figure 14–39 shows a weld that should be rejected.

Figure 14–38. Good weld.

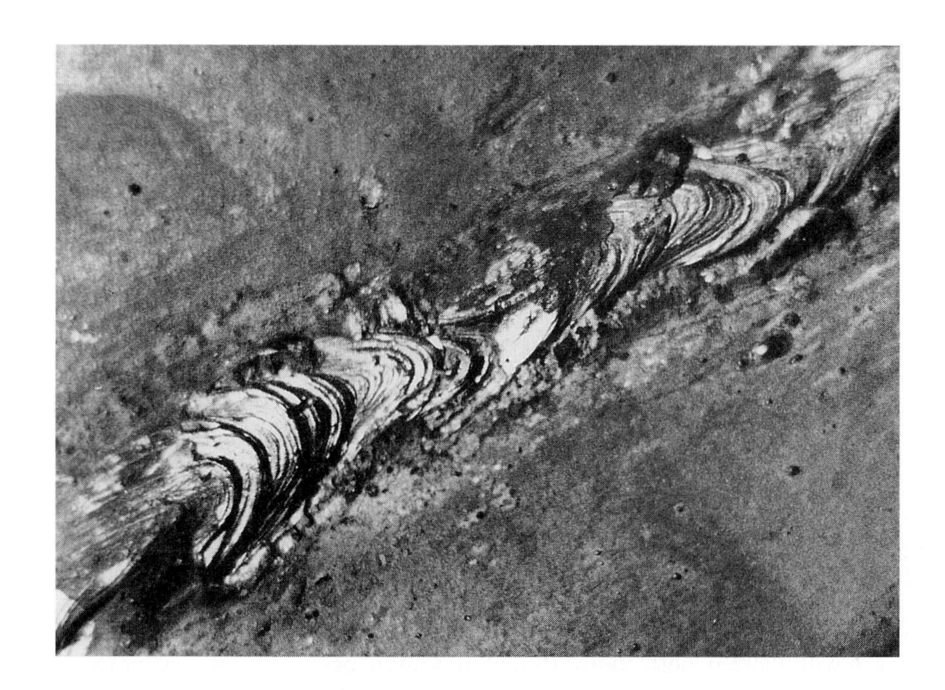

Figure 14-39. Poor weld.

Subsurface Tests

Visual tests detect only flaws on the surface. Subsurface cracks, lack of penetration, and other flaws can be detected by *x rays, gamma rays, sonic tests, magnefluxing,* or *holography.* All of these are nondestructive tests, meaning they do not damage the material. X rays and gamma rays require that a photographic film be laid next to the weld and the rays passed through the weld. Flaws quickly show up on the film. Sonic testers rely on the fact that sound waves, projected through the base metal, are reflected by flaws and their echoes can be seen on an oscilloscope screen. In magnefluxing (Figure 14–40), a heavy magnetic field is placed in the base metal around the weld. Cracks and other flaws will deflect the magnetic field and align it with the crack. A light sprinkling of iron filings follows the lines of magnetic flux and the filings will concentrate in areas of a flaw. The location of the flaw can then be easily seen.

Figure 14-40. Magnefluxing.

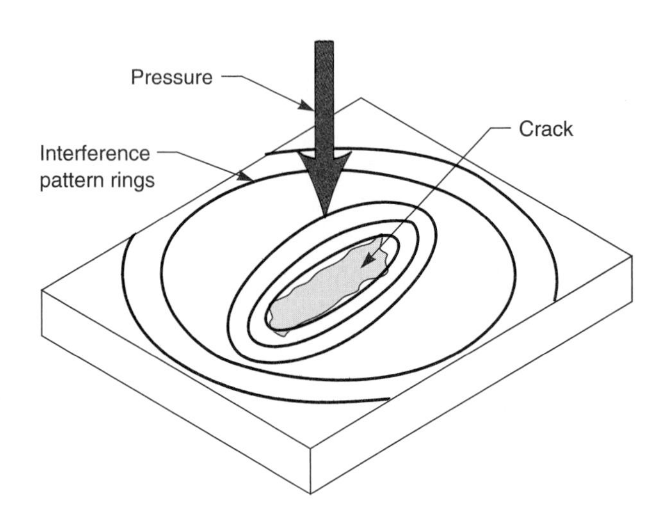

Figure 14–41. Holographic test.

A newer method of nondestructive testing involves holography. A uniform laser-generated holographic pattern is projected on the metal over the weld. A slight load placed against the base metal will deflect the material. The holographic pattern will show this deflection even though it may only be a ten-thousandth of an inch or less. If there is a flaw in the metal, the pattern will not deform uniformly but will distort more along the flaw. Holographic patterns are so sensitive that they can detect a person leaning against a concrete wall from the other side. Figure 14–41 depicts an example of holographic testing.

Mechanical Tests

Guided Bend Test

Welds are destructively tested in several ways. In the *guided bend test,* the weld is placed atop a cylinder having a radius of 1.5 to 2 inches. A good weld allows the metal to bend 180 degrees and not crack or break. Figure 14–42 shows a guided bend test.

Tensile Test

Strips from a vee or butt weld can be placed in a *tensile test* machine and the bar pulled to failure. The bar should break in a place other than the weld. Figure 14–43 is a photograph of a tensile test machine pulling a welded sample.

Impact Tests

Welds are also tested by *impact* machines. As explained in Chapter 2, an impact is the sudden application of a force. Welded parts that are expected to take or receive sudden blows should be tested in impact. Welded Izod or Charpy bars are made with the weld in the notched portion of the bar. The impact machines record the amount of energy (in foot-pounds) that is absorbed by the metal before breaking. If the impact energy of the weld falls below a specified standard, the weld is rejected. Refer to Figures 2–38 through 2–42 for the application of impact.

■ WELD APPLICATIONS

To select the proper weld for a specific application, the welder needs to consider the advantages and disadvantages of each type of weld. Table 14–2 compares several welds in the categories of strength, cleanliness, and cost and provides a good reference for the best applications of different types of welds.

■ WELDING SAFETY

Chapter 20 deals exclusively with industrial safety, but welding has some specific safety problems that we should discuss now. Because all welding, except chemical welding, requires very high temperatures, welders and other workers must be protected from the heat. They should be dressed in leather or other protective garments. These include jackets with sleeves, gloves, aprons, and leggings. Arc and electric welds also generate significant amounts of ultraviolet radiation. Besides the heat, welding produces splatter with liquid metal often flying several feet from the welder.

As with any high-temperature operation, welders and other workers must be protected against the intense heat and bright light of the welding process. For this reason, welders wear leather aprons, arm

Figure 14-42. Guided bend test.

Figure 14–43. Tensile test.
Photograph courtesy of Tinius Olsen Testing Machine Co., Inc., Willow Grove, Pennsylvania.

Table 14–2. Comparison of Welds

Weld Type	Strength	Cleanliness	Cost	Applications
Forge	Good	Poor	Low	Solid steel plates, rods, pipes, bars
Oxy-acetylene	Fair	Poor	Low	General welding
Oxy-hydrogen	Fair	Fair	Expensive	High-melting-point metals
Plasma	Good	Good	Expensive	Ceramics, very high-melting-point metals
Resistance	High	Good	Low	Automobiles, aircraft, blind welds
SMAW	High	Good	Low	Low- and medium-carbon steels
Submerged arc	High	Fair	Moderate	Low- and medium-carbon steels
FCAW	High	Fair	Moderate	General purpose
GTAW	High	Very good	High	Stainless steels, aluminum, magnesium
GMAW	High	Good	Moderate	Steels, continuous beads, robotics applications
Electron beam	High	Best	Very high	Extremely clean, high-quality welds

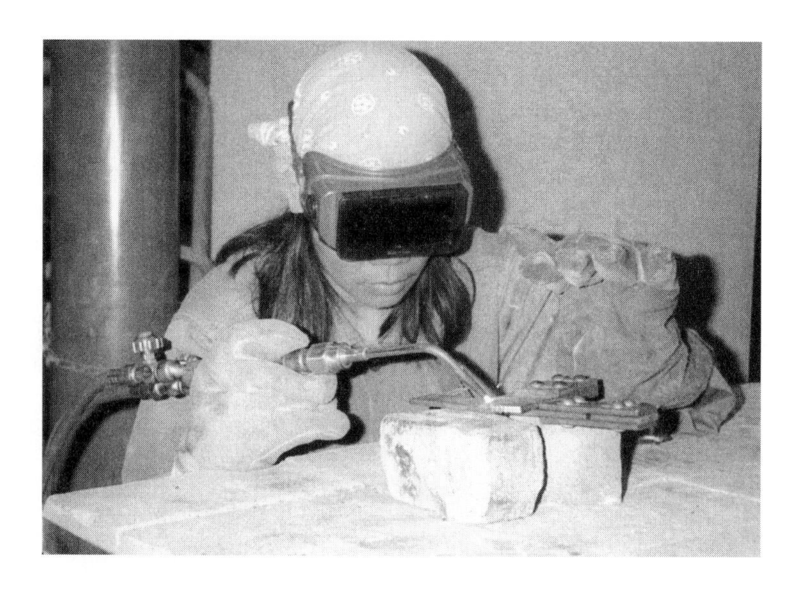

Figure 14–44. Oxy-acetylene gas welder.

protectors, gloves, and possibly leggings. Dark-colored cobalt glasses, face shields, or welding hoods are also mandatory. (Figure 14–44 shows a gas welder at work.)

No one should be permitted anywhere near welding in progress without proper eye shields. The light, especially from arc welding, is so intense that it will cause blindness in a very short time. Even spectators should wear safety garments and eye shields.

People have received "sunburns" on unprotected skin simply by being around arc welding. If more than one welder is working in close proximity, weld screens should be placed between them.

In all forms of electric welding, care should be taken to ground the welding machines. These machines work on 220- to 440-volt lines. They should be inspected periodically to ensure they are in good working order.

Gas welding also poses other problems. Oxygen reacts explosively when in contact with oils and greases. In fact, a case has been recorded in which a person tried to cool himself by running oxygen down his back and was severely burned as the oxygen reacted with the oils in his skin. Remember, oxygen supports combustion. Although oxygen itself doesn't burn, pure oxygen will cause small fires to burn more rapidly. Care should be taken to keep any combustible materials away from the welding area.

Acetylene also requires special precautions. If acetylene is pressurized to more than about 15 psi, it will polymerize, release considerable heat, and cause an explosion. The tanks in which it is sold contain diatomaceous earth or other material to prevent this polymerization from occurring, thereby allowing somewhat higher pressures in the tank. The carbon in the acetylene can also embrittle the tanks, causing them to explode at relatively low pressures. All acetylene and other high-pressure tanks must be pneumatically tested prior to refilling. Acetylene is a dangerous chemical and should be treated as such.

The hazards of using hydrogen are well known. It is one of the most flammable gases known. Its light atomic weight allows it to escape readily through faulty valves. These gases are odorless and colorless and therefore difficult to detect when leaking into the atmosphere. Escaping gas can cause explosions.

Even though the inert gases argon, helium, nitrogen, and carbon dioxide do not burn, they are stored in their containers under very high pressures. The fact that these gases are highly compressed can create safety hazards. Oxygen and other gases are stored in tanks at pressures in excess of 2000 psi. If a control valve is broken off, it can shoot out like a bullet. The flow valves should be released completely prior to turning on the main valve on the tank. Many accidents have been caused by faulty valves and defective tanks.

Many states have laws requiring that all high-pressure tanks be inspected and tested periodically. The dates of these inspections are usually imprinted on the tanks. Users of welding gases should note the date of the last inspection and be alert to cracks or other defects in the tanks. Tanks that have been dropped should be tested prior to refilling. Gas tanks must also be tethered or strapped into a position that will prevent them from tipping. High-pressure tanks are a constant source of danger and should be treated carefully.

All welding operations must be well ventilated. Special fume hoods should be placed at the welding torch or welding gun level, not overhead. The overhead fume hoods draw the fumes from the welding process past the face of the welder on their way to the hood; instead, the fumes should be drawn away from the welder.

PROBLEM SET 14–1

1. Which method of welding uses the least expensive equipment?
2. Which method of welding produces the cleanest weld?
3. Which method(s) of welding must be done in a vacuum?
4. What is an oxidizing flame on a torch? What is it used for?
5. What is a reducing flame on a torch? When is it used?
6. What is the fastest method of making a weld?
7. What are the advantages and disadvantages of gas metal arc welding?
8. What are the advantages and disadvantages of oxygen-acetylene welding?
9. What are the advantages and disadvantages of shielded metal arc welding?
10. In GMAW, what controls the current flow?
11. What variables must be controlled in GMAW?
12. What method of joining would be easiest for a permanent bond of cast iron?
13. List five things that could be visually checked on a weld.
14. What method of welding would be best for high-melting-point metals?
15. What method of welding could be used to fuse ceramics or cermets?
16. What is the tensile strength of a 7018 SMAW welding rod?
17. Given a 7025 SMAW welding rod, what is the tensile strength of the rod? In which positions could it be used? For what polarity (of dc voltage) could it be used?
18. Why must low-hydrogen rods be preheated?
19. Why is the fusing of two pieces of Plexiglas or Lucite together by using a solvent considered to be a weld?

Our discussion on welding not only analyzed the individual types of welding techniques with respect to the strength of the weld, its cost, the materials to be joined, and the position of the actual work, but also discussed the inspection and testing of the completed weld.

The many types of welds processes include the following:

- Forge welding
- Gas welding
- Electrical welding
- Electron-beam welding
- Laser welding
- Friction welding
- Thermite welding
- Chemical welding
- Robotics welding

Each of these types of welding can have many variations.

Welds are specified by the design engineer and must be tested after completion.

PHOTO ESSAY 14-1

Igniting the Oxygen-Acetylene Torch

STEP 1 Select and attach the proper tip for the job to be done.

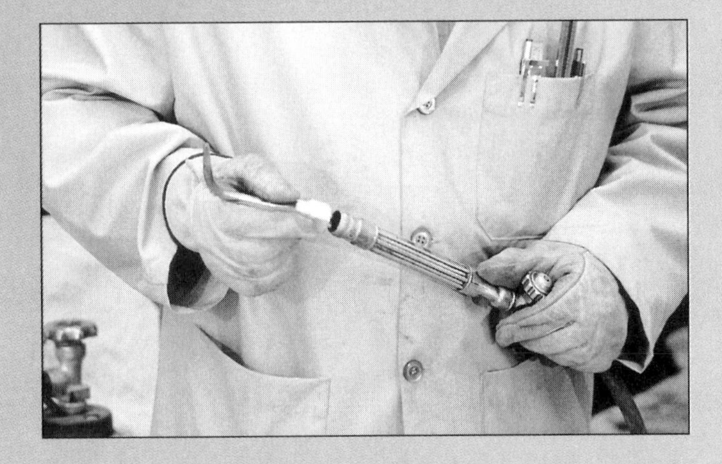

STEP 2 Make sure the pressure adjustment handles are screwed all of the way out (counterclockwise). Then open both of the tank valves at least 2½ turns.

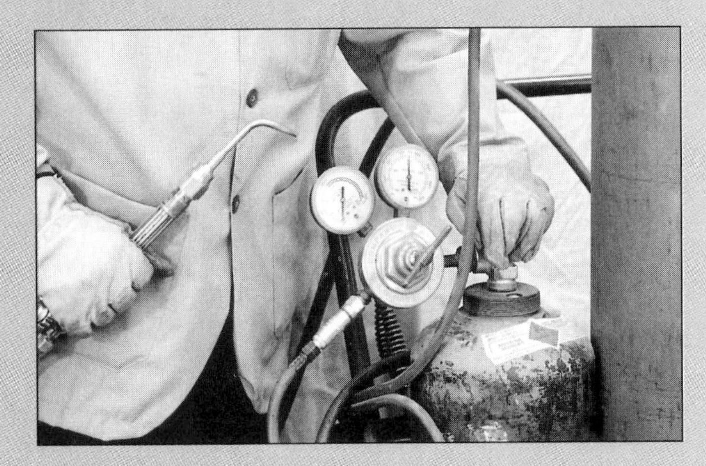

STEP 3 Adjust the gas pressure of both tanks by turning the pressure-adjusting handles clockwise (right) to get the proper pressures for the welding tip in use. The pressures of the oxygen and acetylene are set individually. The pressure should be set with the torch valve open.

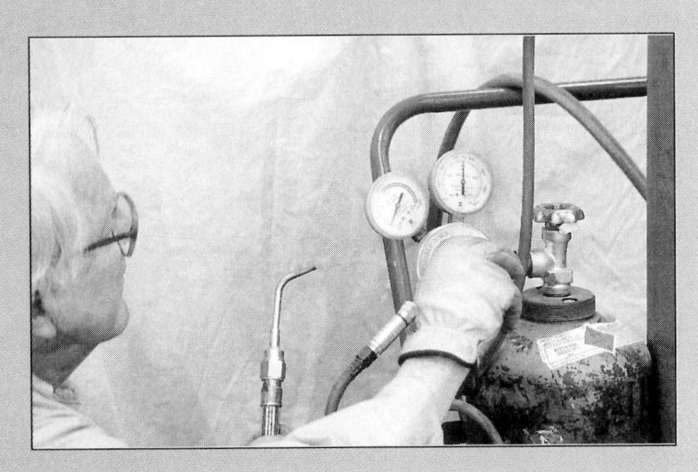

STEP 4 Open the acetylene valve on the torch slightly and ignite the acetylene using a striker.

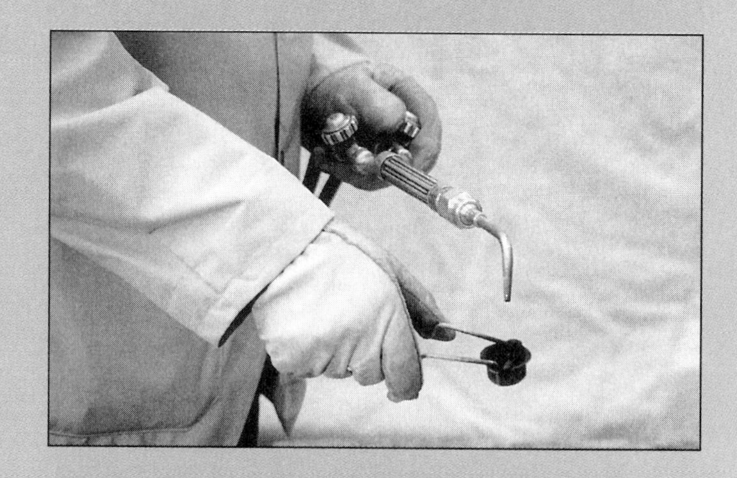

STEP 5 Adjust the acetylene flame (*red hose*) to get a slight "feather" on the flame.

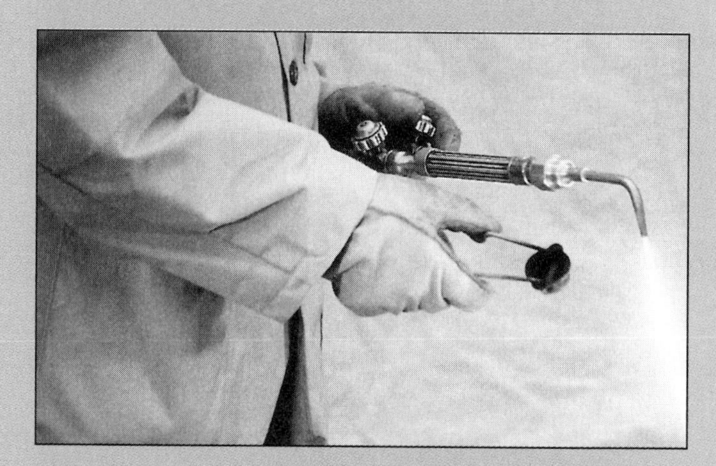

STEP 6 Open the oxygen valve on the torch (*green hose*) and adjust to get a "neutral" flame. The outer light blue cone of flame should be brought back until it just touches the inner bright blue cone.

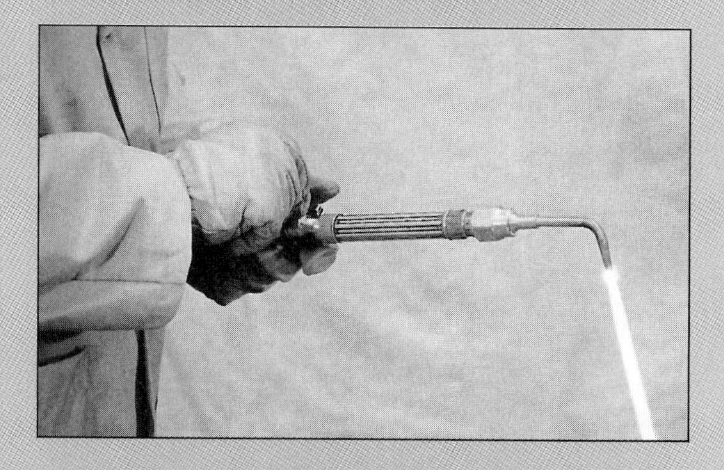

SHUTTING OFF THE TORCH

STEP 7 Always shut off the oxygen first. Shutting off the acetylene produces a loud "pop," which can damage the tip. Turn the oxygen valve clockwise (right) until firmly closed. Only the orange acetylene flame should be left.

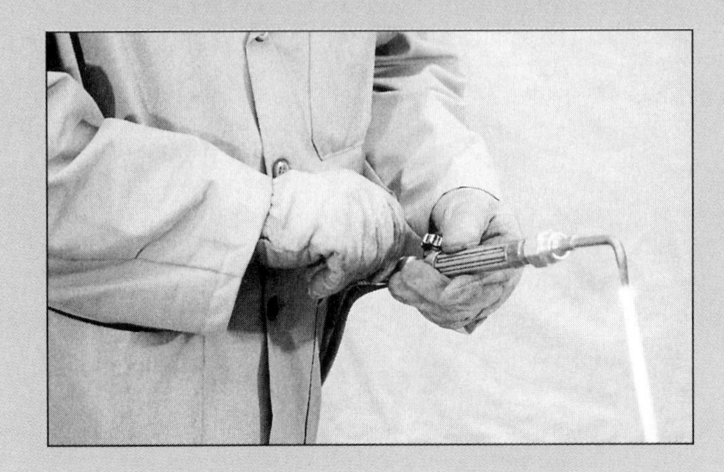

STEP 8 Shut off the acetylene valve. Make sure the acetylene flame is completely out. The acetylene flame may linger a while due to the residual gas in the hose.

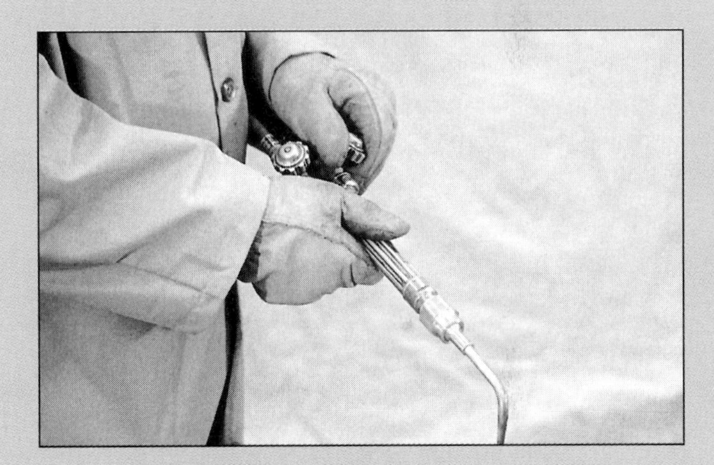

STEP 9 Close the main valves on the oxygen and acetylene tanks.

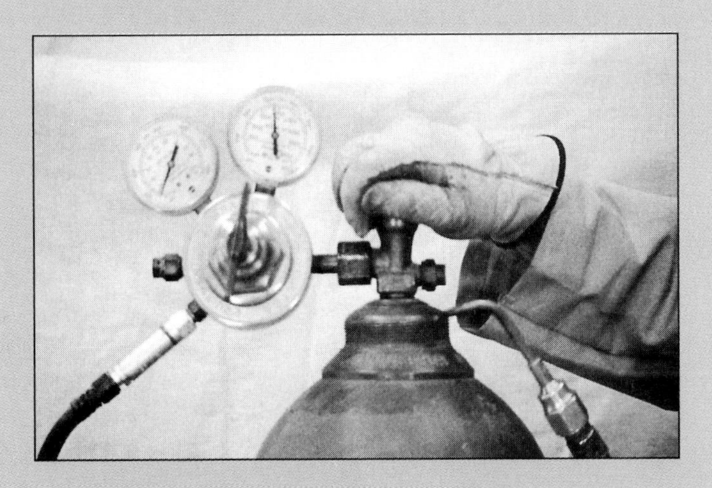

STEP 10 Bleed the lines. Open the valves on the torch to allow the pressure on the pressure-adjusting gauges to read zero. Be sure to reclose the valves on the torch.

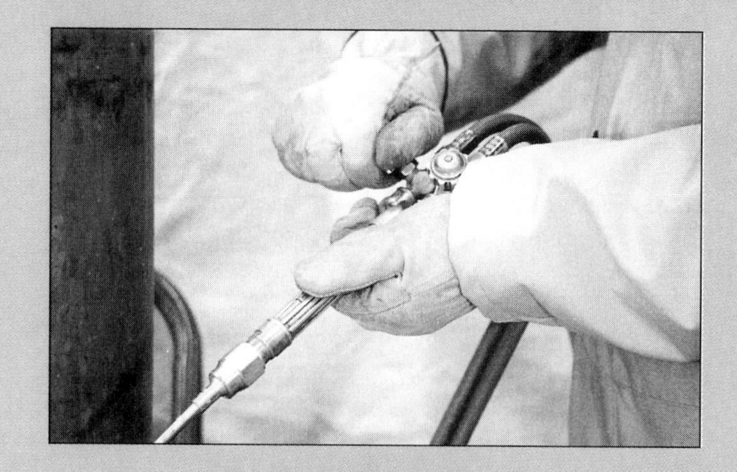

STEP 11 Open the pressure-adjusting handles all of the way, ready for the next opening of the main tank valves.

Chapter 15

Mechanical Methods of Joining

Besides adhesives and welds, parts can be joined by any number of mechanical means. These methods of joining include the following:

- Fastening devices (*fasteners*), such as screws, bolts, nails, staples, pins, rivets, keys, splines, snap rings, etc.
- Crimping
- Making the two parts as one piece.

A few of these methods deserve discussion here.

FASTENERS

Different applications require different types of fasteners. One type of fastener will not work on all applications; and if one kind of fastener won't do for all applications, then we should look at the "rules" for selecting the fastener to match the application. These "rules" come from the needs of the parts to be joined; that is, from the *needs of the application.*

Needs of the Application

Two needs of the application that must be determined are:

1. Is the joint temporary or permanent?
2. Is relative movement permitted?

A third need was mentioned in the preceding chapter on welding:

3. What is the material used to make the parts?

Note, however that the application may have several more needs:

4. In what environment must the fastener survive? (That is, what chemical, thermal, and vibrational conditions might exist?)
5. What tools are required to install it (or to remove it, if it is a temporary fastener)?
6. What skills are required to install it?
7. How much does it cost?
8. What is its availability?
9. How does the finished assembly appear? Will the fastener need to have a special finish?
10. Does the fastener project beyond the part? Will that be a safety hazard?
11. Will the fastener be used for a second function (such as a pivot or a stop)?
12. What strength (or other properties) must the joint have?

And the list goes on! So, given the number of needs that the application may have, it is no simple task to select a fastener. The following discussion cannot cover all possible fasteners; only the most commonly used types are discussed. The material is arranged as follows:

1. Threaded fasteners
2. Nonthreaded fasteners
3. Fasteners for shafts

Threaded Fasteners

Threaded fasteners are the primary type to use when the application requires a nonpermanent fastening. Threaded fasteners include machine screws ("bolts"), wood screws, sheet metal screws, nuts, threaded inserts, studs, and special parts such as screw-hooks or eyes ("eye bolts") and turnbuckles. Note that the parts themselves can be threaded (such as a food container that has a screw-type lid, or a wooden broom handle), but these uses are other *joining methods,* not fasteners.

Threads

Threads come in many types. The most common is the standard 60° vee thread, but some applications require special thread designs. When high strength is required, Acme threads might be used. For applications that require a high force to be applied in one direction, a buttress thread could be used. Square threads are applicable for precision positioning. Knuckle and round threads are often used on glass and rolled metal applications. Figure 15–1 shows a few of the types of threads.

The vee threads have standardized details for the shape, especially how round the top of the ▲ (the *crest*) must be and how sharp the bottom of the V (the *root*) must be. In fact, this became a problem affecting the interchangeability of American and British threaded parts during the second world war, but has since become standardized internationally.

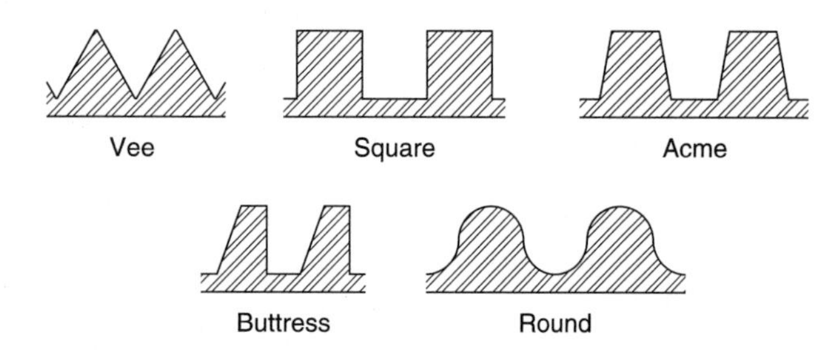

Figure 15–1. Types of threads.

As a threaded part is turned clockwise, it moves either away from you (a *right-hand* thread) or toward you (a *left-hand* thread). Left-hand-threaded fasteners are used when the rotation of the parts would tend to loosen a right-hand-threaded fastener. Generally the fasteners are specially marked if they have a left-hand thread.

Machine screws, wood screws, sheet metal screws, and other threaded fasteners or parts come in standard diameters and number of threads per inch (Table 15–1). The distance between two adjacent ▲'s (peaks or crests on the thread) is the *pitch.* Notice from Table 15–1 that the pitch is given directly in millimetres for the metric threads, but is only inferred in the inch system. For the inch system, the pitch is the inverse of the threads per inch. For example, a *10-24* thread has a crest diameter of 0.190 inch (which isn't related to pitch in any way) and 24 threads per inch. Therefore, the pitch is

$$1 \div 24 \text{ threads per inch} = 0.0417 \text{ in.}$$

For screw threads, each diameter size generally has either two or three pitch sizes: Unified National Coarse (UNC), Unified National Fine (UNF), and possibly Unified National Extra-Fine (UNEF). The coarse thread has more resistance to "stripping" the threads out but less resistance to shearing the bolt due to overtorquing. The fine thread has more threads per inch; it is used to make fine adjustments at the expense of strength. The extra-fine thread is intended for automotive or aircraft applications where the smaller pitch allows more threads to be cut in a thin piece of material.

Threaded parts may fit loosely (and be easy to disassemble) or fit more snugly in high-precision applications. These fits are known as *thread classes* and are specified as follows:

"Loose" fit: external = 1A; internal = 1B (loosest)
"Free" fit: external = 2A; internal = 2B
"Medium" fit: external = 3A; internal = 3B (tightest)

Table 15–1. Screw Threads

Size	Threads per Inch		
Diameter (in.)	UNC	UNF	UNEF
0 (0.060)		80	
1 (0.073)	64	72	
2 (0.086)	56	64	
3 (0.099)	48	56	
4 (0.112)	40	48	
5 (0.125)	40	44	
6 (0.138)	32	40	
8 (0.164)	32	36	
10 (0.190)	24	32	
12 (0.216)	24	28	32
1/4	20	28	32
5/16	18	24	32
3/8	16	24	32
7/16	14	20	28
1/2	13	20	28
9/16	12	18	24
5/8	11	18	24
3/4	10	16	20
7/8	9	14	20
1	8	12	20
1/25	7	12	18
1.5	6	12	18

Metric Screw Threads

Diameter of Screw (mm)	Pitch* (mm)	Diameter of Screw (mm)	Pitch (mm)
6	1.00	12	1.75
7	1.00	14	2.00
8	1.25	16	2.00
9	1.25	18	2.50
10	1.50	20	2.50
11	1.50	22	2.50

*Pitch is the distance between peaks on the threads.

In these designations, the "A" in each case indicates an external thread and the "B" refers to an internal thread. Most purchased threaded fasteners are class 2, although some class 3 fasteners are commercially available. A class 1 fit would be specified if the environment can be expected to clog the thread and interfere with disassembly.

The tool for cutting internal threads is known as a *tap;* for cutting external threads, the tool is called a *die.* Either kind of thread can be cut on a lathe or formed by several other manufacturing processes. Figure 15–2 shows typical taps and dies.

Machine Screws

The great variety of available machine *screws* allows a selection to meet nearly any application need:

- Relative movement is permitted.
- Some (but not all) materials can be threaded easily (to accept the fasteners).
- Fasteners can be designed to resist hostile environments (such as vibration).
- The tools required can be common (if removability is important) or specialized (if limited access is important).
- Little skill is required to install a threaded fastener.
- Threaded T-nuts (which project into material such as wood or sheetrock) eliminate the projecting end of the fastener.
- Fasteners are available with a variety of heads to perform a variety of secondary functions.
- High-strength machine screws are available for applications requiring high fastener strength.

Machine screws are commercially available with approximately 12 standard head shapes and with several standard finishes, so there should be one that is appropriate for nearly every application. A few types of machine screws are depicted in Figure 15–3.

Wood Screws

Wood screws have a tapered thread so as to wedge their way into the grain of the wood. The diameter of the shank of a wood screw should be specified, but the pitch of the thread is not important because it doesn't *thread into* a threaded part.

Sheet Metal Screws

Sheet metal screws are a cross between machine screws and wood screws: They have a tapered thread at the tip (in order to start threading into the pilot hole), but a straight thread for most of their length.

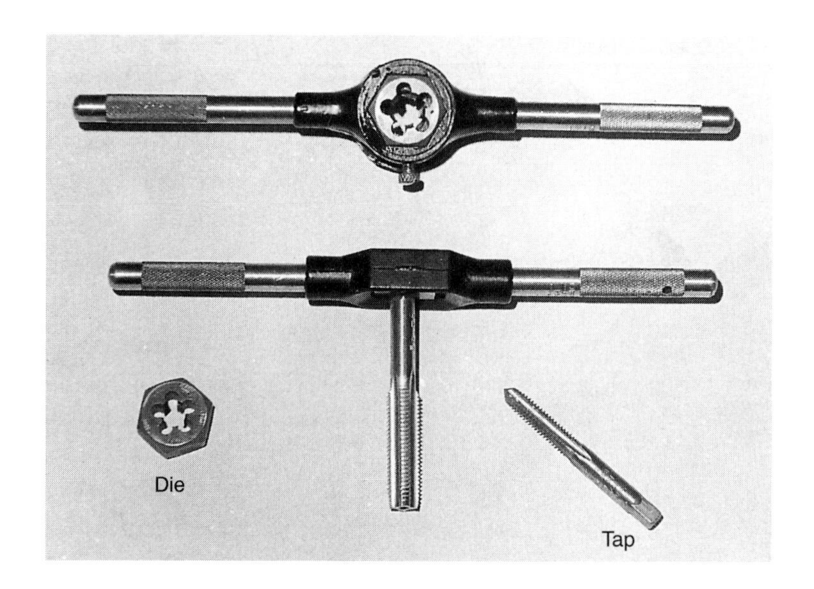

Figure 15-2. Taps and dies.

The tapered thread starts cutting threads (into soft metals such as aluminum) and the straight threads fit into the threads formed by the tip. As with the wood screw, there is no need to specify the pitch of the thread because the screw cuts its own thread into the metal. Self-tapping sheet metal screws have a tip similar in shape to a tap; these are for cutting threads into hard material.

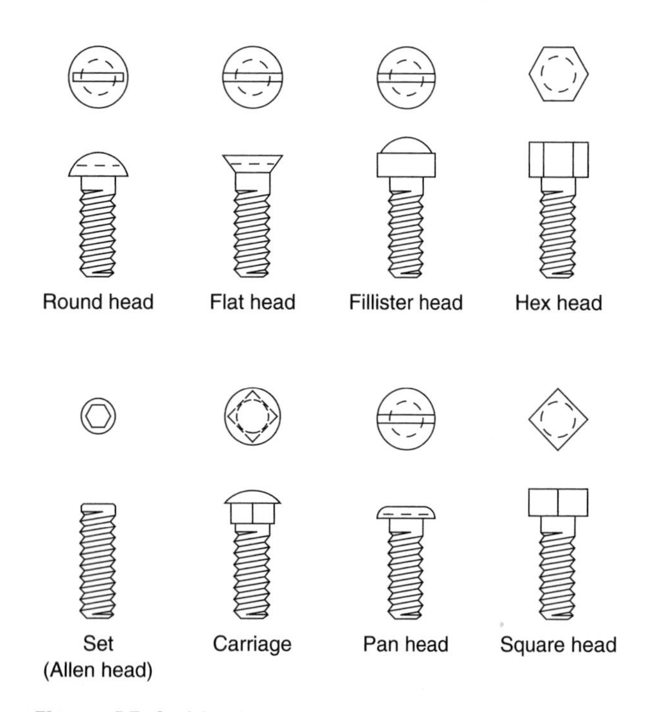

Figure 15-3. Machine screws.

Specifying a Thread

Screws, bolts, and other fasteners are specified by a notation on the drawing. The details of these specifications can be found in most drafting texts. A complete designation for a threaded cap screw might be as follows:

1/4 − 20 UNC2A − LH − Hex Hd. Cap Screw × 2

The translation of this is

1/4 = 1/4-inch nominal diameter

20 = 20 threads per inch

UNC = Unified National Coarse series

2 = class 2 fit

A = external thread

LH = left-hand thread (omit this if right-hand thread)

Hex Hd. = hexagonal head

Cap Screw = type of fastener

2 = 2-inch bolt length

A hole, to be drilled and threaded into a metal block to receive the bolt, would have a designation such as:

7 Drill × 2.5 − 1/4 − 20 UNC − 2B − LH × 2

meaning

7 Drill = use number 7 drill

2.5 = drill the hole 2.5 in. deep (Omit if drilled completely through the piece.)

1/4 = for 1/4-inch-diameter thread

20 = 20 threads per inch

UNC = Unified National Coarse series

2 = class 2 fit

B = internal thread

LH = left-hand thread (omit if right-hand thread)

2 = thread 2 in. deep

Metric Threads

Metric threads have a slightly different method of specification. A typical metric thread note would be

M6 × 1.0

where M = metric thread, 6 = nominal diameter in millimetres, and 1.0 = pitch in millimetres.

There are 25.4 millimetres per inch. A pitch of 1.0 millimetre would have 25.4 threads per inch. Therefore, the pitch of this thread would be between that of a 1/4 – 20 UNC and a 1/4 – 28 UNF thread.

Grades of Bolts

Bolts are specified and made in several grades depending on the strength and hardness of the bolt. The Society of Automotive Engineers (SAE) has devised a grading system by which bolts are designated. Grades run from 0 to 8 with grade 0 being the softest and weakest. A grade 0 bolt has no proof load or specified hardness. Grade 3 and grade 5 bolts are available at most hardware stores. Grade 8 bolts have proof loads of a minimum of 120,000 psi and a Rockwell C hardness of 32 to 38. The grade of a bolt can be seen by the marking on the head. Figure 15–4 shows the markings of some of these bolts. The relative strengths and hardnesses of the various grades of bolts are shown in Table 15–2.

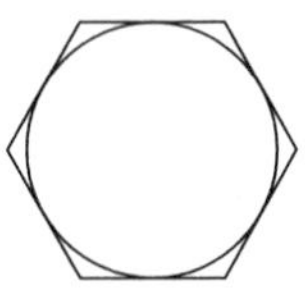

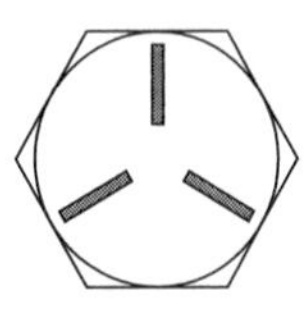

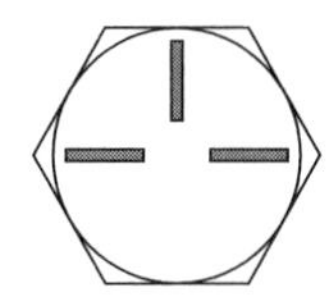

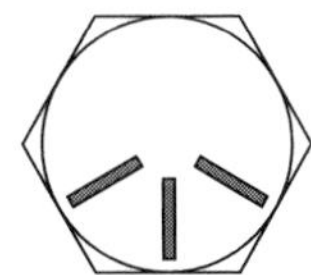

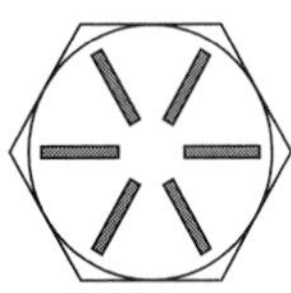

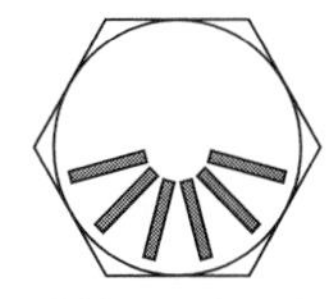

Figure 15–4. Bolt markings.

Table 15–2. Strengths and Hardnesses of Bolts

Grade (SAE)	Tensile Strength (Minimum psi)	Maximum Brinell	Hardness Rockwell
2	60,000	240	100B
5	105,000	285	30C
5.1	120,000	375	40C
5.2	120,000		
8	150,000	350	34C
8.2	150,000		

Nonthreaded Fasteners

There are many types of nonthreaded fasteners, which can be classified as follows:

- Removable with a small effort:
  - rotation allowed: dowels, nails, pins
  - no rotation allowed: staples
- Not easily removable:
  - rivets
  - dowels

Dowels have been used for centuries to join wooden parts, and are still used in furniture, cabinets, and other wood furniture. Although two dowels can be used at angles to each other in such a way that the parts cannot be easily separated, that is seldom done. Instead, the dowels themselves need a "fastener," and an adhesive is generally used to hold the dowel to each of the other parts. (In a sense, the dowel allows more surface for the adhesive to use, and can therefore be considered as an aid to the adhesive solution of joining.)

Nails

Nails are an excellent choice when the application requires the fastener to withstand considerable shearing force but very little tension. Also, we tend to forget that one nail (or screw) still allows for rotation of the parts (around the axis formed by the nail or fastener); a second fastener takes care of that problem if rotation is not desired.

As to which other "needs of the application" nails satisfy, they are *temporary* fasteners, they work in only a few materials (wood, mainly), they can be coated to survive wet environments, and they can be installed with simple tools and with little skill. They do not project beyond the part if properly installed. They are also inexpensive and widely available.

Nails come in a variety of sizes and shapes; popular shapes include box, common, finishing, roofing, and many special shapes such as horseshoe nails. Figure 15–5 illustrates some of the available shapes.

The sizes of nails are given in the penny system. In old England, this designation was the price (in pence) for 100 nails. Of course, the price no longer holds, but the old British notation of "d" for penny is still used for nails (as in a "10d" nail). Nails are sold either in bulk, by the keg, or in 100-pound lots. Nails that are ten-penny or larger are often called *spikes*. The dimensions of wire nails is given in Table 15–3.

Staples

Staples act like two nails side by side, so they solve the revolution problem mentioned earlier. They can be

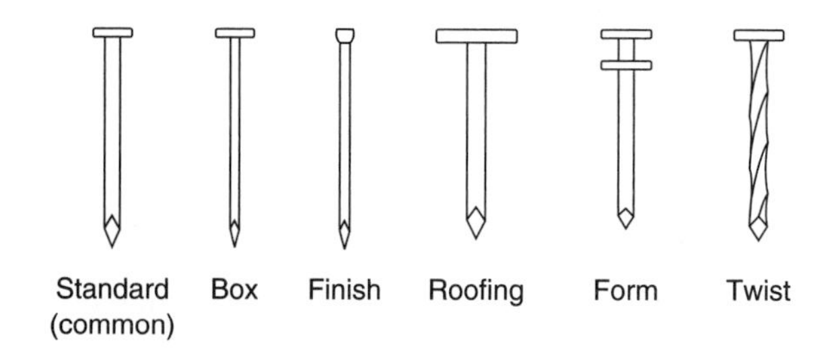

Figure 15-5. Types of nails.

Table 15–3. Sizes of Nails

Size	Length (in.)	Diameter (in.) Common	Box	Finishing	Shingle
2d	1	0.072	0.0672	0.606	0.0915
3d	1.25	0.08	0.076	0.0672	0.1055
4d	1.5	0.0985	0.08	0.072	0.1055
5d	1.75	0.0985	0.08	0.072	0.1055
6d	2	0.113	0.0985	0.0915	—
7d	2.25	0.113	0.0985	0.0915	—
8d	2.5	0.123	0.113	0.0985	—
9d	2.75	0.123	0.113	0.0985	—
10d	3	0.145	0.128	0.113	—
12d	3.25	0.148	0.128	0.113	—
16d	3.5	0.162	0.135	0.1205	—
20d	4	0.192	0.148	0.135	—
30d	4.5	0.207	0.148	—	—
40d	5	0.207	0.148	—	—
50d	5.5	0.244	—	—	—
60d	6	0.2625			

applied at high speed with a powered staple gun for attaching subflooring or insulation in buildings, putting up fencing, crating manufactured products, putting up posters, and a great many other uses. Cardboard boxes, for instance, are assembled using wide staples that are crimped over from the outside using a special tool. As with nails, staples can be used in a limited number of materials. The material must be soft enough to be penetrated by the staple.

Pins

Straight pins, taper pins, split pins, and cotter pins—they all perform the function of resisting shearing forces. A typical application is to join the hub of a gear to a shaft by drilling a hole through both parts and then inserting a pin. The different pins are distinguished somewhat by how they are then held in place (Figure 15–6).

Roll pins are compressed as they are pressed into an undersized hole; the resulting frictional forces hold the pin in place. The ends of split *cotter pins* are bent over once the pin is in place. *Straight pins* are peened over at each end to hold them in place, but *taper pins* need to be peened over only on the small end.

Rivets

Rivets are pins that have a thick diameter compared to their length. They are used (in tension) to hold two parts tightly against each other. They are held in place by having both ends greatly distorted to form a "head" at each end. Typically, one head is formed, the rivet is inserted into the prepared hole, then the other head is formed. Rivets are distinguished by *how* the second head is formed.

If the rivet is installed *hot,* then the second head can be easily formed by forging. Otherwise, the rivet may be split lengthwise for part of its length and then the two tips are separated and the second head formed from them. *Pop* rivets are hollow with a small shaft extending through the rivet. The shaft has a small ball or bulge at the end. Once the rivet is in place, the shaft is pulled so that the bulge spreads the end of the hollow rivet, forming the second head. An explosive rivet uses a small explosive charge to do the same thing. Figure 15–7 outlines several types of rivets commonly in use.

Fasteners for Shafts

Imagine that a belt pulley is to be attached to the shaft of an electric motor. What are the needs of the application? (1) There is a major need to transfer power from the motor to the pulley. (Since this is rotational force, the correct term is *torque.*) (2) The pulley should not work its way off the end of the shaft (that is, no *axial* motion). What choices are available?

If the power to be transferred (that is, the required torque) is not too large, then perhaps all that is needed is a setscrew that passes through the pulley and presses against the shaft. This would be acceptable for a 1/3-hp motor. But what if the motor produces 10 hp? Keys and keyways are one answer.

Keys, Keyseats, and Keyways

A *keyseat* is a slot that is cut into the motor shaft. The *keyway* is cut into the hole through the center of the pulley. Then the slots are lined up and the *key* is fitted into both slots simultaneously (Figure 15–8). As long

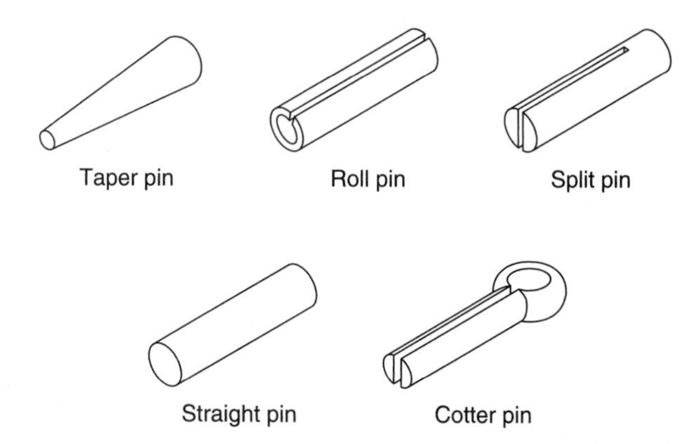

Figure 15–6. Various types of pins.

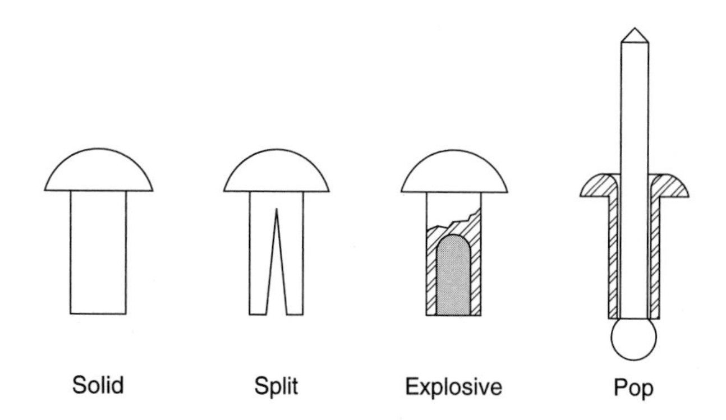

Figure 15–7. Rivets.

as the key is strong enough to resist being sheared by the torque forces, it will continue to transfer torque from the motor to the pulley.

Note in Figure 15–9 that several different shapes are possible for the slots and keys. Yet, they are similar in that each is easy to make with a milling cutter. The straight slot takes a square pin (which could fall out) or a tapered pin (which can be driven in hard, but needs the gib head so it can be removed, just as a common nail needs a head). The Woodruff and Pratt & Whitney keys are first fitted into the slot in the shaft and then the pulley (which has a straight slot) is turned so that its slot lines up with the key and then is pressed or slid onto the shaft.

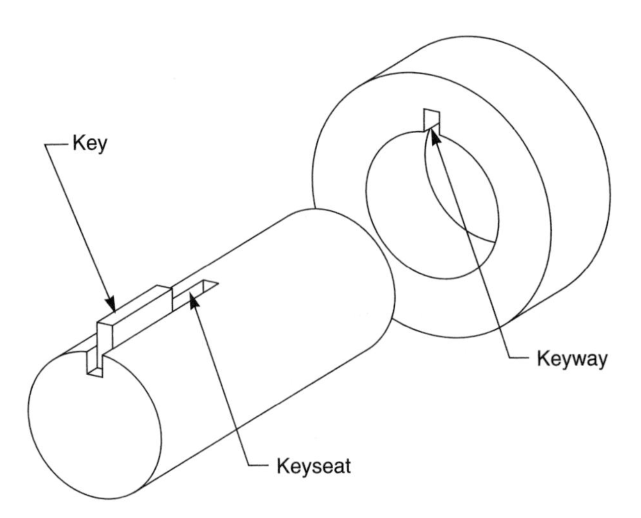

Figure 15–8. Key, keyseat, and keyway.

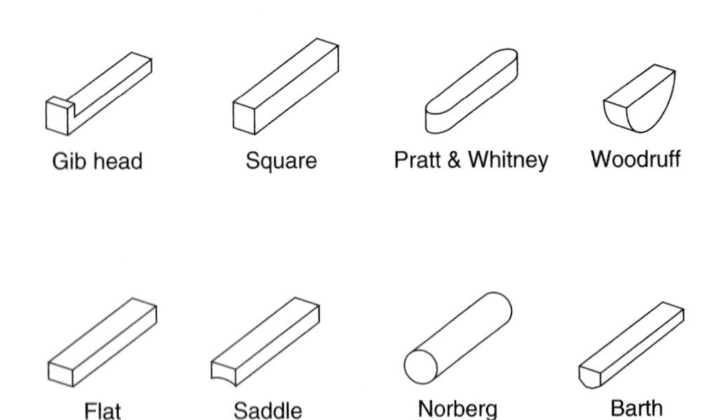

Figure 15–9. Types of keys.

Splines

Splines are toothed external and internal connectors that allow for longitudinal (back and forth) motion on a shaft. In effect, they act as many keys acting together. Splines are excellent for transmitting rotational forces from one part to another and have excellent shear strength. Figure 15–10 illustrates splines.

Snap Rings

In the preceding discussion, reference is made to sliding or pressing a pulley onto a motor shaft. If there is little need to change the pulley, then it is common practice to prevent axial motion of the pulley by designing an interference fit between the motor shaft and the pulley. But what if the application calls for a frequent change of pulley? Then the fit between the shaft and the pulley should be looser. Now it is necessary to prevent the pulley from moving along the shaft. This can be done by tapering the shaft, with a similar taper in the pulley, or with a snap ring. Figure 15–11 shows several shapes of snap rings.

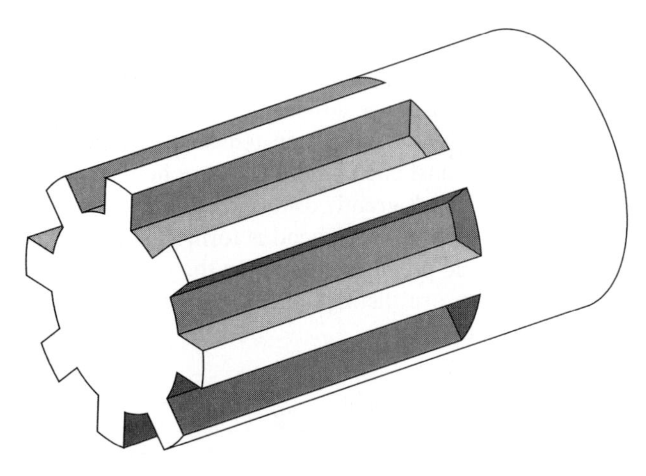

Figure 15–10. Splines.

Figure 15–11. Shapes of snap rings.

Figure 15-12. Crimping.

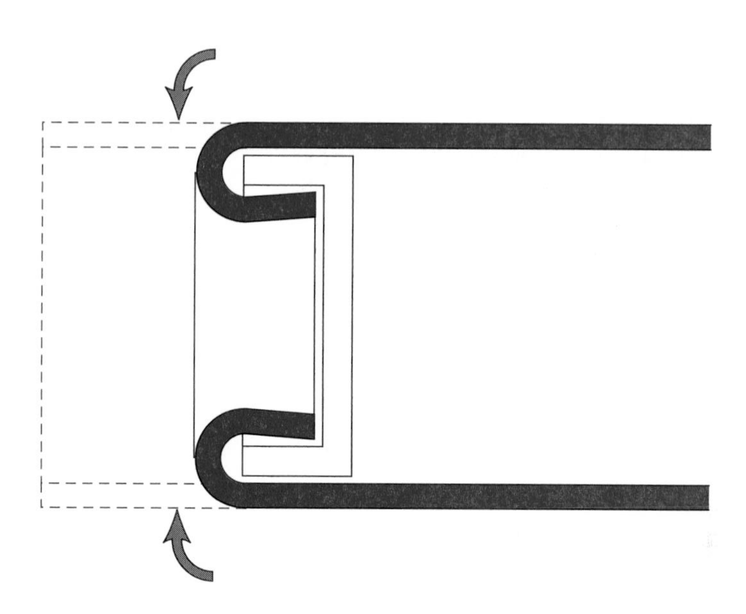

Again, the need for removability (or permanence) arises. If the snap ring needs to be easily removable, then the shape that has holes for snap-ring pliers should be chosen. Otherwise, a plain shape can be selected.

It is customary to cut a shallow groove into the shaft to keep the snap ring from sliding along the shaft, but that may not be necessary in light-duty applications.

■ CRIMPING

Occasionally the simplest fastener is a band of sheet metal (or other shape) that is fitted over the joint between the parts and then bent over or *crimped* to keep it in place. A classic example of crimping is the lid on a tin can. No other fastener is needed in joining materials by this technique. Covers on automotive parts are often attached by crimping (Figure 15–12).

■ MAKE AS ONE PIECE

Perhaps the best solution to many "material joining" applications is to avoid the issue by redesigning the two parts to be a single part. Where that approach is possible (and it isn't always), the cost of production will generally be less and the reliability of the assembly will be improved—and that's tough to beat! Figure 15–13 is an example of how a three-part mechanism might be redesigned and fabricated as a single piece.

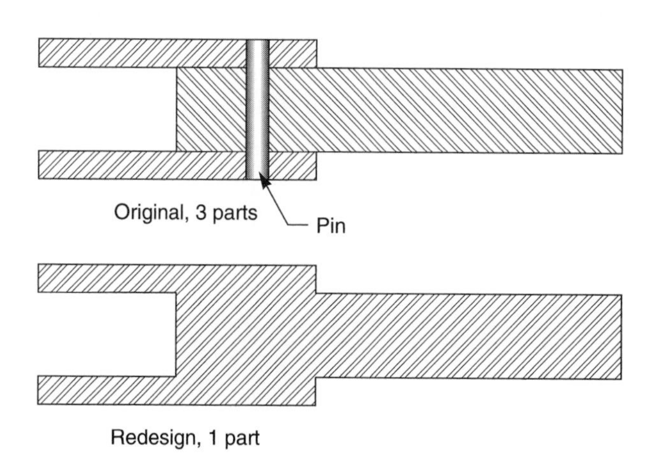

Figure 15-13. Redesigning of a part.

PROBLEM SET 15–1

1. Take a fresh look at the "application" of attaching an automobile wheel to some part of the rear axle. Analyze each of the 12 "needs of the application" for this particular application, and then select a fastening system (or device) that best meets those needs.
2. What is the primary reason to select a threaded fastener instead of a nonthreaded fastener?
3. Two pieces of wood need to be joined permanently. If that rules out nails, then what is a better alternative?
4. If keys and keyways are used to resist shearing forces, and nails are used to resist shearing forces, then why not use nails instead of keys for motor pulleys?
5. A thread is specified as "10 – 24 UNC – 2B." Explain what each part of the specification tells you.
6. In what essential way does the specification of pitch size differ between the inch and the metric systems?
7. For a given thread diameter, why should there be three different pitches of threads available? In particular, explain what UNC, UNF, and UNEF represent, and, for each one, describe what needs of the application would lead you to select a particular one.
8. List examples of six distinctly different heads for machine screws and draw a sketch of each. For each one, describe the *needs of the application* that would lead you to select that particular one.
9. Distinguish among the thread shapes for machine screws, wood screws, and sheet metal screws.
10. Quite a lot is made of the need to select thread pitch and class of fit for a machine screw thread; yet, this is never done for wood screws or sheet metal screws. Why?
11. Recall the history of the use of "penny" to describe nail size, and then explain why the following relationships are reasonable. The lengths of common nails increase 1/4 inch per "penny" through 10d, but they increase 1/4 inch per 5d starting at 20d.
12. What is the primary "need of the application" that pins meet? Distinguish among the four types of pins described in this chapter.
13. "Rivets are pins," but they differ from other pins in some essential features. Describe the similarities that rivets and pins share, and then describe the differences between rivets and the other pins described in this chapter.
14. What was probably the most important consideration in the design of the different keyway slots?
15. Using Table 15–1, give the complete thread notation for a 3/8 hex head cap screw, 3 in. long, having a class 2 fit.
16. If the bolt in Problem 15 is to be screwed 2 in. into a block of aluminum and the hole is to be 1/2 in. deeper than the bolt, give the complete thread specification for the hole and thread.
17. Select any part that has been joined by some fastener. Redesign or sketch how that part can be made as a single piece.

The emphasis in this chapter has been on the analysis of the needs of an application so that the best fastener for that particular application can be selected. A primary consideration is whether or not the fastening should be permanent. Then, should relative motion be allowed? What are the materials to be joined? We discussed 12 needs that should be considered before choosing a particular fastener.

Fasteners were divided mainly into those that are threaded (with considerable discussion on the specification of the thread itself) and nonthreaded fasteners (with a discussion of how to keep the fastener in place).

With sheet metal parts, a solution that avoids the need for a separate fastener is crimping (folding one lip over another).

Finally, the concept of redesigning the assembly to make it all one piece (so that no fasteners would be needed) was introduced as an alternative to the entire chapter! This is a desirable goal, but the nature of the application may not always permit it.

PHOTO ESSAY 15-1

Pop Riveting

STEP 1 Select the proper size rivet for the job. The rivet should protrude approximately 1/8 inch (3 mm) through the workpieces and the backing plate. Drill the holes through the workpieces slightly larger in diameter than the selected rivet.

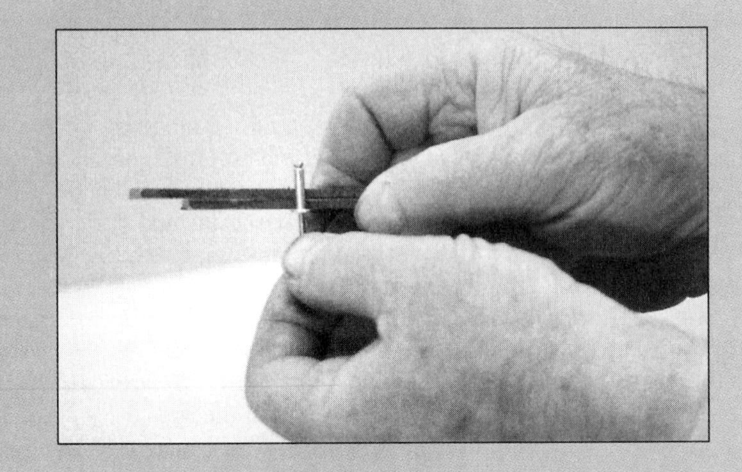

STEP 2 Insert the pop rivet in the riveting tool.

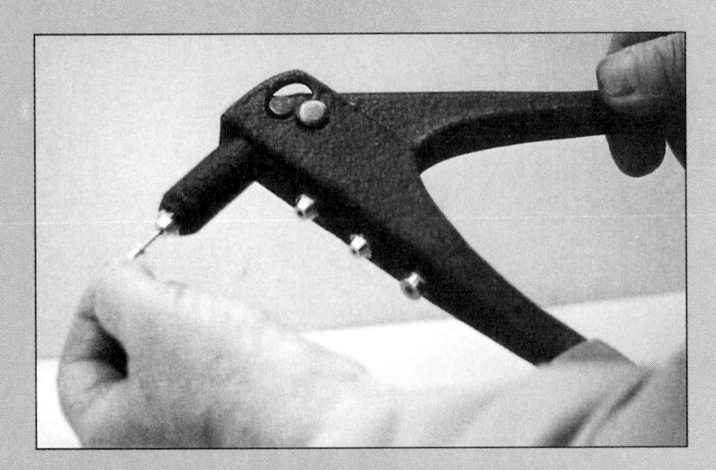

STEP 3 Insert the rivet through the holes in the workpiece, and place the backing plate over the rivet.

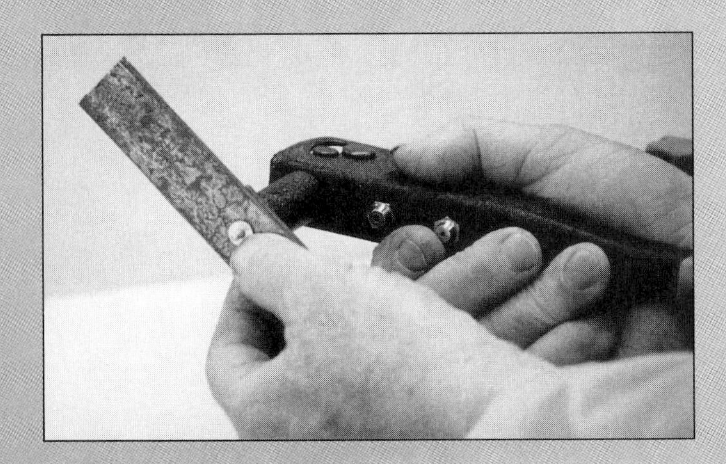

STEP 4 Tighten the rivet by closing the handles on the rivet tool.

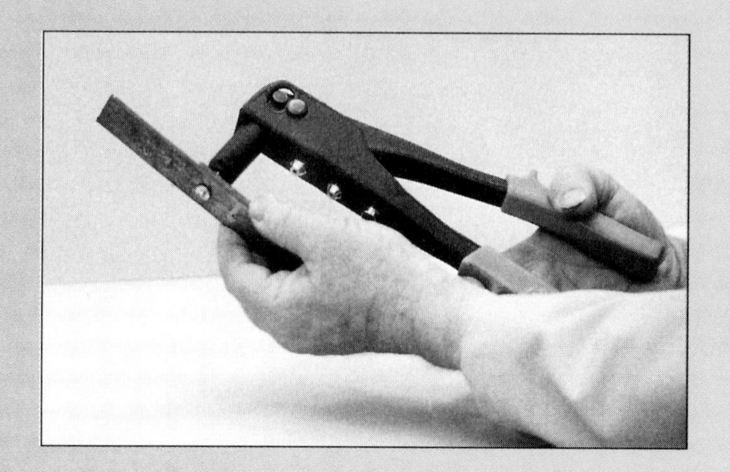

STEP 5 Continue tightening the rivet until the stem breaks off. The "blind" end of the rivet will be spread over the backing plate, locking it into position.

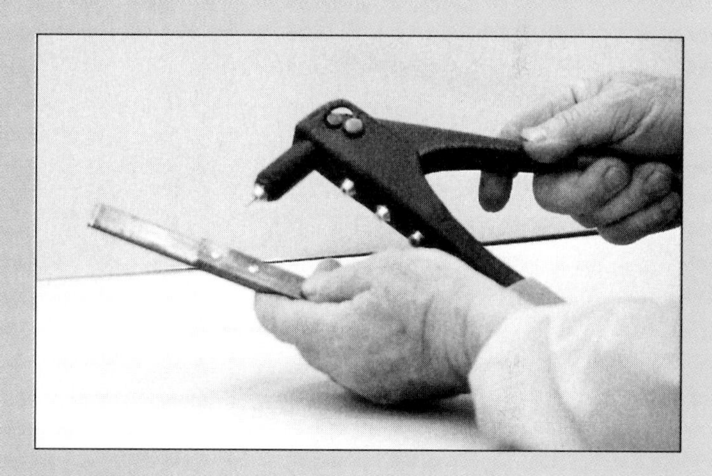

SECTION

VII

Finishing

■ INTRODUCTION

As the sports expression goes, "The game isn't over until it's over." In manufacturing the expression should be modified to "The part isn't finished until it's finished." Finishing is the final manufacturing process before the product is shipped.

Chapter 16

Mechanical Surface Finishing Processes

It is not always clear what distinguishes *finishing* from other manufacturing processes. Finishing affects the surface of the part, but so does grinding (discussed in Chapter 4), plating and other material addition processes (Chapter 8), and case hardening (Chapter 12). The distinction between a finishing operation and these other processes that affect the surface is that *finishing processes are not intended to significantly change the dimensions of a part.* Rather, they are intended to:

- Achieve closer tolerances,
- Provide a protective coating, or
- Improve the appearance of the part.

Finishing processes can be divided into three categories for purposes of discussion: surface preparation, smoothing, and surface treatments; each having several subcategories. A chart of these categories and their subcategories is given here.

Anyone who has repainted a wall knows that first it is necessary to repair nail holes, remove loose paint, clean handprints, grease, and dirt from the surface, and roughen high-gloss surfaces. Without a good job of surface preparation, the finishing coat will not be attractive or last very long. The same principle is true with manufactured parts.

If the workpiece was cast, it may have flashing that needs to be removed. If it was forged, sheared, or punched, burrs may have to be removed. Milled or turned parts may have sharp edges that need to be cut back or *radiused.* Welded and hot-forged parts may have scale, rust, or grease coatings that must be removed prior to finishing. This list of items that must

Surface Preparation	Smoothing	Surface Treatments
Deburring	Honing	Conversions
Hand deburring	Superfinishing	Inorganic coatings
Barrel and vibratory methods	Lapping	Organic coatings
Sanding and belt sanding	Polishing	
Stripping/cleaning	Buffing	
Blasting	Burnishing	
Power brushing	Ballizing	
Ultrasonic cleaning		

be accomplished before finishing is nearly endless. In any case, *before any surface treatment is applied, the surface must be properly prepared.*

SURFACE PREPARATION

Surface preparation processes have been divided into *deburring* processes and *stripping/cleaning* processes, which remove scale, rust, dirt, grease, etc.

Deburring

Burrs are sharp projections that are left on the "back side" (away from the cutting tool) if a shearing, milling, drilling, or other cutting operation doesn't do a clean job. Burrs are such a common problem that a number of processes have been developed just to *deburr* manufactured products. When possible the parts are deburred by automated means, or with the use of machines, but it is sometimes necessary to deburr parts by hand.

Hand Deburring

Many operations in high-tech industries are so demanding that parts must be manufactured to exacting standards and surfaces specified in microinches. Parts for high-performance aircraft and space hardware are often hand deburred by workers using jewelers' files and working with magnifying lenses and jewelers' eyepieces. Parts have been known to require several hours of tedious work to deburr them properly to design specifications. Hand deburring is very expensive and should only be used when no other method will meet the demands of the design. Figure 16–1 shows burrs left by abrasive sawing.

Barrel and Vibratory Finishing

In *barrel finishing,* a number of parts are placed in a drum with an abrasive. The drum or barrel is rotated, tumbling the parts against the abrasive, each other, and sidewalls of the barrel. In this manner, the burrs, sharp edges, rust, scale, etc., are worn away (Figure 16–2). The barrels are generally hexagonal or octagonal to aid the tumbling action. This is the same process used in lapidary work to "tumble polish" rocks except that in deburring, coarser grit abrasives are used.

In *vibratory finishing,* the parts and abrasive pellets are placed in a large "tub," which is agitated by a vibrating mechanism. The resulting jostling of the parts and abrasive against each other produces the same effect as tumbling or barrel finishing. Vibratory finishing is quieter and easier to control than barrel finishing, but it is used for smaller parts than barrel finishing (Figure 16–3).

Either barrel or vibratory finishing can deburr parts very quickly, but neither of these processes works well on internal crevasses or holes. Barrel and vibratory finishing are economical processes especially for exterior surfaces and softer metals.

Figure 16–1. Burrs.

Figure 16–2. Barrel finishing.

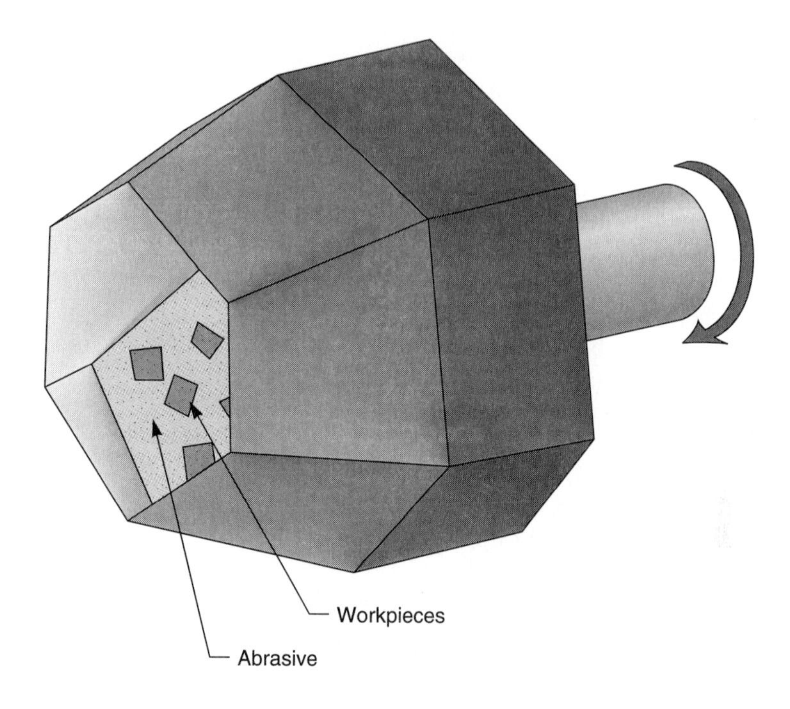

Figure 16–3. Vibratory finishing.

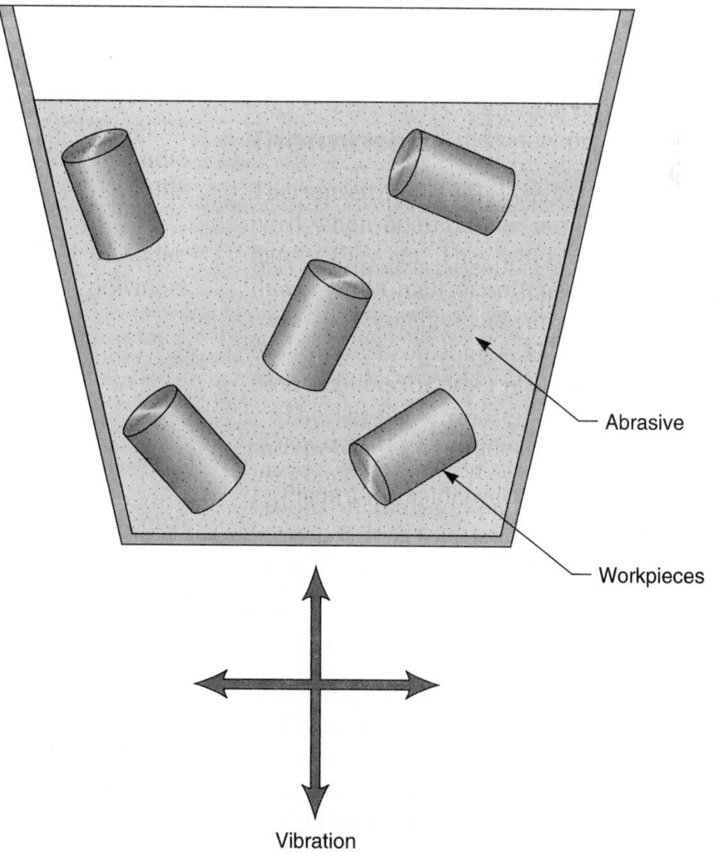

Belt or Disk Sanding

Sanding was discussed as a method of material removal and may seem out of place in a discussion of finishing. However, in those applications where a smooth surface is more important than flatness or cylindricity, parts can be finished very economically by using very fine grit belt sanders. Belt sanders can be either mounted in a fixed position or portable. The belts can be easily changed to fit the needs of the application, both flat and curved surfaces can be sanded, and metals and nonmetals can be sanded successfully. The belt sander is an economical tool for finishing work.

Belt sanding can be done either wet or dry. When very fine grit belts are used, water (or other cutting fluid) should be used to keep the part cool, to carry away the removed particles, and to prevent the abrasive belt from becoming clogged.

Sanding does not leave a bright polished surface. In some instances a finely sanded surface is sufficient to finish the part. In other cases, where a glossy surface is desired, sanding is used prior to polishing, buffing, burnishing, or other finishing operation.

For larger parts, air-powered or electrically driven belt or disk sanders, using medium to fine grit abrasives, can be used to remove flashing or burrs quickly and economically (Figure 16–4). A certain degree of skill is required in the use of sanders to remove burrs. Improperly done, sanding can damage the main body of the part as well as remove the burrs. For delicate workpieces, small disks and grinding stones can be used to remove the burrs.

Figure 16–4. Belt Sander.
(Courtesy of Delta International Machinery Corporation, Pittsburgh, PA.)

Stripping/Cleaning

Sanding is fast and economical, but leaves scratches on the surface. Dry blasting, wet blasting, or ultrasonic cleaning are generally slower, but produce a surface that may be acceptable for the final product. In fact, sandblasted surfaces are occasionally specified for their attractive appearance rather than just a cleaning operation.

Dry and Wet Blasting

Blasting involves blowing abrasive sand against the surface of the part from a high-pressure nozzle. The effectiveness of high-velocity sand as a method of removing paint is apparent to anyone whose automobile has been caught in a severe desert sandstorm. The abrasive force of the individual sand particles, or their ability to remove material, comes from their kinetic energy. This abrasive force is proportional to the square of the speed with which they hit the object. Therefore, to be an effective, economical, commercial process, the abrasive grains need to be carried to the surface of the workpiece in a high-speed stream of either gas or liquid. Both are used.

Fine abrasive particles are usually air-blasted onto the surface. Heavier particles are more effectively carried in a liquid stream. The liquid can be water, a light oil, or other suitable liquid.

The choice of an abrasive is crucial to the success of blasting processes. The abrasive must be hard enough to cut through the material to be removed, but not hard enough to damage the surface of the part. For example, a mixture of walnut shells and detergent in a water stream is used to clean mortar and concrete. The walnut shells are hard enough to cut through grime, grease, algae, and other material on the surface of the concrete, but not hard enough to damage the concrete itself. Sand, silicon carbide, walnut shells, glass beads, and other materials can be used as the abrasives.

The safety of the workers is an essential factor in blasting. The abrasive particles under high pressure can injure people. When blasting exterior surfaces, protective clothing and air filters for breathing must be used by the workers. Small parts are usually blasted in a cabinet or booth that is fitted with protective gloves and viewing screens (Figure 16–5). Besides protecting the worker from the abrasives, the booth makes it easier to recover the abrasive particles for reuse.

The air or liquid is supplied to the nozzles by high-pressure pumps. The sand or other abrasive is mixed at the blasting nozzle or "gun." Since the abrasive would damage the impeller of the pump and erode the piping and hoses of the machine, the abrasive must not be allowed to go through the pumping mechanism. The used abrasive is carefully filtered or separated from the air or liquid and only the fluid is sent back through the pump. Filtering of particles from a liquid involves the use of screens and other filters, which can become clogged. Therefore, it is easier to filter these particles from air by the use of baffles and cyclone separators. Furthermore, liquids can corrode metal parts if the liquids are not carefully selected. Whenever possible, air blasting is the preferred process.

Power Brushing

Power brushing involves the use of stiff fibers mounted in circular brushes and rotated rapidly against the surface to be cleaned. The fibers can be made from metal wire (brass, steel, and aluminum are common), hair, bristles, or fabric. Power brushes can be used to remove either loose particles such as scale and rust, paint, or surface defects in the form of fine, sharp high spots. Brushes are commonly available in a variety of shapes (Figure 16–6). Properly selected and designed brushes can be used to clean irregularly shaped surfaces and difficult-to-reach areas such as slots or inside corners or grooves.

In selecting the proper brush, the hardness of the material to be removed, the hardness of the surface of the part, and the shape of the part must be considered. The fibers must be hard enough to remove the oxide, paint, or other surface coating yet not damage the workpiece. The diameter, shape, and thickness of

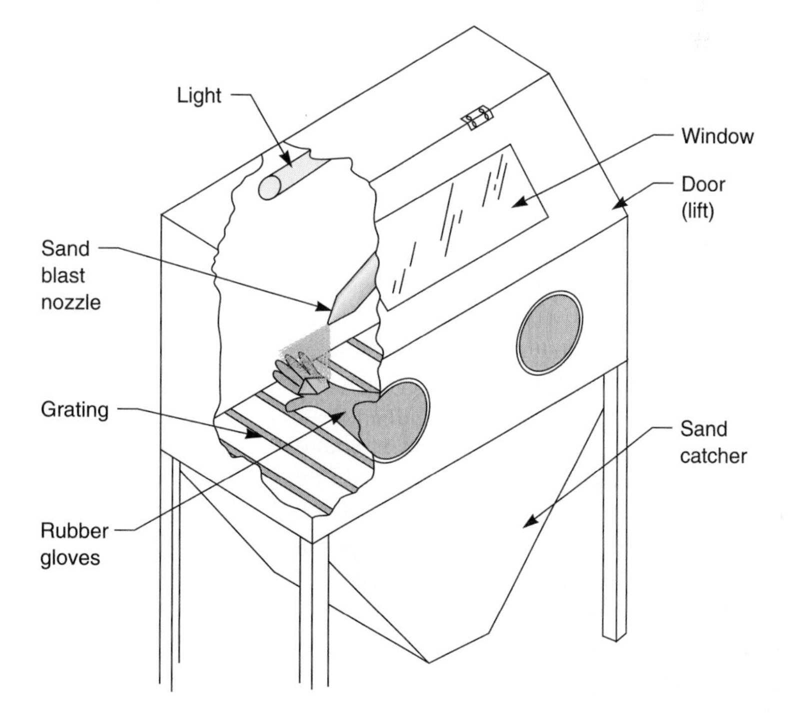

Figure 16–5. Sandblasting cabinet.

Figure 16-6. Types of power brushes.

the brush must allow it to reach into all areas of the part. Very often, technicians will modify the shape of a standard brush to meet individual needs.

Ultrasonic Cleaning

Ultrasonic cleaning uses energy in the form of pressure or sound waves to remove surface dirt, oxides, rust, or scale. Both the amplitude and the frequency of the sound waves should be carefully selected for specific needs. Dirt, oil, and other particles can be removed by relatively low-power, low-frequency sound waves. Scale and oxides require higher powers and specific frequencies, depending on the oxide or scale to be removed. It is a very fast process often requiring three to four minutes to clean a surface.

The parts to be cleaned are placed in the chamber along with the appropriate liquid cleaner. The choice of the cleaner is critical to the effectiveness of the operation but may vary from a detergent to solvents of many types. The high-frequency, high-energy sound waves produce tiny bubbles in the cleaner. The motion and collapse of these bubbles provides a "scrubbing" action. This scrubbing action repeated at frequencies between 10,000 and 100,000 times each second provides a very effective cleaning action that reaches every area of the part.

■ SMOOTHING OPERATIONS

Either the appearance or the application of a part may require that its surface be smoother than that left by the last manufacturing process. This is where smoothing operations come into play. Several factors should be considered when selecting a smoothing operation:

1. The shape of the surface
2. The hardness of the material
3. The dimensional tolerances required
4. Whether or not the entire surface must be flat or just smooth.

Surfaces can be made very flat or cylindrical by *superfinishing* or *lapping*. Accurate cylindrical holes can be finished by *ballizing* and *bearizing*. Sanding, polishing, buffing, and burnishing emphasize smoothness rather than shape.

Honing

Honing is a process that uses abrasive stones. Cylindrical surfaces are honed by a device that has several small abrasive *bars* or *sticks* mounted on a mandrel [Figures 16–7(a) and (b)]. Each stick (hone) is spring-loaded to provide an even pressure against the surface.

Figure 16-7. a. Typical honing tool for interior surfaces. **b.** Photograph of honing tool.

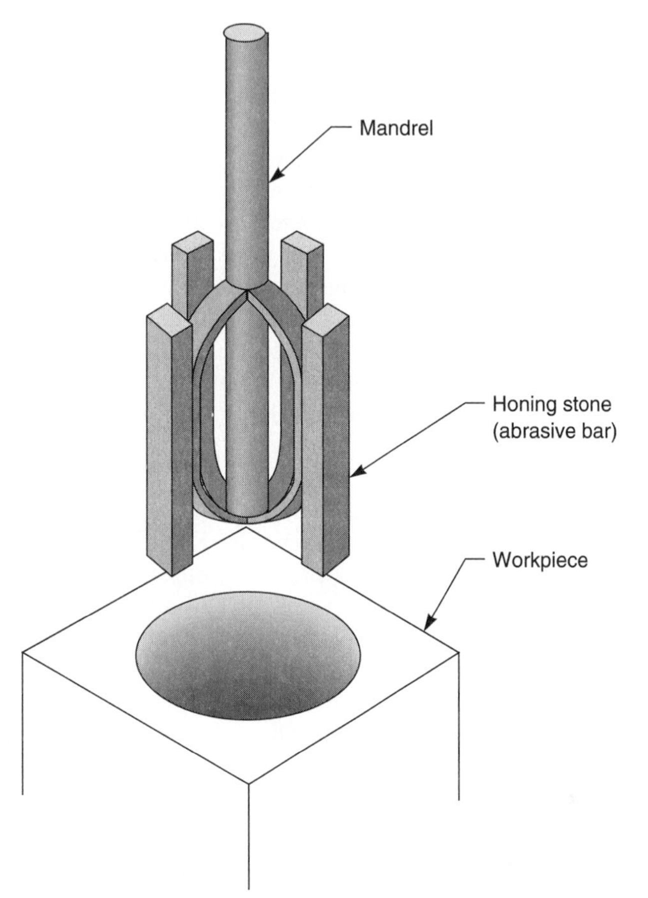

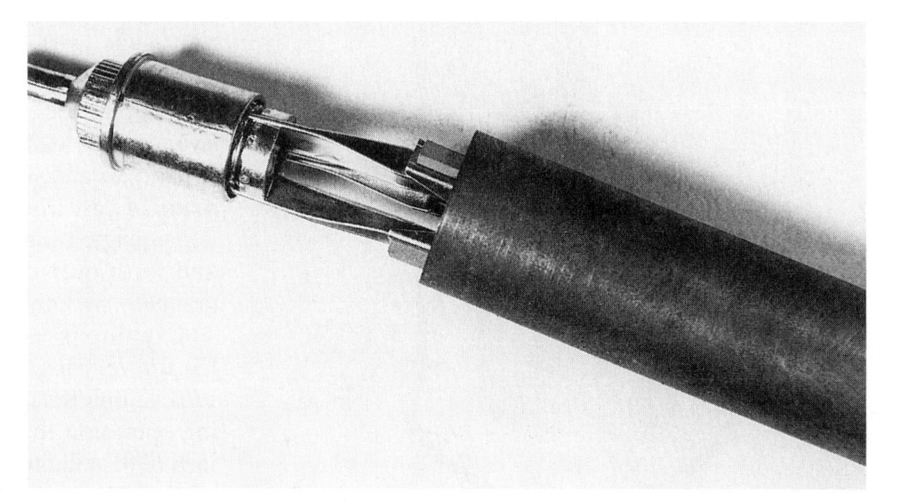

The honing process involves two motions: (1) The mandrel revolves about its axis and, at the same time, (2) the entire honing device is moved along the axis (Figure 16–8). The surface can be made truly cylindrical.

Flat surfaces can also be honed. Flat hones can be moved laterally in several directions across the surface and, if necessary, the hone can be rotated at the same time.

A cutting fluid is generally pumped across the surface being honed to carry away the small chips of surface material. The cutting fluid also prevents the hone itself from becoming clogged with the cut material.

Finishing a surface by honing usually involves several steps. Since finishing hones cut away the surface fairly slowly, the process is accelerated by the use of several different "grits" of stones. Coarse-grit hones with heavy spring pressures are used first to *true-up* the surface, then progressively finer hones and lighter spring pressures are used to achieve the final desired surface.

Honing is a *sizing* process; that is, it is a process used to repair defects left by prior processes. Honing is used to "correct" the size and shape of a surface. For example, we stated in Chapter 4 that due to drill wobble, drilled holes are not perfectly cylindrical or smooth. Honing is one technique that can be used to make the hole very nearly a perfect cylinder. Typically, a honing operation will remove between 0.002 to 0.020 inch (0.05 to 0.5 millimetre) of material to achieve tolerances between 0.0001 and 0.0010 inch (0.0025 and 0.0250 millimetre). Holes as small as 0.060 inch (1.6 millimetres) in diameter to as large as 36 inches (90 millimetres) can be successfully honed. It is considered to be a smoothing process rather than a material removal process. (Material can be removed faster by other means.) With the proper selection of honing stones, almost any material can be honed including metals, ceramics, and some plastics.

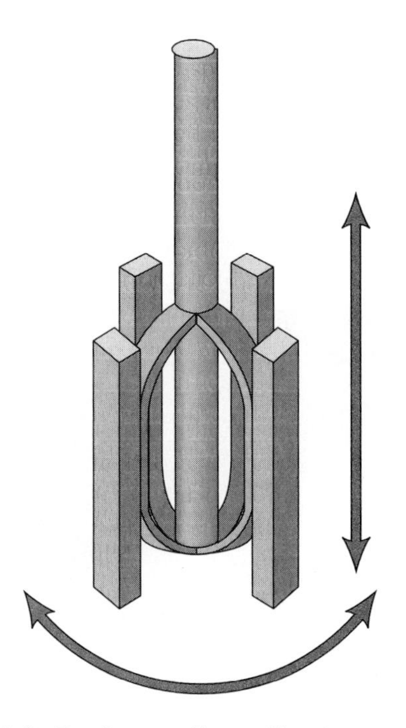

Figure 16–8. The two motions of honing.

Superfinishing

Superfinishing is a proprietary process that was developed by the Chrysler Corporation. This process removes the high spots left over from prior operations. Superfinishing is essentially a honing process that uses lighter pressures and shorter strokes than conventional honing. It can give higher precision than conventional honing and allows two moving parts to fit together nearly perfectly. The result is that the mechanism will last considerably longer than it would without this finishing operation.

Lapping

Lapping is an abrasive process that is used to remove the last minute amount of material from a workpiece. The lapping tool, in contrast to honing, is not an abrasive material itself. The tool, called the *lap,* is designed to hold an abrasive paste or powder. The lap can be made of cast iron, copper, ceramic, leather, or cloth depending on the application. The lap is moved back and forth over the part or rotated against the part gradually smoothing it.

Lapping is used (1) on very hard materials or (2) for very high-precision fits. Lapping is used to remove only small amounts of material. It is a finishing operation that usually removes less than 0.0005 inch (0.01 millimetre) of material. Since a human hair averages about 0.002 inch (0.05 millimetre) in diameter, this amounts to about one-fourth the diameter of a human hair. Lapping can be used on nearly any material. The abrasive used can vary from aluminum oxide (used on steels and other relatively "soft" materials) to silicon carbide and diamond dust for the extremely hard metals, cermets, glass, and ceramics.

Since the abrasive material is held between the lap and the workpiece, both the lap and the workpiece are ground away simultaneously. A modification of lapping is used to fit two moving parts perfectly. For instance, a piston and cylinder can be lapped together to provide a nearly perfect fit. An example of the use of lapping is in the construction of helicopter control mechanisms. As the blades of a helicopter rotate, the pitch of the blade must be changed to provide the lift in the desired direction. In large helicopters, this is done by small pistons inside the rotating shaft. These pistons are fitted to the holes in the shaft by lapping. The surfaces are lapped to within 0.00002 of an inch. (That's 0.0005 millimetre or 0.5 micrometre.) Other examples in which lapping is used include brake cylinders for automobiles, valves of musical instruments, and so forth.

Surfaces can also be made nearly perfectly flat by lapping. To make a perfectly flat surface, three workpieces are needed. Surface A is lapped against surface B until they fit perfectly. Surface B is then lapped against surface C. Surface C is then lapped against A. The process is repeated until all three surfaces match perfectly. The result can be surfaces that deviate from perfectly flat by as little as 0.8 microinch (0.0000008 inch or 0.02 micrometre).

Lapping is an abrasive process. Therefore the force on the lap must be kept constant at all times. Any uneven force between the lap and workpiece will produce an uneven surface, with the result that more material will be removed where there is greater force on the lap. For this reason, most lapping is done by machine rather than by hand. Although small parabolic mirrors for astronomical telescopes are often lapped by hand, the large parabolic mirrors, such as the ones used in the telescopes of large observatories, are ground and lapped by machine to produce a more controllable result.

Polishing

Polishing involves both fine-scale abrasion and *smearing.* Smearing is a process whereby the material from the high points on the surface is pushed over to fill in the low places and smooth out the surface. The purpose of polishing is to remove surface defects, not to change the shape or size of the workpiece. Polishing leaves the glossiest surface possible.

In polishing, a very fine abrasive powder such as jewelers' rouge, Linde-A, rottenstone, or other material is placed on a felt, or other fabric disk or wheel, and the workpiece held against the rapidly turning wheel. The polishing compound is often in the form of a *stick* and is applied by simply rubbing it into the spinning wheel (Figure 16–9). Since the heat generated helps the smearing process, polishing is generally done

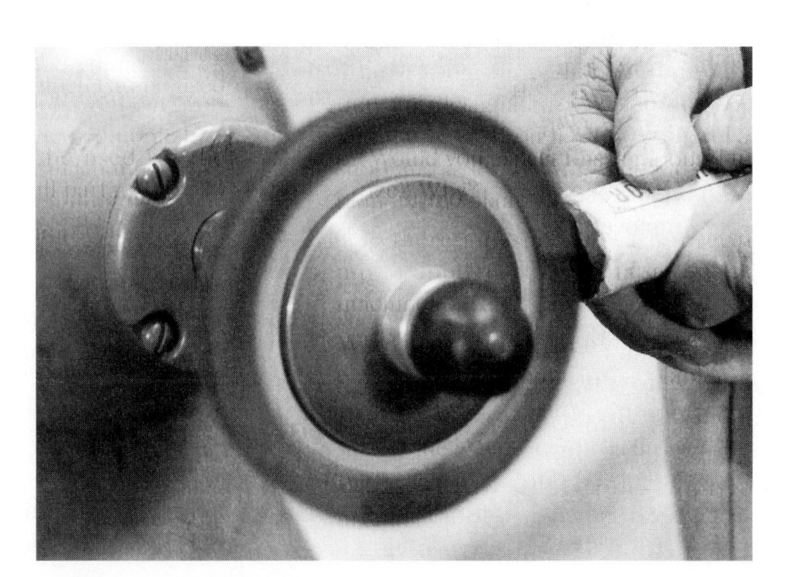

Figure 16–9. Applying polishing compound.

Figure 16–10. Polishing.

dry, without any cooling fluids. Polishing is shown in Figure 16–10.

We can polish surfaces that are flat or large continuous curves or surfaces that have a fairly irregular contour. It is difficult to polish creased or small radius concave surfaces. The polishing of very irregular surfaces requires very small polishing wheels and handwork.

Buffing

Buffing is a process used to produce a high luster by abrading the surface to remove scratches left by a previous process, or to remove an oxide layer. Buffing disks and wheels are made of cloth or leather. The abrasive compounds include Tripoli, pumice, diamond dust, and others. Buffing is done in a manner similar to polishing and often precedes polishing as a finishing process. Buffing removes the scratches and leaves a smooth lustrous surface; polishing puts the shine on it.

Burnishing

Burnishing is a process whereby the surface of the material is brushed or scraped with a wire brush or other burnishing tool. Burnishing leaves a uniform, smooth, but somewhat dull finish to surfaces. Special decorative designs can be applied to the surfaces by imaginative and skillful use of the burnishing tools. The circular marks on gun mechanisms, backs of watches, saw blades, tools, golf clubs, kitchen appliances, and many other devices are made by burnishing. Burnishing can be used with wooden workpieces to align the grains.

Ballizing

Many brittle or ultra-high-strength materials are so sensitive to fatigue and so greatly affected by stress risers that even the interior surfaces of holes drilled through them must be smoothed and finished. The scratches left by machining can cause the materials to fail prematurely. Honing is one method of removing these scratches, but it is relatively slow and costly. A method has been developed that not only removes the scratches, but work hardens the interior surface of the hole. This method is called *ballizing.* In ballizing, the hole is drilled slightly undersize, then a case-hardened steel ball of exactly the correct diameter is forced through the hole (Figure 16–11). Ballizing is a fast and inexpensive technique in which the surface area of the hole is hardened, sized, and smoothed at the same time.

Bearizing

A technique similar to ballizing, and used for the same purpose, is *bearizing.* It is used on holes that are

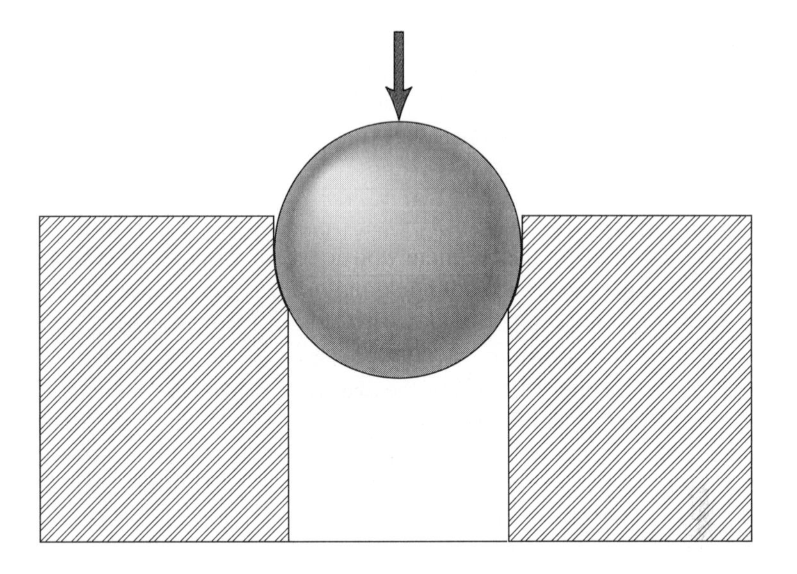

Figure 16–11. Ballizing.

too large to be ballized. Bearizing uses roller bearings mounted on a mandrel. The rollers and mandrel are spun in the hole and the rollers are forced outward into the surface of the metal. This action simultaneously enlarges the hole to the design diameter and work hardens the surface (Figure 16–12).

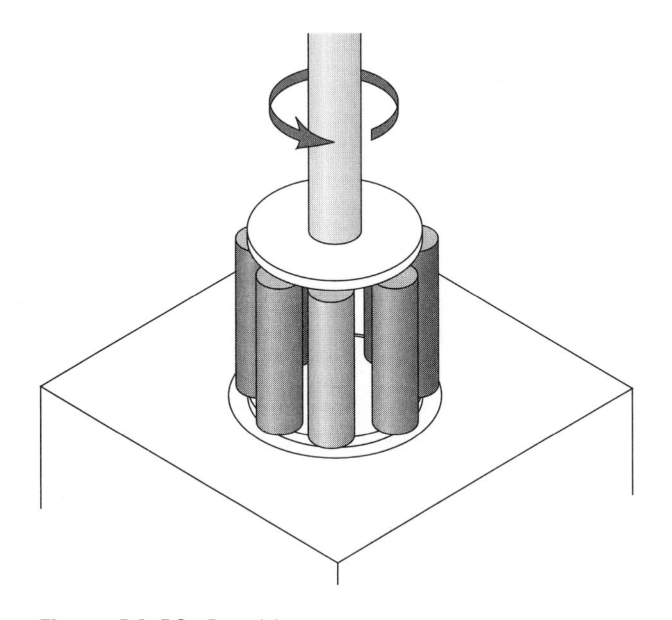

Figure 16–12. Bearizing.

FINISH SPECIFICATION

To properly specify the finish for a part, the following information needs to be given:

- The *process* of the final manufacturing operation (sawing, milling, turning, grinding, etc.)
- The surface waviness, roughness, and lay to be left by the final operation
- The type of process of the surface preparation or finish (buffing, polishing, burnishing, sanding, etc.)
- The type of coating to be applied
- The *process* by which the coating is to be applied.

Surface finishes are specified by giving the desired waviness, roughness, and lay of the desired surface. *Waviness* refers to the long-range undulations in the surface, not necessarily those left by the tool marks. The maximum allowable height (peak-to-valley distances) and lengths of the waves (peak-to-peak distances) can be specified if they are crucial to the proper functioning of the part. Waviness limits can be specified in microinches, microns, angstroms (10^{-10} metres), nanometres, or even as a number of wavelengths of a specified light. For instance, the specification might read "Flat to within 20 helium light bands." Waviness can be measured by dial indicators or by the use of holographic interference patterns, as discussed in Chapter 14. Waviness control is essential to precision machines and instruments. If the bed of a machine is not perfectly flat, the tool post and the cutting tool will be tipped as they move along the bed and an uneven cut will be produced. Industry spends considerable time, effort, and money to keep their machines "true."

Roughness refers to the finely spaced surface texture irregularities. Roughness is usually determined by the tool marks of the final operation. The roughness-height index value is a number that equals the average (mean) deviation of the surface irregularities in millionths of an inch (microinches or 0.000001 inch). Specifying surface roughness is necessary if close tolerances are specified, glossy surfaces are needed, or close fitting of parts is required. The relationship between waviness and roughness is shown in Figure 16–13.

The roughness that can normally be expected from several types of machining operations is shown in Table 16–1. Roughness can be measured by optical comparators, by surface roughness gauges, and by comparison with standard replicas. Cutting tools, torch cutting, and coarse grinding generally leave very rough surfaces, in the range of 125 to 2000 microinches. Finer finishes (32 to 250 microinches) are possible with drilling, boring, fine turning, and fine grinding. Honing, buffing, polishing, lapping, superfinishing, and precision finish grinding can produce fine finishes. Finishes of less than 8 microinches are close to mirror surfaces.

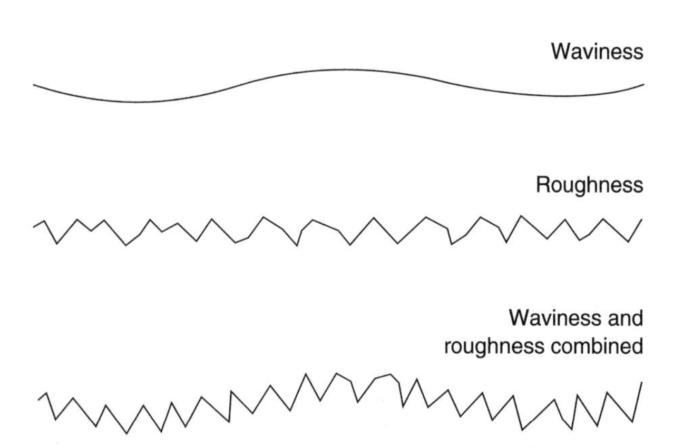

Figure 16–13. Waviness and roughness.

Lay refers to the *direction* of the predominant visible surface roughness pattern. If necessary the lay can be specified as parallel, perpendicular, angular, multidirectional, circular, or radial (Figure 16–14).

Roughness, waviness, and lay are specified by a symbol shown in Figure 16–15(a). An example of this symbol with the numbers properly filled in is given in Figure 16–15(b). Figure 16–16 is a photograph of a roughness comparison gauge. This gauge goes from a mirror-like 2 microfinish polish to a coarse 500 left by a cutting tool. This gauge is the standard against which other surfaces are compared.

The specification for parts should never call for smoother surfaces than are necessary. Making parts smoother than absolutely necessary for the functioning of the part or making it salable often requires more finishing operations, which merely serves to increase the costs of production. These excess costs make the product uncompetitive on the economic market. It has been said that "to overspecify a finish is to sabotage your product." The additional costs of finer finishes must be justified.

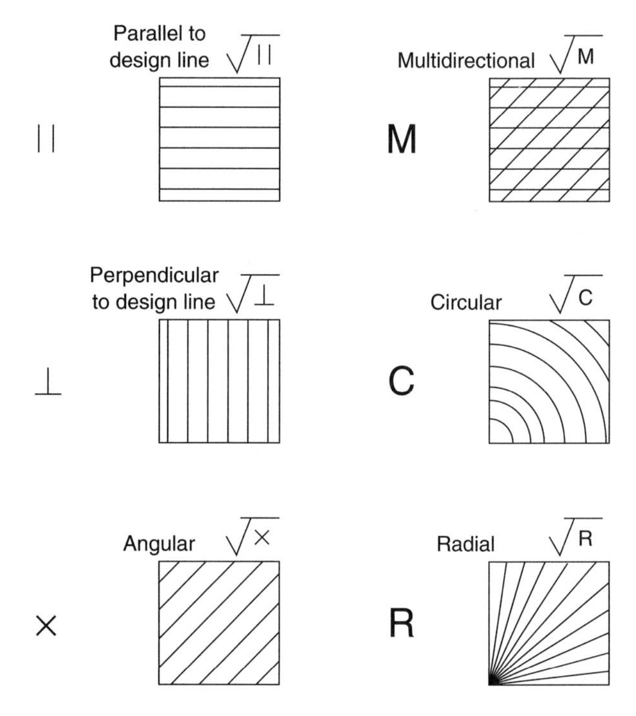

Figure 16–14. Lay.

Table 16–1. Roughness Left By Manufacturing Operations

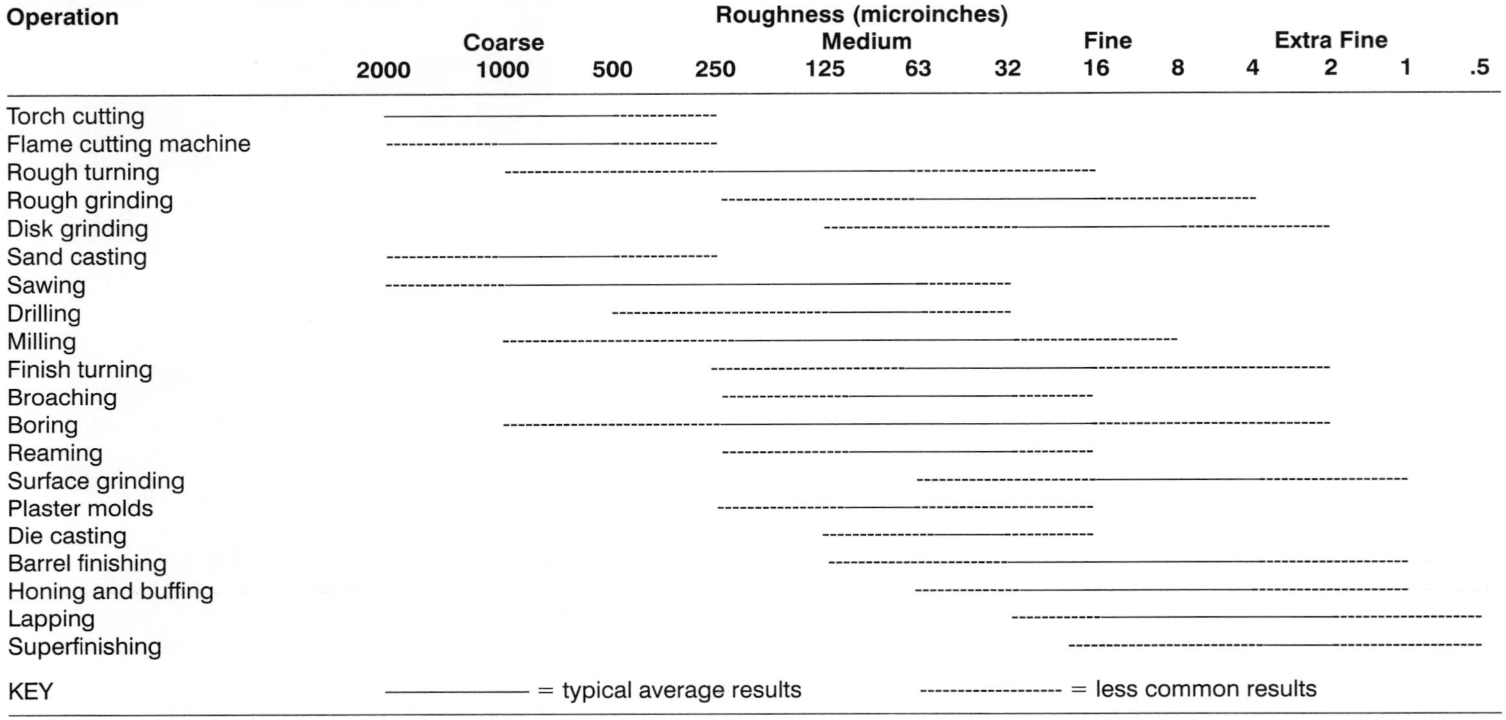

Operation	2000	1000	500	250	125	63	32	16	8	4	2	1	.5
		Coarse			Medium			Fine			Extra Fine		
	Roughness (microinches)												
Torch cutting	———	———	———	- - -									
Flame cutting machine	- - -	———	———	- - -									
Rough turning		- - -	- - -	———	———	———	- - -	- - -					
Rough grinding				- - -	- - -	———	———	———	- - -	- - -			
Disk grinding					- - -	- - -	———	———	———	- - -	- - -		
Sand casting	- - -	———	———	- - -									
Sawing	- - -	———	———	———	———	———	- - -						
Drilling			- - -	- - -	———	———	- - -						
Milling		- - -	- - -	- - -	———	———	———	- - -	- - -				
Finish turning				- - -	- - -	———	———	———	- - -	- - -	- - -		
Broaching				- - -	———	———	———	- - -					
Boring		- - -	- - -	———	———	———	———	———	- - -	- - -	- - -		
Reaming				- - -	———	———	———	- - -					
Surface grinding						- - -	- - -	———	———	———	- - -	- - -	
Plaster molds				- - -	———	- - -	- - -	- - -					
Die casting					- - -	———	———	- - -					
Barrel finishing					- - -	- - -	———	———	———	———	- - -	- - -	
Honing and buffing						- - -	———	———	———	———	- - -	- - -	
Lapping							- - -	———	———	———	———	- - -	- - -
Superfinishing								- - -	- - -	———	———	- - -	- - -

KEY ——— = typical average results - - - - - - = less common results

Figure 16-15. a. Surface roughness and waviness symbol. **b.** An example with the numbers filled in properly.

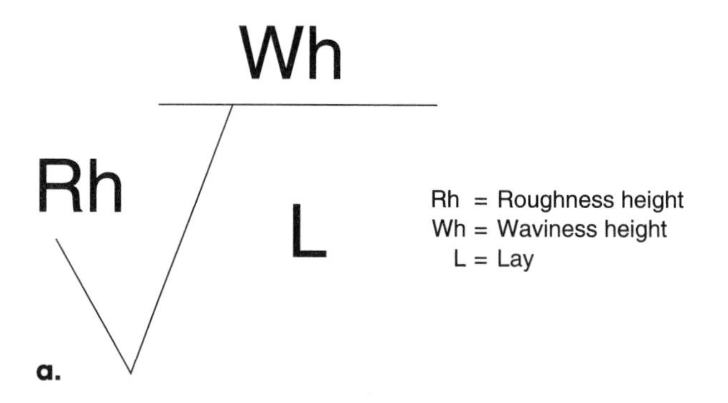

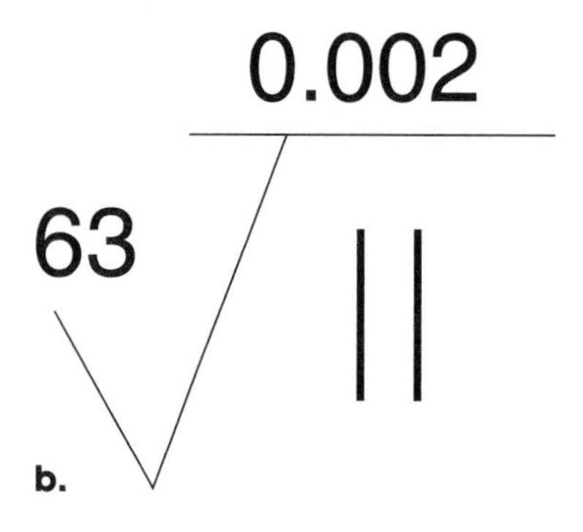

Figure 16-16. Surface roughness comparison gauge.

PROBLEM SET 16-1

1. What are the three goals of finishing operations, as opposed to other operations that change the surface of a workpiece?
2. Define the terms *burr* and *flashing.* Which manufacturing processes produce burrs? Which processes produce flashing? (Flashing was covered in Chapters 9 and 11.)
3. Visit a hardware store. Prepare a descriptive list of the variety of abrasive belts (or disks) that are available for sanders. The list should include a table of (a) sizes of belts and disks available, (b) grits available, (c) types of belts or disks (wet, dry, or both), and (d) the manufacturers of the belts or disks.
4. Prepare a list of types of applications for blasting processes. Check with other trades or crafts including, but not limited to, welders, metalsmiths, artists, dentists, sculptors, builders, and architects.
5. Complete the table on this page comparing the various processes for smoothing a surface.
6. Analyze the following to determine what finishing processes are required, then select a suitable finish for the product:

   Twelve 2-in. by 2-in. blocks have been sheared from a 1/2 in. cold-rolled plate steel. The faces need to be flat and smooth, the edges need to be slightly rounded.
7. What is the difference between polishing and buffing?
8. List three applications for which lapping could be used.
9. What is the difference between honing and lapping?
10. List the steps required to put a highly polished surface on a piece of rough copper.

	Thickness of material removed	Abrasive size and how applied
Honing		
Lapping		
Belt sanding		
Polishing		
Buffing		

SUMMARY

Virtually every surface needs to be "prepared" before a finish treatment is applied. The primary classifications of surface preparation processes are deburring and stripping/cleaning. Deburring can be done by:

- Hand
- Barrel or vibratory finishing
- Belt sanding

The choice of deburring technique is largely determined by the size and shape of the workpiece and the degree of precision needed.

Stripping/cleaning can be done by

- Dry or wet blasting
- Power brushing
- Ultrasonic cleaning

Surface imperfections are removed by smoothing operations such as

- Honing or superfinishing
- Lapping
- Polishing or buffing
- Burnishing
- Ballizing or bearizing

Honing, superfinishing, and lapping emphasize achieving a nearly perfect shape. Sanding, polishing, buffing, and burnishing emphasize smoothness of the surface. Holes can be finished, hardened, and sized by ballizing or bearizing. The choice of method is based largely on the precision required and the shape of the workpiece.

As with most other design criteria, the specification for the finish that a part is to receive depends on its function and the degree of attractiveness the part is to have.

Chapter

17

Surface Treatment Processes and Coatings

One of the stated purposes of finishing is *to provide a protective coating or to improve the appearance of the part.* Chapter 8 discussed methods of adding a metal onto metal using processes such as electroplating or chemical plating. These methods could also be used as a finishing process. If plating is done solely as a finishing operation, then only a thin layer of metal is deposited. All that needs to be done to finish an electroplated part is to buff or polish it. Further discussion of these plating techniques is not needed here.

■ CONVERSION PROCESSES

One of the "tricks" used to prevent corrosion of metals is to oxidize the surface to some chemical compound that adheres to the surface, is aesthetically pleasing, and cannot corrode any further. Iron oxide or rust does not adhere to the surface of the metal. It flakes off, leaving a lower layer of the iron or steel exposed to further corrosion.

Conversion processes are so named because the surface layer of the workpiece is changed chemically to some more stable compound. This is done either to improve the appearance of the part or to make it more resistant to oxidation or corrosion. The major processes include phosphate, chromate, oxide, and cyanide conversions.

All of the coating processes require that the surface of the part be thoroughly cleaned. All traces of dirt, grease, or even fingerprints must be removed. The parts to be coated are then handled by mechanical devices or by hand with the operators using clean rubber gloves.

Phosphate Conversion Processes

The cleaned iron or steel workpiece is immersed for 3 or 4 hours in a boiling solution of a paste made by mixing iron filings and concentrated phosphoric acid (H_3PO_4), then placing it in a weak phosphoric acid solution. The surface of the steel reacts with the phosphate to provide a rust-resistant coating of ferrous phosphate [$Fe_3(PO_4)_2$]. The coating is porous and is an excellent primer coat for later organic coatings. The porous phosphate coating can provide better corrosion resistance if oil is rubbed into it. One method of phosphate coating, known as *Parkerizing,* is used to apply a dull black protective coating on military small arms. With the addition of catalysts (accelerators), phosphate coatings are applied to automotive parts by a processes known as *Bonderizing* and *Granodizing.*

Bonderizing is usually used to provide a base for further painting. Phosphate coatings are also used to make a thin coating to reduce friction for later deep-drawing operations (see Chapter 11). Zinc phosphate coatings used on plain carbon or low-alloy steels leave a spongy, rough surface, which will hold lubricants and allow the part to be easily worked in forming and drawing processes.

Phosphate conversion coatings can be used on aluminum and aluminum alloys. In coating aluminum, it is placed in a warm (110 to 120°F) mixture of ammonium biphosphate ($NH_4H_2PO_4$), ammonium bifluoride (NH_4HF_2), and potassium dichromate ($K_2Cr_2O_7$) for about 5 minutes. The resulting coating will have a bluish color. Phosphate conversion can also be used to protect magnesium parts.

Chromate Conversion Processes

There are two main chromate conversion processes. One uses a mixture of chromic oxide (Cr_2O_3), ammonium bifluoride (NH_4HF_2), and stannic (tin) chloride ($SnCl_4$). The second process uses sodium dichromate ($Na_2Cr_2O_7$), sodium fluoride (NaF), potassium ferrocyanide ($K_4Fe(CN)_6$), and nitric acid (HNO_3). In both methods the part is placed in the solutions from a few seconds to 8 minutes.

Chromate coatings serve a number of functions. They provide good corrosion resistance, good electrical conductivity, good appearance, and can provide a base for later organic coatings. The electrical conductivity of these coatings at radio frequencies makes them ideal for coatings on waveguides for radars and other electronic systems.

Oxide Conversion Processes

Unlike steel, the oxides of aluminum and magnesium adhere to the surface of the parent metal. Aluminum oxide (Al_2O_3) is clay, and magnesium oxide (MgO) is used in making fire brick. The aluminum oxide and magnesium oxide coatings are therefore a very thin layer of ceramic, which protects the metal from further corrosion. Both aluminum and magnesium form these oxide coatings very quickly and naturally, but the appearance is somewhat dull and unattractive. The surfaces can be deliberately oxidized by placing the parts in a hot solution of sodium hydroxide (lye) and sodium nitrate or by *anodizing.*

The anodizing process is similar to electroplating except that the part to be coated is placed on the anode (positive electrode) rather than the cathode (negative electrode) (Figure 17–1). Oxalic acid ($H_2C_2O_4$), chromic acid (Cr_2O_3 in water), or sulfuric acid (H_2SO_4) is used as the electrolyte. The solution should be slightly warmed (100°F or 38°C) and the voltage gradually raised from 0 to 40 volts over a 5-minute period, then held at 40 volts for 30 minutes to 1 hour. The current depends on the surface area to be anodized but should be about 1 to 3 amperes per square foot (11 to 32 amperes per square metre). For instance, a cube, 3 inches on a side, would have a surface area of 54 square inches and the required current would be in the 0.4- to 1.1-ampere range. This can be calculated as follows:

$$\text{area of cube} = (3\text{ in.})^2/\text{surface} \times 6\text{ surfaces} = 54\text{ in.}^2$$

$$= 54\text{ in.}^2/144\text{ in.}^2/\text{ft}^2 = 0.375\text{ ft}^2$$

$$\text{amperage} = 0.375\text{ ft}^2 \times 1\text{ amp/ft}^2 = 0.375\text{ amps}$$

$$\text{to } 0.375\text{ ft}^{2 \times 3\text{ amp/ft}}2 = 1.125\text{ amps}$$

Note that the coating does not add to the dimensions of the part but only reacts with the outer atoms. Whereas the naturally obtained oxide coating on aluminum is only one molecule thick, less than one-millionth of an inch, anodized coatings range from 0.0001 to 0.008 inch (0.0025 to 0.2 millimetre) thick. An anodized coating about 0.0004 inch (0.01 millimetre) thick is necessary for corrosion and wear resistance. The coatings can have any one of several colors depending on which dyes or metals are added to the electrolyte. A chromic acid electrolyte gives a greenish-gray color but red, gold, blue, and some other colors of anodized surfaces are readily available. Colored anodized surfaces are at least 0.0008 inch (0.02 millimetre) thick.

Although anodizing is used mostly on aluminum, magnesium, and magnesium alloys, zinc castings and galvanized (zinc-coated) surfaces can also be anodized.

Anodizing is widely used in industry. Aircraft and automotive parts are anodized to provide corrosion resistance, hardness, and a decorative finish. Even kitchenware is now anodized.

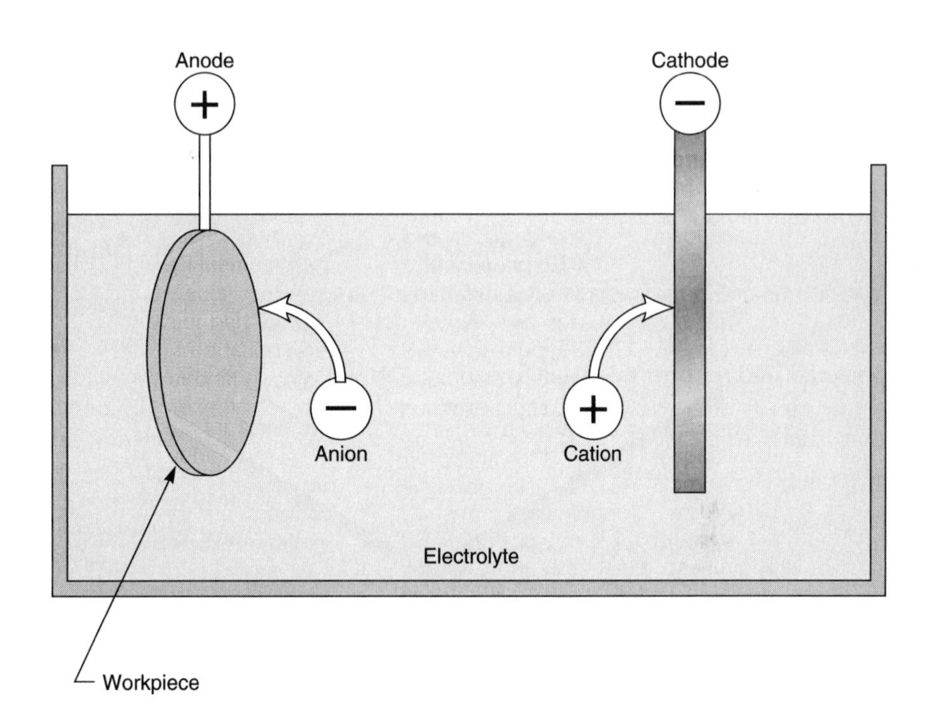

Figure 17-1. Anodizing.

Cyanide Conversion Process

Ferric ferrocyanide ($Fe_4(Fe(CN)_6)_3$), also known as Prussian blue, adheres to the surface of steel and provides an attractive appearance. Since steels contain carbon, a blue surface can be put on steels simply by heating them to about 600°F in air. The nitrogen of the air, reacting with the iron and carbon in the steel, forms the ferric ferrocyanide. Low-carbon steels may have to be buried in bone charcoal while being heated to enhance the formation of the Prussian blue. Steel can also be blued by heating it in a solution of potassium ferrocyanide. This forms the Prussian blue on the surface. Guns have been blued for centuries by this method.

Other methods for bluing steel are available. *Hot bluing* is achieved by placing the parts in a molten solution of sodium nitrate ($NaNO_3$) and potassium nitrate (KNO_3). *Cold bluing* can be done by coating the steel with ammonium polysulfide ($(NH_4)_2S_5$) until the desired color is achieved. A solution of iron chloride ($FeCl_3$), antimony trichloride ($SbCl_3$), and gallic acid has also been used in bluing steels.

■ INORGANIC COATINGS

Inorganic coatings are those that contain no carbon. These coatings are usually metallic or ceramic. Of interest to manufacturing are the metal-on-metal, metal-on-plastics, ceramic-on-metal, and ceramic-on-ceramic coatings.

Metal Coatings

The metallizing, electroplating, and dipping techniques discussed in Chapter 8 as methods of material addition can also be used to coat the surfaces of metals with an extremely thin metallic coating. These thin coatings, used as a finishing process, are intended only to provide corrosion protection or enhance the appearance of the part.

The preparation of the surface for finish metallic coatings may be slightly different than those used for material buildup. Remember that coatings do not remove rough spots on the surface. The coating follows the contour of the roughness. Therefore, the surface must be smoothed and polished before the finish coating is added. It is true that the polished surface must be lightly etched or "pickled" to allow the metallic coating to adhere to it.

Please refer to Chapter 8 for the techniques involved in metallic coatings.

Ceramic Coatings

There are two types of ceramic or *vitreous* coatings: porcelain and refractory. Porcelain coatings are used to create an abrasion-resistant, chemical-resistant surface that is easy to clean. Refractory coatings are used to protect the piece from the effects of high temperature.

Porcelain

Porcelain is a silicate glass with additives that make it opaque and give it color. Glass and the additives are melted together then poured into water to produce fractured particles of glass. The particles are then ground to a powder, called *frit.* The cleaned surface of the metal is treated with a solution of nickel sulfate ($NiSO_4$) to leave a flash coating of nickel, which is a good base to bond to the glass.

The frit is mixed with water to form *slip.* Small workpieces can be dipped into the slip, but the slip is sprayed onto larger objects. The slip is then dried and the coated workpieces fired at about 1500°F (800°C) until the particles fuse into a continuous coating. In coating very large objects, such as bathtubs, sinks, etc., the workpiece is heated and the slip sprayed onto the hot surface. No drying period is needed in this application because the slip will stick, dry, and be ready for firing immediately.

Artists have also used this technique to produce artifacts known as *cloisonne* (kloi′-so-nay). Patterns are outlined on a metal surface by welding wire dams to separate the colors, as shown in Figure 17–2(a). Slips of different colors are applied to the separate areas of the work and the piece is fired. The entire piece is then polished and the part of the wire dams that shows between the colors is plated. An example of the polished and finished product is shown in Figure 17–2(b). Cloisonne is an old manufacturing technique but has many present-day uses.

Porcelain can be applied over steel, aluminized steel, stainless steel, cast iron, aluminum, copper, or other ceramics. Porcelain coatings are used for cooking utensils, laundry equipment, plumbing fixtures, chemical processing equipment, signs, and a host of other applications. The surface is hard, corrosion resistant, and attractive, but it is somewhat brittle.

Refractories

The frit for *refractory* coatings is made from aluminum, magnesium, and chromium oxides. The frit is mixed with only enough water to make a paste, which is troweled or sprayed onto the surface with a plasma spray gun (see the section on metallizing in Chapter 8). Once dry, the workpiece is fired in the same manner as the porcelain.

Refractory coatings are used to protect exhaust manifolds in jet engines and rocket motors, the interiors of ovens and furnaces, heat shields, space reentry vehicles, and for other high-temperature applications.

■ ORGANIC COATINGS

Organic coating materials are those that contain materials which could come from living organisms. They all contain carbon or silicon. With the exception of varnishes, each of the materials is composed of solid particles or pigments suspended in a liquid *vehicle* or *carrier.* Several basic types of organic coatings are used in industry. Some common ones are:

- Oil paints
- Alkyd paints
- Epoxy paints
- Silicone paints
- Varnishes and enamels
- Latex paints
- Casein paints

Oil paint uses a vehicle of natural or synthetic resin plus a thinner and a drying oil. Since the curing time is comparatively long for oil-based paints, their current use in manufacturing is limited.

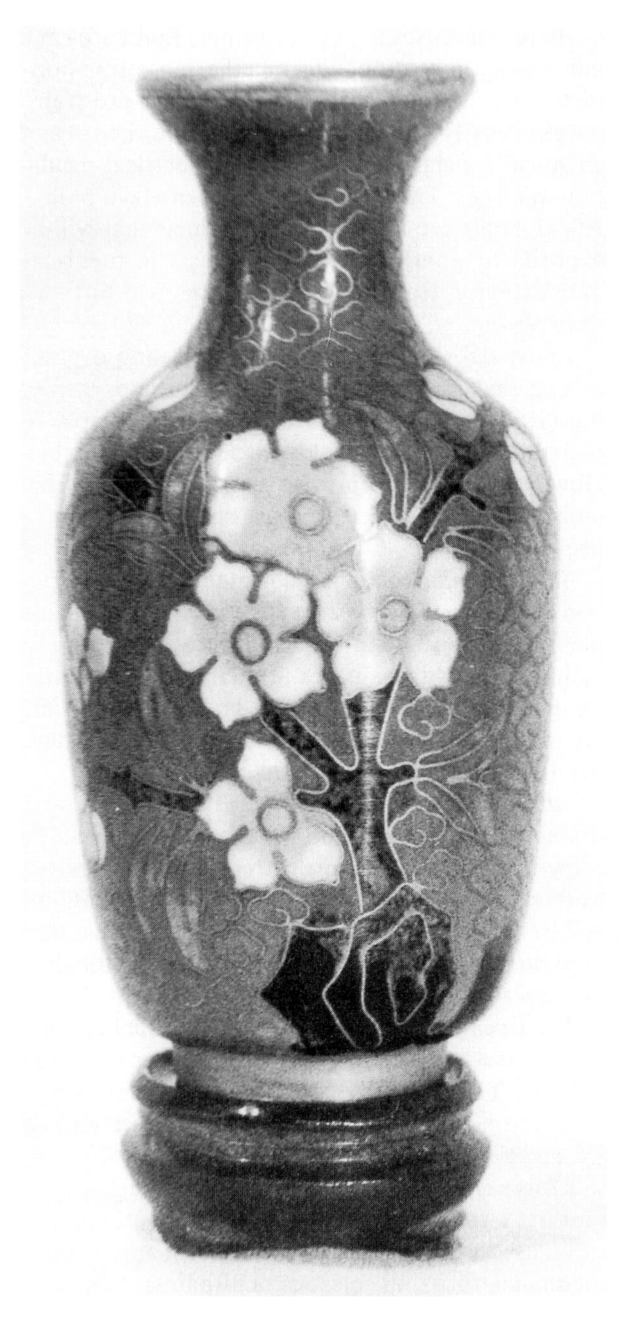

Figure 17-2. Cloisonne metal work.

Alkyd paint uses a polymer vehicle that cures and hardens upon heating to form a thermosetting polymer. These paints adhere well to metals, are transparent, easily colored, tough, flexible, heat- and chemical-resistant, and have good electrical insulation qualities. There are many types of alkyd paints. Alkyd paints are used on steel products that will be exposed to weathering. Lawn and patio furniture and exterior framing for buildings are but two examples.

Epoxy paint uses an oxygen-containing organic vehicle that cannot react further with the oxygen of the air. As such, epoxy-based paints are corrosion-resistant, abrasion-resistant, tough, and flexible. However, one problem with epoxy coatings is that moisture can get under them, causing the coatings to peel off.

Silicone paints contain silicon atoms instead of carbon atoms in the molecule of the carrier. Silicone paints are resistant to heat, moisture, oxidation, or fading. Their disadvantages are that they are brittle, have poor adherence to unprimed metal surfaces, and are more expensive than either the oil-based or alkyd paints.

Varnishes are oil-modified alkyd compounds with drying accelerators in the vehicle. Many varnishes come from naturally occurring resins. Early varnishes were extracted from wood; newer ones use synthetic resins. When pigments are added to varnishes, they become *enamels.* Varnishes are more flexible, tougher, and less affected by moisture than is oil paint.

Latex paints use a variety of synthetic rubber compounds as a vehicle. Many of these vehicles are water soluble. They provide somewhat elastic coatings, which do not crack or peel readily. They are easy to use and cleanup is rapid and easy.

Casein is a white to yellow protein, which is precipitated from milk by a dilute acid. (Vinegar is often used.) Casein has many uses from adhesives to medicines to paper, plastics, and paints. Casein is used as the vehicle for many water-based paints that are not exposed to weather. Casein paints are not waterproof.

The vehicles of these organic coatings do not "dry" after they are applied. Only the thinner or drying agent evaporates. This allows the vehicle to *polymerize* and form a continuous coating on the surface of the material. The effect of this polymerization can easily be seen in older paints that have weathered and split. They can be peeled from the surface in sheets. More discussion of polymerization is provided in Chapter 18.

Organic Coating Processes

A surprisingly large number of processes are available for applying organic coatings. These processes include, but are not limited to, the following:

- Brushing
- Rolling
- Dipping
- Electrocoating
- Spray painting
- Printing
- Curtain coating
- Powder coating
- Heat transfer coating

The choice of method should be determined by:

- The type of coating material to be used
- The size and shape of the workpiece
- The in-house processes that are available

Brushing

Brushing is the process of choice for large flat (or nearly flat) workpieces for which the coating material has a medium to long curing time; when overspray must be avoided at all costs; or when small details need to be done in a variety of colors. Brushing is a relatively slow method of applying coatings, but it can be done in just about any environment, wastes little paint, and cleanup is simple and easy.

Rolling

Rolling is faster than brushing. It covers large areas of flat surfaces quickly. Rolling can selectively coat raised surfaces, such as the raised letters on automobile and truck license plates, without touching the recessed areas. Glues, solder, and other liquids can also be applied by rolling.

Dipping

Dipping is preferred for coating highly irregular shapes with considerable internal detail. Automobile bodies and other large objects are dipped in primer coatings for quick, even, and thorough application. Small parts can be loaded onto trays and dipped. Simple shapes that are to be coated by heavy consistency or viscous paints or that need only part of their surface covered can be dipped.

Electrocoating

In *electrocoating* a static electric charge is applied to the workpiece. This charge attracts the water-soluble paint, causing it to adhere to the surface. Highly irregular surfaces can be given even coatings by this technique. Since the paint is attracted to the surface by the electric charge, little paint is wasted and noxious fumes are reduced to a minimum. Many new automobiles are electrocoated. Electrocoating increases the efficiency of paint use, cuts down on air pollution, and aids in the adherence of the paint to the surface.

Spray Painting

Spray painting is preferred for workpieces that have an irregular, textured, grooved, or concave surface. Spraying can reach areas that are inaccessible to brushing or rolling. It is a fast and inexpensive method of applying a coating. Its disadvantages are that there is some paint lost due to the paint missing the surface (overspray) and that the air is contaminated by the paint spray. Spray booths, exhaust systems, and protective breathing equipment for the workers are all necessary in spray painting operations. Currently, many environmental regulations govern the use of spray painting. The overspray can be reduced considerably by electrocoating.

Printing

Paints can be applied to flat or regular convex surfaces by offsetting. The coating is first applied to a preformed shape, then transferred to the desired surface. Paints can also be applied to metal, paper, plastic, or other materials, which can then be attached to workpieces by adhesives. *Printing* is suitable for a range of applications from applying labels on jars to decorations on aircraft and buildings.

Curtain (or Flow) Coating

Curtain painting involves running the parts on a conveyor belt under a curtain of paint flowing much like a waterfall or under spray nozzles. The excess paint collects underneath the belt and is recirculated. The shape of the workpiece is critical to the success of this process. The gravity flow of the paint would make it difficult to get to interior surfaces. Further, the coating material requires a long drying or curing time or the solvents or thinners must be constantly replenished (Figure 17–3).

Powder Coating

In *powder coating,* the workpiece is heated, then run through a spray of powdered, organic resins. The resin particles fuse to the hot object. As soon as the workpiece cools, it is ready for use. Contact between the workpiece and the resin particles can be accelerated and enhanced by giving them opposite electrostatic charges, in the same manner used in electrocoating. Powder coating has become commonly used in industry since it produces no solvent vapors and little air pollution.

Heat Transfer Coating

A special form of powder coating used primarily to decorate or add writing to the workpiece is *heat transfer coating.* In this process, the powdered resin is attached to a polyester film in the mirror image of the designed pattern. The film is then pressed against the workpiece with the powder side of the film against the workpiece. A hot roller passing over the polyester film fuses the powdered resin into the workpiece. The polyester film is removed, leaving the decoration or writing on the object. This process is a form of offset printing combined with heat transfer and powder coating techniques. Plastic, wood, leather, paper, metal, fiberboard, and polymer objects are commonly decorated by this process.

Figure 17-3. Curtain painting.

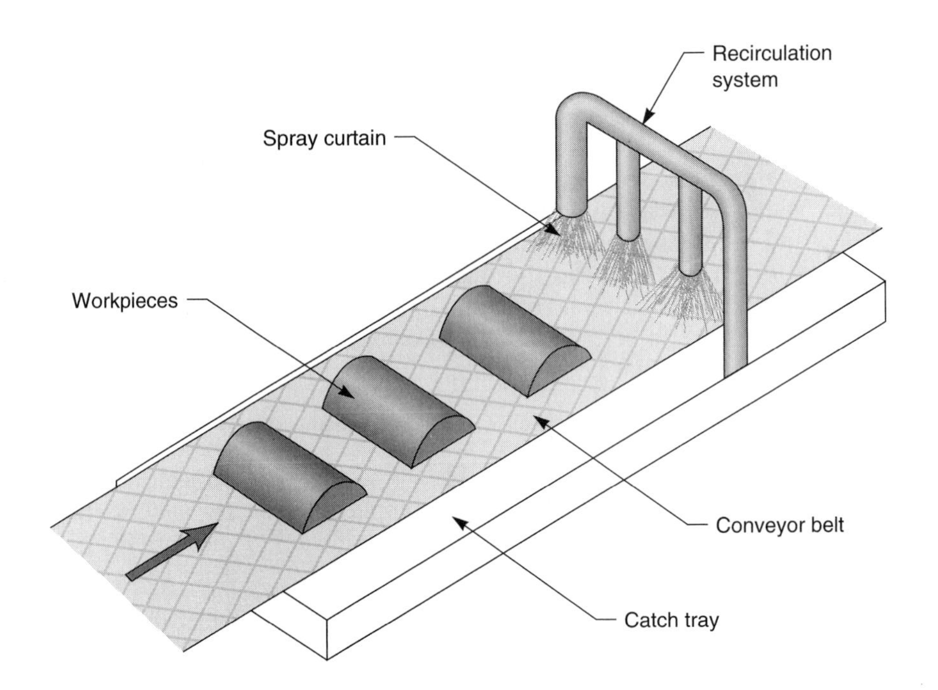

PROBLEM SET 17-1

1. How does a conversion process differ from other finishing processes?
2. How do the metal-on-metal processes in this chapter differ from those of Chapter 8 on material addition?
3. What is the function of a *refractory* coating that makes it distinct from porcelain coating?
4. As presented in this chapter, porcelain coatings sound like the solution to all the world's coating problems. But porcelain coatings have some serious drawbacks. List five of them. (An appliance or bathroom fixture refinisher might be an excellent reference and source of information.)
5. Why aren't porcelain coatings put on plastic workpieces?
6. For each of the coating materials listed in this chapter (counting varnishes and enamels separately), list four objects that would be appropriately coated with that material. Explain the reasons for each choice.
7. For each of the coating processes described in this chapter, list three objects that reasonably could or should be coated by that process.
8. What considerations should be taken into account when selecting a lacquer for a musical instrument?
9. What type of painting is used at one of your local automobile body shops?
10. If you are mass producing a part that must be inorganically coated at the rate of 1000 pieces per hour, what type of coating process(es) could be used?
11. What environmental and safety conditions must be considered when using any organic or inorganic coating?

Surface finishing operations either remove or add insignificant amounts of material to the surface of the workpiece in order to:

- Achieve a closer fit to the ideal shape.
- Provide a protective or functional coating.
- Improve the appearance of the finished product.

The surface treatment processes of importance are:

- Conversion coatings
- Inorganic coatings
- Organic coatings

Conversion coatings change the surface layer of the workpiece to a different chemical compound. The major conversion processes used in industry include:

- Phosphate on iron and steel
- Chromate on aluminum or zinc
- Oxidizing of steel, aluminum, copper, brass, magnesium, etc.
- Cyanide coatings (bluing, etc.)

Inorganic coatings of importance are metal and ceramics. These processes include:

- Metallizing
- Plating
- Porcelain coating (a general-purpose silicate coating)
- Refractory coating (for high-temperature coatings)

Organic coatings include paints (oil, alkyd, epoxy, silicone, etc.) and varnishes and enamels.

Cost, the surface to be finished, and the environment in which the part is to be used all affect the choice of coatings.

Coatings may be applied by several different methods, including:

- Brushing
- Rolling
- Dipping
- Electrocoating
- Spray painting
- Printing
- Curtain or flow coating
- Powder coating
- Heat transfer coating

The size and shape of the object to be coated, along with the properties of the coating material determine the choice of coating process.

SECTION

VIII

Manufacturing with Plastics and Composites

INTRODUCTION

While most products were manufactured in the past from wood, metals, and other naturally occurring materials, plastics and composites may well dominate as the future materials of production. Prior to World War II, the number of known synthetic materials was very small. Bakelite (phenol-formaldehyde) was discovered by Dr. Leo Bakeland in 1907 and was widely used by the 1930s. Celluloid (cellulose acetate), cellulose nitrate, and rayon had been produced, but these were just regenerated natural cellulose fibers derived from cotton and wood. Synthetic rubber had been made in the laboratory as early as World War I, but it was far from a usable product.

World War II gave a great boost to the search for synthetic materials. During the first five years of the 1940s, usable synthetic rubber, nylon, and Lucite (Plexiglas) were commercially developed. That was only the start. Once we understood how polymerization occurs chemically, there was no end to the number of plastics that could be developed. The types and varieties of plastics now number in the thousands.

Plastics have some properties not present in metals and other materials. These unique properties permit the manufacture and use of products heretofore not possible. Plastics do not corrode or fail in fatigue as do metals. Many plastics and composites can be made into finished products using conventional manufacturing techniques, but new methods have been developed for use solely with these synthetic materials.

Chapter 18

Manufacturing with Plastics

The term *plastics* is misleading. One thinks of a plastic as a material that will withstand permanent deformation, such as stretching or bending, without breaking. Yet many of the "plastics" used today are very brittle. Perhaps a better term for these materials is *high-polymer.* However, because the term "plastic" is so common, we will use both the term "plastics" and "high-polymers" or even "polymer" in this chapter.

The Greek word *poly* means "many"; the term *polymer* simply means "many mers." A *mer* is the basic molecule from which a particular plastic or polymer is made. Many thousands of these basic molecules are joined to form a polymer. Chemists classify polymers by the type of chemical reaction used to form them. For instance, there are **addition-type** polymers, **copolymers,** and **condensation polymers.**

Although a complete discussion of the chemical reactions involved in the production of polymers is best left to a course in materials science, it is helpful for the student of manufacturing to know the fundamentals of the chemistry of plastics. Therefore, a brief discussion is in order of the chemical fundamentals of polymerization, the physical and chemical properties of polymers, and the techniques used in manufacturing products from plastics.

■ CHEMISTRY OF POLYMERIZATION

The carbon atom must always have four covalent bonds. (Carbon monoxide is the notable exception to this rule.) The simplest carbon compound is *methane,* which is one carbon atom bonded to four hydrogen atoms (CH_4). The organic chemist would write the structure of methane as:

```
    H
    |
H — C — H
    |
    H
```

Any compound based on two carbons is an *ethane* derivative. Ethane (C_2H_6) has the structural formula:

```
    H   H
    |   |
H — C — C — H
    |   |
    H   H
```

Three carbon compounds are *propanes* (C_3H_8), and compounds having four carbon atoms are known as *butanes* (C_4H_{10}):

```
    H  H  H               H  H  H  H
    |  |  |               |  |  |  |
H—C—C—C—H           H—C—C—C—C—H
    |  |  |               |  |  |  |
    H  H  H               H  H  H  H
```

Propane

Butane

Any of the hydrogen atoms in the organic compounds can be replaced by another atom. For instance, if one of the hydrogen atoms in methane is replaced by a chlorine atom, the result is known as **chloro-methane** having the structural formula:

```
    H
    |
H—C—Cl
    |
    H
```

Chloro-methane

One word of caution about chemical names. One of the most confusing things about chemistry is that the same compounds often have many different names. Prior to the formation around 1932 of the *International Union of Chemists (IUC),* which later became the *International Union of Pure and Applied Chemists (IUPAC),* chemists who produced a new compound could name it just about any way they wished. There was no standard method of naming chemical compounds. Chloro-methane was called **methyl chloride** or even **methylene chloride.** To add to this confusion, many compounds and chemical reactions were given the names of their discoverer then produced under a number of different trade names. There is the Fehlings solution, Sweitzer reagent, Wurtz reaction, and so on. Many chemical compounds still retain these old names as well as the IUPAC standardized name. An example would be **trichloro methane,** which has the formula

```
     Cl
     |
Cl—C—H
     |
     Cl
```

Trichloro-methane, (Chloroform)

The old chemical name for this compound was methylene trichloride, but it is still best known as **chloroform.** One of the early functions of the IUC was to develop a standardized method of naming organic compounds based on the chemical structure of the compound. If the structure of a chemical compound is known, then the IUPAC name can be given to it. Conversely, if the proper IUPAC name of a compound is given, the structure can be derived from the name. The standard IUPAC names are used in this book for the names of compounds. However, the old names of the compounds may be given in some instances to prevent confusion.

Every carbon atom must have four bonds, but it is possible for two or three of the bonds to be between the same two carbon atoms. If two of the hydrogens are removed from the ethane molecule, the bonds that went to the hydrogens will simply form a new *double* bond between the carbons:

```
    |  |
H—C—C—H  ⟶  H—C = C—H
    |  |          |   |
    H  H          H   H
```

Ethene

The suffix for the names of chemical compounds having double bonds changes from **-ane** to **-ene.** Thus the ethane becomes ethene once a double bond is formed.

It is also possible to have **triple** bonds formed in a compound. If two more hydrogen atoms are removed from the ethene, a triple-bonded compound will be formed.

```
H—C = C—H  ⟶  H—C ≡ C—H
    |   |
```

Ethyne (Acetylene)

Triple-bonded compounds have the suffix to **-yne.** The triple-bonded two-carbon compound shown above is known in the IUPAC nomenclature as *ethyne.* This compound is still far better known by its old name **acetylene.**

Any organic compound having double or triple bonds in the molecule is said to be *unsaturated,* meaning that it can take in more hydrogen or other atoms. If there is more than one double or triple bond in a molecule, the molecule is said to be *polyunsaturated.* Any unsaturated compound can be resaturated by reacting hydrogen with it. This is the process by which such synthetic products as margarine and other cooking fats are

```
H   H   H   H        H   H        H   H         H   H   H   H
|   |   |   |        |   |        |   |         |   |   |   |
C = C + C = C  →  —C — C— + —C — C—   →   —C — C — C — C—
|   |   |   |        |   |        |   |         |   |   |   |
H   H   H   H        H   H        H   H         H   H   H   H
```

Figure A

made. One test that can be used to determine if a product is unsaturated is to burn it. Saturated compounds produce no smoke or, at worst, a white smoke when burned. Unsaturated compounds produce a black sooty smoke when burned at one atmosphere of air.

Any unsaturated compound can also be made to react with itself. This is the key to polymerization. A molecule will polymerize **if it has at least two reaction points, and maintains at least two reaction points after each joining of the compound.** Consider ethene for instance. The double bond in two molecules of ethene can "open up" and react (See Figure A).

Because the four-carbon compound has open bonds at each end, it can react with another ethene molecule or with another four-carbon compound, and continue reacting to form chains containing thousands of carbon atoms. This is the process of **polymerization.**

Plastics are usually named for their **mer.** In the case of the polymerization of ethene, the ethene molecule is the mer. The polymerized product of ethene would be called **polyethene** or, more commonly, **polyethylene.** The structural formula for polyethylene is usually written:

```
|   H   H   |
|   |   |   |
|  —C — C—  |
|   |   |   |
|   H   H   | n
```

Polyethylene

Where *n* may be in the thousands.

Polymerization sometimes occurs in unexpected places. Epoxy and other glues polymerize to form the bond that hold the parts together. Gasoline, left over a period of time, will often polymerize to plug up carburetors. Mechanics often note a gummy film on interior engine parts caused by the oil, which is unsaturated, polymerizing. Paint, varnish, and enamels will polymerize once their "thinner" has been allowed to evaporate.

When a hydrogen atom is removed from an organic molecule, leaving an empty or unfilled bond,the product is called a *radical.* The name of radicals end in **-y1.** The methyl radical would be:

```
    H
    |
H — C —
    |
    H
```

Methyl radical

The radical of the ethene molecule is:

```
H   H
|   |
C = C
|   |
H
```

Vinyl radical

The ethene radical is more commonly called the *vinyl* radical. If a chlorine atom was used to fill the empty bond of the vinyl radical, **vinyl chloride** would result:

```
H   H
|   |
C = C
|   |
H   Cl
```

Vinyl chloride

Because vinyl chloride is unsaturated, it will polymerize into **polyvinyl chloride,** often known simply as **PVC.** PVC has been used in many applications, including plumbing.

Replacing two of the hydrogen atoms of the ethene molecule with chlorine atoms produces the **vinylidene chloride** mer. Polyvinylidene is sold under such trade names as *Kynar* and *Saran.*

Methane and radicals can be added to the ethene mer to make polymers with different physical characteristics. One widely used mer has this formula:

Methyl methacrylate

This is the **methyl methacrylate** mer. When polymerized it becomes **polymethylmethacrylate.** Hardly any industry exists that does not use this polymer in its products.

The empty sight on the vinyl radical can also be filled with the cyanide group (–C≡N). This forms the **acrylonitrile** mer.

Acrylonitrile

Polymerization of the acrylonitrile produces a polymer known as *Orlon.*

Other compounds will polymerize as long as there is a double or triple bond in the molecule. Propene can polymerize to form **polypropene,** more commonly known as **polypropylene:**

Polypropylene

Organic compounds can also form *ring compounds.* Benzene, for example, is a six-carbon compound, having only six hydrogen atoms. Its structure is:

Benzene

The circle in the hexagonal molecule indicates that there are three double bonds that are continually shifting positions between the carbon atoms. Rather than writing this complicated formula all of the time, chemists often use just a simple hexagon with a circle inside, to represent the benzene molecule:

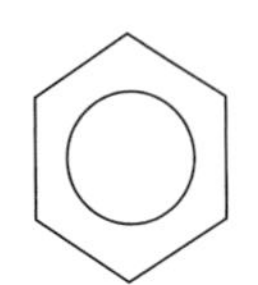

Benzene symbol

If one hydrogen is removed from the benzene molecule, it becomes the *phenyl* radical. The phenyl radical can be added to the ethene mer to produce the **styrene** mer:

Styrene

Polystyrene is an unsaturated compound that produces a black smoke when burned. Polystyrene can be made into a *foam* during its manufacture. The resultant *Styrofoam* is widely used as thermal insulation, in packaging, in roofing, and for many more applications.

If all four of the hydrogen atoms of the ethene molecule are replaced by *fluorine* atoms, the compound becomes tetrafluoroethene:

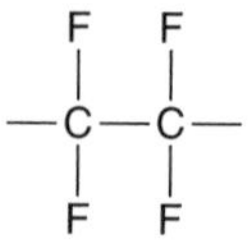

Tetrafluoroethene

Tetrafluoroethene will polymerize into **polytetrafluoroethene,** often abbreviated to **PTFE,** and sold under such trade names as *Teflon.* PTFE has one of the lowest known coefficients of friction and is not affected by acids.

From the examples already given, it is obvious that many more polymers could be made simply by adding different groups to ethene or other unsaturated organic compounds.

All of the polymerizations just discussed occur by simply adding a string of like mers together. These are called *addition type* high-polymers. Two different types of molecules can be made to react with each other to produce a mer. The reaction discovered by Dr. Bakeland reacted phenol with formaldehyde. The formula for the production of this mer is (See Figure B).

In the late 1930s it was found that hexamethylene and adipic acid would react to form a mer that produced water as a by-product. This type of reaction produces a *condensation type* polymer (See Figure C).

This reaction was quickly named **nylon.** Today there are many types of nylons. Any diamine reacted with any organic di-acid, produces a nylon. Nylons are distinguished by the number of carbon atoms in the individual molecules. The example shown above, which is made up of two six-carbon molecules, is *Nylon 66,* which might be used in lightweight cloth. The nylon used in automobile tires and rugs would still have the two amine (NH_2) and the two acid (–CO–OH) groups, but a different number of carbons in the middle of the molecules.

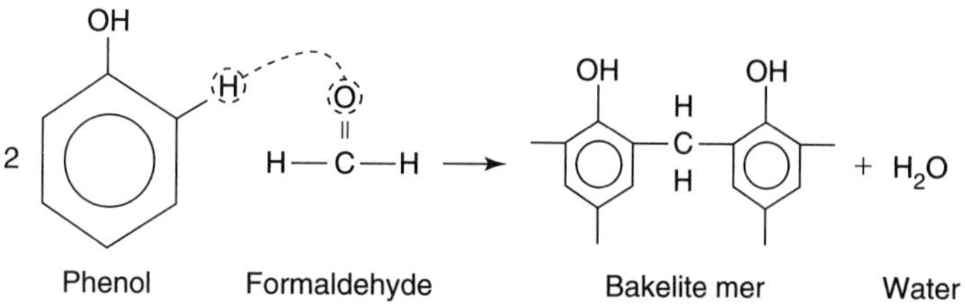

Phenol Formaldehyde Bakelite mer Water

Figure B

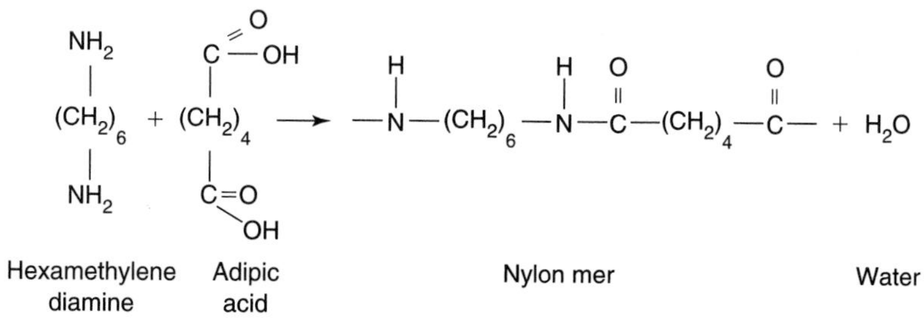

Hexamethylene diamine Adipic acid Nylon mer Water

Figure C

A different type of nylon simply replaces the interior carbon chains with phenyl groups to produce the reaction (See Figure D).

This is the reaction for the production of *Kevlar,* one material used in bullet-proof vests.

Another condensation type mer is formed by the reaction of dimethyl terephthalate and ethylene glycol. Ethylene glycol is an ethane molecule with an alcohol (–OH) group on each end. It is most commonly used as antifreeze in automobile engines. The by-product of this reaction is methanol (methyl alcohol or wood alcohol) (See Figure E).

The end product of the polymerization of this new compound has been given the name *Dacron.* Dacron is used to create durable sailcloth, ropes and other items requiring a strong, weather resistant material.

The dramatic increase of the number of polymers in the last half of the 20th century has produced some rather complicated but useful mers. One such example is polycarbonates, which have the mer:

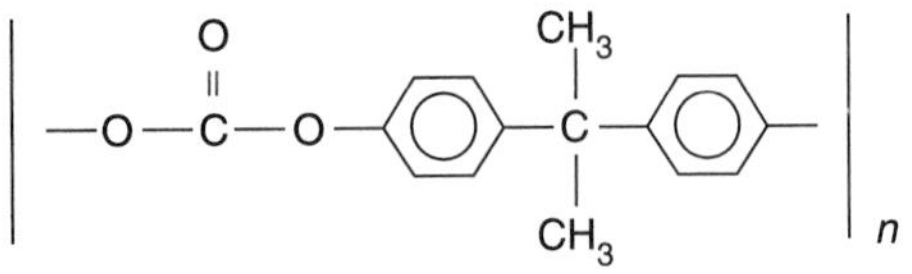

Polycarbonate (Lexan) mer

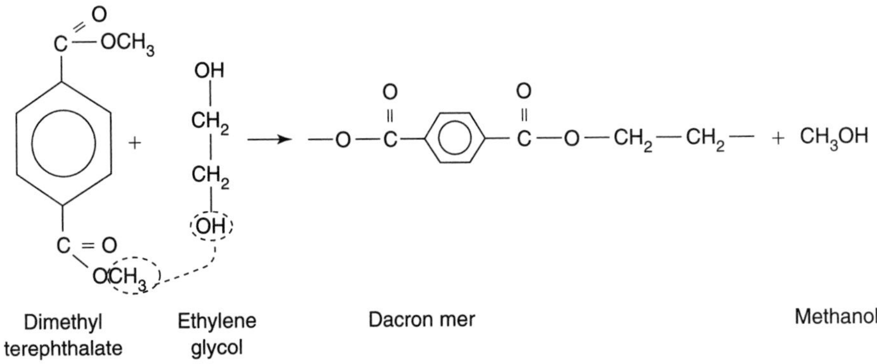

Dimethyl terephthalate + Ethylene glycol → Dacron mer + Methanol (CH_3OH)

Figure D

Kevlar mer + H_2O

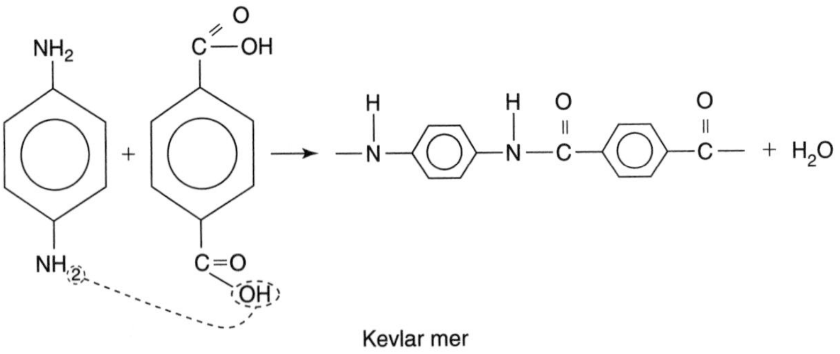

Figure E

A common trade name for this type of polycarbonate is *Lexan.* Lexan is rather soft and can be scratched, but will not break under impact. It has been used to make "bullet-proof" shields, which can be mounted in place of glass.

With our brief chemistry background in polymers, let us now move on to the types of plastics as characterized by their physical properties.

TYPES OF PLASTICS

For the consumer, plastics are divided into three basic types:

1. Thermoplastics,
2. Thermosets, and
3. Elastomers.

Thermoplastics

Thermoplastic high-polymers (called *T-plasts* in industry) get soft when heated. These plastics will soften to a runny liquid, burn, and can be easily molded. They can be reused and recycled simply by melting and remolding them. Obviously, these polymers cannot be used at high temperatures. To make a skillet handle from a thermoplastic material would not be a suitable choice of material.

Most plastics have many trade names for the same polymer. A short list of thermoplastic high-polymers and their trade names would include:

Acrylonitrile
- *Orlon*

Polyamines
- *Nylon*

Polyaramides
- *Kevlar*

Polybutylene

Polycarbonates
- *Lexan*

Polyethylene
- (many brands of trash bags and ropes)

Polyethylene glycol (PEG)

Polyesters
- *Dacron*

Polymethyl methacrylate (PMMA)
- *Lucite*
- *Plexiglas*
- *Acrylic*
- *Acrylan*
- *Acrylite*
- *Methacral*
- *Perspex*
- *Zerlon*

Polypropylene

Polystyrene
- *Styrofoam*

Polytetrafluoroethene (PTFE)
- *Teflon*

Polyvinyl chloride (PVC)

Polyvinylidene chloride
- *Koroseal*
- *Kynar*
- *Saran*

Thermosets

Thermosetting high-polymers (also called **T-sets**) get hard when heated. The more they are heated, the harder they get. They will burn, crack, and char, but they will not melt or soften. Thermosetting materials cannot be remelted and very few methods of recycling them have been found. They can be ground up and used as fillers for other plastics and concrete.

The list of thermosetting materials also includes such polymers and trade names as:

Phenol-formaldehyde
- *Bakelite*

Analine-formaldehydes

Melamine-formaldehydes
- *Formica*
- *Melmac*

Urea-formaldehyde

Polyurethanes

Epoxies

Elastomers

Elastomers are polymers that can be stretched elastically and then return to their original shape and size. Natural and synthetic rubbers are classic examples of elastomers. Elastomers will burn and sometimes soften. However, the softening temperature is often above the temperature at which they ignite. Elastomers often decompose chemically at temperatures below the temperature at which they soften. Natural rubber is polymerized isoprene, which is obtained from the sap of the caoutchouc ("weeping") tree (commonly called the rubber tree). Since the mid-1940s, natural rubber has been largely replaced by a host of synthetic elastomers.

Synthetic rubber was developed by trying to duplicate the molecule found in natural rubber. Natural rubber contains the compound **isoprene,** which has the formula:

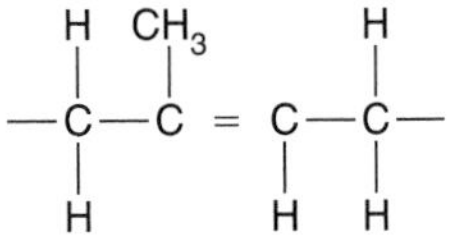

Isoprene mer

One of the first successful attempts to make synthetic rubber simply polymerized a compound called **butadiene.** Butadiene has two double bonds that can "open up" to form a mer:

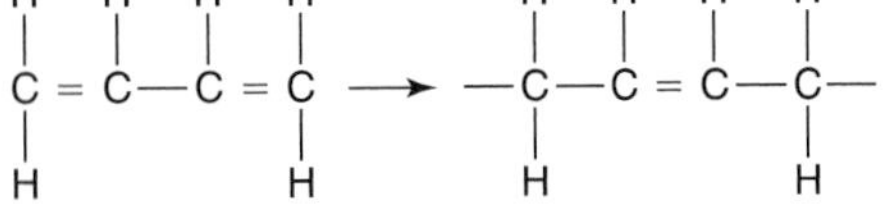

Butadiene → Buna mer

The butadiene rubber is marketed under the name *Buna* rubber.

A partial list of elastomers now available would be as follows:

Butadiene rubbers
Chloroprene
  Neoprene
Thiokol

One problem with rubber is that it quickly decomposes when brought into contact with organic solvents such as gasoline, alcohols, and other petroleum distillates. **Thiokol** rubber was developed to resist this chemical attack. The prefix *thio* in a chemical name indicates the presence of sulfur in the molecule. The chemical structure of thiokol rubber mer is:

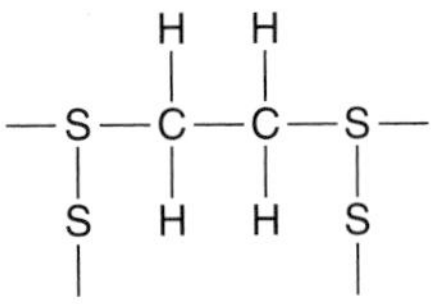

Thiokol

Thiokol rubber is used in fuel lines and gaskets that come into contact with organic chemicals.

Differences in Polymers

The chemical difference between thermosetting and thermoplastic high-polymers lies in the method in which the long-chain molecules react with each other. Thermoplastic high-polymers have no strong chemical bonds between the chains or, at the most, have only van der Waal bonds between the chains. The chains can intertwine and "rub" against each other to provide tensile strength, but many thermoplastic polymers are relatively weak in tension (Figure 18–1). Polyethylene is a thermoplastic polymer used in garbage bags, some rope, and clothes bags. If one simply pulls a polyethylene bag apart, it can be seen that it is very "stretchable" and weak. Thermoplastic high-polymers will get soft and often burn when ignited with a torch or match. An

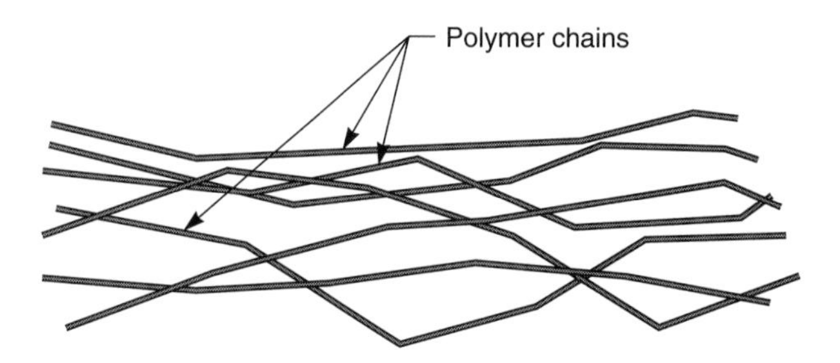

Figure 18–1. Thermoplastic high-polymer.

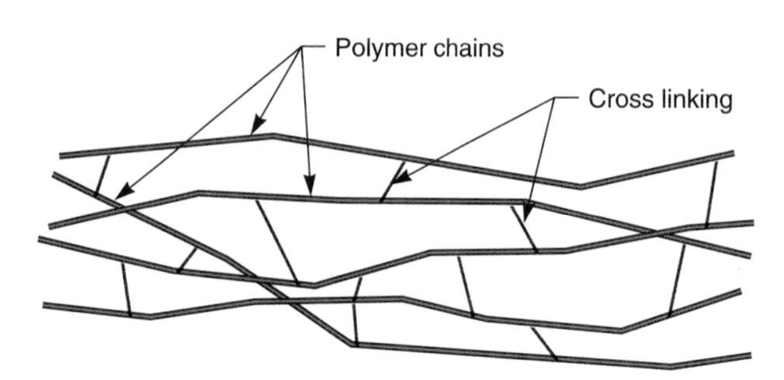

Figure 18–2. Thermosetting high-polymer.

identifying feature of polyethylene is that it will melt and drip burning drops of the polymer when ignited.

Thermosetting high-polymers, on the other hand, form covalent chemical bonds between the long chains (Figure 18–2). These bonds between the chains are referred to as *cross-linking.* When thermosetting polymers are heated, they become more cross-linked and become harder. Thermosetting high-polymers will burn, char, and crack, but will not melt when heated.

The molecular chains in elastomers are in the form of coils (Figure 18–3). When a force is applied to the coils, they stretch like springs. Some cross-linking exists between elastomer chains, but not as much as in pure thermosetting molecules. Some molecules, such as the neoprene found in natural rubber, can form coiled molecules, which are elastic, or straight-chain molecules, which are not elastic. The straight-chain form of natural rubber is called *gutta percha,* but is commonly called hard rubber. In the days before synthetic rubber, gutta percha was used to make items such as billiard balls, bowling balls, canes, pistol grips, and a number of other items.

Other Types of Polymers

Chemists have also combined the molecules of different types of synthetic chemicals to produce specific properties. These plastics are known as *copolymers.* Acrylonitrile, butadiene, and styrene have been joined to produce the ABS plastic used for making sewer

Figure 18–3. Elastomers.

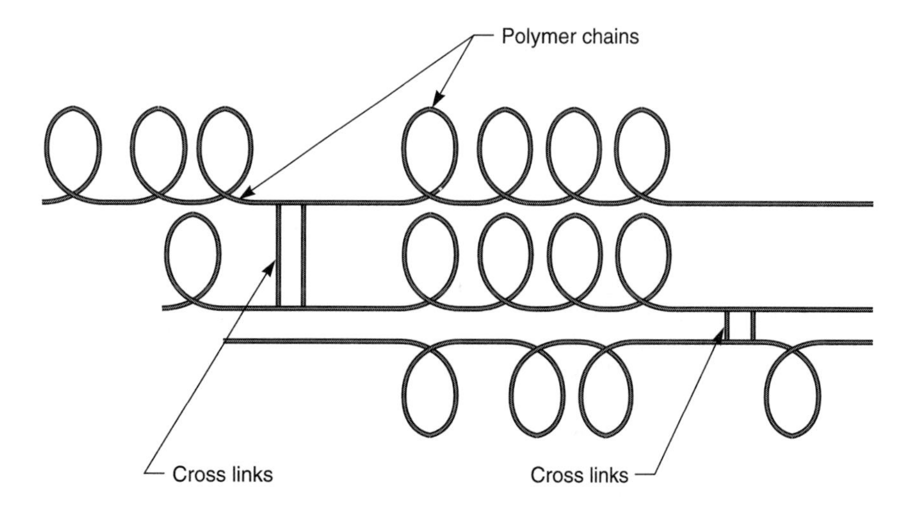

pipes. ABS is affectionately tagged the "Awful Black Stuff" by plumbers. The ABS mer has the structure:

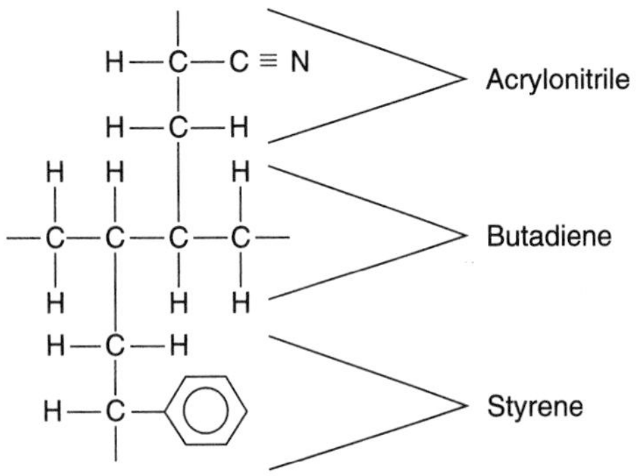

Acrylonitrile
Butadiene
Styrene

Butadiene and acrylonitrile react to produce the elastomer of the Buna-N type rubbers. Similarly, butadiene and styrene combine to make the Buna-S rubbers. The Buna-N rubber is a copolymer of butadiene and acrylonitrile, whereas the Buna-S rubber is a copolymer of butadiene and styrene. The structural formulas of these mers are:

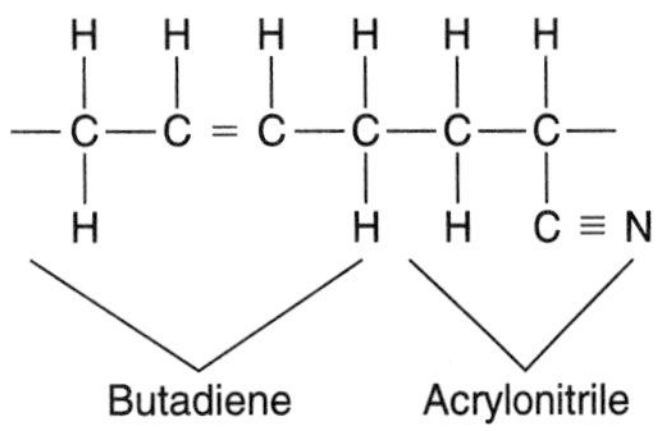

Butadiene Acrylonitrile
Buna-N rubber

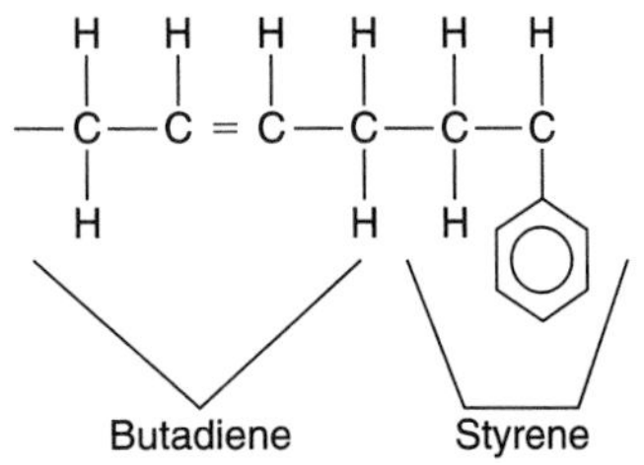

Butadiene Styrene
Buna-S rubber

New polymers are being discovered daily. Chemists and materials scientists now know enough about polymerization to produce a plastic with just about any desired properties. About the only thing polymers cannot do is conduct electricity or heat. They can be used as electrical insulators and heat barriers and heat absorbers. New sources of raw materials and molecules are being found. Polymers nearly as light as air have been produced from seaweed. The future of plastics is unlimited.

Silicone Polymers

Looking at the chemical periodic table, we note that the silicon atom is directly below the carbon atom. The carbon and the silicon atom have similar chemical characteristics. Therefore, anything that can be done with the carbon atom can theoretically be done with the silicon atom.

If the carbon of the methane molecule was replaced by a silicon atom the structural formula would be:

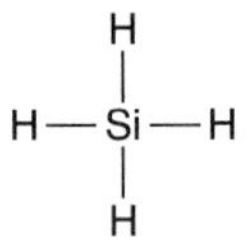

Silane

This compound is known as **silane.** Other silicon-based compounds could just as easily be made.

All of the carbon polymers could, in theory, be converted to *silicone* polymers. In practice, a few polymers have been made using the silicon atom in place of carbon, and have found specific uses in industry. Silicones are a very promising field for the future.

MANUFACTURING WITH PLASTICS

In forming finished plastics, many mers must link together. When the chains grow to between 15,000 and 20,000 mers, the product is an oily liquid and is said to be in its *A stage.* By the time 30,000 to 50,000 mers have reacted, the polymer is a sticky, tacky material much like rubber cement that has been spread on a paper and allowed to set for a few minutes. This is the *B stage* of the polymer. After 70,000 to 80,000 mers have combined, the plastic has fully set and has reached its *C stage.* The basic material of the mer is sometimes called the *resin.*

Plastics can be made into products by means of most of the conventional methods of manufacturing used with metals and other materials. However, plastics differ from metals and other products in that the plastics themselves are often synthesized or formed while the products are being formed. Polymers usually require that a catalyst be mixed with the resin or mer at some point in the molding or forming process. Several molding techniques have been developed for use with polymers.

Resin Transfer Molding

The simplest and cheapest form of molding with polymers is *resin transfer molding (RTM).* Here, the resin and the catalyst are mixed in a container and the mixture then poured into the mold. In the case of composites, the reinforcing fibers will already have been placed in the mold. RTM is used for resins that do not set quickly. The molds are often spun so that the centrifugal force will force out any air bubbles and force the catalyzed resin into the entire mold. Figure 18–4 illustrates RTM.

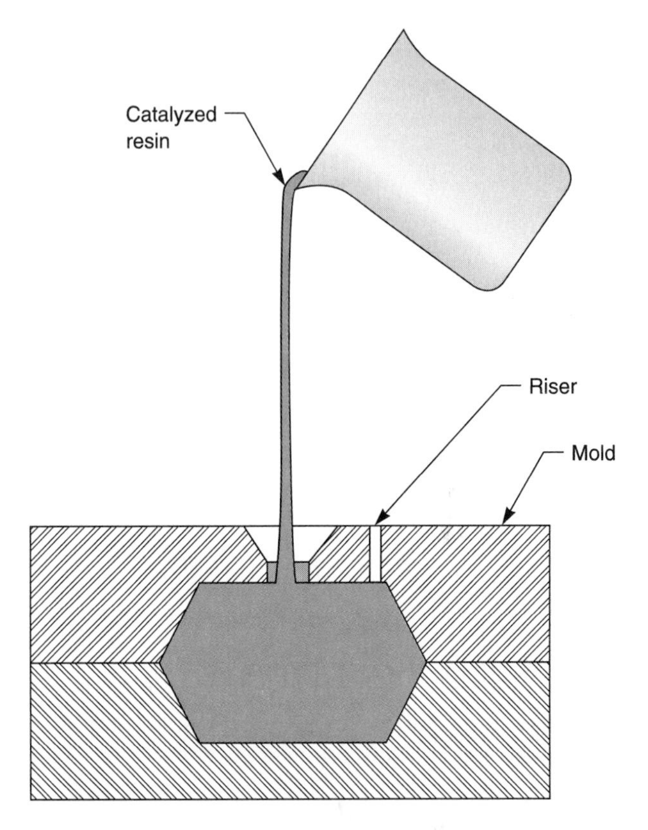

Figure 18–4. Resin transfer molding.

Injection Molding

The difference between RTM and *injection molding* is that in the latter, the catalyzed resin is forced into the mold by an injection pump or piston. This force helps fill the mold completely and gives good dimensional stability to the product. Small gears and other products having delicate surface features are often injection molded (Figure 18–5).

Reaction Injection Molding

Polymers that react quickly with the resin require a faster technique than RTM or injection molding. In *reaction injection molding* (RIM), the resin and catalyst are forced into a reaction chamber where the streams impinge to start the reaction. The catalyzed reaction is then immediately forced into the mold where it is heated and cured to the C stage. A schematic of the RIM process is shown in Figure 18–6.

Figure 18–5. Injection molding.

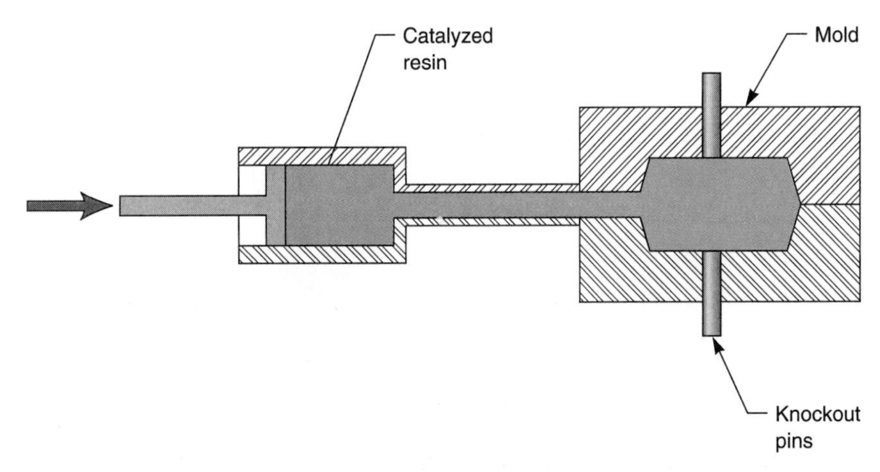

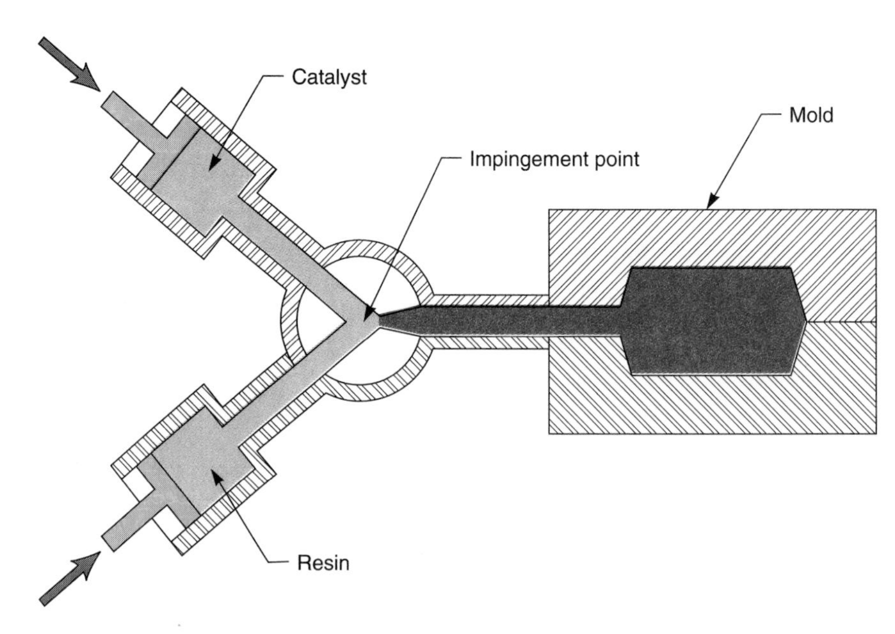

Figure 18-6. Reaction injection molding.

Ultra-Reinforced Thermoset RIM

A method used with thermosetting resins in composite manufacture is that of *ultra-reinforced thermoset reaction injection molding* (URTRIM). In this process, a core is cast of high-temperature epoxy foam mixed with microscopic hollow glass spheres. Foams that makes use of these microspheres are called *syntactic* foams. This technique is also used in manufacturing composites. When used with composites, the core is then wrapped with several layers of graphite or glass fabric and placed in the mold. Epoxy or other T-setting resin is mixed with its catalyst in an injection chamber where it is then forced into the heated mold and cured. A vacuum may also be drawn on the mold to remove any chance for air bubbles getting into the product. URTRIM produces an extremely dimensionally stable part and it is used in the production of aircraft wings, missile fins, and other aircraft-related structures. Figure 18–7 shows this technique.

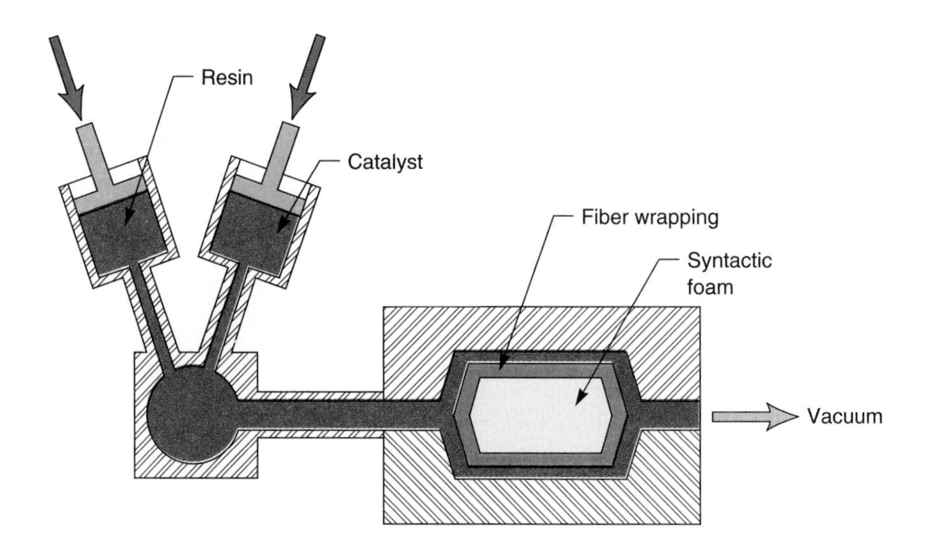

Figure 18-7. Ultra-reinforced thermoset reaction injection molding.

PROBLEM SET 18–1

1. Look around. Make a list of at least three items made of a thermoplastic and three of thermosetting high-polymers.
2. Which of the following items could be easily recycled?
   a. Bakelite electrical plug
   b. Polyethylene garbage bags
   c. Saran wrap
   d. Melmac tableware
   e. Formica table top
   f. PVC pipe
   g. Automobile tires
3. Is a common plastic milk bottle a thermosetting or thermoplastic polymer? (How could you find out?)
4. Used plastic milk bottles are now being remade into plastic shopping bags. How could this be done?
5. What is the difference between resin transfer molding (RTM) and reaction injection molding (RIM)?
6. Cut a strip of a plastic garbage bag, hold it with a pair of pliers, and carefully set it afire. (Do not let any of the molten plastic touch you or your clothing—it is very hot.) Note the color of the smoke, the odor, the color of the flame, and the molten plastic as it drips. Do the same with a Styrofoam cup. Note the difference between the two plastics. This is a standard "flame test," which can be used to identify the type of plastics.
7. Explain how epoxy glue could be used for the "resin transfer molding" of a small part.
8. What makes an elastomer "stretchable"?
9. What is a *syntactic* foam?
10. What is a *copolymer?*
11. If oxygen, chlorine, or sulfur gets into the reaction chamber during a polymerization process, the reaction is said to be poisoned. That is the reaction stops. Why would oxygen or chlorine stop the chemical reaction?
12. Given the following polymers:

    Polyethylene
    polystyrene
    Bakelite
    Buna S.

    Which one would be the most appropriate for use as:

    A skillet handle
    Shoe soles
    Ice box insulation
    clothes line

    Note: There may be more than one answer for each of the above, so explain your reasons for your choices.
13. a. Give the proper chemical name for the following compound.

$$\begin{array}{ccccccc} & & H & & H & & H & \\ & & | & & | & & | & \\ H & - & C & - & C & - & C & - & Cl \\ & & | & & | & & | & \\ & & H & & H & & H & \end{array}$$

    b. Would you expect this compound to polymerize? Why?

Plastics and composites are the materials of the future. They are extremely lightweight, don't corrode, and don't fail by fatigue. There are three types of plastics or polymers: thermosets, thermoplastics, and elastomers. Polymers can be molded using resin transfer molding, injection molding, reaction injection molding, and ultra-reinforced thermoset reaction injection molding.

Chapter

19

Manufacturing with Composites

Composites are defined as a **judicious** combination of two or more materials that produces a **synergistic** effect. The two keywords in this definition are "judicious" and "synergistic." A *judicious* combination implies that the components are carefully selected to provide the desired physical and chemical characteristics to the composite product. *Synergy* (or "synergistic") simply means that the whole product is better than the sum of the individual components. The term *synergy* was coined by Buckminster Fuller. He illustrated his concept of synergy with the following analogy:

If one has 100 individual strings, each capable of lifting one pound, then the maximum load the strings could lift would be 100 pounds. However, if those same strings are twisted, braided, or woven into a rope, the rope will lift far more than the 100 pounds. That is synergy.

Composites are made up of a *fiber* and a *matrix.* Natural composites can be found in such plants as celery, corn stalks, and sugar cane. The long longitudinal strands of these plants are the fiber and the pulp is the matrix. Reinforced and prestressed concretes can be considered composites in that the steel is the fiber and the concrete provides the matrix.

If the fiber in a composite is laid in random directions, or the fibers are very short, the result is often called a *reinforced plastic* or *simple composite.* However, if the fibers are long, continuous, and laid in a given direction, the product is known as an *advanced composite* (Figure 19–1).

Composites can be made using any fiber and any matrix. However, not all combinations are useful. Examples of some of the categories of composite systems include:

Fiber-resin composites
Fiber-ceramic composites
Carbon-metal composites
Metal-concrete composites
Metal-resin composites
Metal-elastomer composites
Fiber-elastomer composites
Wood-resin composites

■ TYPES OF FABRICS

Currently, the most common fibers used in composites are glass, graphite (carbon), and Kevlar. Many others such as boron and silicones have been tried and new ones are being developed.

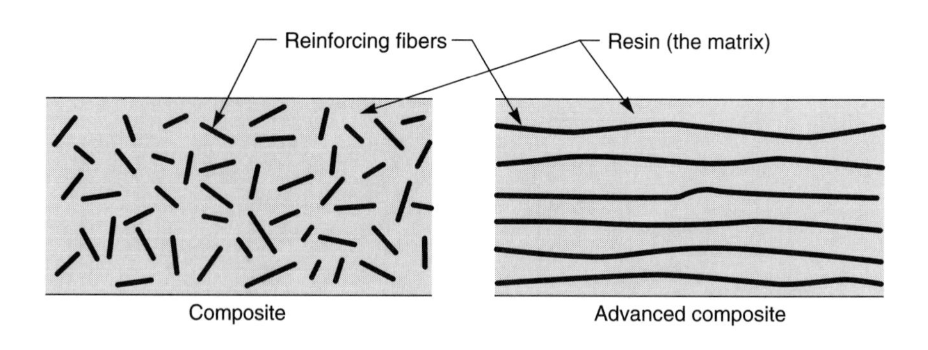

Figure 19-1. Composites and advanced composites.

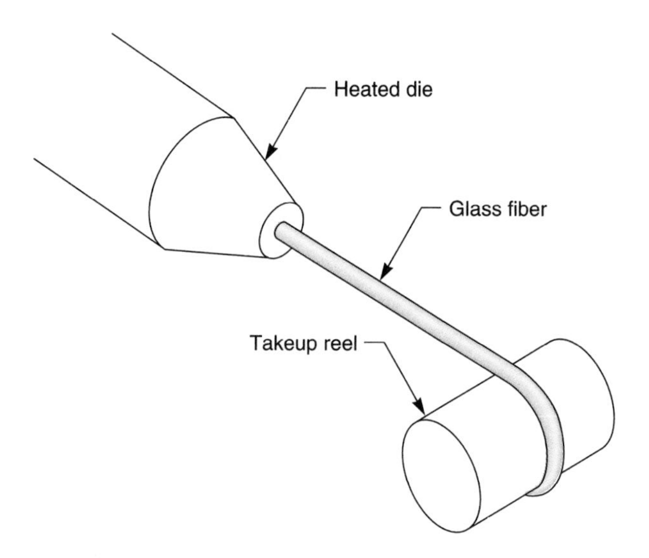

Figure 19-2. Formation of long glass fibers.

If we heat a glass rod to around 1200°F, it can be drawn out into a very thin fiber. Continuous fibers can be made in much the same way. Long, continuous fibers for advanced composites can be made either by extruding molten glass through a die or drawing hot glass through a die (Figure 19–2). Short fiberglass fibers are made by blowing air past a glass rod as it is being melted (Figure 19–3). Many types of glass can be made into fibers. The choice of glass type is determined by the properties needed. Soda-lime (soft) glass fibers are easily made at low cost. Borosilicate (Pyrex, Kymex, etc.) glass fibers are also used. Fiberglass is usually white in color but can be dyed if so ordered.

Kevlar is a polyaramid in the same family as nylon. These fibers are made by extrusion or drawing through a die. Kevlar has a distinctive yellow color.

Graphite fibers are made by several techniques. One of the most popular is to burn rayon, regenerated

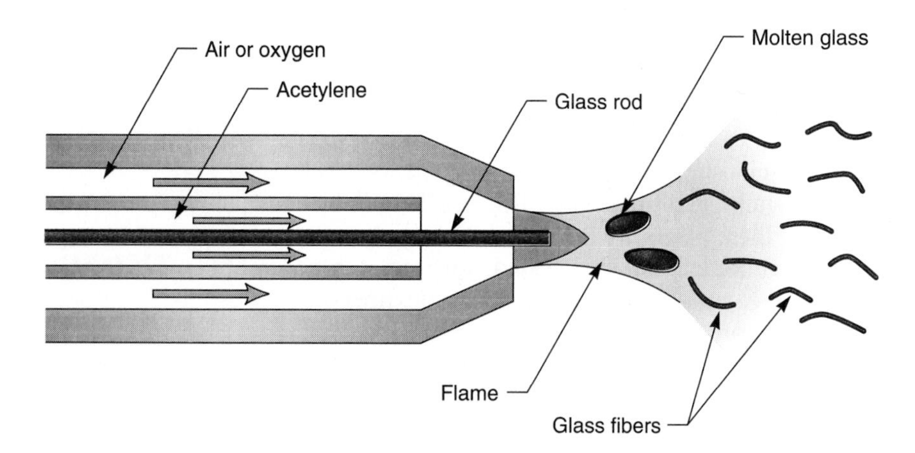

Figure 19-3. Manufacture of short glass fibers.

cellulose, or other materials in the absence of oxygen. The hydrogen and other elements are burned off, leaving only the carbon. Note that the fiber is often called "graphite" but in reality it does not have the crystal structure of graphite. It should be called carbon fiber. In this technique, the fibers are made of the parent material before being converted into graphite. Graphite fibers are always black,.

The properties that fiberglass, graphite, and Kevlar display are remarkable. For example, graphite has a tensile strength three to five times stronger than steel (depending on the type of steel) and has a density that is one-fourth that of steel. Kevlar has about four times the tensile strength of steel and is one-fifth the weight. Similarly, fiberglass has about five times the tensile strength of steel and has a density of about one third that of steel. A comparison of these materials is shown in Table 19–1.

It makes no sense to make fibers stronger than is needed in any direction because this would unnecessarily add weight to the product. In some cases, the fibers need to be quite large in one direction while the perpendicular direction might require only a small fiber or none at all. Some applications may require large fibers in both directions. To obtain a good finish on composites, a fine outer layer of fabric may be needed. As a result, fibers and fabrics are made in many different forms and styles. Figure 19–4 is a photograph of several different types of fibers currently available.

Table 19–1. Properties of Composite Fibers

Fiber Type	Density (gm/cm^3)	Tensile Strength (lb/in.^{2})	Modulus of Elasticity (lb/in.^{2} $\times$ 10^6)
Steel (reference)	7.86	60,000–320,000	29
Graphite			
Low density	1.76	410,000	30–37
High density	1.94	300,000	50–76
Kevlar	1.46	400,000	20
Glass	2.54	500,000	10.5

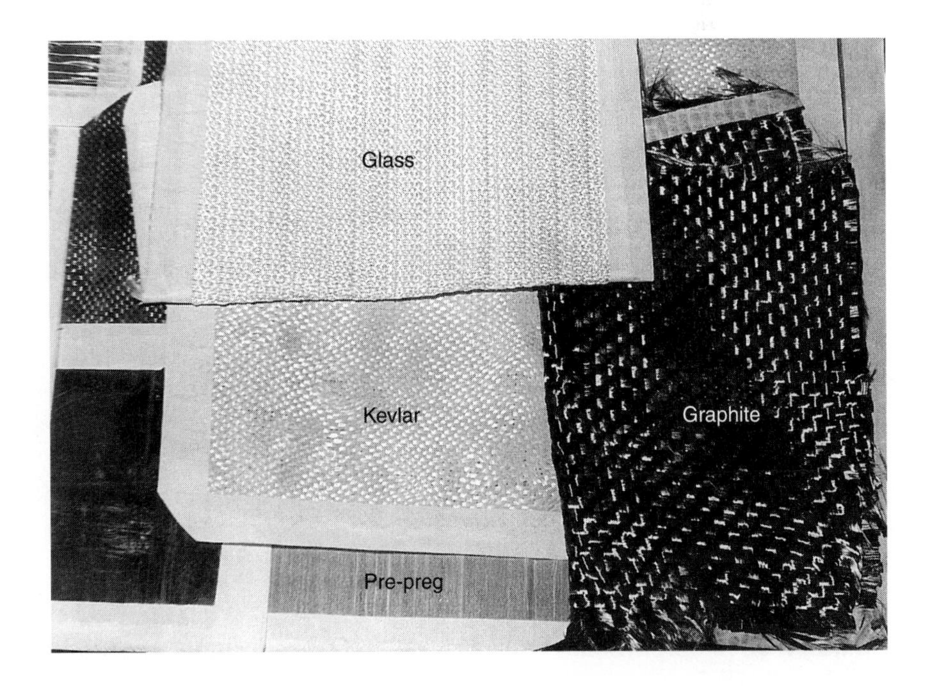

Figure 19–4. Types of fabrics used in composites.

Directions in Fabrics

Stretch any woven cloth in the direction parallel to the fibers and little give will be noted. However, pull on the corners at an angle to the direction of the fibers, and the cloth will be severely distorted. The direction of the fibers in a composite can have a similar effect on its properties. In fabrics, the longitudinal direction, or the direction that will take the most stress is called the *warp*. The direction perpendicular to the warp is the *weft*. The direction midway between the warp and the weft is the *bias* (Figure 19–5). In bias ply tires, the fabric is laid so that the direction of the fibers is at a 45° angle with the tire. Radial ply tires have the warp parallel to the plane of the tire while the weft wraps around it.

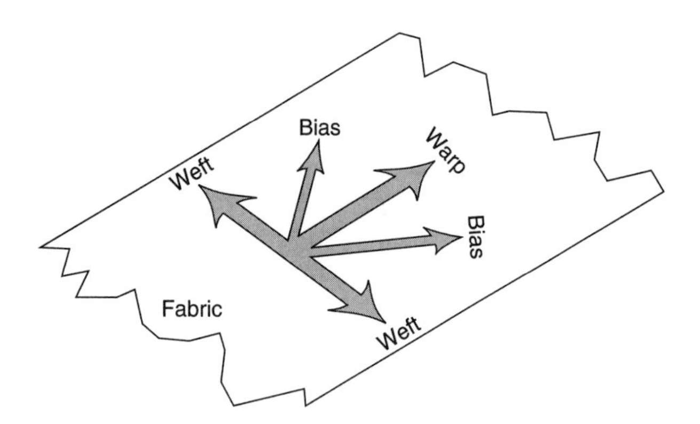

Figure 19-5. Fiber direction.

The properties of reinforcing fibers can also vary considerably. Glass is a strong and cheap reinforcing fiber, but it is twice as heavy as Kevlar and graphite. Fiberglass is best used in compression. Kevlar has good abrasion and puncture resistance as well as tensile strength, but has a relatively low modulus of elasticity. It flexes more than graphite. Graphite is expensive, but gives the best rigidity of any of the fibers. The optimum solution is to mix or *hybridize* the fibers.

Hybrids

A hybrid is a combination of fibers within a single matrix. Hybrids come in several forms. Alternate strands of different fibers in a single layer or ply are *intraply hybrids* (Figure 19–6). If the different plies are of different fibers, it is said to be an *interply hybrid* (Figure 19–7).

In some manufacturing techniques, fibers of different types can be selectively placed (Figure 19–8). In some beams the maximum tensile stress would be on the bottom of a beam, whereas the top might be in compression. Graphite would then be the choice of fiber for the bottom of the beam, whereas glass might be placed on the top. Different layers of fabrics can also be sewn or woven together. This prevents them from moving about in a mold during the curing process. The interlacing of these different layers or plies is known as *interply knitting* (or *interply weaving*). Interply knitting is shown in Figure 19–9.

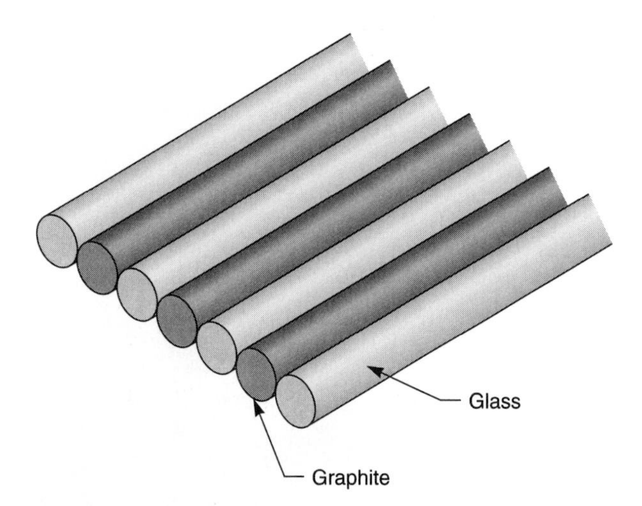

Figure 19-6. Intraply hybrid.

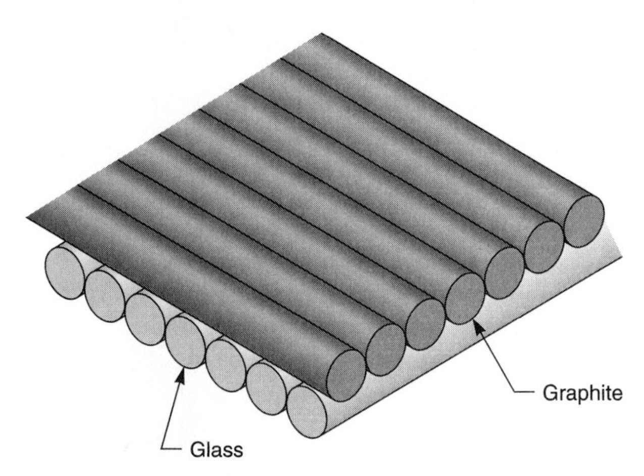

Figure 19-7. Interply hybrid.

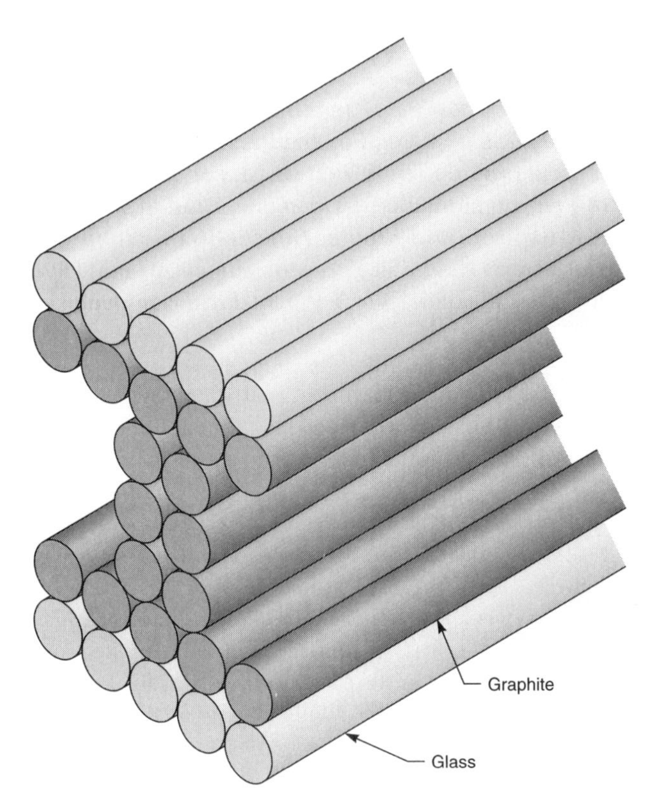

Figure 19-8. Selective placement of fibers.

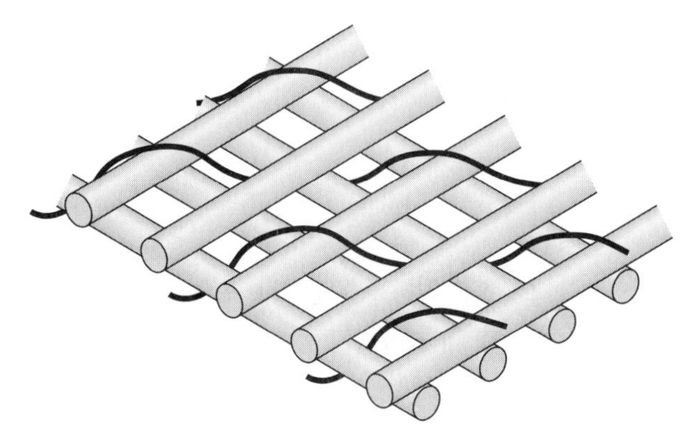

Figure 19-9. Interply knitting.

TYPES OF RESINS

The properties of both the resin and the reinforcing fabrics in composites must be understood in order to design and manufacture a successful product. The resins may be either thermosetting or thermoplastics but they must be compatible with the fibers. For instance, a thermosetting resin should not be used with a thermoplastic fiber nor should a resin that chemically dissolves the fiber be used. Table 19–2 lists some examples of resin types commonly used in composites.

MANUFACTURING TECHNIQUES OF COMPOSITES

Several steps are necessary to manufacture items with composites. The fabric or reinforcing material must be cut to size; it must be properly placed in a mold, pattern, or form; the resin must be applied; and it must be cured. Note that resins do not "dry" but react chemically to polymerize to a final product.

Hand Layup

One of the oldest forms of composite manufacture is the *hand layup* or *hand-lay.* The form is coated with a resin using either a paintbrush, roller, swab, spatula, or other method. The fabric is then pressed into the resin and another coat of resin applied to the outside. Backyard boatbuilders and repair people have used this technique with fiberglass for years. The method is cheap and requires little equipment. It does require some skill on the part of the operator and it is very wasteful of resin. The product is usually quite heavy compared to other techniques. It is used in industry for one-of-a-kind products and prototype products.

Table 19–2. Types of Resins Used in Composites

Thermoplastics	Thermosetting
ABS	Epoxy
PMMA	Bakelite
Fluorocarbon (e.g., Teflon)	Melamine
Nylon	Polyesters
Polycarbonate	Urea-formaldehyde
Polyphenylene sulfide	Urethanes
Polypropylene	Silicone
Styrene	
Vinyl	
Vinylidines	

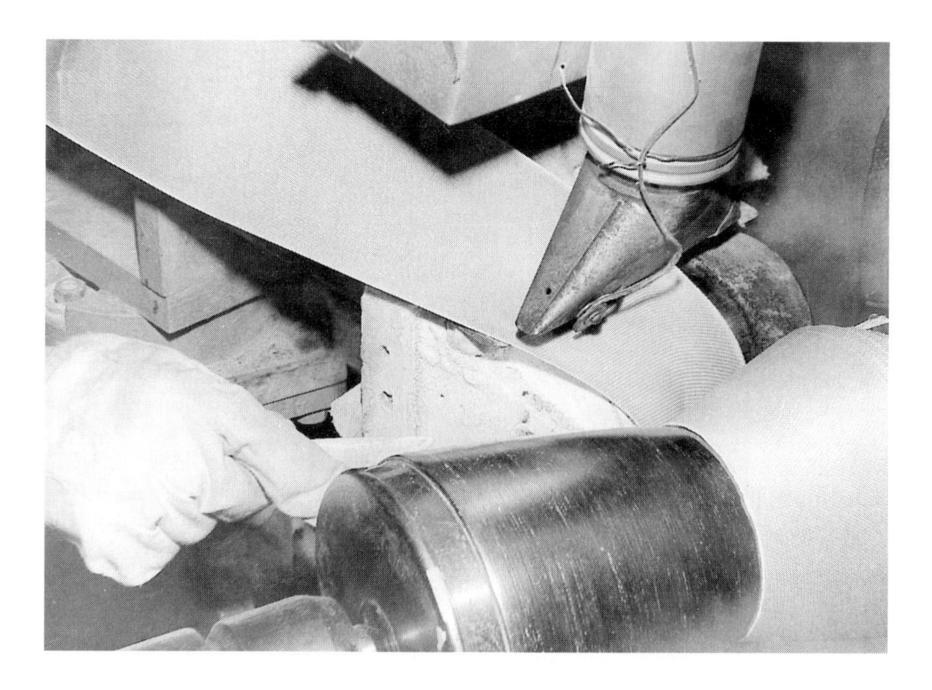

Figure 19–10. Wrapping a pre-preg.

Pre-Preg

Fabric layup can also be done using a *pre-impregnated* or *pre-preg* fabric. In pre-preg, the fabric is saturated with the resin, the excess resin squeezed out by rollers, and the result cured to the B stage. The tacky material can be stored for a short while until ready for use. The shelf life of pre-preg can be extended somewhat if it is kept refrigerated, but generally the shelf life is approximately a week to 10 days.

Pre-preg can be wrapped around a mandrel, or cut by computer-controlled machines and even laid up on forms by laser-guided robots. Once the pre-preg is laid up, it must be put under pressure and cured. This process is discussed later. Figure 19–10 is a photograph of a worker wrapping a pre-preg part.

Filament Winding

One of the problems with using fabrics over patterns and forms is that the fabric often will not readily take the shape of the mold. The result is distortion of the fabric and wrinkles in the resin. If the pattern is a convex shape, having no indentations, a method known as *filament winding* can be used. In filament winding, the individual fibers are drawn through the resin then stretched around a mandrel in the same manner as a thread is wound on a sewing machine bobbin or spool. After the winding is complete, the product is placed under pressure, cured, and the mandrel removed (Figure 19–11). Nose cones for aircraft, radar domes, and missile nose cones and bodies are manufactured by filament winding.

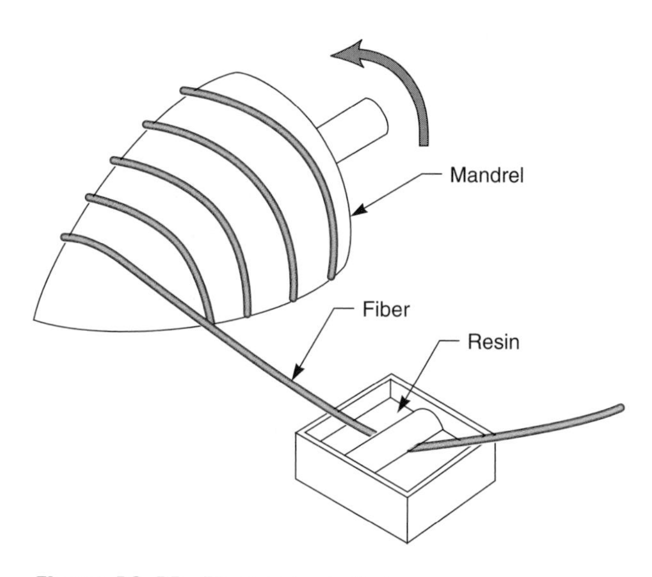

Figure 19–11. Filament winding.

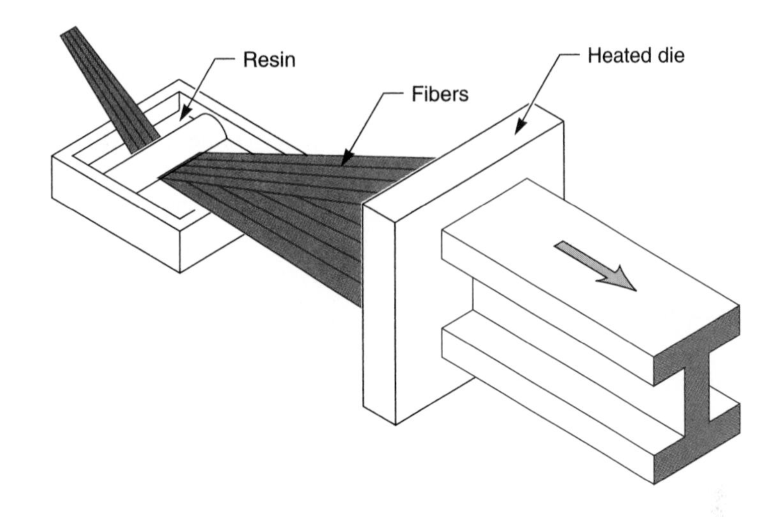

Figure 19-12. Pultrusion.

Pultrusion

A unique technique of composites manufacture that allows for selective placement of fibers is *pultrusion.* In pultrusion, the bundle of arranged fibers is drawn through a resin bath, then pulled through a heated die. The result is a product having the cross-sectional shape of the die. The product is then cured and cut to the desired length. This technique allows for the production of channels, flange beams, T-bars, and other shapes to be produced in very long lengths. Further, the top layer of the beam might be made of fiberglass while the bottom layers could be graphite. Figure 19–12 illustrates this concept.

■ CURING OF COMPOSITES

If air bubbles and delamination are to be avoided in composites, they must be cured under pressure. This pressure can be applied by the use of internal and external forms on the opposite sides of the composite. The forms are then heated to cure the composite. Although this technique could be used in mass production, it requires expensive equipment. Figure 19–13 shows the concept of this method.

Vacuum Bagging

One of the simplest and cheapest methods of applying pressure is *vacuum bagging.* After the composite is laid up either by hand or by computer-controlled robots using pre-preg, the workpiece is placed in a polyethylene, rubber, or other airtight flexible bag. Since air exerts a force of 14.7 pounds per square inch of pressure at sea level, a vacuum pulled *in* the bag would allow this pressure to be evenly exerted on the part. While complete vacuums are not feasible with this technique, pressures of around 12 pounds per square inch are possible. This pressure is often sufficient for many applications. The entire bag and part is then heated to the curing temperature. Figure 19–14 illustrates the vacuum bagging process.

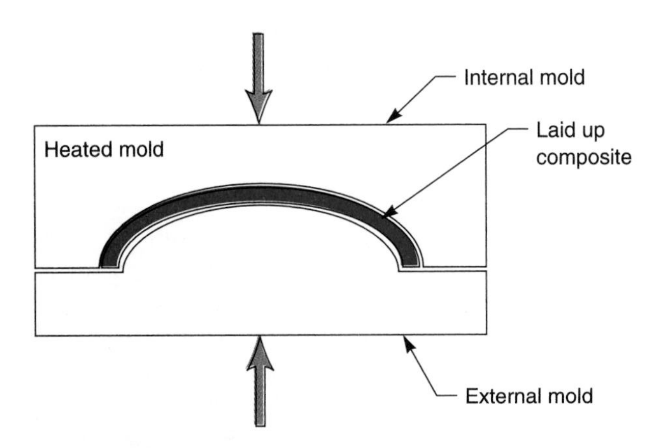

Figure 19-13. Pressure forms.

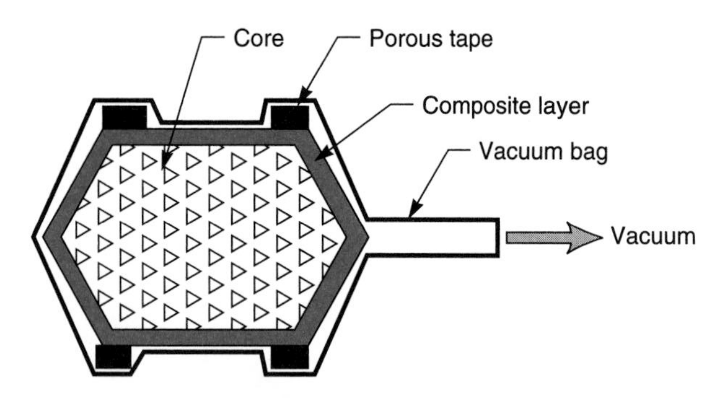

Figure 19-14. Vacuum bagging.

Autoclaving

Extremely complex parts may require more than one atmosphere of pressure. In this case an *autoclave* must be used. An autoclave is an oven that can be sealed and pressure applied by air or other gas. Temperatures of 1000°F (540°C) and pressures of 20,000 lb/in.^{2} are possible in autoclaves.

■ SANDWICHES AND HONEYCOMBS

Glue a couple of pieces of paper on the top and bottom of a small slab of Styrofoam and notice the increase in flexural strength. The same principle can be used with composites. Cores of Styrofoam, syntactic foam, or polyurethane foam can be wrapped with fiberglass, Kevlar, or graphite cloth and the entire product fused together. Many medium-sized sailboats are made of composites laid over a balsa wood core. The core also could be made of a honeycombed aluminum, polymer impregnated paper (Nomex), fiberglass, graphite, or other material. The outer layers of composite are then bonded to the honeycomb. The result is a very lightweight, very strong material. Helicopter blades, truck and aircraft bodies, and some parts of airplane wings and tail surfaces are made of honeycombed composites. Figure 19–15 shows a cutaway section of a honeycombed plate.

Some very recent developments in materials have led to the use of lightweight ceramic cores sandwiched between layers of graphite. Composites with ceramic cores combine the heat resistance and hardness characteristics of ceramics with the tensile strength of graphite. These strong, lightweight, heat-resistant composites have many applications. Figure 19–16 shows this method.

Figure 19-15. Honeycombed sandwich composites.

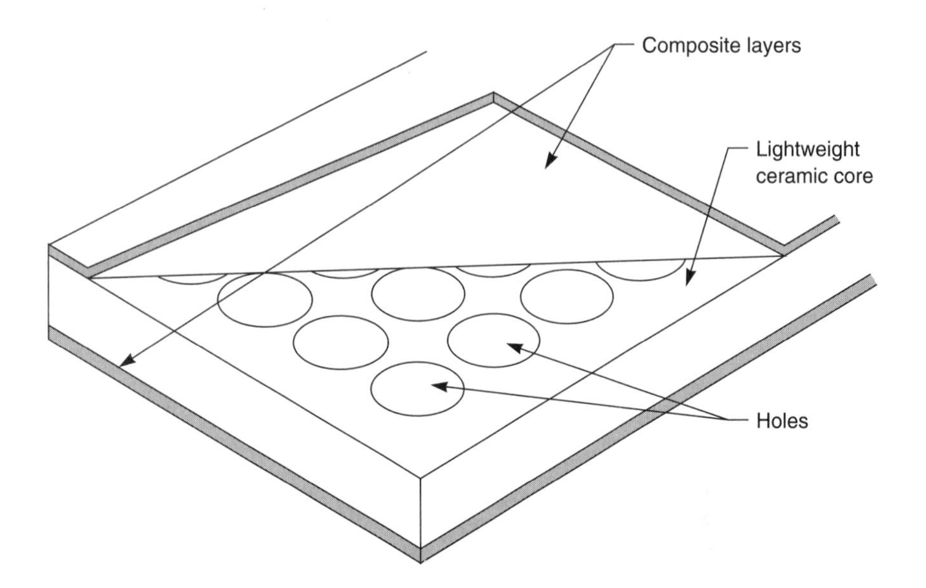

Figure 19-16. Ceramic-composites sandwich.

JOINING OF COMPOSITES

Plastics and composites can be joined by conventional means. They can be held by threads, pins, rivets, and other mechanical means. Further, thermoplastic polymers can be fused by heating in a fusion weld. Both thermosetting and thermoplastic high-polymers can be chemically welded. Many adhesives have also been developed for use with plastics and composites.

ADVANTAGES AND DISADVANTAGES OF PLASTICS AND COMPOSITES

Plastics and composites have many advantages over traditional materials. They do not fail by fatigue nor do they corrode as do metals. Plastics such as polycarbonate (Lexan) will not break. It may be scratched, burned, or sawed, but it will not break on impact. It is even used as bullet-proof automobile and aircraft windows. Composites also have a much higher strength-to-weight ratio than do metals.

Plastics and composites can be machined, drilled, sawed, shaped, or joined by most conventional manufacturing techniques. New methods of manufacture have been developed specifically for composites.

Some recently completed projects would not have been possible without the use of composites. The *Voyager,* the Burt Rutan-designed aircraft piloted by Dick Rutan and Jeana Yeager, which flew around the earth without refueling, was made entirely of composites. The plane, which carried more than 7000 pounds of fuel, two engines, and a crew of two, itself weighed less than 1000 pounds. The flexure of the composite wings would have caused metal wings to fatigue before the plane could have completed its mission.

A second accomplishment chalked up to the use of composites was the flight of the *Daedalus.* The *Daedalus* was a human-powered aircraft, designed and built at the Massachusetts Institute of Technology, which was flown by pedal power alone from Crete to the Greek island of Thios, a distance of 76 miles over water. This plane had a wingspan longer than that of a Boeing 727 yet weighed only 68 pounds.

Racing cars employ composite bodies to lower the weight. Even the leaf springs can be made of composites. Composite bodies and frames, combined with ceramic engines, may soon drastically reduce or even eliminate the use of metals in automobiles (except for electrical conductors).

Then what are the drawbacks to the use of composites?

Too often, the wrong polymer is chosen for a given function. Plastics are sometimes selected on the basis of cost rather than on the properties needed by the part in question. Mention plastics and many people automatically think of Tupperware or trash bags rather than graphite-epoxy.

Composites can **delaminate.** The plies of reinforcing fibers can separate because of improper layup or curing. Air bubbles and other gases can be trapped in the composites during fabrication. Gases trapped near the surface can cause blistering in the composites. Thus a good quality control system is necessary to obtain a good product. Polymers and composites can be attacked by organic chemicals. Oxygen and other atmospheric gases can react with some of the elastomers.

Cutting composites and their reinforcing fibers is often difficult. The glass fibers have a surface hardness greater than many steels. The fibers used in Kevlar and graphite are often so fine that scissors and shears have difficulty cutting them. Special scissors and shear machines have been developed for use with these fabrics, but they are very expensive. Cutting fiberglass and cured fiberglass composites generally requires carbide-tipped cutting tools and hardfaced tools. Water saws and wiresaws have been used to cut composites, and other techniques are constantly being sought.

Perhaps the greatest drawback to the use of composites in present-day products is the curing time required. To make an automobile door or fender from steel or aluminum, a few minutes and a hydraulic press are all that is required. These parts can therefore be produced rapidly and cheaply. To make the same parts out of composites would require in excess of 24 hours. High mass production rates are presently not possible using composites. However, considerable research is being done on this problem and it may soon be possible to cure composites in a very few minutes.

Cost is also a drawback to the use of composites. The raw materials (reinforcing fibers and resins) are expensive. Add to this the manufacturing costs due to the curing time, and composite products become prohibitively expensive, especially for mass production items. An automobile that sells at between $15,000 and $25,000 if conventionally manufactured would be in the $200,000 to $250,000 range if produced in composites. It is hoped that research and new production techniques will reduce these costs. In fact, it has been predicted that within a decade, most commercially produced automobiles will be made almost entirely of composites.

PROBLEM SET 19-1

1. What is the difference between a simple composite and an advanced composite?
2. Make a list of products that are made of composites.
3. Make a list of products for which composites would *not* be a good material of construction.
4. What are the principal advantages and disadvantages of composites?
5. What is *pultrusion?*
6. List three applications in which filament winding could be used.
7. Explain how vacuum bagging is done.
8. Outline a process whereby composite "skis" could be made.
9. Given the fibers Kevalr, graphite, and fiberglass, answer the following questions:
   a. Which one has the highest density?
   b. Which one is the strongest in tension?
   c. Which one is the strongest in compression?
   d. Which one has the best abrasion resistance?
   e. Which one has the highest rigidity?
10. The simplest way to make a composite object is the hand layup method. What are the advantages and limitations of this method of fabrication?

SUMMARY

Composites are strong and lightweight. They can be made into useful parts using hand layup, robotics and computer-controlled laser-guided layup, pre-pregs, filament winding, or pultrusion. Curing can be done using vacuum bagging or autoclaves. Both composites and plastics have advantages and disadvantages, but if properly selected for the job, the limiting factor is the imagination of the engineer, technician, and others who are using these synthetics.

PHOTO ESSAY 19-1 Hand Layup of a Composite

STEP 1 Collect the necessary materials which include the following:

a. Mold release compound
b. Resin and hardener
c. Mixing buckets
d. Acetone or TCE for cleanup
e. Rubber gloves
f. Composite cloth
g. Paintbrusz
h. Mixing stick
i. Squeegee
j. Rag

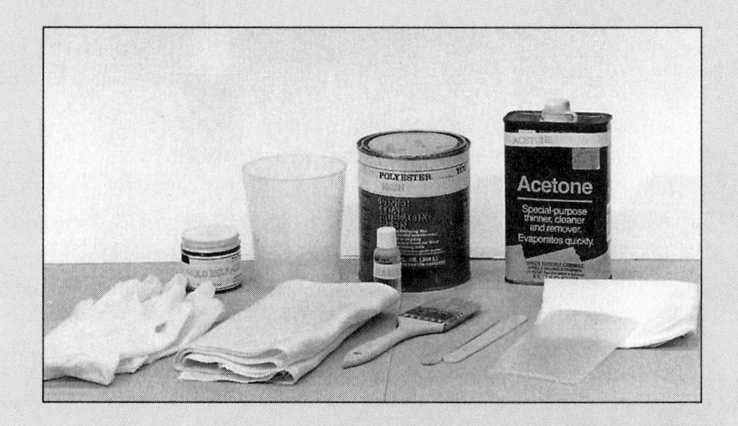

STEP 2 Prepare form. If the pattern is not to be a permanent part of the workbox, the surface must have a "mold release" coating applied to it. The mold release can vary from a thorough coating of wax to "Gel-Coat" or other material that will not bond with the resin of the composite.

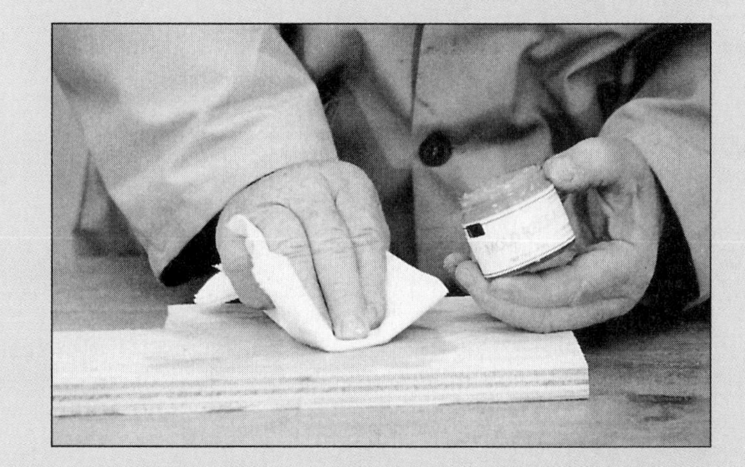

STEP 3 Mix the resin and hardener. It is best to weigh each component carefully since the setup time, curing time, and final properties of the resin often depend on the resin/hardener ratio.

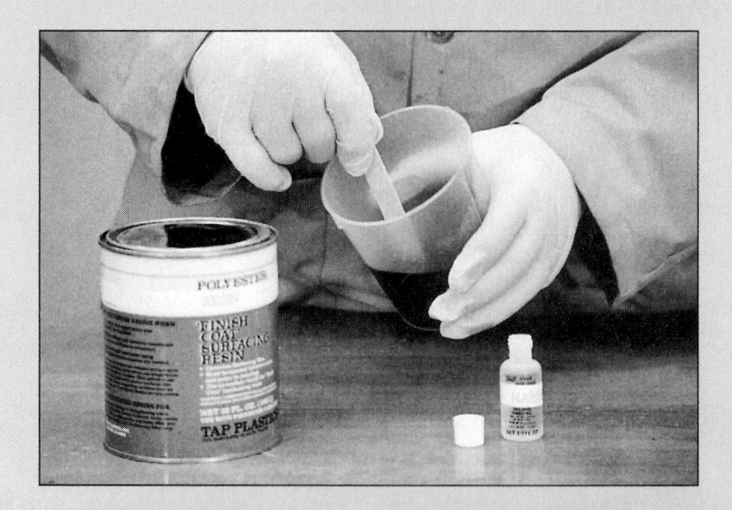

STEP 4 Coat the form, pattern, or mold with a thin but even coating of the mixed resin.

STEP 5 Lay a precut portion of the fiber matrix (usually fiberglass, Kevlar, or graphite) into the resin and "squeegee" the resin through the fibers. Make sure there is an even coating of resin through the cloth.

STEP 6 Recoat the first ply of cloth with resin.

STEP 7 Apply the second ply of cloth. This cloth can be the same material as the first ply or different, depending on the requirements of the final product. In this example, the second ply is being applied on the bias to provide strength in different directions.

Repeat steps 6 and 7 for as many plies as needed for the final product.

STEP 8 Squeegee off all excess resin and make certain there are no air bubbles between any of the layers of cloth. The part is now ready for curing. Curing may be done by allowing the part to stand or by vacuum bagging, autoclaving, or pressing.

After curing, finish the part by sanding, painting, or other technique described in the section on finishes (Section VII).

Note: The type of finish required should be considered in the design of the composite. For instance, Kevlar does not, by itself, produce a very fine finish. If Kevlar is used, it might be wise to apply a very thin fiberglass cloth as the surface layer. This surface layer can then be sanded and coated for a good finish.

PHOTO ESSAY 19–2 Vacuum Bagging

STEP 1 After the part is laid up, place it on a sheet of plastic (polyethylene, polyvinyl chloride, or other plastic that will not react with the resin).

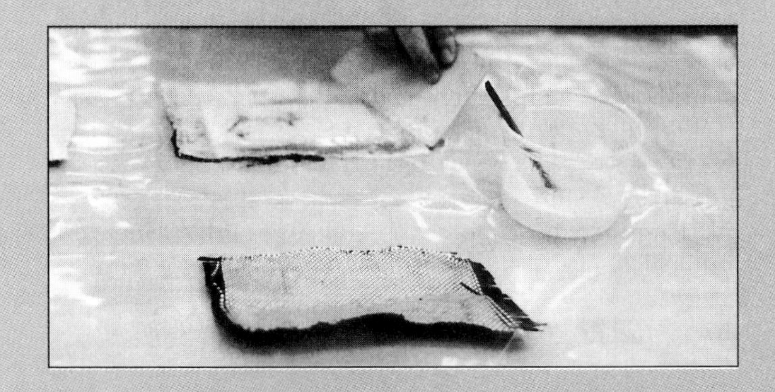

Step 2 Overlay the laid-up part with a porous plastic. This plastic will keep the bag and other material from sticking to the part so it can be easily removed.

STEP 3 Place a porous plastic mesh over the plastic. This mesh will act as a stand-off to allow the excess resin to escape and to provide escape channels for the air.

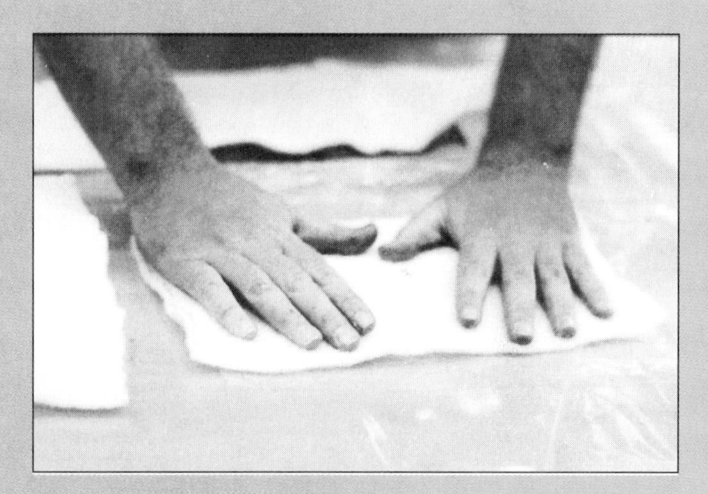

STEP 4 Place sealing tape around the edges of the plastic bag.

Note: Steps 1, 4, and 6 can be accomplished by placing the laid-up part inside a pre-made bag.

STEP 5 Insert the vacuum tube through the sealing tape. (If a bag is used, the tube must be sealed in the opening of the bag.)

STEP 6 Place the top sheet of the bag over the layup and seal it against the sealing tape. It helps to make this sheet fit loosely so as not to block the flow of air to the vacuum.

STEP 7 Apply a vacuum. The air pressure on the outside of the bag draws the plastic bag firmly around the laid-up part. Allow the layup to set while the vacuum is still applied. If the lay up is a pre-preg, the bag and part must be placed in a mold or autoclave for curing.

SECTION

IX

Industrial Safety and Health

Chapter 20
Industrial Safety and Health

INTRODUCTION

Industrial safety and health is a full-time job. It starts with an attitude and continues into practice. Every industry, from the person working in the garage to the multinational institution, must place safety on a par with production. Safety is not just the responsibility of the safety officer or other professional,

> SAFETY IS EVERYONE'S RESPONSIBILITY.

Several organizations—from the Occupational Safety and Health Administration (OSHA), National Safety Council (NSC), and the Environmental Protection Agency (EPA) to local safety councils (and even insurance companies)—publish and enforce rules and guidelines for the safety of workers. But rules and guidelines by themselves will not prevent accidents. No matter how carefully designed and constructed, no machine that is used for cutting, welding, melting, bending, or forming metals or other materials is entirely safe. It is only when the machine is properly maintained and operated that accidents are prevented.

Machines cannot prevent accidents, but they do need to be designed **not** to invite them either. This chapter deals with some of the guidelines by which industrial plants and the machinery in them can be made safer. As such, this chapter will discuss industrial safety in general, equipment safety, personnel safety, product safety, and the human factors (ergonomics) of safety.

Remember,

> SAFETY IS NO ACCIDENT.

Chapter 20

Industrial Safety and Health

GENERAL RULES OF SAFETY

In every industrial plant several basic concepts should be implemented. The ones listed here are the minimum ones. Specific plants may have additional specific needs. Do not hesitate to add to this list.

1. Plan for safety. Do not leave safety to chance any more than you would leave planning a product to chance. Every aspect of the plant needs to be evaluated for safety as well as for production efficiency. It may seem strange, but the two actually augment each other. Investigate the design, construction, operation, and maintenance of the areas listed below for safety. OSHA and many state and federal regulations require that a written safety plan be established for every industrial plant. Don't trust anyone's memory. Put the safety regulations in writing.

A few of the items that must be covered in any safety plan and checked periodically are the following:

- Entrances, exits, windows, and other wall openings:

  Do doors stick?

  Are there doors that stick or window sills or corner places that stick out that could cause head injuries?

  Are there loose panes of glass, etc?

- Floors, walkways, stairs, ramps, and platforms:

  Are there slippery spots on walkways or other places that could cause people to fall?

  Are all rugs and carpets anchored correctly to the floor?

  Are the concrete walkways of a nonskid surface?

- Storage facilities:

  Are the storage rooms ventilated?

  Do they have sprinkler systems and fire alarms?

  Are the storerooms neatly organized?

  Are chemicals, toxic materials, and other supplies properly marked and labeled?

  Are there adequate locks on the storage rooms to prevent unauthorized entry?

- Electric wiring and installation:

  Is the wire size consistent with the circuit breakers or fuses?

  Are all circuits properly grounded?

  Does the installation meet the electric codes?

Are there emergency "cutoff," "kill," or "panic" switches close to areas where machinery is used?

- Illumination:

  Is the lighting in each room of the type and kind required for the job being done in that room?

  Are there glare spots or reflections?

  Is the lighting arranged so that concealing shadows are eliminated?

  Is the light level comfortable?

- Ventilation, heating, and air conditioning:

  Is the ventilation, heating, and air conditioning adequate for the size of room?

  Are the heating and air conditioning facilities adequate for the type of climate in which they operate?

  Are there hot or cold spots in the rooms?

  Are there drafts in the rooms?

  Are all toxic chemical areas ventilated?

  Are all fume hoods and air returns clean of oil, grease, dirt, and debris?

  Are the ventilation and air conditioner motors grounded and properly fused?

- Fire prevention and extinguishment equipment:

  Are the fire extinguishers in working order? Have they been checked and certified within the past year?

  Will the fire extinguishers work on all types of fires (electrical, chemical, and wood or cloth types)?

  Do all gas lines have readily available emergency cutoff valves?

  Does every room have adequate sprinkling systems and do they work?

  Is there an adequate water supply readily available for fighting fires? Are fire blankets available? (See Figure 20–1.)

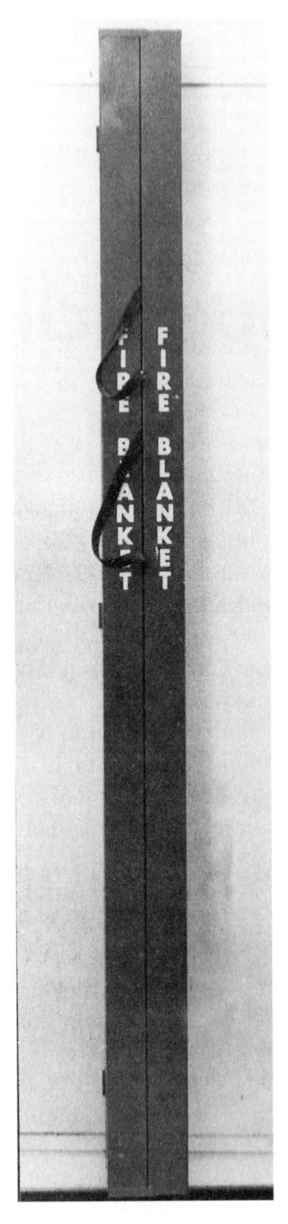

Figure 20–1. Fire blanket installation.

- Elevators:

  Do they work properly?

  Do the doors close properly?

  Do they stop accurately at each floor level?

  Is the lighting sufficient in them?

  Is the ventilation sufficient in them?

  Are there emergency alarms and phones in them and do they work?

  Are they clean?

  Are the floors slippery?

- Water supply:

  Are water fountains of a sanitary type?

  Is the water supply checked periodically for purity?

Are the fountains kept clean?
Is the water level of the fountains the proper height?
Are all water pump motors and refrigeration motors electrically grounded?

- First-aid facilities:

Are well-stocked first-aid kits readily available?
Are CPR, Heimlich maneuver, and emergency procedure posters placed in critical areas?
Are fire blankets readily available? Are there eye wash and emergency shower facilities in areas where chemicals are stowed or handled?
Are protective clothing, hard hats, and eye and noise protection equipment readily available for everyone in the plant?
Are there people trained in first aid, CPR, and emergency procedures in every shop on every shift?

- Boilers and pressure equipment:

Are they adequate for the size of the plant and the job they are required to do?
Have they been checked for corrosion and other maintenance requirements within the past year?

- Personal service facilities:

Are parking lots adequate and well lit?
Are rest rooms and food service areas cleaned daily?
Are all items of equipment (stoves, ovens, commodes, washing facilities, dishwashers, refrigeration units, etc.) adequate for the size of the plant and are they properly maintained and cleaned?

- Waste disposal facilities:

Are there suitable storage facilities for waste metal, glass, rags, chemicals, and toxic materials?
Are hazardous areas marked?

Many of the preceding items are discussed in detail later in this chapter.

2. Designate a safety officer. Every industrial plant, regardless of size, should have a designated person responsible for continually checking on the safety of the facility and its equipment. In small companies this job may be carried out by the manager or president of the firm. In larger companies, an engineer or technician may be given the full-time position of safety engineer or safety officer. The safety officer should be properly trained to assume the duties. Many local safety councils offer short courses for training people in industrial safety. Some colleges and universities offer courses and even degrees in safety. Many professional technical societies such as the American Society of Mechanical Engineers (ASME), the Society of Manufacturing Engineers (SME), the Society of Automotive Engineers (SAE), and the American Society of Safety Engineers (ASSE) (to name a few) also conduct training courses in industrial safety. OSHA standards require larger plants to designate a safety official and periodic reports must be sent in regarding accidents, toxic chemical usage and storage, and many other safety-related matters.

The person managing the safety program should be responsible for the company's required safety courses. That person (or the safety committee under his/her supervision) should also make sure all safety shields and guards are maintained and properly used, and make periodic checks to ensure that the employees are properly wearing eye shields, hard hats, and other protective garments as the job requires. The safety officer should also check periodically that employees have not disabled safety switches and other devices. More about this later.

3. Establish training courses in safety. Safety comes from experience, but it doesn't always have to be an individual or personal experience. Safety can be taught. Every new employee should be indoctrinated in the safety regulations of the plant before being allowed to work in the facility. Every employee must understand that he or she is not only responsible for his or her own safety, but for the safety of those working around them. New or apprentice employees need to be taught safety and the older employees need to be periodically reminded not to take safety for granted.

4. Select the right employee for the right job. It would be ridiculous to assign a 120-pound man to play guard on a professional football team, or to designate a 5-foot, 2-inch tall man as the center on a basketball team. Yet many industrial plants often make employee assignments that are about as ridiculous. To assign to a color-blind person the task of sorting the red and green acetylene and oxygen tanks might pose a problem. Similarly, a person wearing a cast on a leg should not be required to drive a forklift. That is not to say that physically impaired people should not be hired, they just should not be given tasks where their particular impairment prevents them from safely and properly doing their job. It is not always necessary to select the "best" person for the job, but it is necessary to select the "right" person for the job. This is discussed in more detail in the section on Ergonomics.

5. Establish the safety attitude. A good motto for any plant would be "**Machines don't cause accidents, people do.**" It has been well documented that nearly 90% of all accidents are directly due to operator mistakes. The phrase used by the safety professionals is "they did an unsafe act." A rule of safety is usually violated in every accident. In the accidents in which machinery failure caused the accident, most of the failures were due to faulty or inadequate maintenance of the equipment. The machinery was not properly lubricated, electrical wires were allowed to fray or the insulation become brittle, or the tires on the automobile were bald.

People working in the plant need to be encouraged to be alert to safety problems. They should be urged to report improperly maintained machinery, poor lighting, fire hazards, and other safety problems so that they can be immediately corrected. Workers must not be chastised or punished for reporting to the management such items as leaking oil lines, faulty or inoperative safety devices, or other unsafe working conditions.

Distractions should not be allowed in the workplace. There is a place for distracting signs, practical jokes, or celebrations, but it is not on the work floor. Not only will the safety of all personnel working there be impaired, but the quality and quantity of production will also fall.

The attitude of "accidents don't just happen, someone is responsible for the accident" should be taught to the employees. The excuse of "the machine broke" is not acceptable—there was a reason *why* the machine broke. It may be faulty design, faulty maintenance, or faulty operation, but someone's mistake caused the accident. A good corollary to this slogan would also be "Safety doesn't just happen, someone is responsible for safety." That someone is every employee.

6. Keep the morale of the workers high. It has been well established that industrial facilities with high accident rates usually have poor morale. Management and labor must work together in this area. Communication between management and the employees is essential. The employees should have input in the matters of safety and be assigned to the company safety committee. If they help establish the safety rules, they are more likely to abide by them.

7. Establish safety rules and enforce them. Every plant should have definite posted safety rules applicable to their particular industry. Many of these rules are mandated by OSHA and other regulatory bodies. These rules should be as few in number as possible and they should be applicable to the operation of that facility. For instance, rules regarding the safe operation of deep sea diving equipment need not be posted in a plant making rugs. However, fire safety regulations for textile plants would be mandatory. Whatever the rules, **they must be enforced.** Not only should the management, supervisors, and safety officers enforce the rules, but workers should also insist that their colleagues obey the rules.

8. Do not allow untrained personnel to operate any equipment. It would be ludicrous to allow a person who cannot swim to be a lifeguard at the beach. It is just as bad to permit an untrained person to operate a drop forge or a lathe. Beginners and apprentices must be carefully trained in the safe and proper operation of equipment and work only under the supervision of qualified personnel. Even well-trained machinists, welders, and other skilled tradesmen need to be "checked-out" on new equipment before allowing them to run it unsupervised. Untrained personnel not only present safety problems, but they can cause damage to expensive production equipment as well.

9. Implement a planned maintenance program. Defective or poorly maintained equipment is a potential source of accidents. Properly planned maintenance programs should include daily, weekly, monthly, quarterly, semiannual, and annual maintenance schedules. The items on these schedules should include the following as a minimum:

Daily Maintenance

Clean up of all chips, cuttings, filings, and other debris from the machine and the surrounding area.

Oil all critical points if the machine has been used.
Wipe down to remove dust, dirt, and coolant material.
Sharpen cutting tools used on the machine.
Visually inspect equipment for worn or broken parts.

Weekly Maintenance

Grease bearings.
Adjust belts, gears, drive shafts if necessary.
Check alignment of beds, tailstocks.

Monthly Maintenance

Check internal gearing, cams, drive belts.
Replace belts and worn parts as necessary.
Check switches and controls.
Check on safety cutouts, guards, and other safety devices for proper operation.

Quarterly Maintenance

Check calibration of all dials and gauges.

Semiannual Maintenance

Check electrical cords.
Check switches and controls.
Check parts for corrosion.
Check motor mounts.
Check bearings.

Annual Maintenance

Evaluate machine as to need for replacement.
Check bed alignments.
Perform annual calibration check.

Of course, routine maintenance should be performed as the need is seen. Do not wait for the check times to correct any deficiencies that may arise.

EQUIPMENT SAFETY

Safety rules should not only protect the workers, but also protect the manufacturing equipment. While it is true that machines don't cause accidents, their design should not invite accidents either. Each machine has need of special safety rules for its operation. However, there are a few general safety rules for the operation of machinery. The National Safety Council manual recommends the following general rules:

1. Operation, adjustment, and repair of any machine tool must be restricted to experienced and trained personnel.
2. Safe work procedures must be established, with shortcuts and chance-taking prohibited.
3. Management and supervisors must be responsible for the enforcement of the safety policy and for making certain that no deviation from them occurs.
4. When purchasing new equipment, make sure that specifications conform to all applicable regulations concerning safeguarding, electrical safety, etc.
5. New equipment should be inspected and safety innovations made before allowing operators to use the equipment.
6. Attention to the work in progress must be full time. Shut down the operation or move the tool far enough away so that the operator will see where he or she left off if it is necessary to talk to others.

The National Safety Council also lists some specific rules for the operation of machine tools:

1. Machine tools should never be left running unattended.
2. Operators should not wear jewelry or loose fitting clothing, especially neckties and loose sleeves or cuffs of shirts or jackets. Long hair, which could be caught by moving parts, should be covered. (More about this later.)
3. All operators must wear eye and ear protection, as should others in the area. This includes inspectors, supervisors, stock handlers, and visitors.
4. Throwing refuse or spitting in the machine tool coolant should not be allowed—such actions foul the coolant and may spread disease.
5. Manual adjusting and gauging (calipering) of work should not be permitted while the machine is running.
6. Operators should use brushes, vacuum equipment, or special tools for removing chips. (Chips are hot and sharp. They should not be touched without gloves or protection on the hands.)
7. All electrical controls of all machines should be built and installed in accordance with the National Electrical Code, applicable local codes, and recognized industry standards.

8. The machine control panel should contain the main disconnect switch, motor starters, clutch control relays, and control transformers. The control panel should be on or immediately adjacent to the machine it serves, but out of the operator's way during normal operation.

OSHA also has the following requirements:

1. Two-inch-wide yellow lines will be painted on the floor in at least a 3-foot circle around each machine. No one other than the operator(s) of the machine is permitted inside that line while the machine is in operation.
2. Machinery must not be crowded in spaces that hinder the safe operation of the machine.
3. Rotating heads, cutting tools, saw blades, grinding stones, etc., will have safety guards attached.
4. Pulleys, belts, drive shafts, and gears will be completely enclosed.
5. Fans will have guards that will prevent fingers from being hit by the blades.

Drop forges, punch presses, spot welders, and other semiautomatic machines should also require one or more of the following safety devices:

- *Die enclosure guards.* Remember, if hands can reach through, around, over, or under a guard, it is inadequate and not acceptable by itself.
- *Two-hand tripping device.* These are palm buttons, *both* of which must be pressed by the operator to trip or operate the machine. If more than one operator is required to operate the machine, there must be a two-hand trip device for each operator—all of the devices must be depressed in order for the machine to trip.
- *Restraints.* These are devices, worn by the operator, that will automatically pull the operator away from the machine if it is accidentally tripped.

Note, however, that all of these devices can be locked out or circumvented by the operators. The supervisors must be alert that die enclosure guards have not been taken off, two-hand tripping devices have not been taped down, and that restraints are worn. All too often workers become too cocky or careless around the machines, thereby inviting accidents.

Robots, computer-automated machines, numerically controlled machines, and other automatic machines should be fenced off and gates with electric circuit cutoff switches installed to prevent anyone from entering the envelope of the machine. Proximity sensing devices are also widely used for these machines. Proximity sensors will turn off the machines if anyone is detected in the unsafe area.

The National Safety Council also lists many common causes of accidents with specific types of machines. A few of these causes are noteworthy.

Lathes

- Contact with projections on the work or stock, face plates, chucks, lathe dogs, or those with projecting set screws.
- Loose tailstocks.
- Flying metal chips.
- Hand braking of the machine.
- Filing right-handed (with the left arm over the headstock), using a file with an unprotected tang, or using the hand instead of a stick to hold an emery cloth against the work.
- Calipering or gauging the workpiece while the machine is in operation.
- Attempting to remove chips when the machine is in operation.
- Contact with rotating stock projecting from turret lathes or screw machines.
- Leaving the chuck wrench in the chuck. (Use a spring-loaded chuck wrench to eliminate this hazard.)
- Catching loose clothing or wiping rags on revolving parts.

Drill Presses

- Contacting the spindle or the tool. (Do not touch the tool while using a quick-change clutch.)
- Breaking a drill.
- Using dull drills.
- Being struck by insecurely clamped work.
- Catching hair or clothing in the revolving parts.

- Removing long, spiral chips by hand.
- Leaving the key or drift in the chuck.
- Being struck by flying metal chips.
- Failing to replace the guard over speed change pulley or gears.
- Falling against the revolving work or tool.
- Allowing cuttings to build up on the work table.
- Removing chips by hand.

Planers and Shapers

- Running bare hand over sharp metal edges.
- Failing to clamp the work or tool securely before starting the cut.
- Having insufficient clearance for the work.
- Failing, when magnetic chucks are used, to make certain the current is turned on before starting the machine.

Grinding Machines

- Failure to use eye protection in addition to the eye shield mounted on the grinder.
- Holding the work incorrectly.
- Incorrect adjustment or missing work rest.
- Using the wrong type of wheel or disk or a poorly maintained or unbalanced wheel.
- Grinding on the side of the wheel.
- Applying work too quickly to a cold wheel.
- Failure to use wheel washers (blotters).
- Vibration and excessive speed, which lead to bursting of the wheel or disk.
- Use of bearing boxes with insufficient bearing surface.
- Installing flanges of the wrong size, with unequal diameters or unrelieved centers.
- Incorrect dressing of the wheel.
- Contacting unguarded moving parts.
- Using controls that are out of the operator's reach.
- Using an abrasive saw blade instead of a grinding wheel or disk.

Equipment such as forklifts, cranes, tow-bars, and other items used for transporting equipment have special safety requirements. All cabs must have roll bars, protection against falling objects, and backup and warning horns. Operators must be well trained (and even licensed in some cases). No unqualified personnel should be allowed to operate this equipment. Personnel working around these machines should also be trained never to get under the forks and to be constantly alert for falling objects. They should be aware of the position of the forks and hooks at all times. No passenger is allowed to ride the vehicle and no one is ever allowed to ride the forks of a forklift or the hook of a crane.

Safety devices such as saw blade guards, shields, and electric cutouts need to be checked periodically also. They become worn, defective, and broken just as do the machines. If any machine has a defective safety device, the machine should be shut down until the device is repaired. Companies that design and manufacture defective devices can be made to replace them with reliable ones. In most states it is against the law to remove or disable a safety device and allow the machine to be used without it.

There are many more types of machines with specific safety problems for each one. We have just shown a few examples of the safety precautions that are necessary for each machine. Nothing should be left to chance. Every possible cause of accidents or injuries should be anticipated if possible and action taken to prevent those accidents. The best prevention against accidents, however, is a well-trained operator.

PERSONNEL SAFETY

When a person accepts employment at a manufacturing plant or facility, they are bound to follow the regulations of that company. This includes safety and health regulations. Regulations involving the personal safety of employees should include the following:

1. *Dress codes.* Employees should dress for safety. People working around machinery should meet the following requirements.
    - *Hair styles.* Hair should be no longer than shoulder length or it should be tied back and under the hard hat.

- *Pants.* Long pants should be worn on the production floor. Skirts and shorts should be prohibited. The fabrics should not be synthetic types that melt under heat.
- *Shoes.* Hard or steel-toed shoes are preferred. No open top or tennis type shoes should be permitted.
- *Safety clothing.* Specific tasks require specific protective clothing. Welders should wear leather protective garments including aprons, arm protection, and leggings. People dealing with chemicals should wear lab coats and other protective garments as necessary.

2. No person under the influence of alcohol or other drugs should be allowed to operate any machinery. This includes some prescription drugs as well as the illicit ones.
3. Persons who are ill or injured should not be allowed to operate any machinery. If they are not feeling good, they are prone to make mistakes, which could lead to accidents.
4. Rest breaks should be provided for machine operators. Personnel could be rotated so that fatigue does not lead to accidents.
5. Do not allow policies that deter safety. Unreasonably high production quotas and incentives that encourage reckless practices are to be avoided. Set production quotas at the limit at which they can safely be produced.

PRODUCT SAFETY

It may sound trivial, but it is still true: A company is only as good as its product. Besides caring for the safety of the personnel and the equipment, the *product* of the company should also be protected. To do this several precautions could be taken:

1. Establish a good quality control and inspection system.
2. Parts that are dropped at any time during manufacture should be thoroughly inspected for both external and internal damage.
3. Teach pride in workmanship to all employees.
4. Delicate parts should be wrapped or safeguarded at every step during production.
5. Item 5 from the personnel safety section applies here too. Unreasonably high production quotas encourage sloppy workmanship, a poor product, and add to accidents and injuries.

ERGONOMICS

Ergonomics goes under many names: human factors engineering, bioengineering, biotechnology, engineering psychology, and human engineering. The word *ergonomics* is derived from a Greek word meaning "work." Its basic premise is that the machine, the operator, and the environment in which they operate are a single system. The machine is considered an extension of the operator and the operator should be considered a part of the machine. Good design involves a system approach. Good machine design involves designing the machine around the operator, not the other way around. The principles of ergonomics are as applicable to the design of a lathe as they are to the controls of an automobile or an airplane.

It would be senseless to design an automobile where the driver could not see out of the window and could not reach the controls. One would laugh at a computer designed so that the delicate controls would be used by one's feet. Yet sometimes machines are designed that do not fit the operator. Such a design can lead to accidents.

For the purposes of this chapter, let us define machines as any mechanical, electronic, hydraulic, or pneumatic device that can be designed to do specific tasks. These include everything from a simple lever to a computer-automated lathe or a robot. As time progresses, machines are being designed that can do many of the jobs previously relegated to humans and other animals. It may be a blow to the human ego, but machines can do some tasks better than people. Conversely, people do many things better than machines. Good design simply lets each do that which it does best. It is necessary then to understand the strengths and limitations of humans and machines.

People are better than machines at the following tasks:

- *Reasoning.* Although people often rely on machines for data in the form of numbers from calculators, temperatures from thermometers, and the like, the *meaning* of those numbers and temperatures can only be understood by people.

- *Intuition.* People do not have to go through all of the possible results of an action to make a decision. In scientific terms the intuitive approach is called *heuristic logic.* Buried deep in the human experience is often the forgotten basis of making daily decisions. We do things because we feel it is the best course of action.
- *Selecting inputs.* A machine will use any information it is given. People are better at selecting which information is the best and then reacting accordingly.
- *Adapting to unexpected situations.* Machines can be programmed for specific situations. In the case of robots, they can be reprogrammed for other situations, but they cannot react immediately to all of the situations that may be encountered.
- *Reacting to unusual situations and emergencies.* Machines cannot react to situations for which they are not programmed. People can.
- *Judging.* Judging involves a bit of intuition and reasoning. People therefore do judging far better than machines.
- *Supervision.* Machines lack the human skills and characteristics needed to supervise humans. I for one hope I never see the day when my supervisor is a machine.
- *Interpreting data.* Machines provide data, but they have little or no understanding of the *meaning* of the data, or the impact of that data on present and future actions.

People are limited by the following:

- *Emotions.* Machines do not have burnt toast for breakfast or traffic jams on the way to work. Even if they did, it would not bother them. Machines don't cry.
- *Environmental conditions.* People can function only within very limited conditions. People can work only in temperatures between roughly the freezing point of water and 100°F. Even in the extremes of this range, human functions are hampered. People cannot function well around toxic fumes, extremely high audio levels, barometric pressures greatly different from one atmosphere, around high radiation levels, or in the absence of oxygen. Further, people are weak, fragile, and they cut and break easily.
- *Distractions.* People have trouble keeping their minds on the business at hand. They respond to extraneous noises, strange odors, and many other stray stimuli. Distractions are a major cause of accidents.
- *Opinions.* People often depend on their own opinions more than factual data. Machines can easily have their memories or input data changed without generating feedback.
- *Fatigue.* People get tired and cease to function at full efficiency. Unfortunately, people often try to function while fatigued, which produces poor quality of workmanship and also accidents. Machines function at maximum efficiency until parts break, wear out, or the power fails, at which time they just cease to function at all.
- *Speed.* People do not move very quickly. Even hand speed and finger motion are relatively slow compared to that of some machines.
- *Reflexes.* People are limited by a very slow nervous system. Signals detected by any of the senses are slow to reach the brain; the brain is also slow in sending out a response to the muscles and the muscles are even slower in acting on the impulse. Machines can respond many times faster to input signals than humans can.
- *Strength.* Even the strongest of humans can produce forces of only a few hundred pounds. Machines are limited only by their design maximums, which are many times that of mere people.
- *Hunger.* People must stop periodically and eat. Machines will work until the switch is turned off.
- *Training.* Learn by experience. Training a human is a long, slow process. The process must be repeated for each new skill. Machines can be trained by turning a dial or inserting a new program disk.
- *Other limitations.* People have other limitations that are not necessarily bad, just different. Such

matters as one's culture, language, religion, morality, and social ethics affect the way a person thinks and acts. Machines have no such inhibitions.

Machines are better than people at the following tasks:

- *Repetitive operations.* People are easily bored. When it comes to performing the same simple action time after time, design a machine to do it. The machine will do the task faster and make fewer mistakes.
- *High-strength applications.* When forces of more than a few hundred pounds are required for a job, a machine must be used. People are unable to perform high-strength tasks.
- *Speed of operation.* Whether the task is simply moving a part a few inches or many miles, machines can be designed to move faster than people.
- *Precision.* Precision is defined as the repeatable accuracy of a function. Machines can duplicate actions very accurately.
- *Information storage and retrieval.* The memories of humans are limited, and data are often lost and forgotten. Not so with either mechanical or electronic controls and data banks.
- *Monitoring.* For routine monitoring of parts to determine if they meet tolerances, design weights, colors, or other measurements in which parts can be accepted on a go/no go basis, machines can do the inspection faster, more reliably, and for longer periods of time than humans.
- *Poor environment operation.* People cannot work in extremely high or low temperatures, in toxic atmospheres, around radiation, or under very high or low atmospheric pressures. Machines can be built that will function well in temperatures from the cryogenic to red hot range, or from the depths of the sea to outer space.
- *Displaying information.* Analog and digital readouts from clock faces to speedometers are a primary function of machines.
- *Continuous operation over a long period of time.* As long as the machines have a supply of power they will continue to work and to work well. Simple maintenance is all that is needed to keep them going. They can work 24 hours a day if permitted.

Machines are limited by the following:

- *Power failures.* Pull the plug or run out of fuel and the machine stops.
- *New situations.* Machines are designed to do specific functions. One cannot use a sewing machine to do heat treatments. Even robots that are designed to be multifunctional must be reprogrammed to do each task.
- *Insensitive to operator frailties.* An alert human will not swing a hammer if someone is in the way. If a machine arm is programmed to swing through a 180° arc, it will do so whether someone is standing in the path of the arm or not.
- *Reasoning power.* Machines have to be told everything they must do and they will do exactly what they are programmed to do. They do not think or reason whether an action is correct or not—that is left to the operator.
- *Input.* Machines are governed by the old cliche "garbage in, garbage out." The skill of the operator is often the limiting factor for machine operation.

The problem in ergonomics lies in the communication between the machine and its operator. The operator must give the machine inputs that it understands and can respond to, while the machine must give the operator signals and readouts that a human can sense. What are these stimuli that each can answer?

People sense and respond to:

- Light, using the eyes
- Sound, detected by the ears
- Touch, felt as pressure by the skin
- Odors, detected by the nose
- Tastes, picked up by the tongue and the mouth
- Pain, sensed by the nerves and brain
- Heat or cold, picked up by the skin and nervous system

Machines can be made that will respond to:

- Light, detected by photoelectric cells
- Heat, sensed as temperature by thermocouples and thermometers

- Radio waves, picked up on many types of antennas and detectors
- X rays and other radioactivity, sensed by Geiger tubes, scintillation counters, and photomultiplier tubes
- Electricity, sensed and measured by voltmeters and ammeters
- Sound, detected using microphones
- Chemicals, found by other chemicals, pH meters, and gas detection equipment
- Pressure, measured by gauges, manometers, pitot tubes, and strain gauges
- Motion, sensed by gyroscopes and interferometers

The operator communicates to the machine by using the controls on the machine. These controls must be easily manipulated by some part of the human body. Further, certain parts of the human body are better at given functions than other parts. In manual controls the fingers are used for fine or delicate manipulations. Legs and feet are used when large forces or if power is needed. Automobiles are therefore designed such that the switches, steering, and other small controls are operated by the hands, while the brakes are operated by foot controls. The controls of the machine must be designed and built to fit the operator and to respond to the actions of the operator.

Besides the usual manual inputs, machines can now be designed that will respond to voice inputs, eye movements, and even brain waves.

Engineers, technicians, and machine designers must understand the operator as well as the physics of the machine. Besides the physical size limitations of the operator, humans must be studied to determine which stimuli they respond to best.

The machine responds to the operator with analog or digital dials, sounds, odors, and even force against the fingers, hands, arms, legs, or feet. (This is referred to as *kinesthetic* feedback.) Accidents arise when there is confusion in the signals between the operator and machine. Scientists have conducted many experiments to determine which type of information displays are most easily and most accurately read. In general, fewer mistakes are made reading digital displays than analog or dial-type gauges (Figure 20–2). However, when it comes to synchronizing speeds of engines, or matching two dial readings, properly oriented analog or dial indicators are better. Further, the orientation of the dials can aid the operator in proper adjustment. For example, research has found that it is easier to match dial readings if the dials are arranged so that the same readings on each dial can be obtained by making the indicators point at each other rather than having the indicators point upward or be parallel to each other. Two constant speed generators can be synchronized more easily if the "lock-in" speeds are on the line connecting the centers of rotation of the dial indicators so that the indicators point at each other when the desired speeds are attained. See Figure 20–3 for examples of poor and better dial layouts.

From the moment of birth, humans are being programmed. People soon learn that one flips a switch to the up position to turn it on, down to turn it off. Green means go, and red means stop, hot, or danger. Blue usually means cool and yellow is a caution. Moving a lever forward usually makes a machine go forward or down. Pulling it back makes something stop, go back, or go up. Clockwise motion tightens things, whereas counterclockwise motion loosens them. In the United States people drive on the right side of the road and will instinctively turn to the right to escape danger. This fact often causes accidents when drivers used to being on the right side of the road are driving in countries in which the traffic is on the left side of the road.

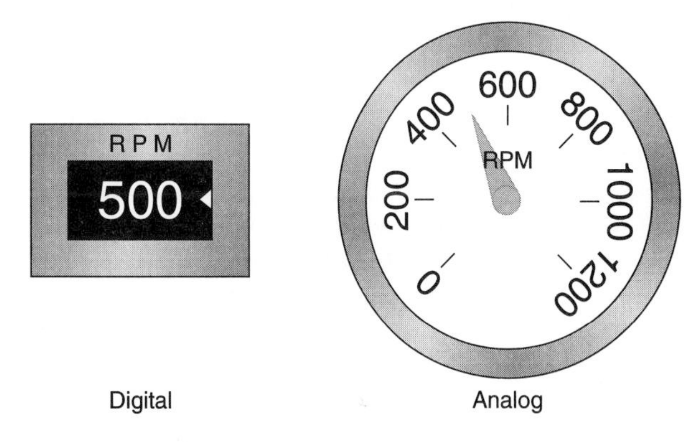

Figure 20-2. Digital versus analog gauges.

Figure 20-3. Matching dials.

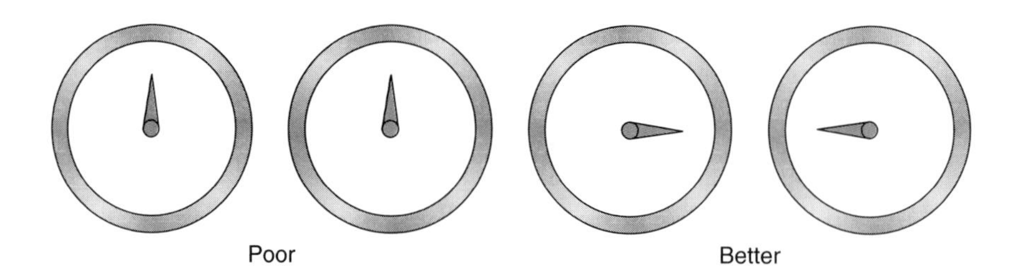

In designing machines for safe operation, every consideration must be taken to prevent the operator from inadvertently giving the machine wrong instructions. This includes such items as turning the machine on when it should be turned off or speeding it up when the intention was to slow it down. Accidents can be caused if machines are designed with the controls in a position contrary to these programmed expectations. Painting the stop button on a machine green would only invite disaster. Similarly, on–off switches oriented with the off position being up could cause inadvertent accidents. Accelerator levers should be designed to increase speed by pushing forward.

Good engineering also dictates that not only should the operator be protected from the machine by the use of guards and shields, but that the machine be protected from poor operators too. Perhaps master switches that require the input of a code given only to experienced operators should be installed. Thermal breakers and cutout switches that respond to overloads and overheating should be used.

OTHER SAFETY CONSIDERATIONS

There are a few items in the design of manufacturing machines that should be checked with respect to safety and health. The Environmental Protection Agency (EPA), OSHA, local codes, and the National Safety Council have many regulations covering machine design. Only a few of these are listed here:

- Start switches or handles that activate machine action should be in depressions or hidden so that they cannot be inadvertently struck and activated.
- Machines should have ample working room about them. They should be oriented in the room in such a manner that the operators have unobstructed views of all of the surroundings.
- Machine operators should not be alone while working in the room. There should always be someone else around in case emergencies arise.
- Emergency alarms should be within reach of every machine operator.
- People seldom look up. Overhead cranes and other machinery can often strike people who may not be aware they are in danger. Design overhead equipment carefully with ample warning devices installed on them.

SAFETY OF THE ENVIRONMENT

Environmental protection is not a recent problem. A lawsuit was filed in Germany more than a century ago by farmers who were seeking an injunction to force a company making bleach from dumping its waste product (hydrochloric acid) into the Rhine River. The acid was damaging the crops and cattle downstream of the plant. (The farmers won and the plant operators soon found that they could turn their waste product into a salable product.) Before the coming of the automobile, large cities had a serious problem with the waste from horse-drawn vehicles. Now, many cities, states, and federal governments have environmental protection agencies with the duties of enforcing strict laws to keep the air, water, and ground free of toxic matter.

Large industrial plants are now required by law to designate an individual as an environmental protection officer for the plant. This person must account for all materials listed as hazardous by the EPA, be responsible for the emission control equipment of the plant, and be in charge of the disposal of toxic wastes.

Industrial plants are given a "quota" of hazardous materials that they can allow to escape into the atmosphere, ground, and waterways. There are heavy fines for plants that exceed the limits. The environmental protection agencies monitor many of these emissions by periodic sampling of the streams and waterways and air and ground around the plant. Groundwater samples are also taken at intervals. For hazardous materials that must be taken to a toxic waste dump for disposal, the accounting method is very simple. The EPA checks the amount of these chemicals that the company bought during the previous reporting period, subtracts the amount still on hand at the end of that period and the amount sent to the toxic waste dumps, and assumes the rest went into the atmosphere and surroundings. If that amount exceeds the allowable amount, fines and cease-and-desist orders may be instituted. Solvents such as 1,1,1-trichloroethane (TCE) and methyl ethyl ketone (MEK) are examples of these types of chemicals. Exhaust gases from smokestacks are monitored for carbon monoxide, carbon dioxide, sulfur and sulfur compounds, nitrous oxides, halogens (chlorine, fluorine, bromine, and iodine), unburned fuels, and many other compounds.

Paint booths and degreasing stations must now be fitted with scrubbers to remove paint and other solvents from the air before the air is released back into the atmosphere. Many of these solvents are being collected and recycled by specially designed equipment.

Spent chemicals used in chemical machining, electrochemical machining, electroplating, and anodizing are also taken to hazardous waste dumps for disposal. Since this adds to the cost of producing any part, considerable industrial research is now being conducted on ways of reducing these chemicals to inert forms that can either be recycled, used for other purposes, or made easier to dispose of. Degreasing solvents are now being filtered to remove the oil and solid particles and reused many times. Before any chemicals are transported to the assigned dumps, they are distilled to remove water. It is often cheaper to remove the water than pay the cost of shipping the added weight to the dump. Industry is also seeking nontoxic and nonpolluting ways of removing grease, paint, and corrosion from their products.

In addition to chemicals, scrap steel, aluminum, copper, and other materials are now being recycled, rather than dumped. This recycling has several beneficial results. Less dump area is needed, the scrap can be sold to recoup some of its cost, and the environment is less polluted.

Storage facilities for hazardous materials, underground tanks, and pipelines also require constant monitoring for leaks. Older tanks are now being replaced with newer, double-walled tanks.

The items mentioned in this short treatise are but a few of the regulations imposed on industry by the environmental protection programs of the government. The persons responsible for each industry's environmental protection programs must be specially trained in courses established or sanctioned by the EPA. Besides just producing a product, industry must now take responsibility for any by-products and waste products produced in making the part.

PROBLEM SET 20-1

1. You have just set up a small plant to manufacture wood vases. You have a wood lathe, a drill press, table saw, band saw, belt and disk sanders, and hand tools. You have rented a small building with a large room for a shop, and three small rooms that are designated a glue room, a paint room, and the office. You have two other workers in the shop and a secretary. Outline a safety policy and plan for the plant.
2. Are there any special safety precautions that might apply to a metal lathe but not to other machine tools?
3. What safety equipment should be available to every shop or manufacturing plant?
4. If you were asked to state the most important safety rule, what would you answer?
5. What specific safety rules apply to grinding machines?

6. What would you do if you had a defective safety guard on a power saw?
7. What special precautions should be taken for safety of personnel in an automobile paint shop?
8. Your plant employs 75 workers. Your product is pistons for automobile engines. You are the safety officer for the plant. Your plant operates a small foundry, machine shop, paint stripping facility, electroplating bath, and a drop forge. Outline a safety training program for the plant.
9. It is often assumed that safety hazards are in the shops. But what safety precautions, if any, would you think necessary for an administrative office in a plant?
10. If a person refused to follow safety instructions in a plant, what steps should be taken?
11. Give three other names for *human factors engineering.*
12. What is the purpose of human factors engineering?
13. How can human factors engineering help prevent accidents?
14. What safety precautions should be taken in one's own home? List at least six hazardous places in the home.
15. What products commonly found around the home should be taken to hazardous waste dumps for disposal?
16. The photograph below shows Mr. Hunt, one of the authors of this book, in a posed picture at a lathe. How many safety violations is he committing in this photo? There are at least nine violations and maybe more. (*Note:* The picture was taken with the lathe turned off, for his protection.)

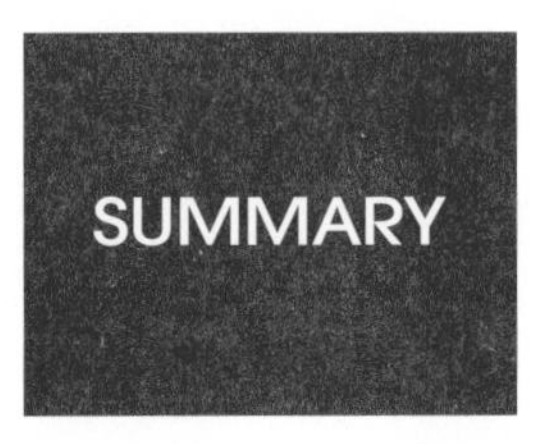

Make the machine as safe and as dependable as possible, but never depend on the machine for safety. The operator, machine, and the surrounding environment are one unit. Design the machine to fit the operator, and pick the right operator for that machine. It may sound trite, but no matter how safe a machine design is, or how many safeguards are placed on it, accidents can happen if the operator is careless.

Environmental protection is now an industrial concern. Companies are now being held accountable for keeping the environment around their plants clean.

SECTION

X

Production Control

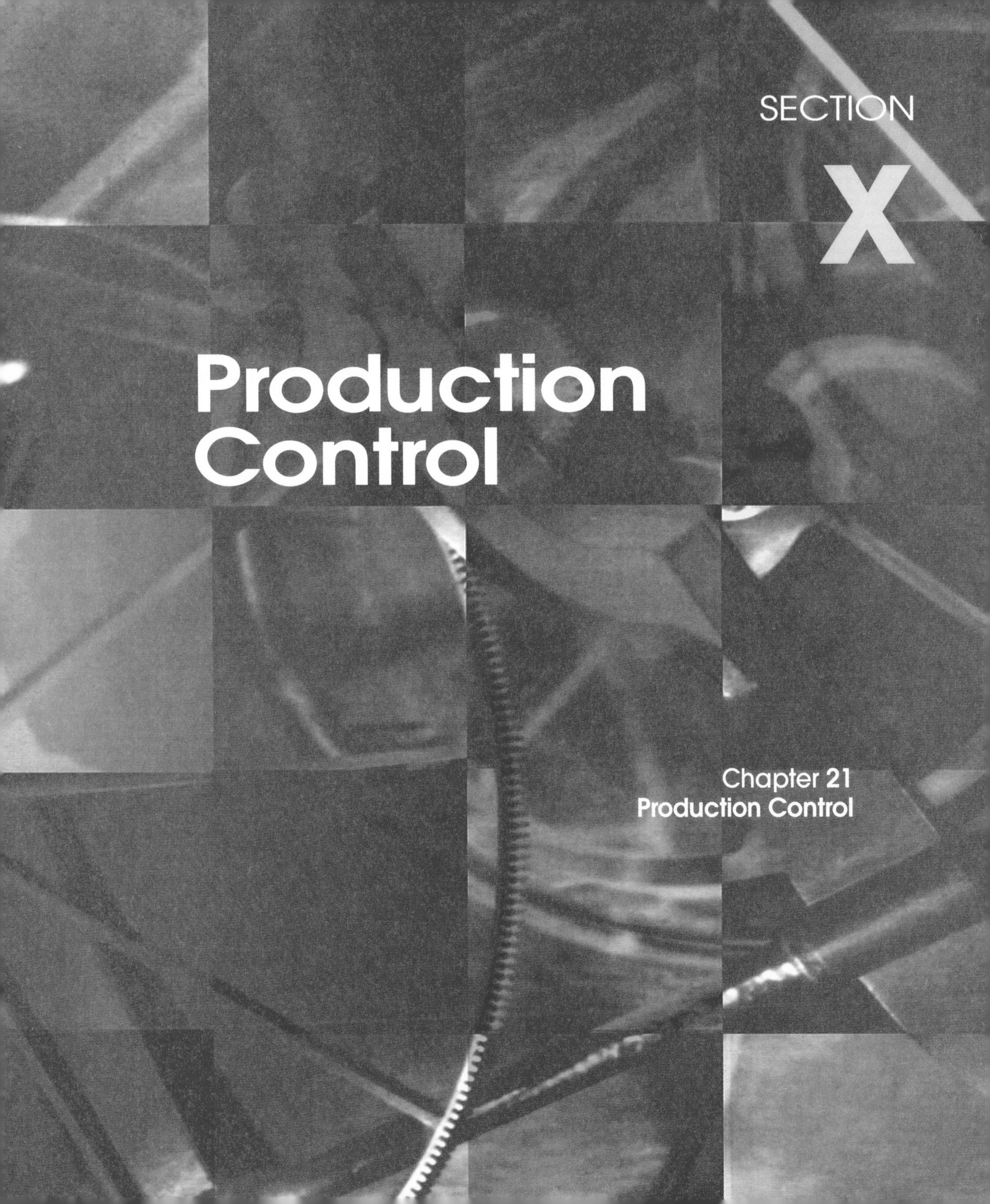

Chapter **21**
Production Control

■ INTRODUCTION

How important is production control? Consider that *the finished product has more value than the value of the raw materials alone.* For this reason, a major contribution to our nation's gross national product (GNP) comes from the act of manufacturing.

The critical issue here is **productivity.** The greater the productivity of a manufacturing facility, the greater the contribution to the general wealth and the higher the standard of living of the employees. To get the maximum productivity from any facility, production must be optimized.

Figuring out how to optimize production is a complicated matter that involves many considerations. Among the considerations that must be taken into account to get the optimum production are the following:

- Design of the product
- Selection of manufacturing processes
- Plant layout
- Time studies of each step in the manufacturing sequence
- Performing required maintenance on production machinery.

Many more considerations may need to be studied for optimum production, depending on the item being manufactured.

Each of the items in the preceding list must also be optimized itself. For instance, several considerations go into the choice of manufacturing processes to optimize the productivity of the manufacturing facility. These include:

- How many different products are being made at one time in the facility?
- For each product, how many are to be made and at what rate of production?
- How large is the production facility (in terms of the number of employees)?

Why is optimization so critical to overall productivity? Consider this example. If $1 is saved in the production of each part and 1,000,000 parts are made, then $1,000,000 is saved in the production run. These savings may make the difference between the product being economically competitive or not selling. In other words, the survival of the company may lie in its ability to optimize all of the parts of production.

Of course, not all productions are a million parts, so just how realistic is a million-parts estimate? There are roughly 100,000,000 (1×10^8) households in the United States today. If each one has a stove with four burners, then that's 4×10^8 burners installed. If each burner lasts an average of 40 years, then

$$(4 \times 10^8)/40 = 1 \times 10^7 = 10{,}000{,}000$$

new burners will be needed each year. While one company may not make all of the burners, one can see from this example (and from the ones that follow) that large production runs are common.

Currently (in round numbers) 7,000,000 vehicles are produced in the United States each year. That requires 35,000,000 new tires each year (counting spares) just for the new cars (not counting replacements for the old ones). These new cars need roughly 140,000,000 new light bulbs each year. So a production run of 1,000,000 is not unreasonable and only a *penny* saved on each one produces an increased potential profit of $10,000 per run.

Optimization of production is a critical part of manufacturing, and the selection of manufacturing processes is critical to that optimization. That's **production control,** the topic of Chapter 21.

Chapter

21

Production Control

Individual manufacturing processes have been described in some detail in the previous chapters. Now it is time to look at them as part of a larger plan. After all, manufacturing processes are only one part of production.

Figure 21–1 shows the organizational chart of a typical manufacturing plant. However, many industrial plants are not "manufacturing" plants. For instance, a food processing plant is an industrial facility, but it doesn't have the Manufacturing block in its table of organization. The three blocks at the bottom of Figure 21–1 have lines leading to the bottom of the figure to indicate that they represent the company's contacts with the world outside of the facility. The purpose of this chapter, therefore, is to place the manufacturing processes, which have already been described, into the context of several different types of manufacturing facilities.

It will become clear that the organization of a production unit is critically dependent on several factors. The most important of these is the rate of production of the product which that unit makes, because that determines which manufacturing processes will be the most efficient for any given product. Therefore, manufacturing facilities will be typed by production rate and the characteristics of these various types compared.

CHARACTERISTICS OF DIFFERENT TYPES OF FACILITIES

A tour of various successful manufacturing plants would show a tremendous variety of manufacturing practices and organizational forms. Why is there so much variety of practice and form?

One answer is that production choices depend heavily on how many parts are to be made, or on how fast they need to be made. This is the point that Table 21–1 makes. For example, a homeowner who is using basic power tools to make a repair part still needs some way to hold the workpiece while drilling or sanding it, and still needs to inspect the workpiece to see if it is going to do the job when it's finished. This is "Piecework," and Table 21–1 shows "jigs and fixtures" and "inspection" as manufacturing concerns for a Piecework type of manufacturing facility.

Once the production rate is increased to the "Small Batch" category, jigs and fixtures and inspection are still needed. But now the production manager has the additional concerns of needing storage facilities and transportation of parts between production and storage. Furthermore, the higher production rates can be

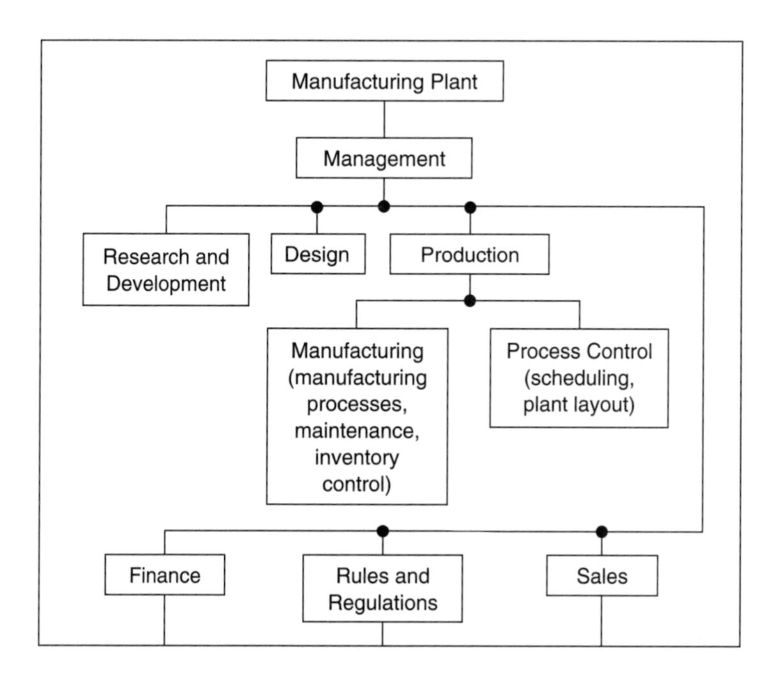

Figure 21-1. Organization of a manufacturing plant.

achieved more efficiently by means of numerically controlled machines rather than hand-operated ones. Therefore, "numerical control," "storage," and "transportation" are covered (in this chapter) under Small Batch facilities because they make more sense in that context. That's also true for the rest of the manufacturing concerns of Table 21–1.

The concerns of a manufacturing facility often depend on the required production rate. For purposes of this discussion, manufacturing facilities will be divided into four groups:

1. Piecework (1 piece per hour or less)
2. Small batch production run (1 to 20 pieces per hour)
3. Large batch or continuous run (20 to 200 pieces per hour)
4. Mass production (more than 200 pieces per hour).

The choices of numbers used as divisions between adjacent "types" of manufacturing facilities are somewhat arbitrary, but the fundamental distinctions are valid. In Table 21–1, no effort was made to include all of the typical products possible or even all of the relevant manufacturing processes appropriate to each type. Only representative examples are shown in the table.

The dividing lines between the types of production are not fixed. One plant producing 50 pieces per hour might be run as a small batch facility, whereas another plant producing only 15 pieces per hour might be operated using the concepts common to a large batch or continuous production run facility. Also, different production areas within the same plant may need to be operated differently, depending on the production rate of the product in each of the areas.

So, there aren't any hard-and-fast rules for optimizing production, but the following discussion should at least set some basic guidelines and propose some general considerations that will help in that optimizing process.

■ PIECEWORK PRODUCTION

Piecework can be defined as the fabrication of one complete unit at a time. Anywhere from one hour to several days or weeks may be required to make a unit or part. Piecework describes (among others) the "handyman" who is making a one-of-a-kind repair part or a carpenter who is fabricating some special architectural detail. Other examples of piecework would be a craftsman who is making a prototype part

Table 21–1. Characteristics of Different Types of Facilities

Type	Quantity	Production Rate	Typical Product	Typical Manufacturing Processes	Manufacturing Concerns
Piecework	1–50	1/week to 1/hour	Experimental or prototype part Replacement part Die sets Major part for low-production-rate aircraft or ship	Sand casting Electric-discharge Machining (EDM) Free-standing mill, lathe, saw Spray coating Explosion-forming	Jigs and fixtures Inspection
Small batch	50–2000	1/hour to 20/hour	Parts for specialty vehicles Major part for commercial aircraft Power-generating equipment	Investment casting Automatic lathe or mill or drill Shearing, punching Brake bending Welding, brazing Heat treatment	Jigs and fixtures Inspection Numerical control (NC or CNC) Storage Transportation
Large batch or continuous	2000–100,000	20/hour to 200/hour	Office machine parts Public utility large parts Parts for police or fire vehicles Parts for small gasoline engines Major parts for home appliances	Stamping, punching Die casting Forging Tumble abrasion Numerical-controlled lathe, mill Roll-forming Extrusion Powder metal forming Powder coating	Jigs and fixtures Inspection Numerical control (NC or CNC) Storage Transportation Just-in-time (JIT) scheduling CAD/CAM Robotics Quality assurance through automatic testing and statistical analysis
Mass production	Over 100,000	Over 200/hour	Zipper teeth Fasteners Automobile parts Home appliance parts Residential architectural detail parts Power distribution hardware Small electric motors Food containers	Computer-controlled welding Continuous casting Roll-forming Fast-cycle forging Finish coat by dipping	Jigs and fixtures Inspection Numerical control (NC or CNC) Storage Transportation Just-in-time (JIT) scheduling CAD/CAM Robotics Quality assurance through automatic testing and statistical analysis Cellular organization Adaptive control

(which will be mass produced at a later date) or making an experimental device for a research project.

This type of work is generally produced for a fixed, agreed-on price or fee (although the price may be based on the number of hours required to make the part). Within reason, productivity is not a concern. Coordination with other workers may not be a concern since one person, or a few people at most, generally do the entire project. Utilization of equipment is not a great concern since the equipment is not usually shared or tied up continuously for the production run. The equipment available generally includes very few specialized tools dedicated to one operation. Lathes, bench grinders, drill presses, and the other equipment generally used in a piecework facility do not require a large capital investment.

Piecework and small production run facilities are often *batch* processes in which products are made individually. Large production run facilities almost always require continuous flow processes where the parts are sent down manufacturing and assembly lines where individual manufacturing operations are done at many stages along the line.

Jigs and Fixtures

A *fixture* is a device that is individually designed to hold a specific workpiece in place. The part may be held in place for a specific machine or machining operation, or held in place relative to other parts for assembly. Its specific job is to facilitate setup or to make holding easier. (Note that a vise for a drill press is not a "fixture" since it is a general-use devise and not designed for a specific workpiece.)

A *jig* is a fixture that also establishes certain critical location dimensions on the workpiece. In fact, dimensioning reference planes for irregularly shaped workpieces are usually the planes on the jig against which the workpiece mounts (Figure 21–2). *Workholder* is the general term for either a jig or a fixture.

Desirable Characteristics of Workholders

There are ten desirable characteristics of workholders. Admittedly, workholders seldom incorporate all ten goals. These are:

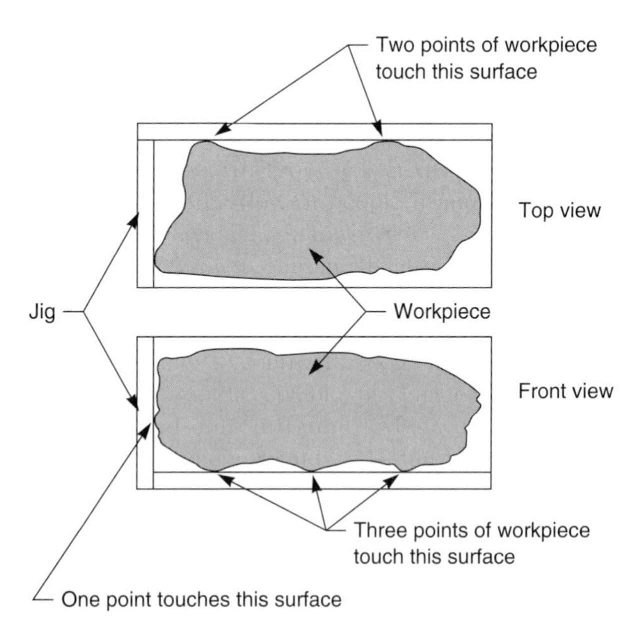

Figure 21–2. Irregular workpiece against jig faces.

1. Positive location for the workpiece without allowing any freedom of motion.
2. Clamping forces sufficient to keep the workpiece in place in spite of machining, vibrational, and other forces that would tend to dislodge it.
3. Easy access for the operator to clamp and unclamp the workpiece and to load or unload the workpiece.
4. Distortion of the workpiece due to clamping forces should be small enough not to cause a problem. This may involve designing pads or flanges on the workpiece so that the clamping forces can be spread over a wide area. These pads must be put in a location on the workpiece that will not be machined.
5. Identical workpieces will be located identically on the workholder. This is essential if the reference planes for the workpiece dimensioning system are the surfaces of the workholder, not the part being manufactured (refer to Figure 21–2).
6. Accept all workpieces that are within the expected range of allowance for the "incoming" part. This may be a significant problem if rough castings or mill-run stock sizes are to be significantly reduced in size.

7. Maintenance of the workholder should be simple and easy to do.
8. The workholder needs to be durable. It should withstand the abuse of loading and unloading workpieces. Parts on the workholder that are subject to wear should be easily replaceable.
9. The workholder should neither extend unnecessarily beyond the workpiece nor create a safety hazard.
10. The design and construction of the workholder should be simple. If possible, it should be made from standardized parts.

If a handyman is cutting a piece of door molding to length, a miter box or a bench vise might be an adequate workholder. An electrical technician who is soldering leads onto a connector would probably use a particular fixture to hold a particular plug at the angle for easiest soldering. A machinist who is drilling precise holes (into a casting) for the shafts of a three-dimensional mechanism would surely have the casting held in a jig, and the jig would have guides for the drills so that the holes would be located accurately. These are all demonstrations of different workholders for different needs.

Inspection/Dimensioning

A clear knowledge of modern dimensioning practices is essential if you are to gain an understanding of the complexities of the manufacturing and inspection of a part.

Types of Dimensioning

There are basically three types of dimensions: shape, size, and location dimensions. A model of nearly every manufactured product could be built by gluing together various shapes or features. These shapes can generally be defined in terms of geometric shapes such as pieces of rectangular solids, cylinders, spheres, cones, etc. Some dimensions specify the shape of the feature while others show its size. Finally, dimensions are required to specify where one feature is placed relative to the other features. In practice, some dimensions actually perform two or even all three functions. In Figure 21–3, the 37° dimension is a shape dimension since it shows the angle at which the part is cut. The "R" in the 0.75R stands for "radius," clearly implying that the *shape* of that part is a circular arc. The same 0.75R dimension is also a size dimension since it specifies the length of the radius. The 0.70 dimension locates the center of the arc with reference to the back of the object, thus it is a location dimension. The 0.56 dimension specifies the thickness of the front "shelf" and can therefore be considered a size dimension. Note that the part is not completely dimensioned since it could not be made from this drawing alone.

The idea that parts can be made by gluing shapes together fails when the workpiece has holes. However, one can think of a hole as a component or feature that is added (or subtracted) at some later stage in manufacturing. Drilling a hole is simply adding an empty cylinder, which is dimensioned in the same manner as a physical cylinder would be.

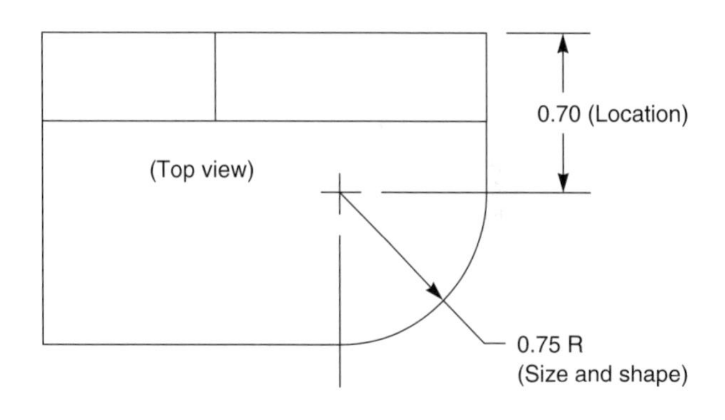

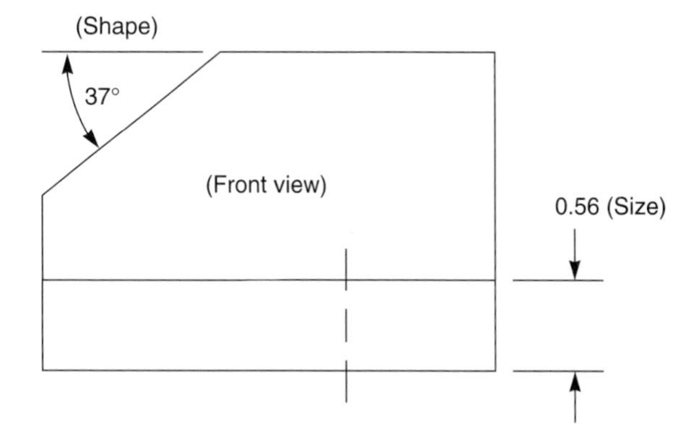

Figure 21–3. Types of dimensions.

Dimensional Tolerancing

No part can be made "exactly" to a dimension. Even though the sizes, shapes, and locations of features on a part may vary by only a few micrometres or millionths of an inch, differences in them can be found. In spite of this, parts can be made close enough to the "exact" specification to be usable and interchangeable. The second level of dimensions is a specification for each dimension of how much a feature of the workpiece may *differ* from the ideal size, shape, or relationship to other features. This is called *dimensional tolerancing.* Tolerancing would include such specifications as surface roughness and flatness (both discussed in the chapter on finishes); how much a hole might deviate from being a perfect cylinder ("cylindricity"); how far apart the center of a hole and its counterbores might be ("concentricity"); the limits to which multiple holes or other features might differ from being parallel to each other ("parallelism"); and the degree to which perpendicular features might differ from the prescribed 90° ("perpendicularity"). In each case, some reference must be specified as the basis from which the dimensions may vary. The reference may be provided either by the workholder (as mentioned earlier) or by some feature of the workpiece, such as the axis of a hole.

Coordinate Systems

If person A tries to tell person B where to find something, the two of them need to have the same understanding of where the starting point is, which direction is North, what "depth" means, the same system of measurements, and so on. In other words they need to have the same *coordinate system.* Although cylindrical and spherical coordinate systems are used for such special applications as navigation and theoretical mathematics, a rectangular coordinate system is by far the one most commonly used in manufacturing (Figure 21–4).

The essential features of a rectangular coordinate system are as follows:

- There are three directions ("axes") at right angles. These are generally labeled *X, Y,* and *Z.* Their common point is the *origin.*
- Units of measure, typically inches or millimetres, are marked on each axis.

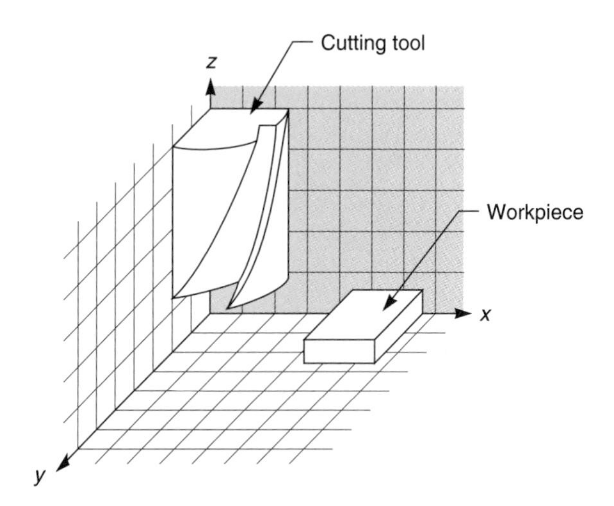

Figure 21-4. Rectangular coordinate system.

- A direction is given for positive numbers along each axis, with the implication of negative numbers in the opposite direction from the origin. Since there is no limit on how large a number can be, each axis can be considered to extend infinitely far in both directions from the origin.

Two intersecting lines define a plane, so that there are three infinite and mutually perpendicular coordinate planes. These are generally shown with a grid formed by extending the unit markings onto the planes. In reality, there is also a three-dimensional grid formed by extending the unit markings onto the space between the planes. These three-dimensional grids are implied rather than shown because the drawing would quickly become unworkably cluttered if the grid were shown. This is especially true if the unit divisions are very small.

The faces of the jig shown in Figure 21–2 are the three planes of a coordinate system. It is a straightforward matter to measure how far a point on the workpiece is from each of the jig surfaces. The three numbers that result from these measurements are the coordinates of the measured point (Figure 21–5). Generally, the coordinates are written in the form (x,y,z), where x is the distance from the *Y-Z* plane, y is the distance from the *X-Z* plane, and z is the distance from the *X-Y* plane.

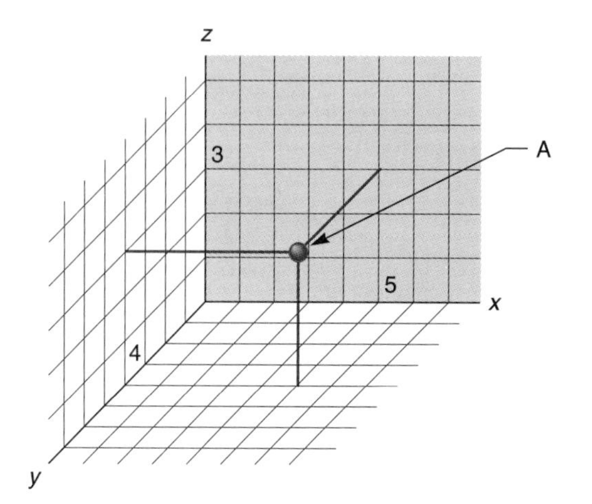

Figure 21-5. Coordinates of a point.

Linear Distances

Once a coordinate system has been established, the distance between two features can be determined by measuring the coordinates of those two features. From several of these linear measurements, the shape of many features can be specified. However, determining the shape of a feature by linear measurements may not always be the best method. For example, if the specifications call for a feature to be a square 15 millimetres on a side, then both the size and the shape can be determined if the distance between each pair of corners is measured (Figure 21–6). (For a square, 6 measurements between points would be required.) But if the shape of the feature is specified as a regular octagon (eight sides), then many more measurements would be required (a total of 28 measurements to be exact). Although some shapes can be defined using only linear dimensions, other shapes cannot easily be defined using linear dimensions. Other methods are discussed later.

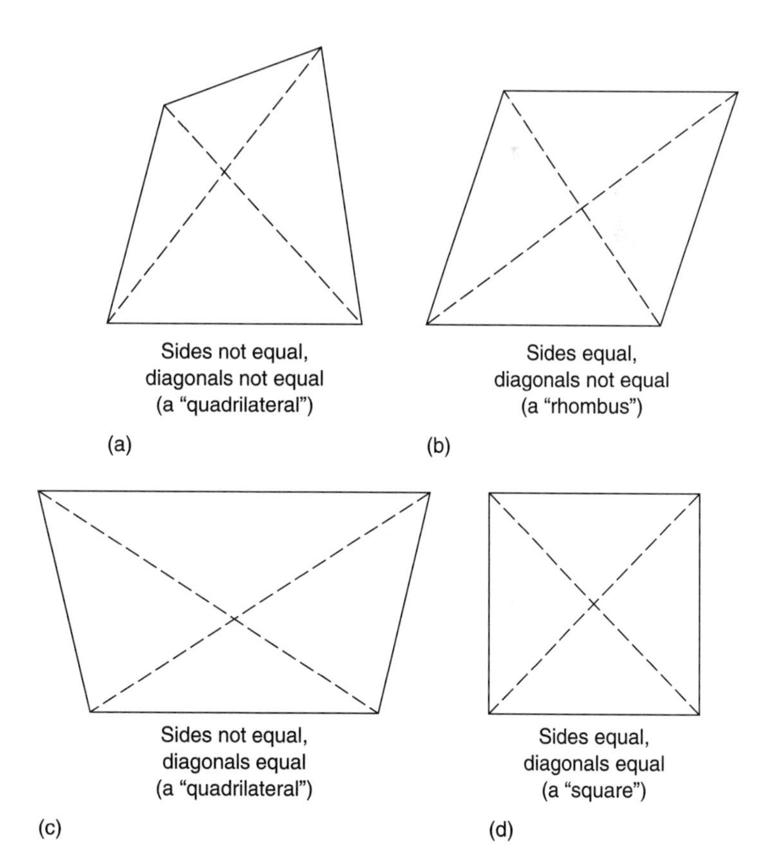

Figure 21-6. Measurements to determine a quadrilateral shape.

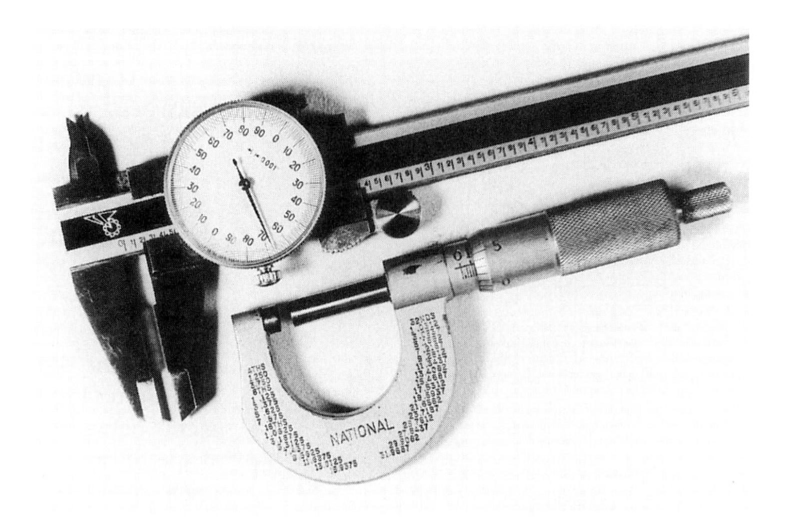

Figure 21-7. Dial caliper and micrometer.

Tools for Measuring Linear Distances. Since linear distances are the most common measurements used, quite a number of instruments have been developed to measure them. Although a steel measuring tape, electronic distance meter, surveyor's wheel, or surveyor's chain can be used for measuring long distances (with some measuring devices being more accurate than others), they are not often used in the manufacturing industry. Among the more common linear measuring devices used in manufacturing are the following:

- Measuring tape, or scale ("ruler")
- Calipers (either simple, used with a scale, vernier, or micrometers)
- Dial indicator
- Measuring telescope
- Interferometer
- Digital height gauge.

The use of steel scales and other small measuring devices is usually adequate for measuring small distances with a precision of 0.01 inch (0.25 millimetre). A dial caliper or micrometer caliper (Figure 21–7) can be used to a precision of 0.0001 inch (1/10,000 inch or 0.0025 millimetre). We must emphasize that there is some skill required in properly using any measuring device. Careful calibration and careful reading at *both* ends of a scale are essential to gain the full precision possible with these tools.

A dial indicator (Figure 21–8) is a device that can easily be used to indicate the *change* in some linear dimension. For example, the thickness of a piece of paper could be measured by the change in reading between the quill when it is resting on the table and when it is resting on the paper. The dial of the indicator can be rotated by hand so that the dial is set at zero for the first reading. This makes any calculations unnecessary, thereby reducing the chance for a mistake. Dial indicators can also be used to measure flatness, make adjustments in lathes and milling machines, and in many other uses.

A measuring telescope is used to determine the linear dimensions of a feature that must be measured from some distance away. This may be necessary because of danger to the measurer or because the workpiece is too fragile to handle, or the product would be contaminated by the act of measurement (Figure 21–9). It is composed of a telescope mounted on a carefully graduated stand. Precision is limited by the ability of the operator to distinguish visually if the crosshairs are centered on the feature to be measured. The precision of this device is limited to about 0.004 inch (0.1 millimetre).

An interferometer is a device that uses the interference of light waves to measure extremely small distances. Interferometers can detect distances at the microinch (0.000,001 inch) level. If a light beam is reflected from two surfaces that are within a wavelength of the light apart, the reflected light waves will

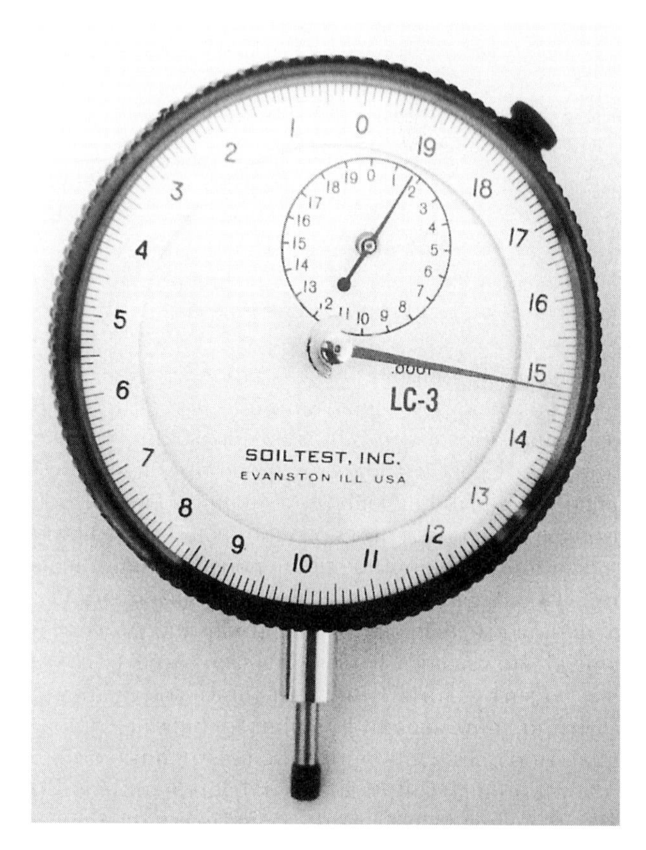

Figure 21-8. Dial indicator.

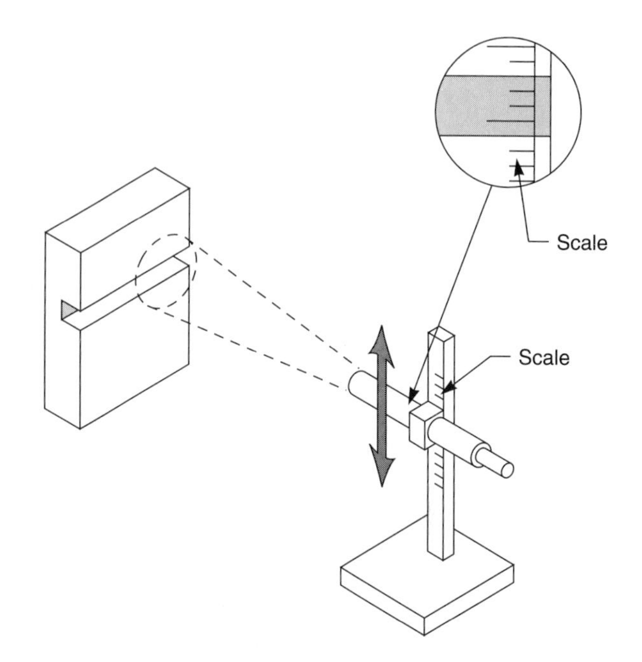

Figure 21-9. Measuring telescope.

View from above

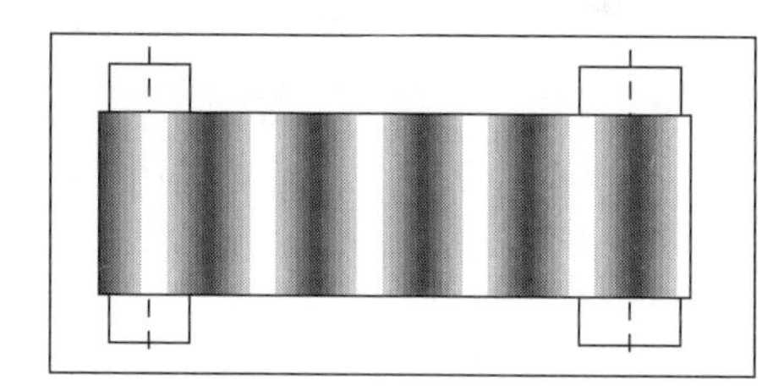

Front view

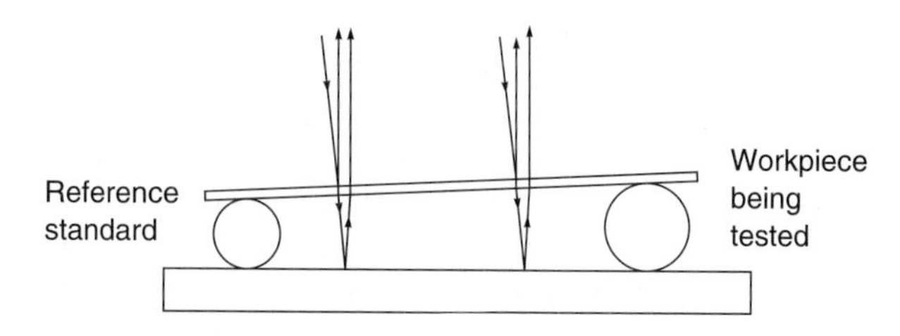

Figure 21-10. Interferometer principle.

be out of phase. These out-of-phase waves will create some regions in which the light is canceled and other regions in which the light wave is reinforced. By measuring the distance between these "light and dark" areas, the difference in the distances of the surfaces that reflect the light can be determined. The closer together the reflecting surfaces, the greater the distance between light and dark interference areas (Figure 21–10). Scales are usually placed on interferometers to indicate the distances directly, eliminating the need for calculations. Interferometers can be used to measure waviness in flat surfaces and the closeness of matching surfaces in mechanical assemblies.

The combination of a height gauge, such as the measuring telescope, with a computer has resulted in a relatively recent device called the *digital height gauge.* The digital height gauge requires that the part

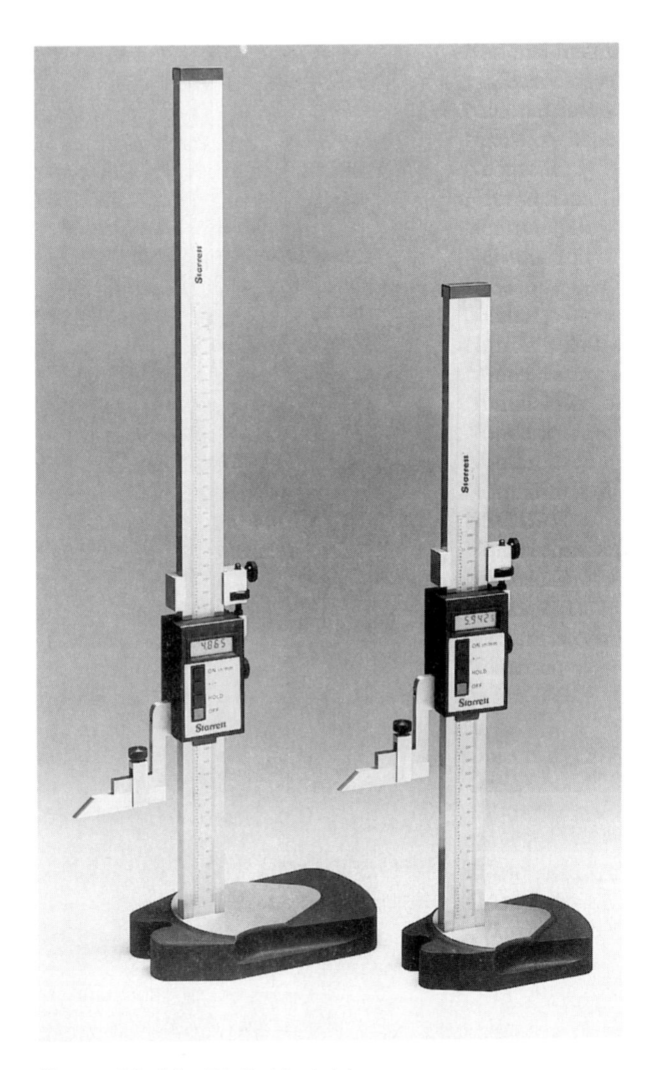

Figure 21–11. Digital height gauge.

be placed on a perfectly flat table. The digital height gauge measures the distance a feature stands above the table as the height of the feature (Figure 21–11).

Tools for Measuring Roundness

If good fits are to be maintained, drilled holes, turned shafts, or other rounded surfaces must not deviate from perfect circles regardless of size. One definition of a circle is *the path of a point moving so that its distance from a fixed point is constant.* Circular measurements, therefore, are a measure of the *deviation* in the distance from that "fixed point" or axis. Dial indica-

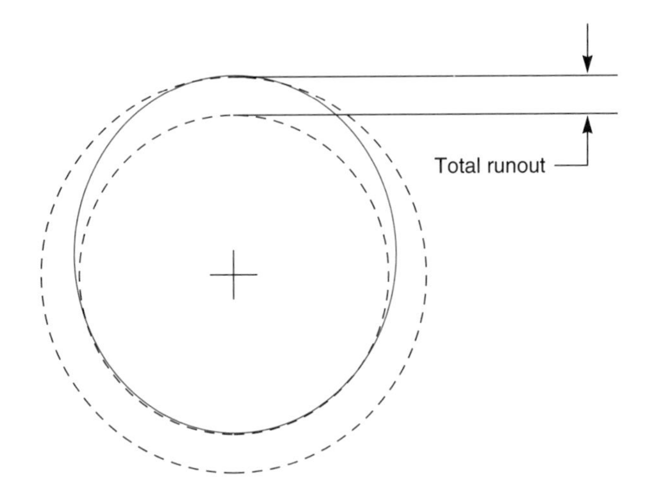

Figure 21–12. Circular runout.

tors are ideal for making such measurements. Deviations from the ideal or specified radius are called *runout.* Tolerances on circularity or cylindricity are given as so much *total indicated runout.* This total indicated runout is the range of dial indicator readings as a part is turned about the axis (Figure 21–12).

Angle Measurements

Angles can be measured either by a vernier protractor or by the use of a sine bar and gauge blocks (Figure 21–13.) Gauge blocks are metal blocks that have been made with very high dimensional precision. As a result, they have polished ends that are so flat that if they are slid together in such a way as to exclude air, the suction created between them will hold them together.

A *sine bar* is a bar that has a cylinder near each end of the bar. These cylinders are shown in end view in Figure 21–13. The cylinders have identical diameters and the distance between their centers is known to a very high degree of precision.

When the sine bar is set up with one cylinder resting on the work table and the other cylinder resting on gauge blocks, the angle of incline can be determined with very high accuracy and precision. Since the height of the gauge blocks and the length between the centerpoints of the cylinders is known, the gauge block height divided by the distance between cylinders on the sine bar is the sine of the angle of inclination. (Hence, the name *sine bar* for this tool.) Measurements of angles

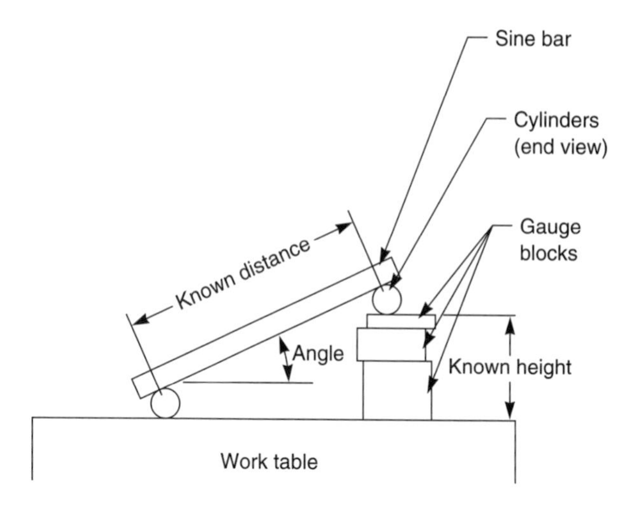

Figure 21-13. Sine bar.

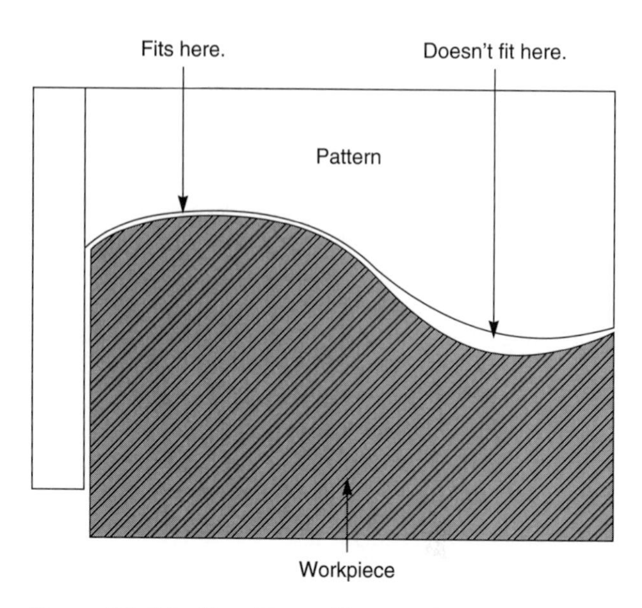

Figure 21-14. Checking a fit against a pattern.

on an object can be made by placing the sine bar on the object and measuring the distance of the centerpoint of the higher cylinder with a measuring microscope, steel scale, dial indicator, or other device normally used for measuring linear distances. Therefore, the use of a sine bar allows angular measurements to be made by measuring only linear distances.

Measurement of Other Shapes

Often in industry, mass production of oddly shaped objects requires innovative methods of measurements. Matching patterns are often made to a high degree of precision and the manufactured parts placed against them to measure how closely they fit (Figure 21–14). The clearances between shafts and bearings are often determined by placing a plastic or wax wire of known diameter between them, then measuring how much the wire is flattened when the parts are tightened together. (Plastigauge is one trade name of this technique; see Figure 21–15.) Sometimes high spots or points of interference between closely fitted parts are determined by painting both surfaces, rubbing them together, then measuring how much paint is worn off. The measurements of irregularly shaped surfaces are often made by techniques designed solely for that application. For this reason, a "general" discussion of these special techniques is not possible.

Figure 21-15. Plastigauge.

Inspection

Inspection is the act of determining whether or not the size, shape, and location of all features of the workpiece are consistent with the drawings or other specifications. No one knows if the specifications have been met until the part is inspected to compare the actual dimensions of the workpiece with the dimensions called for on the drawings for the part. The inspection of a workpiece for conformation to its design specifications can often be done without removing the part from the lathe or other type of machine. In other cases, inspection may require a considerable amount of equipment and consume even more time than the machining did. Parts are inspected by the machinists or other technicians at many stages in the fabrication processes. Micrometer readings are often taken after each cut or pass in machining. Parts can be quickly "roughed out," but in the final stages, finer cuts are made. Machinists and other technicians always try to err on the safe side, keeping the part a little too large or the hole a little too small in the next-to-last pass of the cutting tool. A little more material can always be taken off, but it is more difficult to add material.

The final inspection is always made to determine if all features of the part meet specifications. If the part doesn't meet the design limits, it must be either reworked or scrapped. Scrapped parts represent a considerable loss of manufacturing time and investment. The inspection equipment can be as simple as a set of micrometers but often is very sophisticated. Optical comparators, integrated with computers, are now being used to speed the process of inspection. Optical comparators cast an enlarged shadow of the part on a screen from which very accurate measurements can be made. Figure 21–16 shows a typical optical comparator.

The television camera has been combined with an optical comparator. Now the part is viewed through a television screen. Besides being able to measure the

Figure 21-16. Optical comparator.

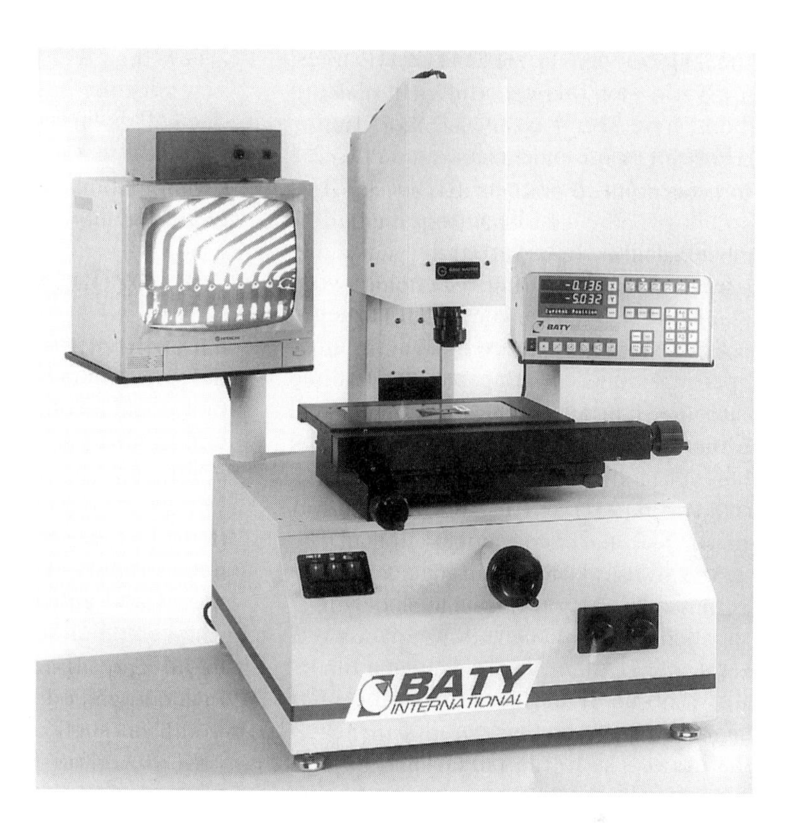

Figure 21–17. Video inspector.

part to see if it meets specifications, the magnified image of the part can be examined for flaws, burrs, and other imperfections. Figure 21–17 is an illustration of a video inspector.

■ SMALL BATCH PRODUCTION

For purposes of discussion, small batch production can be considered to be the manufacturing of from 1 to 20 pieces per hour. Small batch production poses some problems that do not occur at the piecework level. At this rate of production, completed parts begin to take up a significant amount of space before they can be shipped; raw materials are needed at a faster rate than they are with piecework. Management and engineering people must now plan to keep supplies on hand, storing them and later transporting them to the production areas on demand. Moreover, the transporting and storing of the completed parts away from the production machinery must be considered in the planning stages if they are to be accomplished efficiently.

Another factor that occurs in mass production, even at the small batch rate, is that since one worker generally makes the same parts over and over, *boredom* becomes a concern. Boredom not only affects the quality of the product, but safety is also affected. For these monotonous tasks, it would be nice if the worker could "tell the machine how to do it" then simply supervise production. This would allow the worker to be a part-time supervisor and release her or him to do other tasks that machines cannot do very well. The worker could now spend some time on maintenance of the machinery, inspection of products, or a myriad of other tasks that could relieve the monotony and resulting boredom. Techniques have been developed whereby machines take over these repetitive tasks.

Numerical Control

Wood lathes that could copy chair legs, spindles, or other complicated wood parts have been around since the turn of the 20th century. Turret head metal lathes, which can perform several operations in making screws and bolts, have also been used in industry for some time. However, these machines were dedicated (one-function) machines. *Numerical control* (NC) of the machines allowed them to be reprogrammed to produce many different types of parts.

In numerical control, machine commands were standardized and given numerical codes. The sequence of codes that described the sequence of steps required to make the part was punched into paper tape and the tape loaded into the machine's tape reader. Then, all the operator had to do was to load the raw material into the machine, start the machine, and come back later to collect the completed parts.

With the first NC machines, the numbers representing the codes were expressed in binary form and put on a paper tape in the form of punched holes. The holes corresponded to a number; a hole would equal 1, no hole was a zero (Figure 21–18). The tape reader interpreted the holes as instructions, which the machine then used to perform the operations.

One of the drawbacks to the paper tape NC system was that even the slightest change to the sequence of operations required that a complete new tape be punched. If a mistake was found in the tape, a new tape had to be made (with the chance that a new mistake could be created while fixing the old one). This lack of flexibility made the process slow and cumbersome. Yet the paper tape did have its positive points. It was an important step on the way to automation.

Computerized Numerical Control

Computerized numerical control (CNC) was the next development in the automatic control of machines. Not only could a computer be programmed to control several production machines, but it could also store programs and data that would allow it to keep track of the wear of the cutting tools and compensate for that wear in the instructions to the machine. The computer could detect trends in the final dimensions of the product pieces and anticipate service needs of the machine. The computer could also be programmed to respond to tool breakage and other malfunctions, keep track of the progress of the order, call for a new supply of raw materials, and keep track of maintenance needs. One computer could control a plotter that was drawing the plans for a part at the same time that it was controlling the machine that made some other part. Automatic control of machines had arrived.

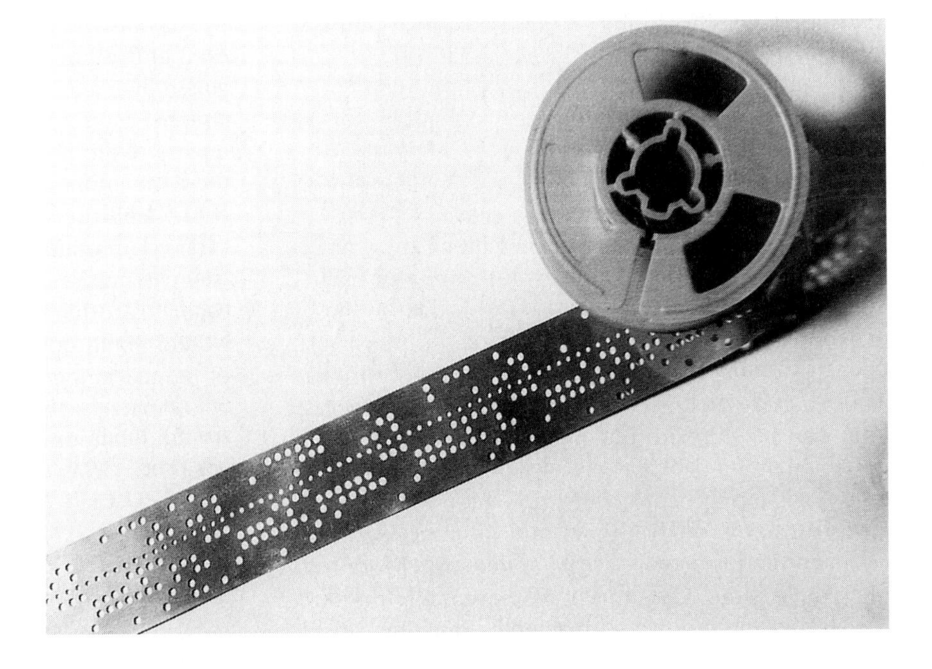

Figure 21–18. Punched paper tape.

Of course, the revolution in the control of the machines required changes in the machines too. There is only so much that can be done using a drill press with a quarter-inch drill. A worker operating a drill press not only controlled the speed and feed rate and the depth of the drill, he or she changed the drills to the correct size, or even changed from a drill to a countersink (or to a counterbore tool or a tap), checked for flawed tools, adjusted the coolant flow rate, mounted the part in the holding fixture, and even oiled and cleaned the machine. If CNC were to exploit the complete range of possibilities of any machine, then it needed to be able to do these functions in addition to operating the machine. This required a change in the design of the machines. Machines were designed with a tool carrier so that tools could be changed by computer control. Feeds and speeds now had to be controlled electronically rather than mechanically.

One result of the change from manual control to CNC was in the safety of operation. Since the operator no longer had to have immediate access to the workpiece to make repeated measurements and adjustments, shields and other safeguards could be set up about the entire machine. Trying to get a picture of the workpiece in a CNC machine is almost impossible since the entire operation is almost hidden from view. Figure 21–19 is a picture of a machine with the workpiece in place, and therefore almost completely hidden.

Some Problems of CNC

The economics of CNC are treacherous. The new machines and control systems are very expensive. Many small job shops that do piecework and small batch projects cannot justify the additional cost of these machines. Furthermore, not all of "manufacturing" is machining.

Figure 21–20 illustrates graphically the results of a study done by the Cincinnati Milacron Corporation regarding the amount of time that a workpiece spends in the various stages of manufacturing. It is a shock to discover that 95% of the time a workpiece is being made it is either being transported or is waiting to be machined. Only 5% of the time is it actually on the machine. This downtime does nothing to add to the

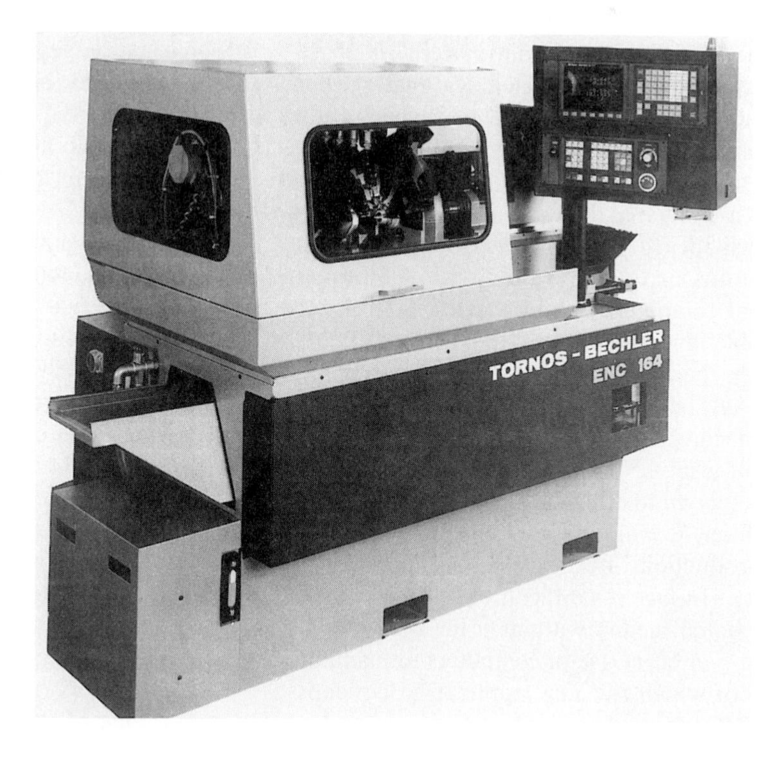

Figure 21–19. CNC machine.

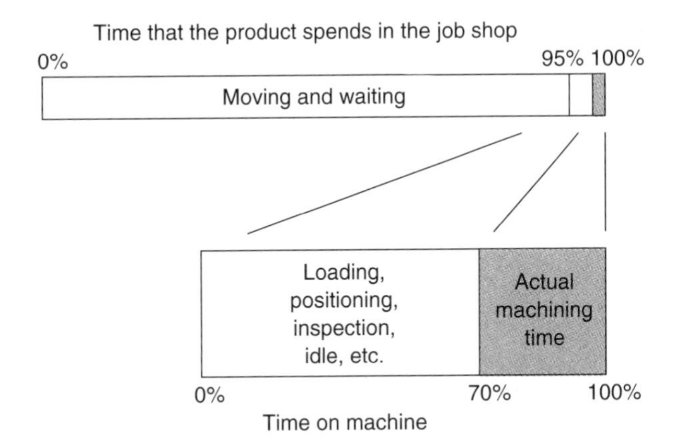

Figure 21-20. Utilization of equipment.

value of the part. Furthermore, of the 5% of the time that the part is actually on the machine, it is actually being machined for only 30% of that time. The rest of the time is used in setup or inspection. Since 30% of 5% is 1.5%, the part is only being machined 1.5% of the time it is in the facility. Therefore, if all that can be done to improve the process efficiency is to improve the efficiency of the 1.5% of the time, little overall improvement of productivity can be expected. Fortunately, computers can also be used to improve the efficiency of the rest of the time, as discussed later.

There is a second problem with the introduction of CNC. Does it really free the worker to do other, more profitable activities? If it frees him or her to supervise more machines or to spend time with the human activities that computers cannot do well, then it may be a profitable investment. But if the worker needs to sit and watch the machine work, then there may not be much, if any, benefit in having CNC, especially in a small batch facility.

In spite of what CNC did (and can) accomplish, it does not even *begin* to utilize the capabilities of even a small computer. Computers are capable of handling the large production rates that characterize a large batch facility. (In fact, it's difficult to imagine operating a large batch facility without using computers.) But the first significant use of computers in manufacturing, out of which the new application concepts grew, was CNC for small batch processes.

LARGE BATCH OR CONTINUOUS PROCESSES

For purposes of discussion, assume that a large batch facility is one that produces from 20 to 200 parts per hour. In addition to the problems it shares with facilities of smaller production rates, an entire new set of problems arises that is associated with mass production of this magnitude. No longer does one person see the part from start to finish. Instead, several workers work on it, each doing a single operation (or a few operations) on the part before it is passed on to the next step. The process is changed from a batch system to a continuous-flow system. The logistics and supply of raw materials, components, and tools to the manufacturing and assembly lines create many problems. Methods of transporting parts from one manufacturing station to the next must be devised. Henry Ford's solution to mass production lay in one of his mottos: "Keep it off the floor." In other words, no part is to remain stationary. Parts are manufactured and assembled as they are moving down the line. Any lack of communication between workers at different parts of the production line can easily create problems.

Scheduling is an important part of manufacturing. For instance, suppose that a completed assembly has only two parts, but one of the parts takes 30 minutes to make and the other part takes only 10 minutes for fabrication. To keep the assembly line supplied with parts, three machines making the 30-minute part are required for every machine making the 10-minute part. This assumes that 100% of the parts coming from each machine pass inspection, that they are usable, and that none of the machines breaks down. Pretty risky assumptions!

One can easily see how this problem would be magnified for the production of an automobile engine requiring hundreds of parts for assembly. One solution to this problem is to acquire a large stockpile of parts and tools for the assembly line; a newer solution (just-in-time scheduling) is discussed later.

Some years ago a film titled *The Tyranny of Large Numbers* reported on the difficulties the International Resistance Corporation faced when they tried to increase by a large amount the rate at which they produced resistors (a small component of electronic circuits). One of the major issues facing them was the delay between detecting a production error and correcting it. For example, suppose that 6000 resistors per hour (100 per minute) were coming from the production line. If an error was detected and it took five minutes to correct it, then they had 500 resistors to scrap, plus all of those that were in the production line and hadn't yet reached the point of inspection. Considering that one of the steps required a four-hour curing time, it is clear that the number of parts that would have to be scrapped as a result of a manufacturing error would be unacceptably large. They were forced to redesign the entire fabrication method to avoid such a long delay. Since the per-piece profit is small on items such as these, the expense of scrapping unusable parts seriously reduces the profit and could, in extreme cases, cause the company to fold.

One of the major solutions to the preceding problem was to inspect the parts frequently so that they could detect trends in terms of parts not meeting standards and take corrective action before the parts were so far from the specifications that they had to be scrapped. With the implementation of computers, each part can be inspected at several points in the manufacturing process and minute adjustments made immediately instead of waiting until the parts become so bad that the entire line must be shut down.

Computer-Aided Drafting/ Computer-Automated Manufacturing

Computer-aided drafting (CAD) has been available for a number of years. It started out as little more than an "electronic drafting board" that had the capacity to do perfect lettering, draw circles and lines with perfect line weight, repeat features, change the scale of a drawing, and move views around with ease. Now it has evolved from the original passive drafting aid into more of an interactive design aid (computer-automated engineering, or CAE). Where once the drafter worked to create three views of an object, the designer now works in a pictorial mode and the computer draws the three views when asked to do it. But that is not all that a computer can do. Once the design is completed in the computer, the computer can:

- Determine the volume and weight of the part and locate its center of mass.
- Do a complete structural analysis of the product calculating the stresses and deflection resulting from any applied forces.
- Determine clearances and allowances between matching parts.
- Devise the optimum steps for the manufacture of the product.
- Create the commands for the computer-controlled machines that make the product.
- Keep the inventory records and other documentation for the product
- and more.

CAD is now being used even by small industries and in many different types of production situations. (It should come as no surprise to the reader that all of the drawings in this book were done using a computer.)

As impressive as all of this is, it's all history. The application of computers has passed through the realm of computer-aided drafting and computer-automated engineering and is now into the realm of *computer-automated manufacturing (CAM)* and, in some cases, even beyond that! Entire books have been written on the topic of CAM, so only a few highlights are touched on here.

Two concepts regarding the use of computers are crucial to the understanding of their application to manufacturing:

1. Computers come in all sizes. A successful CAM facility will have many computers scattered throughout the plant, preferably all of them interconnected with the same data line. Information ("data") needs to be processed in some way to make it useful. One computer can be used to preprocess the data and then it can be the data source for other computers. The total body of information about a product (the "database") may be used in the drafting department to create the drawings and assembly plans, by the production department (on the manufacturing floor) to create an applications drawing in the sales department, or by the engineering department to do a flexibility study. Each of the small computers has access to the database computer and in each case the information comes from the central computer in the same form. The local computers process the data differently to meet the local needs.

2. It is important to distinguish between data flow and control flow. On the production floor, the typical sequence is that first a control instruction is fed to a machine. The control instruction may be to start the machine, move the tool in some direction, speed up the machine, turn on the coolant fluid, or any other single step required to make the part. In response to that instruction, data about the status of the machine is fed back to the control source (computer) to let it know whether or not the machine is ready to respond to the instruction and perform the action. If the machine is ready for the instruction, data are fed back to advise the computer that the operation has begun. When the control instruction is completed, data are fed to the computer that it is ready for the next step, and on it goes until the part is completed. Basically the computer and the machine have to "talk" to each other. Remember that the machine does nothing until it is "told" to do so, and then it does only what it is told to do. It will follow an erroneous control instruction as well as a correct one. Often the computer can be programmed to detect if a machine is following an erroneous instruction and issue another control command to prevent the machine from damaging a part.

Data flow and control flow can also be applied to the entire plant management. Figure 21–1 showed the general organization of a manufacturing plant. It is not difficult to imagine that all of the boxes are connected by data flow and control lines. Control flows from the management to production and on to process control. Control lines also flow from management to research and development, design, sales, and the other departments. Data must also flow back from process control and manufacturing to production and on to management. The communication lines must be kept open in both directions in everything involved in manufacturing. This applies to the machines as well as the personnel involved.

Typical Computer Operations in CAM

Besides the use of computers in CAD, CAM, and CAE, which have already been discussed, computers have many more applications. A few of these would be to:

- Control the inventory of raw materials, supplies, completed products, tools, and other necessities of manufacturing.
- Direct the transportation of raw materials, completed products, and parts at different stages during manufacture.
- Control the collection and distribution of recyclable scrap.
- Produce daily reports on the status of each product line.
- Produce end-of-shift reports on tools needing sharpening or repair for the next shift.
- Create machine maintenance schedules.
- Control the automatic inspection of each part during and after each step in manufacturing.
- Analyze the inspection results to determine whether or not any modification of control instructions is required.
- Conduct performance tests on completed products.
- Maintain documentation on each workpiece.
- Maintain documentation on each machine.
- Control the actions of robots.
- And on and on.

Even this incomplete list should convey some sense of the variety of applications in a CAM facility where computers have been given so many of the tasks formerly done by humans.

Figure 21–21 is an example of a computer-controlled cutting torch. This type of torch can now cut a plate of steel much faster than sawing. Combine this type of cutting with a plasma torch, and the cutting rate is further increased and the cut is nearly as smooth as that achieved by mechanical sawing. Figure 21–22 shows a computer-controlled system whereby one person can control the drilling of large holes in up to six steel parts at one time.

Robotics

This topic has been labeled *robotics* rather than *robots* because *robot* refers only to the machine. There is much more to the use of a robot than just the device itself. The robot must interact not only with a human programmer/operator but also with a central production–control computer, the environment, and the workpiece. Therefore, a considerable

Figure 21-21. Computer-controlled cutting torch.

Figure 21-22. Computer-controlled multiple drill.

amount of information processing and communication must go on continually if robots are to be used to maximum effect. This communication is done by external devices, and the entire functioning complex of equipment is referred to as a *robotics system* or *robotics technology.*

A robot has been defined by the Robot Institute of America as *a reprogrammable, multifunctional manipulator designed to handle material, parts, tools, or specialized devices through variable programmed motions for the performance of a variety of tasks.* The keywords in the definition are "reprogrammable" and "multifunctional." Dedicated machines such as turret-head lathes can do many tasks, but they are not reprogrammable. Tape-controlled machines might be reprogrammable but not multifunctional. Therefore, neither machine qualifies as a robot.

Robots have five essential components:

1. Manipulators, which consist of the "arm" and the "hand." The manipulator is the part of the machine that actually handles the tool, material, or part.
2. Actuators on the arm and hand respond to the signals from the controller, causing the manipulator to perform its assigned function. The manipulator can hold and move the part through six degrees of freedom (translation in the x, y, and z axis directions and rotation about each axis). The manipulator is driven by electric, hydraulic, pneumatic, or magnetic motors through a series of gears, chains, screw jacks, or other mechanisms.
3. Sensors detect the positions of the various components of the arm and hand and feed data back to the controller.
4. Controllers are the computers that send control signals to the manipulators and respond to the data fed back to them by the sensors. These controllers are programmed for the task assigned. In the older robots, these controllers were mechanical devices but now they are almost exclusively electronic.
5. Power supplies provide the energy that drives the controllers, sensors, manipulators, and actuators.

Use of Robots

Much has been written about how all of the drudgery will (someday) be taken out of everyday living by the use of robots, on the assumption that they will be as common as television sets are today. But there is a lot more to the use of industrial robots today than just avoiding "drudgery." Robots are often cheaper to operate when compared to the cost of hiring humans. They can operate in hazardous environments and at temperatures, pressures, and atmospheric conditions that would be fatal to humans. They tend to perform their function with consistent precision beyond the capability of human beings. Their performance capabilities can be greater than that of a human in strength, size, speed, and accuracy. They can be reprogrammed almost instantly by a central computer or operator. They can work 24 hours a day, never need coffee breaks (well, maybe for maintenance), and don't argue with the boss.

Robots are now widely used in a variety of industrial applications such as materials handling, tool use, finishing processes, welding, assembly, inspection, and data collection. They are used in processes such as casting, forging, heat treating, press work, molding, adhesives application, coating processes, and cutting. The composites industry also relies heavily on robots for layup, cutting, resin application, and painting of products.

That is not to say that robots are perfect. Some situations benefit from human presence. For example, it is difficult to program a robot to handle an unexpected emergency or solve an unanticipated problem. That's where a human worker has the advantage. Vision, especially *pattern recognition,* is still at a relatively primitive level in robots. It is true that much research and development is currently going on to find a solution to that limitation.

One other dilemma of robots is purely economic. The term *robot* is derived from the Czechoslovakian word meaning "compulsory service" or "worker." One robot can replace several workers doing these repetitive jobs. Although robots are expensive to install, it is sometimes cheaper to use a robot than to hire several human workers to do the same job. Labor now has a competitor. It has been said that every $1 per hour increase in the cost of labor makes 1000 robots economically feasible. Yet robots do not earn a salary, buy the goods they produce, or pay taxes. It is possible to imagine the extreme case where everything is done by robots, but no humans would have a job or the income with which to buy the robots' products.

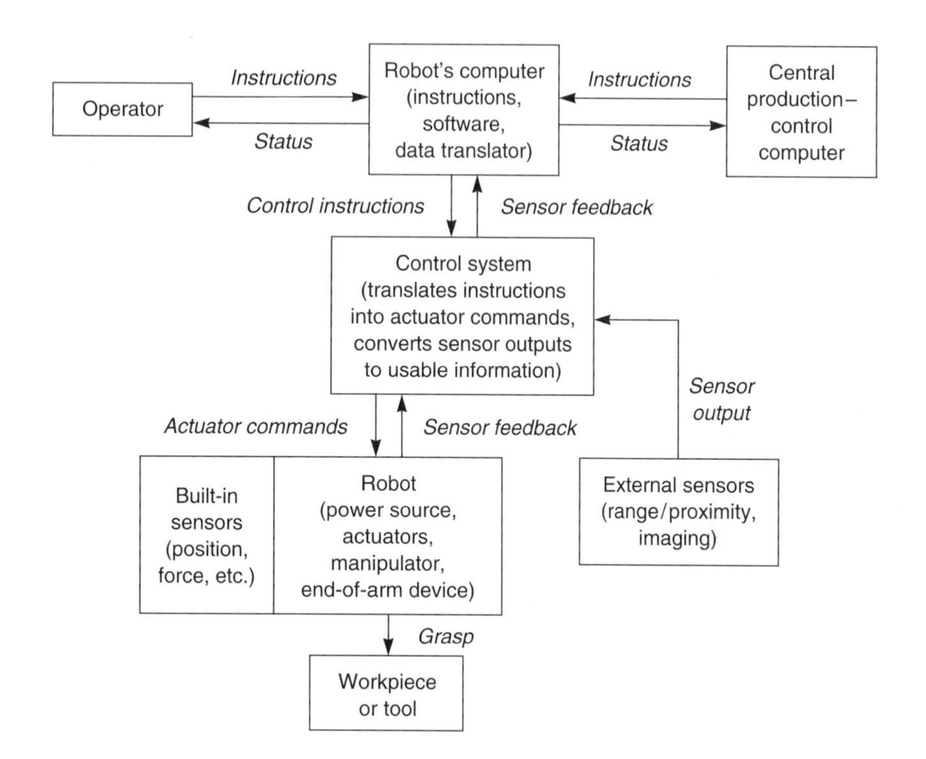

Figure 21-23. Diagram of a robotics system.

Figure 21–23 is a diagram of a robotics system as it interacts with the operator, the central production–control computer, and the workpiece or tool. The operator describes to the robotics system the task at hand. Software in the computer translates this information into a sequence of instructions that details positions, movements, and actions to be taken by the robot. For example, "locate the lower-right corner of the workpiece, move up 3.0 inches, and spot weld." These instructions are then translated by the control system into specific commands to particular actuators and the robot begins to move. Sensors in the robot send information on arm position and motion back to the computer so that the instructions can be updated. Once the sensors signal that the mechanism is in the desired position, signals are sent to the actuators to stop there and proceed with the next instruction. The computer indicates the status of the robot on some readout device in a format the operator can understand and also sends status information to the production–control computer in a format that it can understand.

Manipulators. The robot *manipulator* must duplicate essentially every movement of the human arm and hand. If one analyzes the movements of an arm and hand, the elements of the movement would be as follows:

- *Arm sweep.* The right-left movement of the arm from the shoulder. For a human, sweep is approximately 270° (without turning the body). A robot arm can sweep through 360°.
- *Shoulder swivel.* The up-and-down motion of the arm, pivoting at the shoulder.
- *Elbow extension.* The lengthening of the arm by pivoting the forearm at the elbow.
- *Wrist roll.* The rotation of the wrist as in the action of turning a doorknob. This is limited to about 270° in the human arm, but is possible through at least 360° with robots.
- *Wrist pitch.* The up-and-down action of the wrist as in knocking on a door or rapping on the table.
- *Wrist yaw.* The right-left action of the wrist with the arm held rigid.

The movements just described are illustrated in Figure 21–24. This type of mechanism, which duplicates the motions of the human arm and hand, is called an

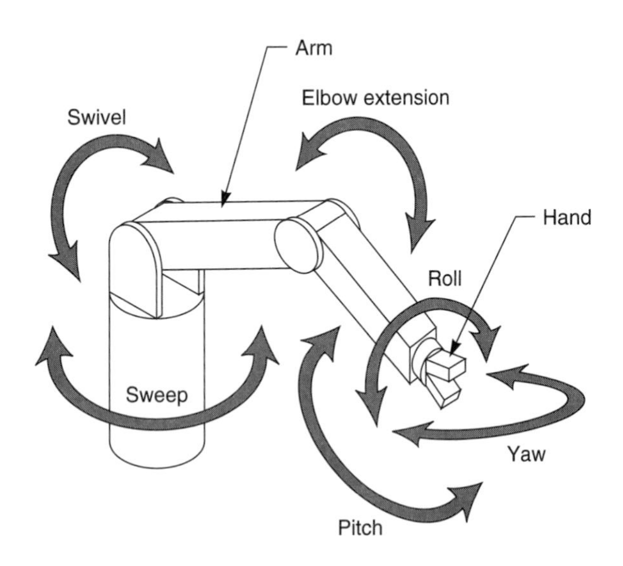

Figure 21-24. Anthropomorphic movements of a robot.

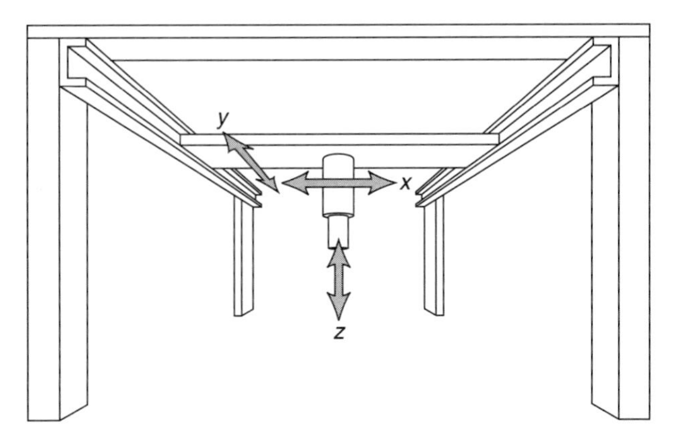

Figure 21-25. Gantry robot system.

anthropomorphic system. Some of these motions may not be needed in certain industrial applications. An anthropomorphic system is expensive to duplicate so if some motions are not required, a number of simpler, less costly designs are available.

In the Cartesian mechanism, the "hand" is simply moved independently in the *X, Y,* and *Z* directions without rotation. In this system the arm is supported above the work surface by a four-legged frame. Since this is the same form as a "gantry" crane, it is generally referred to as a gantry mechanism. It is simple, sturdy, and capable of high speeds and accurate work (Figure 21–25).

Cylindrical and polar mechanisms involve one or two pivot axes so that they are capable of more complex motions than the gantry mechanism. Their greater complexity makes them more expensive than the gantry mechanism, but they are still less costly than the anthropomorphic design. Figure 21–26 depicts the possible movement of a *polar* robotic system and Figure 21–27 depicts the movements possible in a *cylindrical* robotics system.

Each mechanism type has a fixed volume within which it can work. This volume is called its *working envelope* (Figure 21–28). Since a robot is limited to that volume, all functions that it is to perform must be within its envelope. If a robot is doing a pick-and-place task, for instance, the point where it will pick up the workpiece and the point where the workpiece will be placed must both be inside its working envelope. If all of the workpieces and tools cannot be fitted within that volume, then either a more complex robot, or a larger one, must be used.

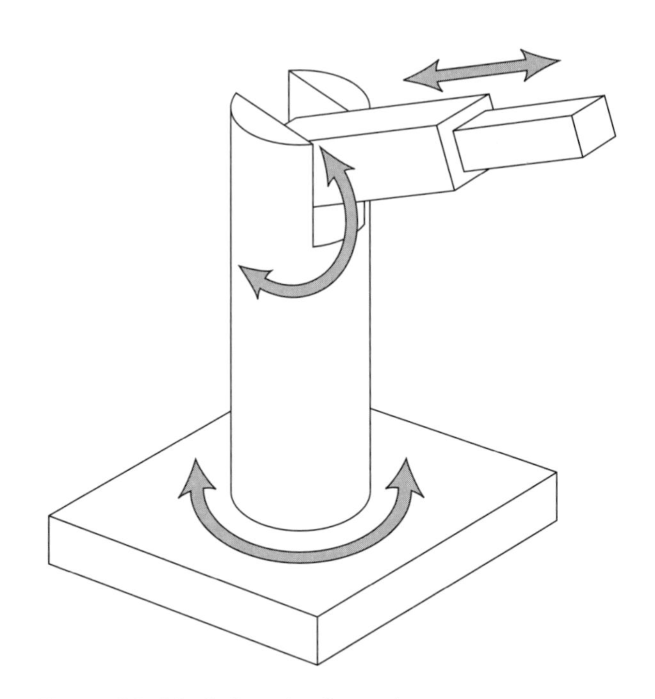

Figure 21-26. Polar robotics system.

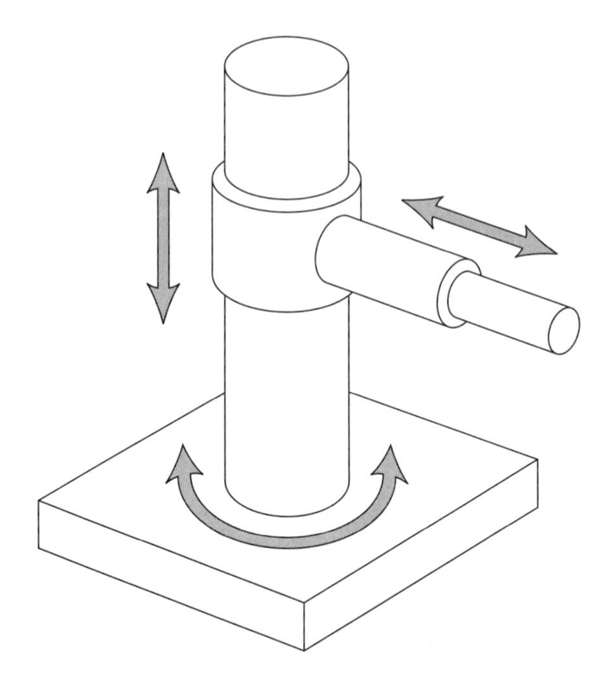

Figure 21-27. Cylindrical robotics system.

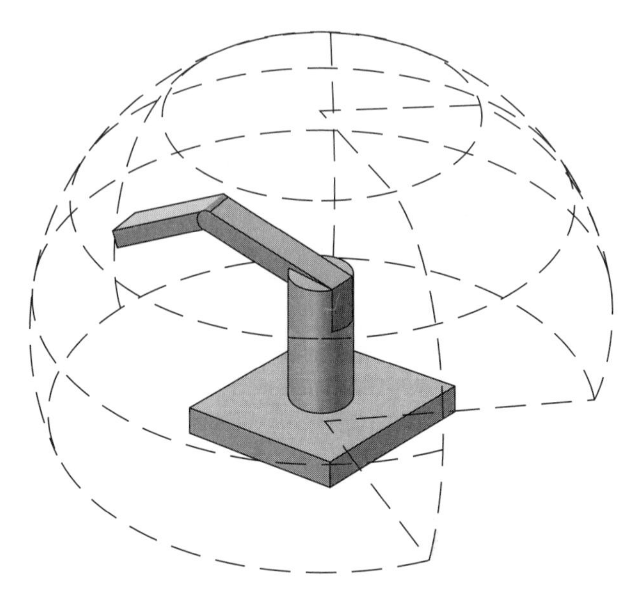

Figure 21-28. Working envelope of a robot.

Actuators. The "muscles" of the robot mechanism are called *actuators.* These may be powered by air, hydraulic fluid, electricity, magnetism, or a vacuum. There are advantages and disadvantages to each type of actuator.

Air-operated and vacuum-operated (pneumatic) mechanisms are usually small and limited by the amount of force they can exert, but they are inexpensive to construct and operate.

Hydraulic mechanisms are capable of exerting large forces, are high powered, and are amazingly quick for their size. They can be made very large and can be safely and reliably used in hazardous environments. They do tend to be messy due to the inevitable leakage of hydraulic fluid.

Electrically driven mechanisms are appropriate for light duty. They are very accurate, precise, and clean.

Magnetically driven actuators such as solenoids are very quick but are limited by the amount of force they can exert on the workpiece.

Computer Functions. The "brain" of the robot is the computer that controls the robot. The computer has three major functions:

1. The computer must contain a software program that can interact with the programmer to convert the programmer's wishes into instructions that will cause the robot to perform the assigned task. Instructions can be given to the computer by digital input from a keyboard, or through analog input from a joystick, levers, knobs, or switches. Some robot computers are programmed simply by setting the computer in a learning mode then letting a skilled technician guide the arm through all of the motions to complete the task. The computer memorizes all of these motions and will duplicate them exactly for each successive workpiece.

2. The computer must carry out the instructions for the current application. Since robots are reprogrammable manipulators, the computers can also store sets of instructions for alternative uses of the robot.

3. The computer must translate data from the rest of the components of the robotics system. If the instructions from the operator are given to the computer in an analog form, then the computer must translate those data to a digital form, which will control the manipulator and actuators. The computer must also be capable of receiving instructions from the central production–control computer and converting those instructions into commands its own manipulator can use. This may involve making bit-rate changes, synthesis of a number of data into a summary form, translation and interpretation of coding schemes, etc. In addition to these three major functions, the computer's software must also be able to:

- "Model the world"; that is, it must describe the known positions of object obstacles within its envelope. It must also keep track of tool positions, wear on the tools, and the condition of its own mechanisms.
- Generate the details of the motion of the robot's arm from one point to the next, considering the obstacles, the power available, and the shortest clear path from one point to another within the envelope.
- Analyze sensor data to detect errors in measurements, tool failures, power failures, hydraulic fluid pressures, air pressures in the pneumatic systems; to identify workpiece features; and to determine the positions of all parts of the mechanism.

Control System Functions. If the computer is the "brain" of the robotics system, then the control system is its "nervous system." For example, the control system takes the instructions for a change in the position of the hand (including the details of the path, the velocity to be achieved, and the acceleration or rate at which the velocity should change) and converts them into commands to the actuators, which power the various motions of the mechanism. Then it takes the outputs of the various sensors and converts them into a form that the computer can use to determine how well the instruction is being followed and when the instruction has been completed.

Often the configuration of the robot doesn't have the same motions as envisioned by the operator or the programmer. The program may have originally been written for an anthropomorphic arm and is to be used on a gantry mechanism. In that case, the control system will combine the output of several sensors to calculate the motion in the type of coordinate system being used.

Just as the human hand has a variety of types of sensors, so does the robotics system. Sensors may be either a contact (kinesthetic) type, which rely on touch, force, torque, position, or change of position for their input, or a noncontact type, which use optical, sound, proximity, heat, x-ray, or even gamma-ray input.

Robot Functions. Through its end-of-arm device ("hand"), the robot either picks up the workpiece and does something with it, or picks up a tool and performs some function with it. For example, robots can either transport workpieces between machines or hold the tool for some manufacturing process. Drilling, grinding, spray painting, and spot and arc welding are common examples of work often done by robots. Inspection of parts is also done by robots. Thicknesses and diameters of parts can easily be done by automated calipers attached to the robot hand. In fact, there are documented instances where one robot is programmed to pick up a part, often at very high temperatures, and hold it up in position for a second robot to measure it. If the part passes inspection, the part is placed on a conveyor belt by the first robot; if the part fails to meet the design specification, the first robot discards it.

Quality Assurance Through Automated Inspection

Quality control (QC) is concerned with ensuring that the products from a factory meet their design standards before leaving the manufacturing plant. In the past, QC has meant that the workpiece was inspected after it had been completed. If a particular workpiece failed to meet the design criteria, it was reworked (if possible) or scrapped. Scrapping wastes both material and the manufacturing investment and is to be avoided if possible.

Today's philosophy in quality assurance is to inspect to *prevent* defects from happening rather than inspecting to *find* defects. This preventive inspection is done continuously if possible or by periodic sampling and inspecting of parts along the manufacturing line and is called *statistical quality control (SQC).* Statistical quality control attempts to detect trends that would ultimately lead to the production of rejects if left uncorrected. SQC has several goals including the following:

- In-process gauging and inspection, rather than postprocess gauging if possible

- Automated inspection of 100% of the workpieces if at all possible
- Statistical analysis of the results of the inspections to detect trends and problems.

In-process gauging means, for example, that a part being turned on a lathe is measured as it is being machined, so that the part has been inspected even before the lathe is turned off and the part is removed. SQC is discussed in more detail in a later section.

Automated inspection is a computer-controlled activity that is capable of automatically measuring size, surface roughness, temperature, force, power requirements, vibration, tool wear, and the presence of defects in a part being manufactured. It uses a computer-driven *coordinate measuring machine* (CMM), which is essentially a device capable of moving a probe and accurately establishing the coordinates of selected points between the probe and the part. CMMs can be used either for in-process gauging during an operation or for postprocess gauging of a finished part. Figure 21–29 is a photograph of a coordinate measuring machine.

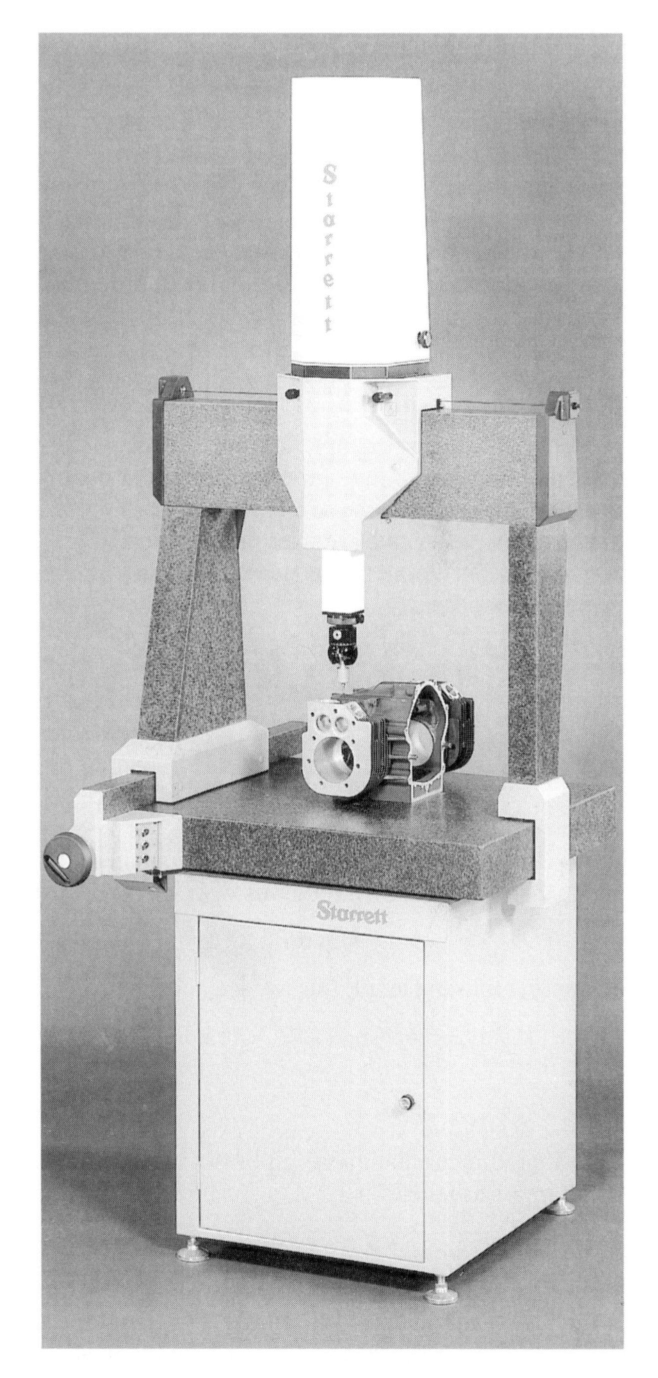

Figure 21-29. Coordinate measuring machine.

Causes of Product Variation

Suppose that a 5-millimetre hole, 15 millimetres deep, is drilled into a bracket using a drill press. The process produces 250 brackets per hour (about 4 per minute). How reliably can the depth of the holes be controlled?

First of all, there must always be some variation in the output of a machine/machinist combination. So there will be some variation in the depth of the holes due to *inherent variation,* and the variation will be random from one sample to the next.

On the other hand, some of the variation will be caused by changes in the properties of the raw materials, some due to changes in the machining environment (workpiece loose in the jig, tool wear, deterioration of the cutting fluid, etc.), and some due to operator error (perceptual limitations, inattentiveness, etc.). These are *assignable causes;* they need to be detected and corrected.

A manufacturing process is said to be "in control" when the assignable causes have been detected and corrected so that only the inherent variation is left. But if the holes will naturally vary in depth due to inherent variation, then how is it possible to detect variation due to assignable causes? SQC is the answer.

Statistical Quality Control

In the simplest possible approach, the depth of each hole would be measured, the result would be graphed as a function of time, and the graph would be analyzed to see if any trend was developing. But statistical analysis asserts that a sample of the proper size can give a reliable measure of how the trend is going, and that's all that is needed.

Consider another example. At some regular interval, take a *sample* of the output, measure the depths of the holes, calculate the average value for the sample, and plot the sample averages on the graph.

Assume that the measured depths of the samples came out as follows:

#1	#2	#3	#4	#5	#6
15.01	15.11	14.95	15.06	15.01	14.94
14.96	15.03	14.96	14.91	14.97	15.12
14.92	15.07	14.93	14.88	14.92	14.99
15.06	15.10	14.92	15.12	15.02	15.07
15.00	15.04	14.94	15.03	14.93	15.03

From Chapter 3, the average (or "mean") value for each sample is calculated as follows:

$$x = \Sigma x/n$$

Therefore, for sample #1, the average hole depth is

$$x = \frac{(15.01 + 14.96 + 14.92 + 15.06 + 15.00)}{5}$$

$$= 14.99$$

Repeating this calculation for all of the samples gives the averages as follows:

#1	#2	#3	#4	#5	#6
14.99	15.07	14.94	15.00	14.97	15.03

If the individual readings for sample #3 were plotted on a graph and compared to a plot of the readings in sample #4, it would immediately be obvious that there is another measure to be considered: the range of the readings. The *range* of a set of readings is simply $R = x_{max} - x_{min}$. It is important to calculate (and graph) the range of each sample as well as the average value of each sample. Therefore, the information that would be of value from these samples is:

Sample Averages:

#1	#2	#3	#4	#5	#6
14.99	15.07	14.94	15.00	14.97	15.03

Overall average of the sample averages: 15.00

Sample Ranges:

#1	#2	#3	#4	#5	#6
0.14	0.08	0.04	0.24	0.10	0.18

Average of the sample ranges: 0.13

These averages are needed so that upper and lower limits for the average readings and the ranges can be calculated (from sampling theory, based on the value of R and the number of pieces in each sample). Once that's done, the actual graphing can be done; the result is shown as the first six points in Figure 21–30(a). At the beginning of the chart, the readings were widely variable from one sample to the next, suggesting that the process was only barely under control. (Perhaps the machinist was new to this machine.) Then the amount of variation became smaller (near the middle of the chart), suggesting that the process was under good control. But the last third of the points began to show a trend that would soon exceed the upper limit. At this point someone would check with the machinist to see if the drill was getting dull or if the workpieces weren't fitting correctly in the jig, or whatever.

It is important to distinguish between (1) the process being "in control" and (2) the process producing parts that are close to the "ideal" size. Figure 21–30(b) shows a process whose range lies well within the expected limits, but it is producing holes that are too shallow. The fact that the readings all lie close to one horizontal line shows that the machine is accurately producing the result that it was instructed to do. Its instructions need to be changed slightly, that's all. Then the completed pieces should all be close to the new setting.

Figure 21–30(b) also shows that the separation between the upper and lower limits is related to the mean value of R. Since the mean value of R is smaller in Figure 21–30(b) than it is in Figure 21–30(a), then the upper and lower limits must be closer to the

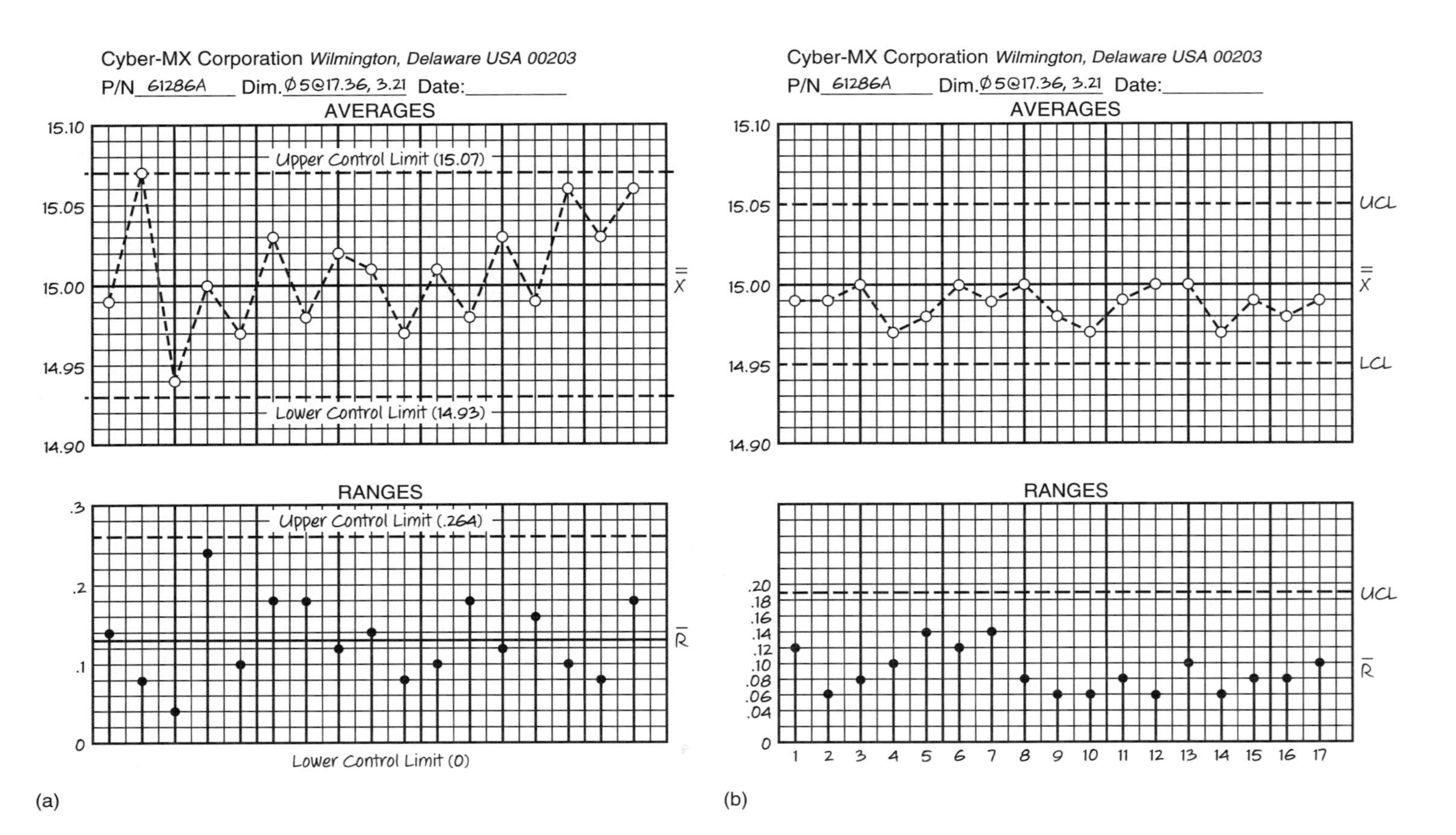

Figure 21-30. Control charts.

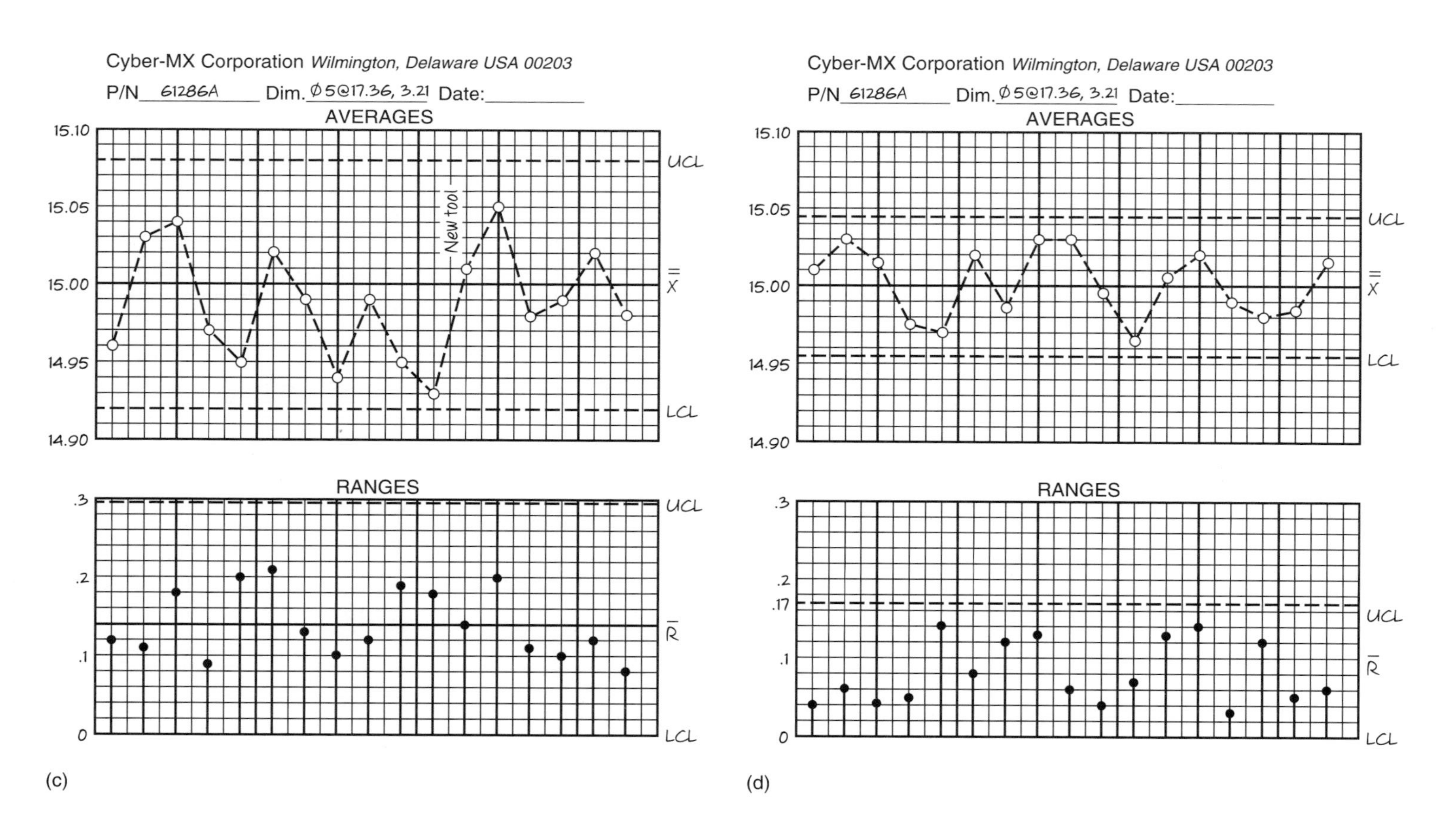

Figure 21–30. *continued*

mean-value line. But this should not be difficult to achieve if the range of values is small for each sample.

Figure 21–30(c) shows that a problem was detected and corrected, producing workpieces well under control. Note that at no time did the depths of the holes exceed the limits so that no pieces had to be scrapped. Figure 21–30(d) shows a process that is "well under statistical control"—always the goal.

The Role of Computers

Note that the preceding discussion related to the measurement and control of *one single feature!* Imagine how many people would be required to do this by hand for every measurement in every product in a large manufacturing plant. Therefore, this has become practical only now that computers are commonly available.

Meeting Drawing Tolerances

All that has been described thus far in this section relates to whether or not the process is "under control"; nothing has been said about meeting the tolerances on the drawing. Here is where the designer and the production control supervisor need to be communicating well. If the designer asks for tolerances that are smaller than can be achieved with this machine, then either the tolerances need to be "loosened" or else the process needs to be done on a different machine.

Just-in-Time Scheduling

Traditionally, large stocks of parts were either produced or purchased, then stored or "warehoused" for use on the assembly lines. This was done for a variety of reasons:

1. The supplier of the parts was able to produce the parts for an assembly line more economically in large batches and in a short time, so that several months' worth of parts were available at once. These parts had to be stockpiled.
2. The large store of parts served as a buffer in the event that the supply was unexpectedly interrupted.
3. If there was no assurance that all stored parts would meet specifications, then the rejection of some parts would not slow down the assembly line if extra parts were in the warehouse.

However, this approach wasted materials, machines, time, and human resources. Furthermore, it required a significant investment of money to maintain the inventory and the storage places for the parts. In the 1950s, the Toyota Motor Company developed what has come to be called the *just-in-time (JIT)* production scheduling.

In JIT scheduling, purchased supplies are delivered to the production area, daily or more often, *just in time* to be used. Parts are produced *just in time* to be included into subassemblies. Subassemblies are produced *just in time* to be included into finished products, and finished products are delivered *just in time* to be sold. The coming of computers and communications systems that can keep track of all of the parts all of the time has made JIT scheduling a workable reality.

One consequence of this approach is that the responsibility for quality control is shifted to the supplier. (The supplier might be an in-house subassembly line.) In the past, many corporations had quality assurance programs that tested incoming parts to make sure that they met specifications. With JIT, the delivered parts must be tested by the supplier since the parts will be used before there is time for incoming inspection. This means that the suppliers probably will have to set up some sort of automated inspection and SQC process to inspect 100% of the parts they are supplying.

In spite of the increased premium that JIT places on cooperation between the supplier and user, and on the need for the supplier to be extremely reliable, dependable, and adaptable, the popularity of JIT scheduling continues to grow in large industries.

The savings produced by JIT scheduling (by not having to build large storage facilities and maintain inventory records) should be obvious. But JIT has other, less apparent, advantages as well. JIT scheduling has been shown to:

- Reduce production costs by 20% to 40%.
- Reduce inventories by 60% to 80%.
- Reduce rejection rates by up to 90%.
- Reduce lead times by 90%.

- Reduce scrap, rework, and warranty costs by 50%.
- Increase direct labor productivity by 30% to 50%.

The obvious drawback to JIT is that the production line is shut down if even one part does not come in "just in time." This high-risk approach to manufacturing is not suitable for every product or for every company, but for those capable of controlling the uncertainty associated with JIT, it does offer considerable savings.

MASS PRODUCTION

Entire courses have been offered in both engineering and business schools in solving the problems unique to mass production. Sometimes the problems are very subtle and may not be anticipated. This section is not designed to delve deeply into mass production problems, but a few of the problems associated with mass production might give an insight into the complexities of that type of manufacturing.

When producing more than, say, 300 pieces per hour, mass production has its own set of problems and advantages. At 300 pieces per hour, a saving of 20 cents per piece amounts to a reduction in cost of $1 per minute. A savings of *one dollar* per piece is a cost reduction of $300 per hour, $12,000 per 40-hour work week, or roughly $600,000 per year. In this competitive world, this could mean the difference between the company making a profit or not. It makes economic sense to "count the pennies" when manufacturing any product. For instance, ordering the minimum size stock required to produce a part could reduce the cost in several ways. First of all, the smaller stock costs less when bought on a size or per-pound basis. Second, there is less waste. (The collection and disposal of waste material is expensive.) Next, less machining time is required to reduce the stock to the size of the final product. Consider the labor cost alone of removing the extra material. If it takes one minute to make one pass or cut on a lathe, and the cost of operating that machine (labor, power, overhead, etc.) is $100 per hour, then by eliminating that one pass, $1.66 has been saved. (Note that it may occasionally be necessary to order larger than minimum size stock in order to control the material properties, surface flaws, and other properties of the part to make it meet specifications. But even then, the smallest size raw materials consistent with the quality of the product should be ordered.)

In addition to ordering the proper size of stock, parts, and raw materials, two other ways to reduce costs have been developed: adaptive control and cellular organization of production. These are covered next.

Adaptive Control

Adaptive control is an exceptionally powerful application of the general concept of feedback control. What makes it unusual is that the system can respond to unexpected changes that may come from factors *outside* of the process itself. A conventional feedback control system can only compensate for changes within the process. Before examining adaptive control, a look at the concepts of conventional feedback control are in order.

The principles of feedback control become apparent when driving an automobile. Normally, an automobile will not travel more than a few hundred feet down the road before its direction and/or speed need some slight correction. The word *correction* implies that there is some "correct" path and speed for the vehicle. The assertion that some correction is needed comes from the driver's observing the vehicle's position with respect to the lane markers, another vehicle or other condition ahead, and the direction and speed of his vehicle, and then comparing those observations with a set of expectations as to the "correct" path of the automobile. With the information gained from the observations and the comparisons, the driver makes corrections to the course and speed of the car. This is an idealized example of the application of feedback control principles. Figure 21–31 shows the elements of a conventional feedback control system. Note that there is an "input," which is some statement of what action should be produced. It might be an instruction to a radar antenna directing it to point 5° to the left, or to a CNC drill press that it should extend the drill 15 millimetres. In the case of the automobile, the driver may send an instruction through the steering wheel to keep the vehicle 4 feet to the right of the lane markers. The feedback system now must have two other key elements: (1) some "monitor" of the "output" (or action produced) and (2) some way to compare the output

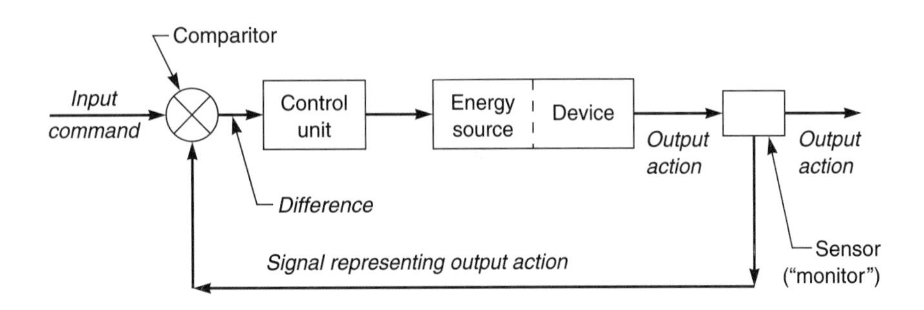

Figure 21-31. Conventional feedback control system.

with the input. If there is any difference between the input and the output, a command is given to the energy source, which will produce an action that will reduce that difference.

For example, suppose that the CNC drill press has just reached the correct position to drill a hole in the workpiece. The command is given to drill the hole. The command probably instructs the drive motor to turn on and the depth actuator to lower the drill until it has a depth position 15 millimetres lower than it now has. The first question is where is the drill (in terms of depth) at that moment? If the sensor that measures the depth of the drill indicates that the drill is in the "retracted" or fully up position, the motor on the depth lead screw is activated to lower the drill. As the drill moves, the depth sensor indicates when the drill is approaching 15 millimetres. At a depth of, say, 14 millimetres, the signal to the actuator is changed to slow it down. At a depth of 14.8 millimetres, the signal to the actuator stops the depth motion altogether. After verifying that the drill is not quite at the 15.00-millimetre depth, the signal to the actuator directs the drill to move very slowly so that it can be stopped at the correct position within 0.01 millimetre.

The system in this example is not very sophisticated, but it will do the job as long as nothing goes wrong. But what if the drill has become dull or broken; what if there is an undetected defect, such as a cavity, in the workpiece; or what if the workpiece has become mislocated in the jig? What if everything seemed to go well but, on inspection, the hole was not drilled to the correct depth? The advantages of adaptive control become clear at this point. Figure 21–32 shows an adaptive feedback control system. Assume that, in addition to all of the "normal" position sensors, the system has sensors that measure the downward force of the drill against the workpiece, the speed of the drill, the power it requires, the vibration of the workpiece, the sounds generated by the process, the temperature of the drill or workpiece, and other parameters that may be specified. A dedicated computer can then be programmed to make sense of all the data and either stop the process or override the automatic commands in order to produce an acceptable workpiece. As long as everything goes as planned, the automatic commands control the process. But as soon as something doesn't go as planned, the computer can adapt to the new circumstances by generating new control commands. In this way the production rate can be maintained, which is a critical consideration for high-production-rate jobs.

As a result of adaptive control, machining time can be reduced by as much as 50% since the system continuously optimizes feeds and speeds for the given cutting tool and material being cut. Other benefits of adaptive control include closer tolerances, reduced scrap rates, improved surface finish, and reduced dependence on operator skill.

Needless to say, all of this comes at a price. Adaptive control systems are expensive and complex and therefore more prone to malfunction. But they do make possible high production rates of fairly complex products, especially when used in conjunction with a manufacturing cell.

Cellular Organization

In response to the external demands for a greater variety of new products of consistently better quality and the internal demands for constantly improved productivity, production control staffs have more and more gone to computer-controlled equipment to meet

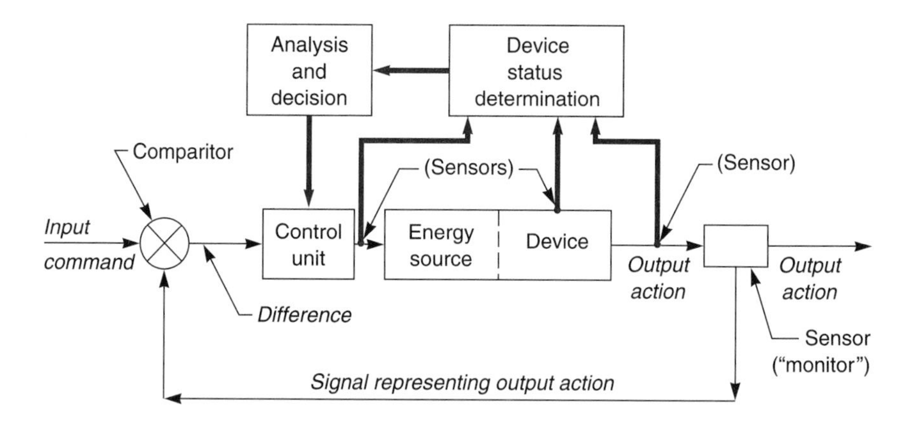

Figure 21-32. Adaptive feedback control system.

these demands. Moreover, a special effort has been made to reduce sharply the amount of transportation and waiting time for each workpiece.

In the traditional methods of manufacturing, the workpiece (such as a rough die casting) has been moved from one special-purpose machine (such as a lathe) to a second special-purpose machine (such as a milling machine), to a third (such as a drill press), to a fourth (such as a grinder), etc. At each machine, the workpiece needed to be mounted into the jig for that particular machine, adding repeated setup times to the other lost hours for that workpiece. *If* it were possible to design *one* machine that could do all of these functions, then there would be only one setup time involved and transportation time would be eliminated. But the catch is that such a machine would need to be able to change tools, rotate and translate the workpiece (that is, the jig) accurately, etc. In short, it would need to be a "smart" (that is, computer-controlled) machine.

And that's exactly what a *machining center* is. A machining center can be fed palletized workpieces by a robot, read the part number from the pallet, "look up" the instructions for machining that part, carry out those instructions, request that the robot remove the completed workpiece, and then start on the next part—if need be, all in the dark! There's no need for human intervention except for one shift a day. During that shift, completed workpieces are removed, new stock is loaded into the "queue" or feed line, tools are sharpened and replaced, and other maintenance is done. In these plants, the machines can work in the dark, with only one worker around to handle any unexpected problems.

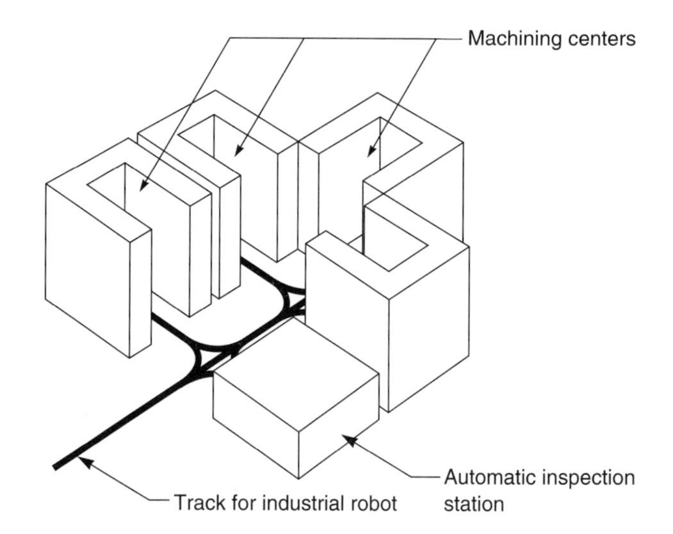

Figure 21-33. Machining cell.

The concept of a *machining cell* is that several machining centers (along with an automatic inspection station) are grouped around a pick-and-place robot, which keeps each of the machines busy most of the time. If the machines are CNC machines, then a variety of products can be made by the cell in a random order. The robot can detect when a product is completed and which product is next in line and advise the computer (which then selects the appropriate set of commands to continue production). Figure 21–33 shows a machining cell.

Flexible Manufacturing Systems

The greatly increased productivity and the versatility of machining cells have made them the basic building block of a larger scale manufacturing organization—the *flexible manufacturing system.* Its basic concept is the same as that of the machining cell except that the elements of the flexible manufacturing system are machining cells rather than machining centers.

If a fairly complex product is to be built, its design (and fabrication) would normally be organized into major assemblies, each of which would be composed of several subassemblies, and so on until the organization reaches the level of the individual parts (Figure 21–34). Then the organization of the machining cells into the flexible manufacturing system follows the reverse order of organization: One or more machining cells create the parts that will be assembled into subassemblies by the next higher level of cells, which in turn feed subassemblies to the next higher level of cells, etc., until the highest level of cells produce completed products.

That's about as far as one can go in organizing the manufacturing organization in order to reduce the cost penalty for product diversity for products of medium production volumes and rates. The next step would involve organizing all phases of the entire manufacturing facility. Is that possible? Perhaps.

Computer-Integrated Manufacturing

Up to this point, the discussion has been about how computers can be used to facilitate productivity in this or that phase of manufacturing or this or that subsystem. Is it possible to integrate *all* of these subsystems by the use of some supercomputer? And even if it is possible, would the system actually achieve its potential? Some specialists in the field believe that it is possible to create such a system. It's called *computer-integrated manufacturing* (CIM).

Note that CIM is a *methodology* and a *goal;* it is *not* a collection of equipment and software. Its method is

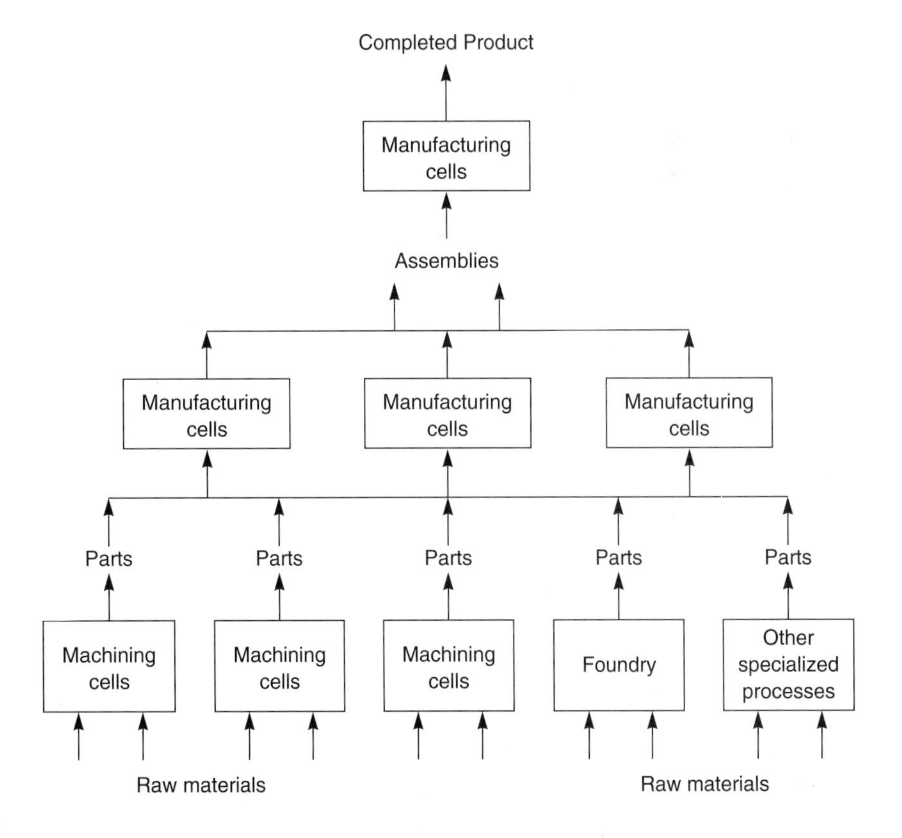

Figure 21–34. Product and machining cell organization.

to integrate all aspects of design, planning, manufacturing, distribution, and management so that it becomes possible to *transform product designs and materials into saleable goods at a minimum cost in the shortest possible time.* Quite a tall order!

In the past, manufacturing has been thought of as a sequential activity: first, do this; then, do that. However, modern manufacturing is not that one dimensional. In the first place, many different activities are going on simultaneously; worse yet, they probably are going on independent of each other, even though each one will generally affect the others. In the second place, making one change (in the allocation of time or resources, for example) will require many other adjustments to be made to accommodate that change (the "ripple effect"). Many of these consequent changes will not be anticipated, and several of them are likely to be disastrous! How is it possible to keep all of these activities constantly in touch with each other in real time? Answer: computer-integrated manufacturing.

Computer-integrated manufacturing would treat manufacturing processes and plant operations and their management as a single *system,* as an integration of the subsystems of:

- *Business planning and support.* Demand forecasting, cost analysis, purchasing, distribution and service analysis, sales and marketing, inventory management, etc.
- *Product design.* Computer-aided design and analysis, manufacturability analysis, prototype testing, etc.
- *Manufacturing process planning.* Selection of manufacturing processes consistent with the nature of the workpiece and the anticipated availability of machines, production scheduling, etc.
- *Process control.* Supervision of manufacturing and inspection, quality control, inspection of incoming parts, collection and disposal of scrap, etc.
- *Shop floor monitoring systems.* Monitors the status of manufacturing activities, sends status reports to central control, etc.
- *Process automation.* Design of jigs and fixtures, programming of robots, scheduling of raw materials delivery, scheduling of storage space and transportation, etc.

The keyword is *integrated.* With some effort, the behavior of each of the subsystems, and of the overall system that combines them can be predicted by a mathematical and/or logical model, which can be "run" much faster than real time. In fact, a computer can run the model so fast that the potential consequences of a proposed manufacturing decision can be anticipated in seconds. If the decision seems to be advantageous, the computer can make all of the scheduling (and other) changes that need to be made in order to implement the decision. If the decision would probably be a disaster, then it isn't made, no harm is done, and an alternative decision is made and evaluated.

Only a few companies have implemented CIM. Why not more? There are several reasons that become clearer if we compare converting to CIM with converting the United States to metric measurements, or converting all European countries to driving on the left side of the street:

- The number of detailed changes that needs to be made is staggering. CIM requires an entire new way of thinking, an entirely new philosophy of making business/manufacturing decisions. How many members of management would resist the new approach, either because of habit or because they honestly believe CIM to be a mistake?
- Most companies are organized into relatively autonomous departments. How many department managers would be willing to give up their control to a computer? And how many managers, having to manage the details of their department for years, have the broad vision to appreciate the benefits of CIM?
- Great care must be taken in the organization of the systems and subsystems to ensure that the failure of a single computer or sensor or program does not bring the entire operation crashing down. The difficulty of making a system fail-safe (or even free from internal contradictions) rapidly increases with each increase in the complexity of the system. It's not clear how to prevent a rare (but catastrophic!) system failure from destroying all of the benefits that a CIM approach would bring to the company.

So there generally is a "go-slow" reaction to the implementation of CIM. But market pressures and international competition may force companies to take the computer-integrated manufacturing approach to production in order to stay competitive and thereby stay in business.

PLANT LAYOUT

Although there are few hard-and-fast rules to plant layout, a few principles should be considered when designing a manufacturing facility. The key consideration is **flow**—flow of materials, utilities, people, and information. This flow must be efficient and must not conflict with any process involved in the manufacture of the product.

Flow of Material

Materials include all items of:

- Raw materials
- Parts between production operations, subassemblies, completed assemblies
- Fasteners
- Finishing materials, paint and varnishes, paint thinners, solvents, sand paper, steel wool, other expendable supplies
- Tools to be sharpened, tools which have been sharpened, tools for temporary use
- Air
- Fumes
- Trash, recovered scrap
- Food
- Safety equipment, protective clothing, medical supplies, emergency equipment, etc.

Flow of Utilities

The flow of utilities includes such items as:

- Electricity
- Fuel gases
- Oil
- Steam
- Water: chilled, cold, hot, sterile
- Pressurized air and other gases

Flow of People

People are an essential consideration in plant layout. Items that concern the flow of people include:

- Workers: equipment operators, supply movers, foremen, inspectors
- Next change of shift
- Safety personnel
- Custodians
- Shop stewards
- Maintenance/repair personnel
- Engineers
- Technicians
- Visitors
- Couriers
- Administrative personnel

Flow of Information

These items include:

- Instructions for fabrication
- Change orders
- State and local codes
- Machine operation instructions
- Repair orders
- Time cards
- Production schedules
- Workpiece status reports
- Machinery status updates
- etc.

In practical terms, providing for the flow of information and utilities is a fairly simple matter. Providing for the smooth flow of people and material requires more thought and planning.

Information

The primary consideration in planning for the flow of information is that it not be contaminated by "noise" from the environment. If two people are trying to talk in a room with a high background noise, the noise can interfere with the reception of the words. In the realm of computers, the "noise" could be false data, interference from electrical distortion provided by the machinery, or any number of items that interfere with the transmission of data from sensors to the computer or one computer to another. As production facilities are increasingly controlled by computers, the need for uncontaminated data becomes more urgent. Expressed another way, the external "world" that the computer "sees" is a world of data. If those data are false, then the computer bases its decisions on a false view of the world and might make decisions that would be inappropriate for the real world.

For this reason, data are transmitted between the sensors and the computer, and between the central computer and the machines, by video signals through fiber optic cables. No matter what transmission system is to be installed, components of the highest possible quality should be used, because the cost of contaminated data could be ruinous to a facility. Cables should be mounted in an overhead cable rack to provide for both easy access and flexibility.

Utilities

Flexibility is the key consideration for the flow of utilities. However, the problem of providing for smooth flow of utilities is such a common problem that it has long been solved satisfactorily. Most utility installations are either overhead or through channels in the floor, with lift-off covers for instant access.

People and Materials

To some degree, it is necessary to consider the flow of people and the flow of materials together, because both require a considerable volume or amount of room to accommodate. Neither can generally function well traveling up a steep slope, such as stairs. Both people and materials need access to the machining equipment and other facilities of manufacturing.

To understand how the two needs conflict, consider a distribution warehouse where the merchandise is temporarily stored in high bays, loaded and unloaded by a forklift device, and where the outgoing shipments are made up on (and delivered to the outgoing bays by) a waist-high conveyor system. That works well for the materials, but the conveyor creates a formidable barrier to the free flow of people in much the same way as a freeway divides a town.

Conflicting flow requirements are an old problem, therefore solutions abound. Pedestrians and vehicles use the city streets by time-sharing the intersections with traffic lights to control the flow. Where the resulting delays are not acceptable, passengers and vehicles are separated onto different levels with pedestrian overcrossings, underground traffic tunnels, or elevated roadways. Airport terminals are a good study in the separation of aircraft, land vehicles, baggage, and pedestrians. But access to the separate levels may require elevators, space-consuming ramps, or other such solutions that delay the flow of material or people.

People generally are easier to move vertically. The use of a freight elevator for materials fails when 20-foot lengths of pipe or steel beams need to be moved from one level to another. How much simpler it would be to have the material receiving area at the same level as the manufacturing floor and let the people generally inhabit the other floors.

If it simply is not practical to separate the material and the people into separate levels, then other solutions, while probably not as satisfactory, are possible. The model here is the "lunch counter" approach. In a lunch counter there are two populations (the employees and the customers) who need to keep out of each other's way at the same time that the opportunity to interact (order and be served) should be maximized within a limited floor space. If the lunch counter floor plan has "finger" counters alternating with access aisles for the customers, the counter acts as a barrier between the two populations and an efficient operation results. Much the same concept can be used to organize the production area with the equipment forming a "leaky barrier" between material and people.

Figure 21–35 shows a floor plan of one possible idealized layout of a medium-sized flexible manufacturing facility. Its basic organization is made up of ten U-shaped manufacturing cells, each composed of from 6 to 11 machining centers (the circles) along with automatic inspection stations (the rectangles), and all joined

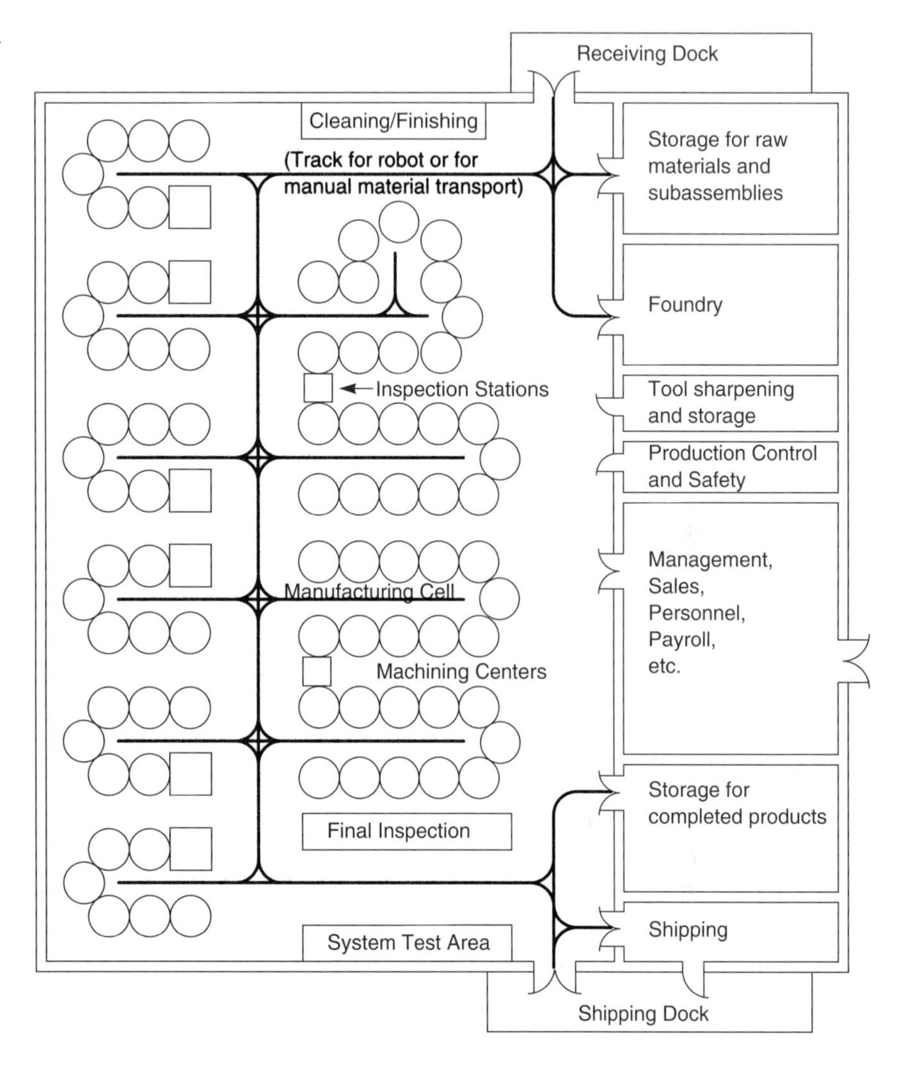

Figure 21–35. Flexible manufacturing facility using manufacturing cells composed of machining centers and inspection stations.

by tracks for materials handling robots. Figures 21–36 and 21–37 show the flow of materials, people, utilities, and information in a flexible manufacturing facility.

Note that the material flows into the top of the floor plan and down into the top quarter of the right side, fills the left half of the floor space and the bottom quarter of the right side, and flows out at the bottom. The "people" part of the floor plan is roughly the center half of the right side, where the people are bounded on three sides by the "materials" part. The people have easy access to the spaces between the manufacturing cells, crossing the robot tracks in only two or three places. Thus the manufacturing cells act roughly as the "finger" counters of the lunch counter model.

Note also that the tracks for the industrial robots contain a large number of switching points so that the delivery route can go between nearly any pair of points. Also, the tracks need not be as "cast in concrete" as one would think. Systems are available wherein the track layout can be changed easily as the manufacturing cells (and sequences of operations) are changed.

As long as the flow of each of the items listed here is taken into consideration when planning the layout of the plant, then it doesn't matter what kind of facility is being planned. It should turn out to be a smooth and productive operation, and probably a pleasant place to work as well. What more could one wish for!?

Figure 21-36. Flexible manufacturing facility showing flow of material and people.

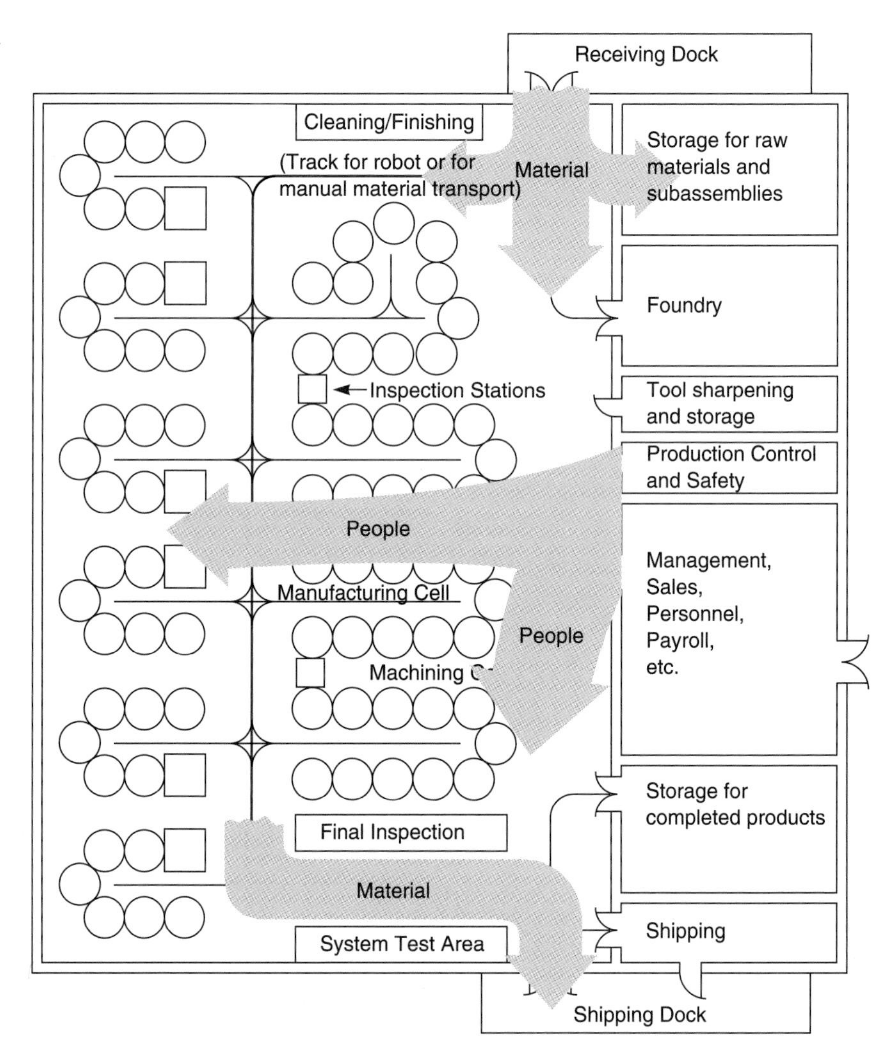

Figure 21-37. Flexible manufacturing facility showing flow of utilities and information.

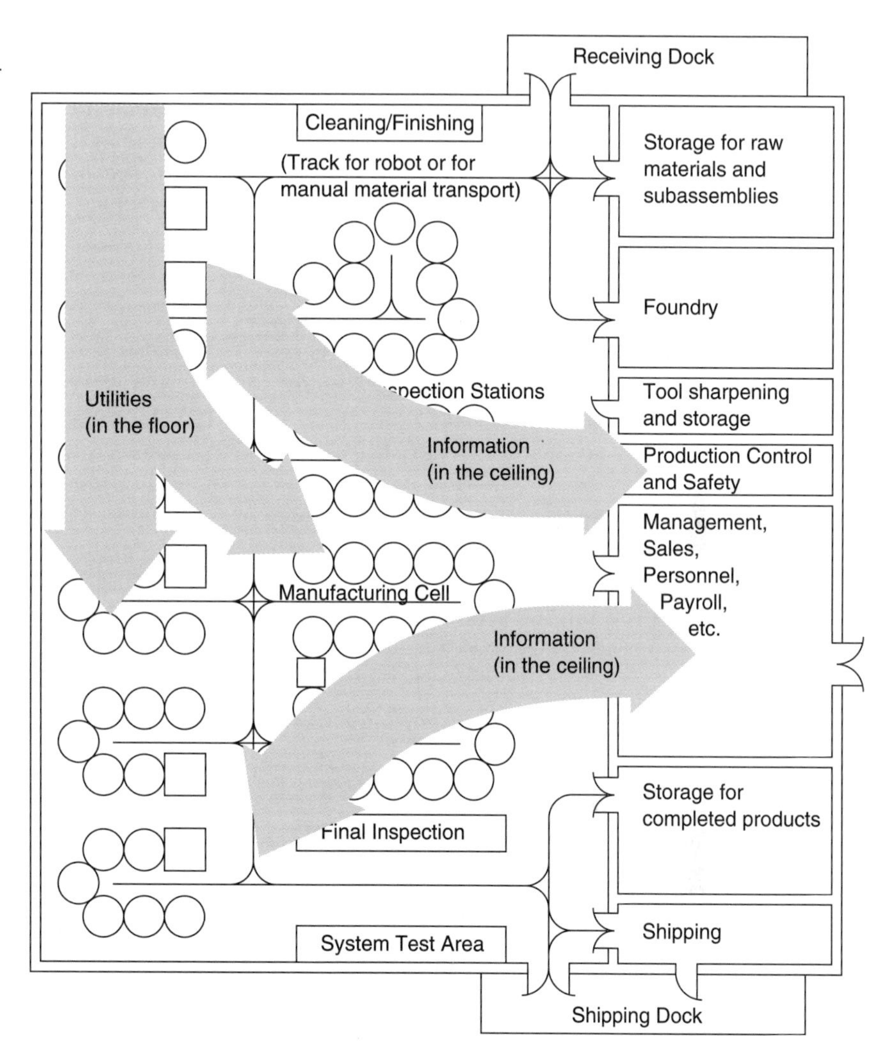

PROBLEM SET 21-1

1. Define *manufacturing facility* and distinguish it from a *processing plant.* Give an example of each.
2. Explain why the issue of productivity (or efficiency) is so central to production control.
3. Assume that you have decreased the net cost of manufacturing a mass-produced part by $0.40 per piece, that the production rate is 500 parts per hour for one 40-hour shift per week, and that there are 50 working weeks per year (allowing 2 weeks for vacation). How much does this reduced cost save your company per hour? Per day? Per week? Per year?
4. Four types of facilities were introduced in this chapter: piecework, small batch, large batch, and mass production. Each added specific concerns appropriate to the rate of production. List the concerns appropriate to each of the four facility types.
5. Distinguish among the terms *jig, fixture,* and *workholder.*
6. Dimensions can serve one or more of three functions. Name each function and create a dimensioned drawing that shows an example of each one.
7. Explain how a dial indicator would be used to determine total indicated runout.
8. In the discussion on measuring linear distances using a tape or a scale, there is a note that "careful calibration and reading at both ends of a scale are essential to gain the full precision possible with these tools." Explain why this is so.
9. Explain how gauge blocks and a sine bar would be used to create an accurate angle of 60°.
10. Describe the advantages and limitations of simple numerical control (NC) of process equipment.
11. Describe the functions that a CNC computer can perform and an NC computer cannot. (That is, in what ways is a CNC setup an improvement over an NC setup?)
12. Some years ago, Cincinnati Milacron did a study of exactly what happened to a workpiece during the time that it was in the shop. (Never mind the storage time!) The results of that study were mentioned in this chapter. What fraction of the total time in the shop was devoted to actual machining of the workpiece? How was the rest of the time "spent"?
13. The text discusses the transition from CAD to CAE to CAM. Explain what each of these terms means.
14. Distinguish between data flow and control flow; give an example of each.
15. Describe three circumstances in which it makes sense to use a robot instead of a human worker. Then describe three circumstances in which it makes more sense to use a human worker than a robot.
16. Several classes of manufacturing processes are listed in which robots are used today. List five of them. For each of the five, explain how the robot would be used.
17. Draw a diagram to show the components of a robotics system. Explain the functions of each component.
18. The text names six possible motions of an anthropomorphic robot: two at the shoulder, one at the elbow, and three at the wrist. Name each of the six motions and give an example of how it would be used.
19. Several different sources of "muscle power" (for robots) were described in this chapter. Name three and explain the advantages of each.
20. Explain what is meant by just-in-time scheduling, and describe both the advantages and the disadvantages of this system.
21. Draw a diagram of a conventional feedback control system and explain how it operates. Use a specific example to illustrate your explanation.
22. Explain how an adaptive control system differs from a conventional feedback control system.
23. Plant layout should be based on the flow of four classes of items. Name each of the four and give several examples of items for each class.

Manufacturing processes are only a part of the "manufacturing" facility. It is the manufacturing activity that distinguishes a manufacturing facility from other types of facilities.

Productivity is the key consideration when judging the contribution of an individual worker to the company, or of the company to the gross national product. Productivity can be optimized at two levels: the productivity of each item or product and the overall plant productivity. The production rate is probably the most significant distinction among different types of manufacturing facilities, because it determines the level of complexity or sophistication of the plant's operation.

The major considerations of a piecework facility are inspection and jigs and fixtures. Inspection requires an understanding of dimensioning practices and coordinate systems. Sometimes the coordinate systems are defined by the jig rather than by any features of the workpiece. Inspection involves determining linear distances, circular dimensions, and angles.

The major considerations of a small-batch facility include those of the piecework facility, plus numerical control and computerized numerical control of process equipment and attention to the problems of storage and transportation.

Given that up to 95% of the time that a workpiece is in the shop is taken up with moving and waiting, major improvements need to start with improved organization of work.

The major considerations at a large batch (or continuous) process facility are those of the small batch facility plus just-in-time delivery of materials, parts or assemblies; computer-automated manufacturing; robotics; and quality assurance through automated inspection and statistical analysis. Clearly, large batch processing requires a major increase in sophistication over lower production rate facilities. Computers of all sizes and functions are located throughout large batch facilities. Most of the computers are linked to a central production–control computer. Some of them control robots, which are mechanized in a variety of ways to serve a variety of functions.

Automated inspection attempts to anticipate potential defects and prevent them from affecting production through in-process gauging and through the use of process charts to detect trends.

Mass-production facilities add the concerns of the cellular organization of process equipment and its control by an adaptive control approach. Adaptive control goes beyond simple feedback control by compensating for factors that can be external to the system. It does this by the use of a variety of sensors that feed data into a computer model of the manufacturing process.

Plant layout focuses on the flow of materials, utilities, people, and information.

Appendix

A

Glossary

A word about definitions. A definition should be a complete sentence that includes everything pertaining to the term and excludes everything that does *not* apply to the term. An automobile should not be defined merely as a "vehicle." Vehicles would also include buses, bicycles, airplanes, motorscooters, etc. Complete sentences require a subject, verb, and predicate. Many dictionaries leave out the verbs to save space, but it is poor practice. Furthermore, definitions should not use the term or its derivatives in the definition. An engineer should not be defined as a person who does engineering functions. What have you learned from such a definition? Use synonyms that are well known in your definitions. Be specific in definitions. Vague definitions are a waste of time.

In the following definitions the authors have attempted to comply with these rules. Once you understand the meanings of the terms, you can probably improve on the wording of these definitions. The purpose here is to convey the idea or concept of the terms.

Note: Capitalized terms in **BOLD TYPE** are related glossary entries.

A-stage is the point in the curing of a **POLYMER** at which it is an oily liquid.

Abrasive Saw is a machine that cuts a metal **WORKPIECE** by **GRINDING** through it with a thin blade made of corundum or other similar material.

Absorptance is the fraction of the electromagnetic radiation falling on a body that is retained by the body. Also called the *coefficient of absorption.* It is the reciprocal of reflection.

Accuracy is the property of a **MEASUREMENT** that specifies how close it is to the true or actual size.

Actuator is a device that converts energy (from a power source) to mechanical motion. Examples are an animal muscle or a hydraulic piston on a **ROBOT.**

Adaptive Feedback Control is a type of **FEEDBACK CONTROL** that senses changes outside of the **FEEDBACK** loop and changes the instructions (to the **ACTUATORS**) so that the desired result is still accomplished in spite of the exterior change.

Addition-Type High-Polymers are those in which a single compound joins to like molecules to form a chain of **MERS.**

Adhesive Bond is a mechanical bond between two pieces of material.

Adhesives are glues or bonding agents that do not react chemically or thermally with the pieces they hold together.

Advanced Composite is a **COMPOSITE** that has long, continuous reinforcing **FIBERS.**

Age Hardening is the natural process of hardening of a metal either because the atoms of an impurity or the atoms of a second material become evenly distributed throughout the metal. These impurities migrate from the **GRAIN** boundaries and pin the slip planes in the metal, thereby hardening it. Also known as precipitation hardening or *crystallizing.*

Alkyd Paint is a coating that has urethane or melamine as a polymerizing agent.

Alloy is the combining of two or more metals to produce an **INTERMETALLIC** compound having characteristics that are different than those of the parent metals. Alpha brass and steel are examples of alloys.

Alpha Iron is pure iron for all practical purposes. It can have no more than 0.025% carbon at 1333°F. Also known as **FERRITE.**

Analog is a related property or function that can be substituted for another to produce the desired results. For instance, the mechanical motion of a pendulum can be used to measure time.

Anion is a negatively charged atom, or radical.

Annealing is the softening of a metal followed by very slow cooling. For steels, the metal must be heated into the austenitic range and cooled very slowly.

Anode is the pole in an electric cell to which the **ANIONS** are attracted. In a cell of a battery, the anode is the negative pole. In a plating cell, the anode is the positive pole.

Anodizing is an electrical **CONVERSION** process in which the **WORKPIECE** is attached to the **ANODE** of a plating cell and the **ANION** of the plating solution attaches to the surface metal of the part to make it resistant to wear and further corrosion.

Anthropomorphic Robot is a **ROBOT** whose mechanism has been designed to have the form and motions of the human **SHOULDER, ELBOW,** and **WRIST.**

Arm Sweep is left-right side-to-side motion (as from the **SHOULDER** with the arm held rigid). Used to describe one motion of an **ANTHROPOMORPHIC ROBOT.**

Assignable Cause is an act or activity that can be identified as the reason why a **WORKPIECE** deviates from the ideal dimensions.

Atom is the smallest part of a sample of an element. Atoms are made of electrons, neutrons, and protons, which can also be broken into other subatomic particles.

Atomic Number of an element is the number of **PROTONS** in the nucleus of the **ATOM.**

Atomic Weight of an element is the number of **PROTONS** plus the number of **NEUTRONS** in the nucleus of the atom.

Austempering is a heat treatment of steel in which the steel is quenched to just above the **MARTENSITE** start temperature and held there for several hours before lowering the temperature to room conditions. (See also **QUENCHING.**)

Austenite is the face-centered-cubic structure of steel. It is nonmagnetic.

Austenitic Stainless Steel is a high-chrome stainless steel that has a face-centered-cubic structure. It is nonmagnetic.

Autoclave is a high-temperature, high-pressure oven.

Average is a measure of central tendency. Also called the *arithmetic mean.* Its value is found by dividing the sum of all scores by the number of scores.

B-stage is the intermediate stage in the curing of a **POLYMER** at which it is viscous and sticky.

Ballizing is the finishing of a hole in a **WORKPIECE** by forcing a hardened steel ball through the hole. Ballizing makes the hole the exact size needed, work hardens the surface of the hole, and smooths the hole.

Band Saw is a power saw that uses a continuous ribbon-shaped blade.

Barcol Impressor is a **HARDNESS** tester used primarily on plastics.

Barrel Finish is a tumbling process that polishes or removes **BURRS** from a part.

Basic Size is the dimension from which the allowances are added or subtracted to produce the **LIMITS** on the design dimension of a part.

Batch Production is the complete processing of one group of **WORKPIECES** before starting on another group (as opposed to doing one operation on one piece at a time).

Beading is the forming of a rolled edge on a piece of sheet metal.

Bearizing is the finishing of a hole by rotating roller bearings against the inside of the hole.

Bed is the long smooth "bar" on a **LATHE** or other machine that carries the **HEADSTOCK, TOOL POST,** and **TAILSTOCK.**

Bessemer Converter is a method of converting iron into steel by blowing air from the bottom of the crucible through the molten iron to remove the excess carbon from the iron.

Bias is the angled direction in a fabric that is halfway between the **WARP** and the **WEFT.**

Billet is the bar or **WORKPIECE** undergoing a **FORMING** process. Also, in steel production, a billet is any large thick slab of steel.

Bioengineering is another name for **ERGONOMICS.**

Biotechnology See **ERGONOMICS.**

Blanchard Grinding is a method of **GRINDING** flat surfaces by moving the grinding wheels over the surface of the **WORKPIECE.**

Blister Copper is 99+ percent pure copper obtained after the **MATTE COPPER** is processed through a **CONVERTER.**

Boiling Point is that temperature at which a **LIQUID** will turn to a gas. Technically, the boiling point is that temperature at which the vapor pressure of the liquid, created by the molecules of the liquid escaping from the surface of the liquid, equals the pressure of the surrounding atmosphere. The boiling points of liquids are generally given referenced to one atmosphere (760 mm of mercury or 14.7 psi) pressure.

Bolt is a cylindrical, externally **THREADED** fastener that must be attached by rotating it into an internally threaded part (such as a nut).

Boring is the cutting of a hole by rotating the **WORKPIECE** over the cutting tool.

Box Annealing is a process of softening steel by placing it in an oven, heating the steel into the austenitic temperature range (see **AUSTENITE**), and then allowing it to cool in the oven after the oven has been turned off.

Brake is a machine or device used to bend sheet and light-gauge metal.

Brass is an **ALLOY** (or mixture) primarily of copper and zinc. Some brasses have minor amounts of other elements.

Brazing is an **ADHESIVE** process of joining two pieces of metal by melting a third metal through and around the joint. Usually the brazing metal is a bronze or brass.

Breaking Strength is the **STRESS** at the rupture point of a material.

Bright Annealing is a method of softening a steel by heating it into the austenitic temperature range (see **AUSTENITE**) and cooling in an inert gas atmosphere. Bright annealing does not corrode the surface of the steel.

Brinell Hardness is a measure of the surface resistance to plastic deformation by impression of a steel ball under a specified load.

Briquetting is the process by which powdered material is compressed into the desired shape prior to the **SINTERING** process.

Brittle Material is a material that breaks at its **TENSILE STRENGTH.**

Broaching is the cutting of an irregularly shaped hole by the action of a toothed file-like cutting tool made in the desired shape of the hole.

Brush Plating is an **ELECTROPLATING** process in which the plating solution is painted on the surface of a metal by means of a brush connected to the positive pole of a supply of direct current.

BTU or British Thermal Unit is the heat required to raise one pound of water one degree Fahrenheit.

Buffing is a process used to produce a high luster by smearing or moving the metal from high points into low points on the surface of a workpiece.

Burnishing is a **FINISHING** process in which the surface is scraped to align surface fibers or scratch marks to produce a smooth but dull finish.

Burr is the jagged, partially cut projection left by a cutting tool.

Butane is an organic compound having 4 carbon and 10 hydrogen atoms.

C-stage is the final cured stage of polymerization of a plastic or **COMPOSITE.**

CAD See **COMPUTER-AIDED DRAFTING.**

CAE See **COMPUTER-AIDED ENGINEERING.**

Calorie is the heat required to raise one gram of water one degree Celsius. More specifically, it is the heat required to raise one gram of water from 14° to 15° Celsius.

CAM See **COMPUTER-AUTOMATED MANUFACTURING.**

Carbide is iron carbide (Fe_3C); also known as **CEMENTITE** in the iron-carbon **PHASE DIAGRAM.**

Carbon Arc Welding uses a carbon rod to strike an arc with the **WORKPIECE** and the filler rod is melted in the arc.

Carbonitriding is a case-hardening technique in which the steel is immersed in sodium cyanide (NaCN). The carbon and nitrogen diffuse into the outer layers of the steel. The outer layers can then be hardened by **QUENCHING.**

Carbonizing is a case-hardening technique in which the steel is buried in graphite or other carbon-containing compound. The carbon diffuses into the outer layers of the steel, increasing the carbon content, thereby allowing it to be hardened by **QUENCHING.**

Cartesian Coordinates make up a system in which a point can be located in space by specifying the distance from a point (origin) in three mutually perpendicular directions (*x,y,z* directions).

Case Hardening is any of several methods of hardening the surface of a steel. A low-carbon steel is immersed in a **COMPOUND** containing carbon, nitrogen, or hydrogen, which diffuses into the outer layers of the steel. The steel surface can then be hardened by **QUENCHING.**

Casting is the process of pouring or forcing a liquid into a mold.

Cathode is the negatively charged electrode in a plating cell but the positively charged electrode of a battery cell.

Cation is a positively charged **ION.**

Cellular Organization is a way of organizing a **MANUFACTURING** facility so that most of the processes are carried out by a small group of machines (the "cell").

Cementite is iron **CARBIDE** (Fe_3C).

Center Hole Drill is a tool designed to drill a hole in the **STOCK** that will fit around the **DEAD-CENTER** or **LIVE-CENTER** in the **TAILSTOCK** of a **LATHE.**

Centrifugal Casting is a casting process in which only the **MOLD** is rotated about an axis. The centrifugal force of the spinning mold forces the liquid metal into the mold.

Centrifuged Casting is a casting process in which the molten metal and the mold are spun about an axis thus allowing gravity to force the metal into the mold. It is often used as a part of the **LOST-WAX** casting process.

Ceramic Mold Casting is an **INVESTMENT** casting similar to **PLASTER MOLD CASTING** in which the **PATTERN** is coated with a ceramic or refractory slurry. The slurry is then fired to make the **MOLD.**

Ceramics are anything made from aluminum oxide (Al_2O_3).

Ceramic Tools are tools that have a ceramic (aluminum oxide) cutting tip. Aluminum oxide has the same **HARDNESS** as corundum. Ceramic tools can cut almost any metal and withstand high heats, but they are very brittle and will not take shock.

Chain Saw is a power cutting tool made up of teeth on a chain, which is driven around an elongated bar. Chain saws are used mainly in the lumber industry.

Change of Condition is any treatment of a metal that changes its internal structure or changes the physical properties of the material.

Change of Form is any process that changes the shape of a material.

Charpy is an **IMPACT** test that uses a notched bar, held horizontally in the test machine and struck in the middle.

Chasing Dial is the part on a **LATHE** that rotates past an indexing mark to allow the **TOOL POST HOLDER** to be engaged at exactly the same point for each pass in relation to the rotating **WORKPIECE.**

Chatter is the uneven cutting of a tool caused by the blade skipping on the material. This skipping causes internal vibrations resulting in a loud squeal or noise. It should be avoided because it causes uneven cutting.

Chemical Machining is any method of removing material by chemical attack.

Chemical Plating is the deposition of a metal onto the surface of a more chemically active metal by the more active metal replacing the lesser active metal in solution. Copper can be easily chemically plated onto iron by this process.

Chuck is the rotating part on a machine tool that holds the **STOCK** or cutting tool. Chucks are adjustable to hold different sizes of stock or cutting tools.

Circular Saws are saws in which the cutting teeth are set in a flat disk.

Closed Forging is the use of **DIES** in the forging process to produce a desired shape or pattern in the **BILLET.**

CNC Computer numerical control; see **NUMERICAL CONTROL.**

Cogging is a forging process that reduces the thickness of a single **BILLET** by small increments.

Cohesive Bond is the **FUSION** of two pieces of like material together into one continuous piece. No interfaces are left after a cohesive bond occurs. All true welds are cohesive bonds.

Coining is a **FORGING** process that uses **DIES** to produce a low bas relief pattern on a **BILLET.** It is a process by which coins are produced.

Cold Working is the breaking of **GRAINS** by mechanical action. The result is that the metal is both hardened and embrittled.

Collet is the rotating part on a **LATHE** or other machine tool designed to hold one size of **STOCK** or cutting tool.

Composite is the judicious combination of two materials to produce a synergistic effect.

Compound is the product of joining two or more chemical elements by electronic action into a new material. The elements in compounds cannot be separated by mechanical means.

Compression is the application of **FORCES** on an object directly toward each other.

Computer-Aided Drafting is a method of drafting that uses a computer to draw on the screen as instructed by an operator. The computer acts only as an "electronic drafting board"; it does not do anything on its own.

Computer-Aided Engineering is a system that uses a computer to make design drawings (as in **COMPUTER-AIDED DRAFTING**) but which additionally is capable of doing engineering analysis of the designed object.

Computer-Automated Manufacturing is an approach to organizing a **MANUFACTURING** plant so that all possible analyses, decisions, and actions are assigned to a master computer.

Condensation-Type High-Polymers are those in which the **MER** is made from two different molecules which react to produce water or alcohols.

Continuous-Pour Casting is a method of casting used in the steel industry in which the liquid steel is poured through a cooled **MOLD** allowing the hot but solid ingot to come out the bottom of the mold. The size of the ingot is limited only by the amount of metal in the crucible.

Controller is that part of a **ROBOTICS SYSTEM** that acts as the "interface" between the incoming instructions and the mechanical part of the **ROBOT.**

Conversion Processes are methods of changing the surface of a metal to a chemical compound that will resist corrosion and adhere to the surface.

Converter is any method of converting iron into steel. The common converters are the **OPEN HEARTH, BESSEMER, ELECTRIC ARC,** and **OXYGEN LANCE.** Converters are also used to remove the impurities from **MATTE COPPER.**

Coordinate is a general term referring to some measure of the status of an object or system. It could indicate linear position, rotation, temperature, etc.

Coordinate System is a set of independent **COORDINATES,** which together adequately describe the status of an object or system.

Cope is the upper part of a casting **MOLD.**

Core is a preformed part of a pattern that is placed in a **MOLD** to make internal cavities in a part.

Core Print is an indentation in a **MOLD** used to hold a **CORE.**

Cotter Pin is a type of pin formed by folding a piece of half-round rod around a small mandrel and folding the rest against itself to form a split pin with a head. The head keeps the pin from falling out one way and the free ends are separated to keep the pin from sliding out the other way.

Covalent Bond is a bond between atoms caused by the sharing of **ELECTRONS** between the atoms.

Creep is the permanent elongation or deformation of a part under **STRESS** over a long period of time.

Crimping is a method of joining materials by bending or folding one part over the other.

Critical Fatigue Point is the **STRAIN** at which cyclic bending of a material will start to cause the material to fail. Below this strain, cyclic bending has only a small effect on the failure of the material.

Cross Feed is the direction of movement (of the cutting tool on a **LATHE**) that is perpendicular to the longitudinal axis (rotation) of the lathe.

Cryolite is sodium aluminum fluoride (Na_3AlF_6), which is used to dissolve bauxite, or aluminum ore, to allow the aluminum to be plated out electrochemically.

Crystal is an orderly array of particles.

Crystal Imperfection is anything that upsets the orderly array of particles in a crystal.

Crystal System is any of seven different possible ways that particles can join in an orderly array.

Cubit is an ancient measurement defining the distance from one's elbow to the tips of the fingers.

Cure Time is the length of time required for a glue, paint, or resin to attain its design strength or hardness.

Cutting Speed is the velocity by which a cutting tool passes the surface it is cutting. The surface can be moving past the tool, or the tool past the surface.

Cyaniding See **CARBONITRIDING.**

Cylindrical Coordinates is the system of locating points in space by specifying one linear distance, one perpendicular distance, and one angle from a starting point (origin) (x,y,θ).

Dead-Center is the tool that holds the end of a **WORKPIECE** in a **LATHE.** It is opposite the chuck. The dead-center does not rotate and the stock must rotate over it.

Deburring is any process that removes the sharp projections (**BURRS**) left by cutting tools.

Defect is anything that alters or upsets the orderly arrangement of a material. Also called an **IMPERFECTION.**

Density is the mass per unit volume of a material. Density always has units associated with it. The density of water is 1 gram per cubic centimeter in the **SI** or **METRIC SYSTEM** or 62.4 pounds per cubic foot in the English system.

Depth of Cut is the width of a ribbon-chip (measured perpendicular to the **FEED** direction).

Destructive Test is a test of a material or part that damages it to the point that it cannot be used for the purpose it was intended.

Detonation Spraying is the **MATERIAL ADDITION** process whereby a powdered material is fed into an oxy-acetylene or oxy-hydrogen flame then "blasted" onto the surface of a workpiece by an electric arc.

Diamond Pyramid Hardness is the measure of surface plasticity as determined by the **VICKERS HARDNESS TEST.** It is the result of impressing a diamond-shaped pyramid point into the material and measuring the diagonal of the resulting indentation.

Die is a tool with a cavity into which or through which a **BILLET** is forced.

Die (thread cutting) is a **TOOL** for cutting external threads on a shaft.

Die Casting is a casting process that uses a metal permanent mold and in which the liquid metal is forced into the **MOLD** under high pressure.

Dielectric is a material that does not conduct electricity.

Die-Sinking EDM See **RAM EDM.**

Diffusion is the migration of atomic or molecular size particles throughout a material.

Dimensional Analysis is the use of the units or dimensions of a quantity to make sure that these units are the same on both sides of a mathematical equation.

Dimensional Tolerance is the total amount by which a specific dimension is allowed to vary. It is the difference between the largest acceptable size and the smallest acceptable size.

Dimpling refers to the cylindrical **FLANGES** placed in sheet metal parts.

Dipping is any method of **MATERIAL ADDITION** whereby a solid is immersed into a liquid. The liquid coats and solidifies on the solid, thereby increasing its size.

Double Shear is the cutting of a piece in two places at the same time by slightly offset opposing **FORCES.**

Draft Angle is the angle of taper on a **MOLD** used to allow the casting to be removed from the mold.

Drag is the bottom part of a **MOLD** used in casting.

Drawing (forming process) is the pulling of a bar through a **DIE** to reduce the cross section. It is a process by which wire can be made.

Drawing (thermal process) is the process of heating a metal and slowly cooling it to slightly reduce the hardness of the metal.

Drilling is the machine cutting of a hole by a rotating bit or solid tool.

Dry Sand Casting is a casting technique that uses sand **MULLED** with oil or anything but water. Dry sand castings are often baked to give them more strength and rigidity in the **MOLD.**

Ductile Material is a material that will endure considerable **PLASTIC DEFORMATION (PLASTICITY)** before rupturing.

Dye Penetrants are very low **VISCOSITY** liquids that will seep into the surface flaws of a part. The flaws can then be easily seen with the naked eye or with low-power magnification.

ECG See **ELECTROCHEMICAL GRINDING.**

Edge Angle is the angle between the pair of faces that forms an edge of a cutting tool.

EDG See **ELECTRICAL DISCHARGE GRINDING.**

EDM See **ELECTRIC DISCHARGE MACHINING.**

Elastic Limit is the exact point on a **STRESS-STRAIN CURVE** where the curve ceases to be a straight line.

Elastic Region is the area under the straight-line portion of the **STRESS-STRAIN CURVE.**

Elastomer is a stretchable **POLYMER.**

Elbow Extension is the motion of a human **ELBOW** (or the elbow of an **ANTHROPOMORPHIC ROBOT**) that is possible with the shoulder and wrist joints locked.

Elbow (Robot) is that part of an **ANTHROPOMORPHIC ROBOT** that mimics the motions of the human elbow.

Electric-Arc Converter is a method of converting iron into steel by burning out the excess carbon by means of an electric arc from huge carbon electrodes being brought near the surface of the molten iron.

Electric-Arc Spraying is a material addition process that melts a metal rod in an electric arc and then sprays the atomized metal onto the surface of a **WORKPIECE.**

Electrical Discharge Grinding (EDG) is a method of grinding in which the grinding wheel has been replaced by an electrical discharge wheel to remove the metal.

Electric Discharge Machining is a method of **MATERIAL REMOVAL** by use of an electric arc.

Electrical Resistance is the obstruction of the flow of **ELECTRONS** through a metal. Electrical resistance is measured in **ohms.**

Electrochemical Grinding (ECG) is the grinding of a metal under a fluid through which an electric current is passed.

Electrochemical Machining is a **MATERIAL REMOVAL** process that is the opposite of **ELECTROPLATING.** The part to be machined is placed on the anode or positive pole of a plating cell.

Electroforming is a method of **MATERIAL ADDITION** in which metal is electroplated onto a metal-coated form and the form later removed after sufficient thickness is plated.

Electrolyte is any solution that will conduct electricity. Electrolytes for plating cells must contain the metallic ion that is to be deposited.

Electrolytic Tough Pitch (ETP) Copper is copper that is nearly pure and is produced from the electroplating of **BLISTER COPPER.**

Electromotive Series is the list of the chemical elements in order of their chemical activity, or ease with which they react with acids and water. Lithium, sodium, and potassium are the most active elements; copper, silver, platinum, and gold are among the least active. Hydrogen is considered the neutral element in the series.

Electron (imperfection) is an extra electron (particle) in the crystal lattice.

Electron (particle) is a negatively charged, particle in the atom. It is outside the nucleus and has a mass about 1/1836 that of a **PROTON.**

Electron Beam Machining is a form of **MATERIAL REMOVAL** by cutting a metal with a beam of electrons (in a vacuum).

Electron-Beam Welding uses a beam of electrons to melt the surfaces of a metal together. Because the weld is done in a vacuum, it is one of the cleanest welding techniques available.

Electroplating is the depositing of a metal on another negatively charged metal by means of a direct electric current.

Element is a fundamental type of material from which everything in the universe is made. There are 92 naturally occurring elements and several more artificially made elements.

Enamels are **VARNISHES** with a pigment added.

End Milling is the removal of material by use of the end of a rotating cutting tool.

Energy is the ability to do **WORK.**

Engineering Design Strength is the maximum stress for which a structure is meant to survive. Often a **SAFETY FACTOR** is applied that sets the engineering design strength at much more than the maximum stress the structure will ever have to take.

Engineering Psychology See **ERGONOMICS.**

Engineering Stress is the load applied to a test specimen divided by the original cross-sectional area of the specimen.

English System is the system of measurements that uses inches, feet, yards, miles, pounds, ounces, pints, quarts, gallons, and other nondecimally related units. Sometimes called the inch system.

Epoxy Paint is a coating that uses epoxy ($-C_2O$) as a polymerizing agent.

Ergonomics is the science of designing equipment to fit the user.

Error is the inaccuracy in any **MEASUREMENT.**

Ethane is a two-carbon organic compound having six hydrogen atoms.

Eutectic is the lowest melting composition and temperature of a system that is formed of two or more components.

Eutectoid is the composition and temperature at which a single-**PHASE** solid will change directly into a two-phase solid without passing through a two-phase region.

Evaporative Pattern Casting See **LOST-WAX CASTING.**

Expanded Polystyrene Casting See **LOST-FOAM CASTING.**

Explosion Forming is the process by which a **WORKPIECE** is shaped by blasting it into the cavity of a **DIE.**

Extrusion is the process of forcing a material through a **DIE** to produce a very long **WORKPIECE** of constant shape and cross section.

Facing is the turning of a flat surface (perpendicular to the rotation axis) on a **LATHE.**

Fathom is six feet of water depth.

Fatigue (in metals) is the weakening or breaking of a material by repeated or cyclic **STRAINS.**

FCAW See **FLUX CORE ARC WELDING.**

Feed is the movement of a cutting tool longitudinally along a piece of stock, or the movement of a piece of stock past a cutting tool.

Feed Rate is the speed at which a cutting tool passes longitudinally along a **WORKPIECE.**

Feedback is the act of sending a report back to the **CONTROLLER** regarding the effect of some action.

Feedback Control is a control system that compares the input instruction with the reported "output" and uses the difference to change the output to reduce the difference.

Ferrite also called **ALPHA IRON** is, for all practical purposes, pure iron. It can have up to 0.025% carbon at 1333°F. Ferrite is body-centered-cubic in structure.

Ferritic Stainless Steel is any of a group of low-carbon chromium-iron alloys that are very ductile and highly resistant to corrosive action by acids.

Fiber is any filament used in cloth or as a reinforcement for **COMPOSITES.**

Filament Winding is the layup of **FIBERS** by wrapping resin-coated fibers around a mandrel.

Fillet is an internal "corner" that is "filled in" to reduce **STRESS RISERS.**

Fillet Weld is a method of **WELDING** two plates of metal together whereby the weld metal forms a concave surface between the plates.

Finishing is any **MANUFACTURING** process designed to give closer **TOLERANCES,** provide a protective coating, or improve the appearance of a part.

Fixture is a device designed to hold a specific **WORKPIECE** so that a **MANUFACTURING PROCESS** can be performed on it.

Flame Hardening is a method of hardening only the surface of a high-carbon steel by means of a gas flame jet followed by a spray of water or oil.

Flame-Spray Metallizing is the process whereby a metal rod is run through an oxy-acetylene or other flame, melted into droplets, and sprayed on the surface of another metal.

Flange is a part of the **WORKPIECE** that protrudes from the sides of the part.

Flaring is the bending of a sheet or the expanding of the walls of a cylinder of metal at an angle of less than 90 degrees.

Flashing is the excess material squeezed out from a **BILLET** in a **CLOSED FORGING** or stamping process.

Flask is the container used to hold the sand or other material when making a casting.

Flexure is the bending of a material.

Flux Core Arc Welding (FCAW) uses a flux-filled hollow wire as the arc welding rod with no inert gas.

Foil is a metal sheet thinner than 0.0003 in. (0.008 mm).

Foot is (a) twelve inches or (b) the bottom end of a base.

Force is energy applied in a single direction against an object.

Forge Weld is the **FUSION** of two pieces of metal by heating and pounding them together.

Forging is the delivering of a force or impact to a **WORKPIECE.**

Forming is any manufacturing process in which an external force is used to shape the **WORKPIECE.**

Frank-Read Source is a mechanism by which new **DISLOCATIONS** are generated in a material under cyclic stress that eventually "pile up" to start a crack in the material. This leads to the failure of the part.

Friction Weld is the **FUSION** of two pieces of stock together by rapidly rubbing them together, which generates enough heat to melt the contact surfaces.

Frit is a glass and additive that have been melted together then poured into water to produce fractured particles, which are then ground into a powder.

Full Annealing is the heat treatment of a metal that places it in the softest condition possible.

Full-Mold Casting is the same as **LOST-FOAM CASTING.**

Furlong is one-eighth of a mile or 660 feet.

Fusion is the joining of two pieces of material by the mixing of the surface atoms.

Fusion Weld is the melting of two metal surfaces together without the addition of any added rod or material.

Galvanized Steel is steel that has been dipped in molten zinc. The zinc protects the steel both by providing a physical barrier and by electrolytic action.

Gamma Iron is the same as **AUSTENITE.**

Gamma Rays are ultra-high-frequency and energy electromagnetic radiation. Gamma rays have more energy than **X-RAYS.**

Gantry is a type of crane that can reach any point within a rectangular volume. It does this by using two side rails, which run the length of the volume (along the top edge of the sides); a cross-rail, which rolls on the side rails; and a hoist, which rolls along the cross rail.

Gantry Robot is a **ROBOT** whose mechanism has been designed to mimic a **GANTRY** crane rather than the human **SHOULDER/ELBOW/WRIST.**

Gas is that state of matter that has no structure at all.

Gate (in casting) is a restriction in the **RUNNER** used to control the flow rate of the material being cast.

Gear Hob is a machine that cuts teeth in gears.

Gib Head Key is a **KEY** that has a small hook or "L" shape on the end for ease of removal.

Glass. Originally glass was any amorphous material made from fused silicates. Glasses now include amorphous materials made from fused borosilicates, phosphorous oxides, lead oxides, and germanium oxides. Glass does not have a long-range crystal structure and thus no definite melting point.

GMAW is gas metal arc **WELDING.** Formerly known as MIG or metal inert gas welding.

Grain is a single **CRYSTAL** in any material.

Green Sand Casting is a casting made using a water **MULLED** sand as a bonding agent.

Grinding is the removal of material by abrasion.

Grit is the size of particle of an abrasive. The grit is measured by the sieve used for sizing the abrasive, and is equal to the number of openings per linear inch in the sieve. The higher the number of the grit, the finer the abrasive.

GTAW is gas tungsten arc **WELDING.** Formerly known as TIG or tungsten inert gas welding.

Gunier-Preston (GP) Zone is the volume around a defect in a crystal in which the imperfection is "felt," that is, influences the properties of the crystal.

Gutta Percha is hard rubber.

Hacksaw is a saw usually used to make downward cuts in a piece of metal. Hacksaws can be either hand or machine operated.

Hand (robot) is the part of an industrial **ROBOT** that holds the tool.

Hand-Lay is the application of reinforcing **FIBERS** or cloth and the resins by **HANDWORK** alone.

Handwork is any **MANUFACTURING** technique whereby the tool is held and guided by humans.

Hardenability is the depth to which a steel can be hardened.

Hardfacing is the addition of a hard material to the surface of a softer material.

Hardness is the resistance of a material to deformation.

Heading is the forging process that "upsets" the metal to form heads on nails and similar products.

Headstock is the part of a machine that holds or rotates the cutting tool or stock.

Heat of Fusion is the heat required to change a unit mass of a solid at the melting temperature to a liquid at the same temperature. The heat of fusion of water is 79.7 **CALORIES** per gram, or 144 **BTU** per pound.

Heat of Vaporization is the heat required to convert one unit of mass of a liquid at the **BOILING POINT** to the same unit of mass of gas. In the **SI** or metric system the units of the heat of vaporization are **CALORIES** per gram; in the English system the units are **BTUs** per pound.

Hemming refers to the folding of a piece of sheet metal or other material back on itself.

High-Polymer (Hi-Polymer) is an organic compound that has linked many thousands of times to similar **COMPOUNDS.** Commonly but often inaccurately called **PLASTICS.**

Hogshead is an ancient unit of volume equal to 63 U.S. gallons.

Holes (imperfection) are missing particle electrons from their proper location in a crystal.

Hole Saw uses a cylindrical cutting tool to cut holes in the **WORKPIECE.**

Holography is the science of using interference patterns of light for very precise **MEASUREMENTS** or other uses. Lasers are often used in holography. Holography can also be used to produce true three-dimensional photographs.

Honing is a sizing process that uses fine grinding stones to repair some of the size and shape defects left by previous processes.

Horizontal Mill is a machine used to produce flat surfaces. It turns a cutting tool about a horizontal axis.

Horsepower is a unit of power required to lift 550 pounds one foot in one second. It is equivalent to 746 watts or 2544 BTU per hour.

Hot Working is the deformation of metals at temperatures that allow the crystals to rapidly regrow to a large size. This is usually close to the "red hot" temperature of the metal.

Human Engineering is another name for **ERGONOMICS.**

Human Factors Engineering is an old name for **ERGONOMICS.**

Hybrid is a mixture of different types of **FIBERS** in a **COMPOSITE.**

Hydroforming is the forcing of a **BILLET** into a mold with a bag of liquid.

Hydrostatic Head is the pressure of a liquid on anything below the surface of the liquid.

Impact is the sudden application of a **FORCE.**

Impact Strength is the energy a material will absorb before breaking when struck by an instantaneous blow.

Imperfection is anything that upsets the uniform structure of a material.

Inch is one-twelfth of a foot.

Inch System See **ENGLISH SYSTEM.**

Inclusions are macroscopic flaws in the crystal structure. Inclusions may be foreign matter or even holes in the structure. Inclusions can be seen by optical microscopes.

Induction Hardening is the hardening of a high-carbon steel, heated by inducing an electric current in the surface of the part and immediately **QUENCHING** in oil, brine, or water.

Inherent Variation is the unavoidable variation in the dimensions of a part due to such causes as **FLEXURE** of the cutting tool, looseness of the bearings, etc.

Injection Molding is a form of casting in which the liquefied material is forced into a mold by pressure. Used for plastics, ceramics, and some metals.

Inspection is any observation of an object to determine if it meets all previously specified criteria.

Intermetallic Compounds are metals that react chemically to form new metallic materials. The elements of intermetallic compounds cannot be separated by mechanical means.

International Union of Pure and Applied Chemists (IUPAC) is the international organization that sets the standards for chemical research.

Interply Hybrid is a combination of two or more layers of **FIBERS** having different fibers in each layer of a **COMPOSITE.**

Interply Knitting is the weaving or sewing of two or more plies or layers of fabric in a **COMPOSITE.**

Interstitial in a crystal is a particle in between the lattice points.

Intraply Hybrid is a combination of two or more types of **FIBERS** within a single layer of a **COMPOSITE.**

Investment Casting is the process by which a pattern is coated with a refractory material to make the **MOLD.**

Ion is a charged atomic particle.

Ionic Bond is a bond between atoms created by one **ELECTRON** being passed from one **ATOM** to another. This creates positively and negatively charged atoms, which attract each other forming the bond.

Iron Carbide is **CEMENTITE** (Fe_3C).

Isotope is an **ELEMENT** having the same number of **PROTONS** as the parent element but a different number of **NEUTRONS.** Isotopes of an element have the same **ATOMIC NUMBER** but different **ATOMIC WEIGHTS.**

Izod is an **IMPACT** test that uses a vertically held notched test specimen.

Jig is a general-purpose device used to hold a **WORKPIECE** so that a **MANUFACTURING** process can be performed on it. A jig differs from a **FIXTURE;** a fixture is designed for a specific workpiece.

Jigsaw is a reciprocating saw usually having a thin blade used to cut intricate patterns in a **WORKPIECE.**

JIT See **JUST-IN-TIME SCHEDULING.**

Jominy Test is an end quench test of steel to determine its **HARDENABILITY** or depth to which it can be hardened.

Just-in-Time Scheduling is a type of production organization in which the material for a part or assembly is delivered just in time to be used, rather than being stored in advance of being needed.

Kerf is the volume swept out by a saw blade.

Key is a piece of material fitted between a shaft and either a wheel, gear, or pulley to prevent rotation between the outer piece and the shaft.

Keyseat is the hole in a shaft where material has been removed to allow for the insertion of a **KEY.**

Keyway is the hole where material has been removed from the inside of a wheel, gear, pulley, or similar piece for the insertion of a **KEY.**

Knurl is a pattern created on a **WORKPIECE** resulting from the upsetting or raising of the material on the surface of the stock by forcing a tool into the material. Knurl patterns range from diamond shapes to straight lines and circles. The purpose of knurling is to provide an attractive, high-friction surface or to expand the surface of the metal.

Lapping is a process designed to remove the last minute amount of material by use of a fine loose abrasive rubbed between a tool and a part or between two matching parts.

Laser is a light source. LASER is an acronym for *l*ight *a*mplification by *s*timulated *e*lectromagnetic *r*adiation. Lasers produce single-frequency beams, which can be focused to a very fine point.

Lathe is a machine used for making cylindrical parts. Flat surfaces can be made on lathes if the proper attachments are used.

Lead (in threads) is the distance a threaded part will advance in one complete rotation of the part. In single **THREADS,** lead and pitch have the same value.

Lead Screw is the long **WORM GEAR** that moves the **TOOL POST HOLDER** along the **BED** of a **LATHE.**

Ledeburite is iron with 4.3% carbon. It is the **EUTECTIC** composition of the iron-carbon system.

Limits are the dimensions within which acceptable parts must be made.

Line Defect is an imperfection that occurs along a plane in the crystal. These include **LINE DISLOCATIONS** among others.

Line Dislocation is an extra partial plane of atoms in a crystal.

Liquid is that state of matter which has a random atomic structure.

Liquidus is the line on a **PHASE DIAGRAM** between the liquid and liquid-solid two-phase region.

Liter is a metric system unit of volume equal to 1000 cubic centimeters. It is equal to 1.057 quarts.

Live-Center is the part of a **LATHE** that holds the free end of the stock and rotates with the stock.

Lost-Foam Casting is a process of **CASTING** in which the **PATTERN** is made of styrofoam or other foamed polimer. The foam is invested in a refractory material and then burned out to make the **MOLD.**

Lost-Wax Casting is a process of **CASTING** in which the **PATTERN** is made of wax. The wax is invested in crystobalite or other refractory material and then melted away to make the **MOLD.**

Machining is the removal of material by powered mechanical action.

Machining Cell is a **MANUFACTURING** unit in which several **MACHINING CENTERS** are grouped together with an automatic **INSPECTION** station and a pick-and-place **ROBOT.**

Machining Center is a machine that can turn a **WORKPIECE** into any position necessary and automatically select (and install) any tool needed to perform nearly any machining operation on the workpiece, all without any human intervention.

Magnafluxing is a method of **INSPECTION** in which a magnetic field is placed through the part and iron filings are sprinkled on it. The filings will pile up along any flaw in the part.

Magneforming is the process in which a sheet of metal is forced into the cavity of a **DIE** by a creating a magnetic field in the sheet, which repels the sheet into the die.

Magnetic Test is a method of detecting flaws in a material by detecting the alteration of the magnetic lines of flux in a magnetic field. A strong magnet is attached to the part and a finely ground iron powder is sprinkled on the surface of the part. If a flaw is present, the filings will line up along the flaw instead of the regular lines of flux. Other methods of detecting the alteration of the magnetic lines of flux can also be used.

Manipulator is that part of an industrial **ROBOT** that actually handles the tool, material, or part.

Manufacturing is the activity of making products.

Manufacturing Cell is another name for a **MACHINING CELL.**

Manufacturing Facility is the physical building (or complex of buildings) where **MANUFACTURING** is done.

Manufacturing Process is any method used to convert raw materials into useful products.

Martempering is a heat treatment process in which a steel is quenched from the austenitic temperature to just above the **MARTENSITE** start temperature, held there for a few seconds to a few minutes, and then quenched. It is used to provide an even-sized martensite throughout the part.

Martensite is the hardest form of steel. It is body-centered-tetragonal in structure.

Martensitic Stainless Steel is a chrome steel that is body-centered-tetragonal in structure.

Mass Production refers to any high-production-rate operation, and especially to production where extremely close control of the **MANUFACTURING** processes is necessary in order to avoid the expenses of many scrapped parts if anything goes wrong.

Material Addition is any **MANUFACTURING PROCESS** in which material is built up on a part or **WORKPIECE** or stock.

Material Joining is any **MANUFACTURING PROCESS** that puts two parts or pieces of material together either temporarily or permanently.

Material Removal is any **MANUFACTURING PROCESS** that reduces the size of a part.

Matrix is (a) the parent material (in **PHASE DIAGRAMS**) of a system of two or more components or (b) the material that holds the **FIBERS** in a **COMPOSITE.**

Matte Copper is the material obtained after copper ore is smelted. It is about 30% pure copper.

Mean is the measure of central tendency found by adding the scores and dividing by the number of scores. Also known as the *arithmetic mean* or **AVERAGE.**

Measurement is the act of determining the dimensions of some feature of an object.

Mechanism of Slip is a direction in which one plane of **ATOMS, MOLECULES,** or other particles in a crystal can slide over another plane. The more mechanisms of slip a crystal has, the more ductile the material will be.

Median is the middle score of a series of data arranged in order from maximum to minimum.

Melting Point is the temperature at which a solid will turn into a liquid.

Mer is the basic or unit **MOLECULE** from which a plastic or **POLYMER** is made.

Metallic Bond is a bond between atoms that have an excess of **ELECTRONS** over the stable state. This creates an electron "cloud" to which the nuclei are attracted.

Metallizing is the addition of a molten metal to a **WORKPIECE** by spraying.

Methane is an organic compound having one carbon and four hydrogen atoms. It is a gas at room temperature.

Metre is the basic unit of length in the metric system. It is equal to the length of the path traveled by light in a vacuum during a time interval of 1/299,792,458 of a second. One inch equals *exactly* 0.0254 metre. One metre is approximately equivalent to 39.37 inches.

Metric System is a system of measurements based on the decimal or base-ten system.

Midrange is the value found by adding the maximum and minimum scores and dividing by two.

MIG Welding See **GMAW.**

Mile is an English system unit of length equal to 5280 feet.

Milling is the removal of material by a rotating cutting tool to make a single flat surface.

Mistake is any wrong action. Not to be confused with **ERRORS,** which are inaccuracies in **MEASUREMENTS.**

Mixture is a combination of materials in which there is no chemical reaction.

Mode is the most frequently occurring value in a collection of data.

Modulus of Elasticity is the change in **STRESS** divided by the change in **STRAIN** while the material is in the **ELASTIC REGION.** Also known as **YOUNG'S MODULUS.**

Mohs Hardness Test is a method of measuring the resistance to surface abrasion. It is primarily used by geologists.

Mold is the container into which liquid metal, plastic, or other material is poured in making a casting.

Molecule is the smallest particle to which a compound can be reduced without changing the chemical structure of the material. Any further reduction would destroy the compound.

Mulling is the process of mixing the sand used in making a **MOLD** for a casting with water or other binding agent.

Neutral Flame is a flame obtained by combining oxygen with any flammable gas to provide the exact amount of oxygen needed to burn all of the gas.

Neutron is a particle with no electric charge in the nucleus of the atom. Its mass is about the same as the **PROTON.**

Newton is a unit of **FORCE** in the **METRIC SYSTEM.** It is equal to one kilogram-metre per second squared.

Nitriding is a method of hardening the surface of a low-carbon steel by the diffusion of nitrogen into the surface of the metal. Ammonia gas is often used as the source of nitrogen and the steel must be heated into the austenitic region (see **AUSTENITE**) and quenched to produce the maximum **HARDNESS** after nitriding.

Nondestructive Test is any test that does not damage the part to the point that it cannot be used for its intended purpose.

Nonferrous Metal is any metal that does not contain iron.

Normalizing is a heat treatment used to give a steel an even **GRAIN** size. It is used prior to **MACHINING** or other heat treatments.

Nose Radius is the radius of curvature on an edge of any cutting tool.

Notch-Sensitive materials are brittle materials that have a tendency to fail by **STRESS-CORROSION** fatigue due to stress risers being placed in the part.

Nucleation is the initiation of a **CRYSTAL** by the joining of the first few **ATOMS** or **MOLECULES.**

Numerical Control is a way of controlling the action of machines by giving them a set of instructions in numerical form.

Numerics are numbers that have no units associated with them in an equation. Examples are pi, $\sqrt{2}$, and coefficients.

Objective Test is a test in which the operator or the person conducting the test has no influence on the results of the test.

Oil Paint is a coating formulated of a natural or synthetic resin, a coloring agent, and a thinner.

Open Forging is the pressing or hammering of a **BILLET** between two flat surfaces to reduce its thickness.

Open Hearth Converter is a method of converting iron into steel by blowing air across the surface of the molten iron to burn out the excess carbon from the raw iron.

Organic Chemistry is the chemistry by which the carbon atom forms molecules.

Overcut is the difference in size between the size of the cut and the size of the electrode in **EDM.**

Overhead Saw See **RADIAL ARM SAW.**

Oxidizing Flame is a gas-oxygen flame in which there is more oxygen in the mixture than is necessary to burn all of the gas. The excess oxygen can react chemically with the metal on which it is used and corrode the metal. Oxidizing flames are used in cutting torches. Oxidizing flames are not as hot as neutral flames.

Oxygen Lance Converter is a method of converting iron into steel by blowing pure oxygen through the molten iron to burn out its excess carbon.

Parkerizing is the **CONVERSION** process of creating an iron phosphate coating on the surface of steel. It provides a dull black surface that resists corrosion.

Parting Plane is the plane of separation between the halves of the **MOLD** in a casting.

Parting Sand or Compound is a very finely powdered white sand or other material that is used to lubricate a **PATTERN** and allow it to be removed from the **MOLD** without damaging the mold.

Pascal is the **METRIC SYSTEM** (SI) unit of pressure or **STRESS** equal to one **NEWTON** per square metre.

Patenting is a heat treatment of steel in which the steel is heated into the austenitic region (see **AUSTENITE**) then quenched in molten lead. It is used to give steel wire ductility.

Pattern is the shape that is to be cast, about which the **MOLD** is made.

Pattern Maker is a skilled craftsman who makes the patterns for castings.

Pearlite is the **EUTECTOID** composition of steel that has approximately 0.8% carbon.

Penny System is the system used for weighing nails. Originally it was the cost in pence for 100 nails of that size.

Permanent Mold Casting is a casting process that employs a **MOLD** which can be reused many times.

Phase is a macroscopically physically distinct and separable portion of matter.

Phase Diagram is a graph of the parameters over which given **PHASES** exist. The more common phase diagrams are a graph of temperature versus composition.

Phenyl Radical is a benzene ring that has lost one hydrogen **ATOM.**

Piecework is that form of **MANUFACTURING** where one product is completed before work is started on the next one.

Pigs or Pig Iron are ingots cast directly from the refining of iron ore.

Pilot Hole is a small hole drilled in a **WORKPIECE** to guide a larger drill.

Pinned Dislocations and Planes are atomic planes that have one end locked in a particular location in the crystal. The rest of the plane of particles can move, but the pinned points cannot.

Pins are nonthreaded fasteners used to join two pieces of material. Pins provide good **SHEAR** strength but usually little **TENSILE STRENGTH** to the joint.

Pitch (in threads) is the linear distance from the crest (top peak) of one thread to the crest of the adjacent thread.

Plane Defects involve three-dimensional imperfections in a crystal. These include **TWINNING** and **INCLUSIONS.**

Planing is a **MACHINING** technique in which the stock moves past the cutting tools.

Plasma is an ionized **GAS.**

Plasma-Arc Spraying is the **MATERIAL ADDITION** process whereby a powdered material is fed into a **PLASMA** flame, which sprays the molten material onto the surface of a **WORKPIECE.**

Plaster Mold Casting is an **INVESTMENT CASTING** process in which both halves of the **MOLD** are made of gypsum or plaster of paris.

Plastic Region is the portion in the stress-strain graph that is beyond the **ELASTIC LIMIT** in which permanent deformation occurs.

Plasticity is the amount of permanent deformation that a material will undergo before rupturing.

Plastics See **HIGH-POLYMERS.**

Plate is a metal sheet thicker than one-quarter inch.

Plumbers Lathe is a **LATHE** designed specifically to cut, and make **THREADS** in pipe.

Plunge EDM See **RAM EDM.**

Point Defect is an imperfection that affects only a point in the crystal lattice. Point defects include **VACANCIES, INTERSTITIALS, SUBSTITUTIONALS, HOLES,** and **ELECTRONS.**

Polar Coordinates is the system of locating one point from another point (the origin) by specifying the distance and two angles from the origin (r, θ, ϕ).

Polishing is a process that removes surface defects without changing the size or shape of the part.

Polymer is the product of joining many of the same type of **MOLECULES.**

Polyunsaturated means that the organic compound has many double or triple bonds in the molecule.

Porcelain is an opaque, colored glass.

Pot Life is the length of time a glue or paint will remain usable after it is mixed or left open.

Powder Forming is a process whereby a finely powdered material (metal, metal oxide, ceramic or the like) is packed into a **MOLD,** placed under pressure, and the particles fused together in a process called **SINTERING.**

Powder Metallurgy See **POWDER FORMING.**

Pratt and Whitney Key is a **KEY** that is rectangular in cross section but has rounded ends.

Precipitation Hardening See **AGE HARDENING.**

Precision is the degree of closeness of several **MEASUREMENTS** to each other.

Pre-Preg is reinforcing **FIBERS** or cloth in a **COMPOSITE** that have been precoated with resin and cured to the **B-STAGE.**

Propane is a three-carbon organic compound having eight hydrogen atoms.

Proportional Limit is the arbitrary point on a stress-strain curve which indicates that the **ELASTIC REGION** has been passed. The proportional limit is usually found by the intersection of the **STRESS-STRAIN CURVE** itself and a line, offset by 0.0002 inch/inch (0.02%) and parallel to the straight-line portion of the curve.

Pultrusion is a method of forming **COMPOSITES** by pulling preimpregnated **FIBERS** through a heated die to form the desired shape.

Punch is a tool that delivers a force or impact to a small area on the **WORKPIECE.** Also see **RAM.**

Punching is the forming of holes or other shapes in a sheet of material by the **SHEARING** action of a punch into a die.

QA See **QUALITY ASSURANCE.**

QC See **QUALITY CONTROL.**

Quality Assurance is the **INSPECTION** and testing of materials and products coming into a facility to ensure that the parts meet specifications.

Quality Control is the **INSPECTION** and **TESTING** of products as they leave the assembly or production line to ensure that the parts meet design specifications.

Quenching is the rapid cooling of a metal to harden it.

Radial Arm Saw is a circular saw that has the motor and saw blade suspended from a slide track. The entire motor and saw are drawn across the fixed **WORKPIECE.**

Radial Forging See **SWAGING.**

Radical (chemistry) is any molecule that has an unfilled bond.

Rake Angles are the cutting-tool face angles that direct the motion of the chip.

Ram is the part of a forge used to deliver the force to a **BILLET.**

Ram EDM is the electric discharge **MACHINING** in which a shaped tool is used to cut a desired shape into or through a piece of metal. Sometimes called *plunge* or *die-sinking EDM.*

Range is the difference between the maximum and minimum values of a collection of data.

Reaction Injection Molding is a method of casting plastics and **COMPOSITES** in which the resin and the catalyst are mixed in a chamber and immediately forced into the mold.

Reaming is the removal of material by use of a straight fluted tool, which rotates to scrape or cut the inside portion of a hole.

Reducing Flame is a flame that has too little oxygen to react completely with the gas. The result is that the gas can react with corrosion on the metal and prevent further oxidation. Reducing flames are used in the melting of metals to prevent oxidation. Reducing flames are not as hot as neutral flames.

Refractory is a material that will withstand high temperatures and will not transmit heat very well.

Reliability is the repeatability or consistency of a test.

Relief Angles are the angles on cutting tools that tilt the noncutting part of the tool away from contact with the stock.

Resin Transfer Molding is a casting technique used for plastics and **COMPOSITES** in which the catalyzed resin is poured into the mold.

Ribbon Weld is a resistance weld that forms a continuous seam between two pieces of metal.

Riddle is the screen through which the sand used in a **CASTING** is sifted in making a **MOLD.**

Ring Compounds are organic molecules in which the carbon atoms are joined in a semblance of a circle.

Ripping is the cutting of wood parallel to the grain.

Risers are vertical holes in the **MOLD** going from the casting to the surface of the mold that allow excess material to rise to the surface of the mold. Risers keep air from being trapped in the casting, allow the workers to see when the mold is filled. They also create a **HYDROSTATIC HEAD** to keep pressure on the casting.

Rivet is a type of permanent **PIN** that has heads or enlargements formed at the ends after insertion.

Robot is a reprogrammable, multifunctional manipulator designed to handle material, parts, tools or specialized devices through variable programmed motions for the performance of a variety of tasks.

Robotics System is an industrial **ROBOT** along with all of its control equipment.

Rockwell Hardness Test is a measure of surface plasticity by measuring the depth a point is driven into a material under a specified load.

Roll is a **WRIST** motion (the motion used to turn a doorknob).

Roll Forming is a process by which sheet material can be bent a continuous cross-section shape.

Rolling is the process whereby a **BILLET** is reduced in cross section or changed in shape by squeezing it between rotating solid cylinders.

Roughness is the short range or close together undulations in the surface of a part or **WORKPIECE.**

Rounds are external "corners" that are curved to reduce **STRESS RISERS.**

Routing is the removal of material by moving a rotating, shaped tool past a piece of stock.

Runners are the horizontal holes in a **MOLD** that go between the **SPRUE** and the **CASTING.** Also refers to the horizontal holes that run between sections of multiple castings.

Runout is the deviation from the roundness or specified radius or a circular or cylindrical part.

Rupture Strength is the **STRESS** at which a material breaks.

Saber Saw is a power saw that has a reciprocating blade anchored at only one end.

Safety is freedom from danger or harm. Industrial safety is the effort made to protect people and equipment from harm by a conscious and deliberate analysis of hazards and a resulting change of conditions, activities, attitudes, and habits.

Safety Factor is the yield **STRESS** divided by the maximum design stress on a part.

Sand Casting is a casting process that uses sand to form the **MOLD.**

Sanding is the removal of material using loose or lightly bonded sand or other abrasive material.

Sawing is the parting of a piece of stock by removing material from the middle of the stock.

Scale is the oxide on the surface of a metal caused by heating.

Screed is a leveling device used to smooth the surface of a **MOLD** or concrete.

Screw is a fastener that has a spiral groove (a **THREAD**) that fits into the grooves of another piece of material.

Screw Dislocation is the "twisted" offset of one plane of atoms on another plane of atoms in the crystal.

Seaming is the joining of two pieces of sheet metal or other material by **HEMMING.**

Sensors are devices that send a signal indicating the amount of whatever they are designed to measure (position, temperature, sound, etc.).

Set (in saw teeth) is the angle to which the teeth protrude from the side of the blade. The set provides the clearance for the blade to pass through the material.

Shaping is the removal of material by passing the cutting tool over the stationary stock.

Shaw Process See **CERAMIC MOLD CASTING.**

Shear is the deformation of a material by slightly offset, opposing **FORCES.**

Sheet is a flat metal plate with a thickness between about 0.0003 and 0.25 in.

Shelf Life is the time a product is usable after it is made. In the case of glues or paints, it is the length of time the unopened container of the material will still be usable.

Shell Mold Casting is an **INVESTMENT CASTING** process in which the pattern is dipped in sand that has a resin binder. The pattern and accumulated sand are then heated and the sand forms a shell around the pattern to form the **MOLD.**

Shore Durometer is a **HARDNESS** test used on plastics and other soft materials.

Shore Scleroscope is a **HARDNESS** test that measures the surface elasticity of a material by bouncing a plummet against the surface.

Shoulder (Robot) is the part of an **ANTHROPOMORPHIC ROBOT** that is designed to mimic the motions of the human shoulder.

Shoulder Swivel is the up-and-down motion of the arm, pivoting at the shoulder.

Shrink Rule is a scale or ruler that is made oversize to allow for the amount a metal casting will shrink during solidification and cooling.

SI or Système International is the unified system of measurements adopted for worldwide use in Paris, France, in 1960. Also referred to as the *metric system.*

Side Milling is the removal of material by using the *side* of the milling or cutting tool.

Silicone Paint is a coating that uses silicon-based compounds as polymerizing agents.

Silicone Polymers are polymers based on the silicon atom instead of the carbon atom.

Sine Bar is an **INSPECTION** tool used to measure angles accurately.

Single Shear is the cutting of a single surface by means of slightly offset, opposing **FORCES.**

Sintered materials are those that have been powdered then packed in a mold and the edges of the particles fused together. The end product is porous but strong enough for many uses.

Skimming is the **POLISHING** of parts using **WIRE EDM.**

Slip Planes are planes in a **CRYSTAL** that slide over each other.

Slush Casting is a process whereby the molten metal or other material is poured into a **MOLD,** the surface of the casting allowed to solidify, then the rest of the cast material poured out or removed. This leaves a hollow casting.

SMAW is shielded metal arc **WELDING,** generally known as **STICK WELDING.**

Soldering is the **ADHESIVE** joining of two metals by means of a low-melting-point metal alloy.

Solid is that condition of material which has a definite long-range **CRYSTAL** structure.

Solidus is the line on a **PHASE DIAGRAM** between the solid phase and the liquid-solid two-phase region.

Solution Heat Treating is the acceleration of **AGE HARDENING** by holding the stock at a temperature below the **MELTING POINT.**

Sonic Testing is the detection of flaws in a material by sending sound waves through and detecting the echo of the sound waves reflecting from the **IMPERFECTIONS.**

Span is an archaic measurement equal to the distance between the tips of the fingers of opposite outstretched arms.

Specific Gravity is the **DENSITY** of a material divided by the density of water in the same system of units. The specific gravity of a material is the same in all measurement systems.

Spin Forming is a forming process in which a sheet of metal is held to a mandrel, rotated, and forced onto the mandrel to shape the sheet.

Spindle Speed is the rotational speed of the **CHUCK** of the lathe, milling machine, drill, or other machine tool. Commonly just called the *speed* of the machine.

Splines are ridges parallel to the axis on a rod used to transfer rotational motion from one shaft to another slipped over the splines.

Spot Weld is a resistance weld applied to a point between two pieces of metal.

Sprue is a hole in a **MOLD** through which the molten material is forced or poured in making a casting. After the casting has solidified, the material of the sprue must be removed.

Sputtering is a method of **MATERIAL ADDITION** in which **ATOMS** are dislodged from a source material by bombarding the source with ions of an inert gas at low pressure.

SQC See **STATISTICAL QUALITY CONTROL.**

Standard Deviation is a measure of distribution of data in a collection. In a normal distribution, 68% of all data would lie between one standard deviation below and one standard deviation above the **MEAN.**

Staples are basically two **FINS** side by side, which are joined.

Statistical Quality Control is the control of the quality of the parts produced by an analysis of the numerical data derived from measurements taken on those parts.

Statistics are the numerical data that describe the status of a system.

Steady-State Test is a test in which the load is applied slowly and evenly throughout the test.

Steel is an iron-carbon **ALLOY** having less than 2% carbon. Steel can also be defined as an iron-carbon alloy that can be heated entirely into **AUSTENITE.**

Stick Welding is a common name for SMAW welding.

Stock is any material in the form and shape in which it comes from the supplier.

Stone is an old unit of weight, still used in Scotland, equal to 14 pounds.

Strain is the amount of deformation per unit length a material undergoes with the application of a load.

Strain Rate is the speed at which the load is applied to a test specimen. Units are usually inches per minute or metres per minute.

Stress is the load or **FORCE** per unit of cross-sectional area applied to a material. The units are usually Pascals or pounds per square inch.

Stress-Corrosion Cracking is a method of **FATIGUE** in metals caused by the repeated opening of a scratch or crack allowing air to get into the crack and break the bonds between the atoms by corrosion.

Stress Risers are points on an object at which **FORCES** are concentrated. Stress risers can lead to the failure of parts. Sharp corners, cracks, or even scratches on the surface of parts can become stress risers.

Stress-Strain Curve is a graph of the **STRESS** applied to a specimen versus the **STRAIN** resulting from that stress.

Subjective Test is a test in which the person conducting the test can influence the results of the tests. Most tests are subjective to some extent.

Sublimation is the transformation of a **SOLID** directly to a **GAS** without passing through a **LIQUID** state.

Submerged Arc Welding is a **WELDING** technique in which a pile of flux is laid over the intended weld joint and the **WELDING** rod is fed through the flux to strike an arc. The layer of flux protects the weld surface from the air.

Substitutionals in a crystal are particles (**ATOMS, MOLECULES, IONS, RADICALS,** etc.) other than the parent material at a lattice point.

Superfinishing is a process designed to remove high spots so that two parts can fit together nearly perfectly.

Surface Feet per Minute is the number of feet of stock per minute that passes a cutting tool, or the number of feet per minute by which the cutting tool passes the stock.

Surface Treatment Process is any **FINISHING** process designed solely to provide a protective coating or to improve the appearance of a part.

Swaging is the forging process by which a hollow cylindrical part is forced tightly around a rod or wire to permanently attach the two parts. It is also known as **RADIAL FORGING.**

Synergy is the effect of combining two or more materials to produce better results than the sum of the two materials in the uncombined state.

Syntactic Foam is a resin in which tiny, hollow glass spheres are embedded to form a foam.

Tailstock is the part of a **LATHE** that holds the free end of the piece of stock.

Tap is a **TOOL** for cutting internal threads in a hole.

Tempering is the removal of internal stresses in a metal by heating.

Tensile Impact is the sudden application of a **FORCE** in outwardly opposite directions.

Tensile Strength is the maximum **STRESS** a material will withstand prior to failure when tested by applying outwardly opposing **FORCES** in a steady-state test.

Tensile Stress is the **FORCE** per unit area that results in the application of **TENSION** forces.

Tension is the application of **FORCES** in outwardly opposite directions.

Testing is any method used to determine if any product meets the minimum acceptable standards.

Therm is 100,000 **BTUs.** Gas prices are often stated in therms. One therm of natural gas is the volume of gas needed to produce 100,000 BTUs of heat.

Thermal Conductivity is the ability of a material to conduct heat.

Thermite is a mixture of aluminum powder and iron oxide that produces iron, aluminum oxide, and a tremendous amount of heat upon ignition. Thermite is used in some **WELDING** applications.

Thermoplastic Polymer is a **POLYMER** that gets soft upon heating.

Thermosetting Polymer is a **POLYMER** that gets harder upon heating.

Thio (prefix) simply means that there are sulfur atoms in the molecule.

Thread (Machine) is a helical ridge and groove on a cylindrical axis, used to hold two parts together.

Tolerance is the total or maximum permissible variation in the size of a part.

Tool is any device designed to perform a specific function in the manufacture of a product. These can be cutting tools, shaping tools, bending tools, etc.

Tool Post is the part on a **LATHE** or other machine that holds the cutting or other type of **TOOL.**

Tool Post Grinder is a powered abrasive wheel that is mounted in the tool post holder of a **LATHE** or other machine tool.

Tool Post Holder is the part on a lathe that rides along the **BED** and holds the **TOOL POST.**

Torsion (torque) is the application of **FORCES** that twist an object.

Torsion Strength is the maximum twisting stress a material can withstand.

Triple Plating is the process by which steel is electroplated by copper, then nickel, then by chrome to finish the product.

Triple Point is the temperature and pressure of a single component system at which the solid, liquid, and gaseous states are in equilibrium.

True Stress is the load applied to a test specimen divided by the instantaneous cross-sectional area of the specimen at the time the load is measured.

TTT Curve is a graph of the temperatures and times at which **AUSTENITE** changes to **PEARLITE** or **MARTENSITE.** Also known as the time-temperature-transformation curve, the Bain S curve, the isothermal transformation curve, and the I-T curve.

Tukon Hardness Test is a measure of surface plasticity determined by applying a load on an elongated pyramid-shaped diamond indenter known as the Knoop indenter.

Turret Lathe is a lathe that holds several tools in a rotatable tool holder. The turret lathe can then do several functions sequentially on the **WORKPIECE.** Turret lathes are often automated to turn out a complete part without shutting down.

Twinning in a crystal is the formation of a mirror image crystal. Twins can be formed during the solidification and cooling process of the crystal (thermal twins) or during **COLD WORKING** (mechanical twins). Twins can occur on either the macroscopic or microscopic scale.

Twist Drill is a **TOOL** used to cut a hole in a piece of **STOCK.** The twist drill has cutting edges on the end and long helical grooves to carry away the cut material. It looks like, and was originally made from, a piece of twisted metal.

Ultimate Strength is the same as **TENSILE STRENGTH.**

Ultrasonic Cleaning is the removal of tarnish or other surface material by the use of ultra-high-frequency sound waves.

Unit Analysis See **DIMENSIONAL ANALYSIS.**

Unit Horsepower is the **HORSEPOWER** required to remove one cubic inch of material per minute.

Unsaturated (chemical) is an organic molecule having double or triple bonds.

Upset Weld is a resistance weld between raised contact points of two metal pieces.

URTRIM is ultra reinforced thermoset reaction injection molding. In this process a **FIBER**-wrapped **SYNTACTIC FOAM** core is placed in a mold and then a **THERMOSETTING** resin is injected around the core and cured.

Vacancies in a crystal are missing particles from a lattice point.

Vacuum Bagging is the application of atmospheric pressure to a **COMPOSITE** while it is being cured, by placing the laid-up composite in an air-tight bag and removing the air from it. Close to one atmosphere (14.7 pounds per square inch) can be applied to the part by this technique.

Vacuum Casting is a casting process in which the molten metal or other material is drawn into the **MOLD** by a vacuum.

Vacuum Coating is the distribution of a vaporized metal on the surface of another material in the absence of air.

Validity is the characteristic of a test that determines the degree to which the test measures the property it is intended to measure.

Van Der Waal Bond is a bond between atoms caused by the asymmetry of the electrons and nucleus. This asymmetry causes a "polarity" in the atom with slightly positive and negative sides. The slightly positive side of one atom or molecule is attracted to the slightly negative side of the next, thus creating the bond.

Variance is the square of the **STANDARD DEVIATION** of a statistical calculation.

Varnish is an oil with modified alkyd or synthetic resin as polymerizing agents. Varnishes have no pigmenting agents.

Vents are holes placed in a **MOLD** to allow steam and other gases to escape from the mold without damaging the mold. Vents do not come in contact with the casting as do **RISERS.**

Vertical Lathe is a lathe that holds the **WORKPIECE** vertically and rotates it about a vertical axis.

Vertical Mill is a machine that uses a rotating cutting tool to make flat surfaces and which rotates the cutting tool about a vertical axis.

Vibratory Finish is a **POLISHING, BURNISHING,** or **DEBURRING** process in which the part is placed in a container with abrasive pellets and mechanically vibrated at a relatively high frequency.

Vickers Hardness Test is a measure of surface plasticity that uses a load applied by a diamond pyramid indenter.

Viscosity is the resistance to flow of a **LIQUID.** Liquids with a high viscosity are thick and flow very slowly. A unit of viscosity is the "poise" or "centipoise."

Visual Inspection is the determination of the acceptability of a part by the naked eye or by optical magnification. **MEASURING,** weighing, and scanning for defects are visual inspections.

Vitrification is the process by which **ABRASIVE** wheels are made. An abrasive is mixed with glass or porcelain, molded into the wheel, and then heated to form a glass-like wheel.

Warp is the direction in fabrics that is parallel to the length of the fabric.

Water Saw is a machine that cuts material by means of a high-pressure stream of water with a very fine abrasive.

Waviness refers to the long-range undulations in a **WORKPIECE.**

Wear Ratio is the volume of the **WORKPIECE** removed divided by the volume of the electrode removed per unit of time at any given voltage and current.

Weft is the direction in a fabric perpendicular to the **WARP.**

Welding is any method of joining two materials by a cohesive **FUSION** of the materials (see **COHESIVE BOND**).

Wire EDM is the cutting of a metal by use of a wire having a voltage applied across a dielectric fluid. The arc between the wire and the metal **WORKPIECE** erodes its way through the metal. EDM stands for **ELECTRIC DISCHARGE MACHINING.**

Wire Saw is a machine that cuts material by drawing a diamond-dust-imbedded wire across the stock.

Woodruff Key is a method of preventing one part from rotating on a shaft by using a half-circular-shaped piece of metal, a matching **KEYSEAT** in the shaft, and a **KEYWAY** in the part.

Work is **FORCE** times a distance. Work is not done unless a force moves something a distance in the direction of the force.

Workholder is any device that holds a **WORKPIECE** so that a **MANUFACTURING** process can be performed on it.

Workpiece is a general term referring to any material stock, part, or assembly.

Wrist (robot) is that part of an **ANTHROPOMORPHIC ROBOT** which mimics the motions of the human wrist.

Wrist Pitch is that wrist motion which makes the hand go up and down.

Wrist Roll See **ROLL.**

Wrist Yaw is that wrist motion which makes the fingers move from left to right while the **ELBOW** and **SHOULDER** joints are locked.

Wrought Iron is iron that has been cold worked to produce a fine grain structure.

X-Ray is very short wavelength electromagnetic radiation. X-rays are invisible to the human eye but can be detected by photographic or electronic means. X-rays can be used to detect flaws in materials.

Yard is 36 inches or 3 feet.

Yaw is a general term for left-right motion.

Yield Point is the **STRESS** at the engineering design strength of any material.

Yield Strength is another name for **YIELD POINT.**

Young's Modulus See **MODULUS OF ELASTICITY.**

Appendix

B

Derivation of Equations

Derivation of the Brinell Hardness Number

The Brinell hardness number, Vickers hardness number, and the Tukon hardness number are all derived as the pressure withstood by the surface of the metal. Pressure is the load divided by the area of contact or

$$\text{HN} = P/A$$

where HN = hardness number, P = load, and A = contact area.

Because the Brinell test uses a spherical ball, the Brinell hardness number is equal to the load divided by the area of the spherical sector of the indentation (see Figure B–1). The area of the spherical sector is

$$A = 2\pi Rh$$

where A = the surface area of the indentation, R = the radius of the sphere, and h = the depth of the indentation.

Therefore the equation for the Brinell hardness number becomes

$$\text{BHN} = P/2\pi Rh$$

The radius of the sphere can be divided into the depth of the indentation and the distance of the center above the indentation x or

$$R = h + x$$

then

$$h = R - x$$

From the Pythagorean theorem we know that

$$R^2 = x^2 + r^2$$

where r = the radius of the indentation. From the Pythagorean equation, a little algebra leads us to the equation

$$x = \sqrt{R^2 - r^2}$$

Substituting this into the equation for the depth of the indentation, we arrive at

$$h = R - \sqrt{R^2 - r^2}$$

We now have the equation for the Brinell hardness number as

$$\text{BHN} = \frac{P}{2\pi R(R - \sqrt{R^2 - r^2})}$$

But remember that

$$R = D/2$$

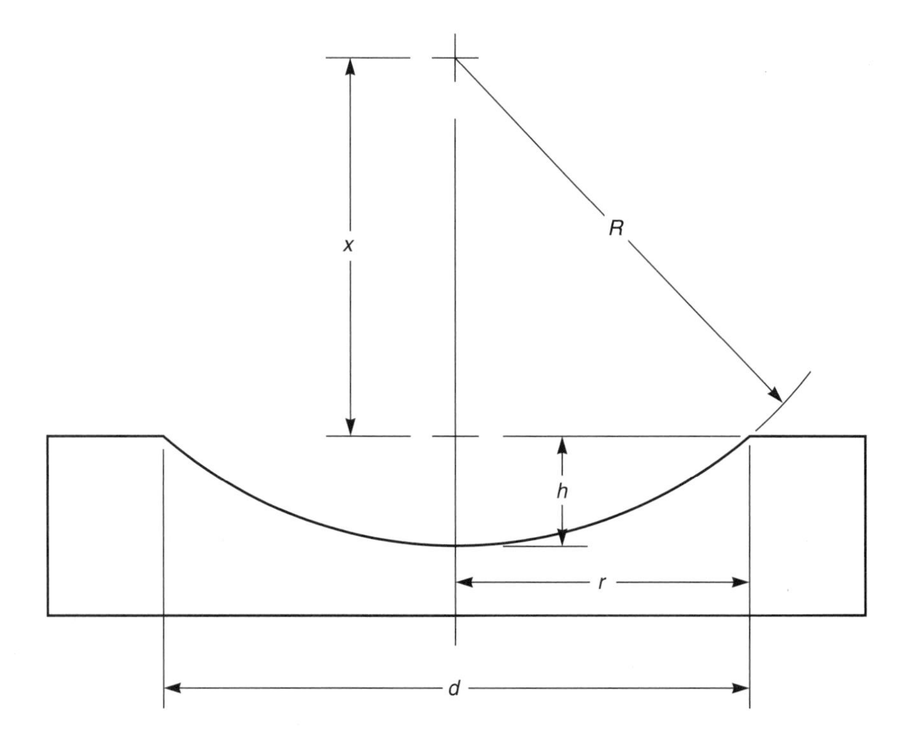

Figure B–1. Brinell hardness diagram.

and

$$r = d/2$$

where d = the diameter of the indentation and D = the diameter of the ball. Therefore, the Brinell hardness number becomes

$$\text{BHN} = \frac{P}{2\pi \frac{D}{2}\left(\frac{D}{2} - \sqrt{\left(\frac{D}{2}\right)^2 - \left(\frac{d}{2}\right)^2}\right)}$$

This equation reduces to

$$\text{BHN} = \frac{P}{\pi D\left(\frac{D}{2} - \sqrt{\frac{D^2}{4} - \frac{d^2}{4}}\right)}$$

Of course, the square root of 4 is 2 so

$$\text{BHN} = \frac{P}{\pi D\left(\frac{D}{2} - \frac{1}{2}\sqrt{D^2 - d^2}\right)}$$

By factoring this equation and inverting the 1/2 in the denominator we arrive at the final equation for the Brinell hardness number:

$$\text{BHN} = \frac{2P}{\pi D(D - \sqrt{D^2 - d^2})}$$

QED*

Derivation of the Vickers Equation

The diamond pyramid hardness (DPH) number derived from the Vickers test is also the load divided by the surface area of the indentation:

$$\text{DPH} = P/A$$

As seen from Figure B–2, the surface area of the indentation is four times the area of each triangular face.

*The term *QED* stand for "quod erat demonstrandum," which is Latin for "that which has been shown." QED is the traditional end for all mathematical proofs.

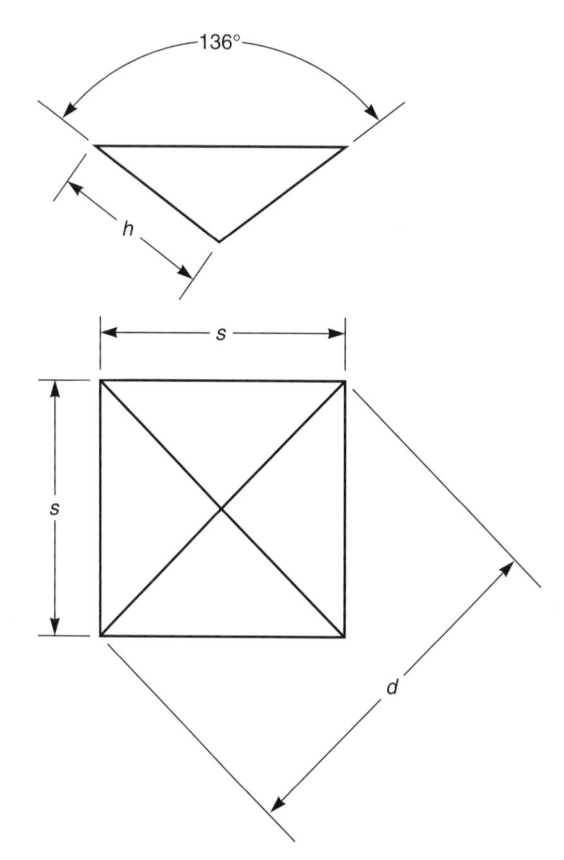

Figure B–2. Diamond pyramid hardness number.

$$A = 4A_t = 4(0.5\,sh) = 2sh$$

Using the Pythagorean theorem,

$$s^2 = (0.5d)^2 + (0.5d)^2 = 0.5d^2$$

$$s = 0.7072d$$

With the help of a little trigonometry

$$h = s/(2 \sin 68°) = 0.53927s$$

Therefore, the area of the indentation is

$$A = 2(0.7072d)(0.53927s)$$

or

$$A = 2(0.7072d)((0.53927)(0.7072d)) = 0.5393d^2$$

The DPH number is therefore calculated as

$$\text{DPH} = P/0.5393d^2$$

or

$$\text{DPH} = 1.8544\,P/d^2$$

QED

Derivation of the Knoop Hardness Number

Unlike the Brinell and Vickers tests, the Tukon equation is derived from the projected or flat area left by the indenter, not the actual surface area. However the Knoop hardness number is still the load divided by the projected area or

$$\text{KHN} = P/A_p$$

The projected area is the area of the rhombus left on the surface of the metal. Refer to Figure B–3 for the symbols used in this derivation.

$$A_p = 2c(L/2) = cL$$

The length L is measured directly by a microscope but a rather roundabout route must be taken to calculate c:

$$L/2d = \tan 86.25°$$

Therefore,

$$d = L/2 \tan 86.25° = 0.03277L$$

But

$$c/d = \tan 65°$$

So

$$c = 0.03277L \tan 65°$$

or

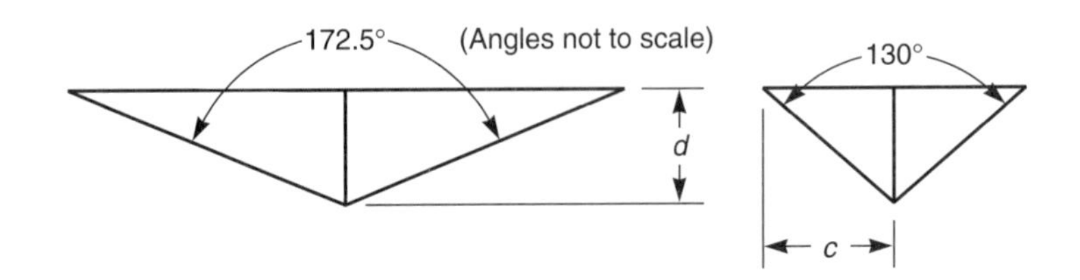

Figure B–3. Knoop hardness number.

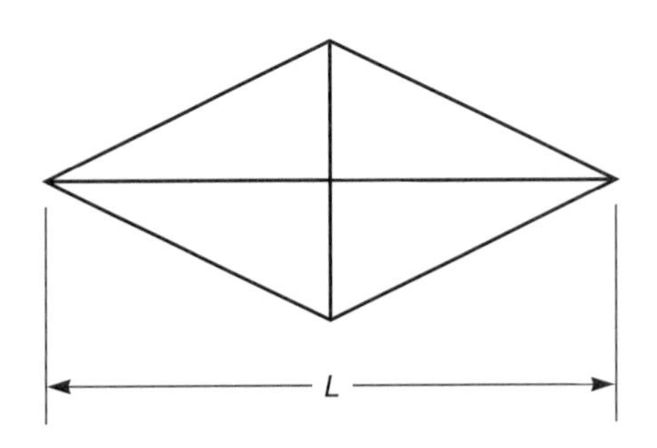

$$c = 0.07028L$$

Then

$$A_p = 0.07028L \times L$$

or

$$A_p = 0.07028L^2$$

Therefore,

$$\text{KHN} = P/0.07028L^2$$

Giving

$$\text{KHN} = 14.3P/L^2$$

QED.

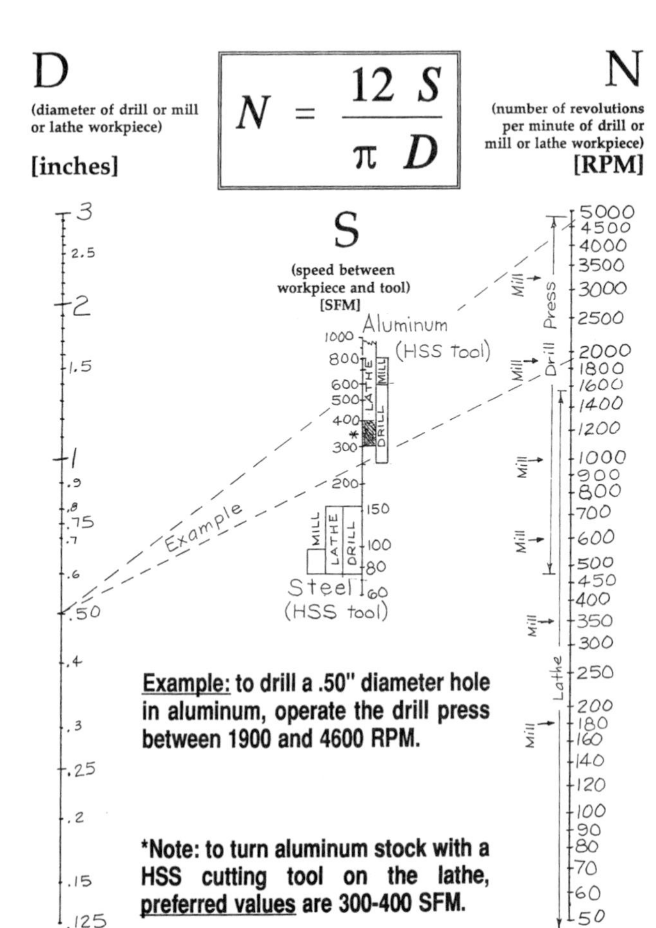

Appendix C

Formulas and Equations

Areas

Circle	$A = \pi r^2$
	$A = \pi d^2/4$
Triangle	$A = bh/2$
	$A = \sqrt{s(s - a)(s - b)(s - c)}$ where $s = (a + b + c)/2$
Trapezoid	$A = h(B + b)/2$
Rectangle	$A = Lw$
Square	$A = s^2$
Sphere	$A = 4\pi r^2$
	$A = \pi d^2$
Spherical sector	$A = 2\pi rh$
Right circular cone	$A = \pi rs$ where s = slant height
Cylinder (side)	$A = \pi dh$
	$A = 2\pi rh$
Cylinder (total)	$A = 2\pi r\,(r + h)$
Pyramid	$A = ps/2$ where s = slant height and p = perimeter
Rectangular parallelepiped	$A = 2(Lw + Lh + wh)$

Volumes

Rectangular parallelepiped	$V = lwh$
Pyramid	$V = Bh/3$ where B = area of the base
Cylinder	$V = Bh$
Circular cylinder	$V = \pi r^2 h$
Cone	$V = \pi r^2 h/3$
Sphere	$V = 4\pi r^3/3$
Pythagorean theorem	$c^2 = a^2 + b^2$

(The Pythagorean theorem applies only to a right triangle.)

Trigonometry

For a right triangle	Sine = sin = opposite/hypotenuse Cosine = cos = adjacent/hypotenuse Tangent = tan = opposite/adjacent $\sin^2 A + \cos^2 A = 1$

Materials Testing

Stress	S = load/area
Strain	e = change in length/original length
Modulus of elasticity	E = stress/strain (in the elastic region)
Impact strength	$I = \pi RL(\cos\theta_2 - \cos\theta_1)$
Brinell hardness number	$\text{BHN} = \dfrac{2P}{\pi D(D - \sqrt{D^2 - d^2})}$
Vickers hardness number	$\text{DPH} = 1.8544\, P/d^2$
Tukon hardness number	$\text{KHN} = 14.3P/L^2$

Manufacturing

Cutting Speed	$N = C_r/2\pi r$ $N = C_r/\pi D$ where C_r = recommended cutting speed in feet per minute, D = diameter of the stock or the diameter of the cutting tool in feet, and N is in revolutions per minute.
Material removal rate	MRR = $12fDC$ where f = the tool feed rate in inches per revolution, D = depth of cut in inches, and C = cutting speed in SFPM.
Power requirements for cutting	$P = P_u \times \text{MRR}$ where P_u = unit horsepower requirement.

Appendix

D

Answers to Odd-Numbered Problems

Section I

Chapter 1

Problem Set 1-1

1. a. Rub the metal piece against a rough concrete sidewalk, brick, or rough stone. No tools required.
   b. File the metal. File required.
   c. Sand the metal. Sandpaper required.
   d. Remove the metal with a whetstone. Whetstone required.
   e. Melt the metal off with a torch. Cutting torch required.
   f. Saw the metal off. Hacksaw required.
   g. Remove the metal by turning. Lathe required.
   h. Place the piece in a drill, start the drill, and hold a file against the metal. File and drill required.
   i. Remove the metal by milling. Milling machine required.
   j. Grind the metal off. Grinder required.
   k. Wet the metal, let it rust, then rub off the corrosion. No tools required.
3. A simple belt would be made of leather with the end looped through a buckle and the loop riveted. Holes punched in the belt for the buckle.

   Step 1. Obtain a calf hide or other sheet of thick leather. Assume the tanning of the hide has been done. Also obtain a belt buckle.

   Step 2. Cut the sheet into strips 1 inch wide and about 8 inches longer than the finished belt. This could be done with a power shear.

   Step 3. Punch the holes for the buckle to fit into. Use power punch or even a gang punch.

   Step 4. Loop the other end of the belt approximately 1 inch back and punch the two layers of belt for a rivet.

   Step 5. Loop the belt through the buckle and insert a pop rivet to lock the buckle in place.

   Step 6. Color the belt with a dye, and wax to provide a luster.

   Note: All of the preceding steps could be done using either power machinery or with hand tools. There are also many other schemes whereby belts could be made.
5. A screwdriver could be made by the following method:
   a. Start with a cylindrical piece of metal. Use a lathe to turn it down to the desired size of the shank of the screwdriver.
   b. Shape the "wedge end" of the shank by milling, grinding, or filing.

c. Mill, grind, or file a flat side to the shank on the handle end. This will prevent the screwdriver blade from turning in the handle.
d. Heat the tip of the blade into a red hot condition then plunge it into oil to harden the tip.
e. Reheat the tip to a "blue" condition and air cool it to temper the blade.
f. Place a cylinder around the "handle end" of the shank. Seal the end of the cylinder against the shank. Lock the shank and cylinder in a vertical position, probably in a vise.
g. Mix an epoxy resin with the hardener, then pour the epoxy around the blade in the cylinder.
h. After the handle has cured, it can be turned and finished in a lathe.

A knife can be made from a file, or large hacksaw blade by the following steps:

a. Heat the file or blade to a bright red condition and allow it to cool very slowly. The high-carbon steel will be too hard to work if this step is omitted.
b. Shape the file into the blade by grinding and filing. The blade will be soft enough after annealing to be ground, milled, filed, or drilled.
c. Drill holes in the handle end to allow a handle to be bolted onto it.
d. Reheat the blade end of the knife to a red heat either in an oven or with a "rosebud" torch.
e. Quench the blade in oil.
f. Reheat the blade to a "blue" condition and air cool to restore the toughness of the blade.
g. Fabricate and attach a wooden handle or cast an epoxy handle to the blade.
h. Remove the oxide by sanding. Progress from coarse sandpaper to finer sandpaper to establish a smooth finish.
i. Hone the blade to sharpness.

7. Examples of environmental problems from production can include the following:
   a. Items requiring electroplating. (Chemical disposal is a problem.)
   b. Engines might be designed that require less lubrication. (Disposal of lubrication oils is a problem.)
   c. The internal combustion engine is a polluting engine since it burns its fuel at several atmospheres of pressure. At these high pressures, the nitrogen in the air starts to burn, producing nitrous oxides. Perhaps the external combustion engine (Sterling cycle) or electric engines should be researched more completely and used if possible.
   d. Furnaces that use a high-sulfur-content fuel could be redesigned to use other, cleaner fuels.
   e. Solvent-free methods of removing greases, paints, and varnishes are being developed.
9. a. Box wrench—material joining
   b. Crosscut saw—material removal
   c. Hammer—material joining, change of condition, change of form
   d. Drill press and drills—material removal
   e. Screwdriver—material joining
   f. Oxy-fuel torch—change of condition, change of form, material removal
   g. Slip joint and channel-lock pliers—material joining, change of form. Material removal if wire cutters are on the pliers.
   h. Soldering iron—material joining
   i. Grinder—material removal
   j. Vise grip pliers—material joining, change of form
11. Many items have become obsolete. These include but are not limited to such things as slide rules, IBM punch cards, key punches, Morse code, mimeograph, ditto machines, 78-, 45-, and 33-rpm records, 8-track tape recorders, Beta TV recordings, 8-mm movie cameras, 5-1/4 inch floppy disks for computers, etc., etc., etc.

Chapter 2

Problem Set 2–1

1. 1130 lb/in.^{2}
3. 6340 lb/in.^{2}
5. $D = 0.65$ in.
7. No, stress = 181,000 lb/in.^{2}
9. 78 lb/in.^{2}
11. 0.61 in.
13. 120 ft-lb
15. 3 ft
17. 53,000 lb/in.^{2}
19. 0.42 in.
21. 15×10^6 lb/in.^{2}
23. 0.005 in.

Problem Set 2–2

1. a. 206
   b. 60
   c. 103
   d. 3.26 mm
   e. 464
   f. 25
   g. 18.5
   h. 1240 kg
   i. 155
   j. 0.65 mm
3. No, DPH = 148
5. No, hardest Rockwell C scale for steel is 68.
7. No, the scleroscope hardness number is a measure of surface elasticity, whereas the Brinell hardness number is a measure of resistance to surface plastic flow.
9. Cutting tools: Any situation involving possible friction wear. Plow blades, bearings, bushings. Hardness must also be known in order to select the tools that will cut the material.
11. Density is the mass per unit volume and must always be quoted with units. Specific gravity is the ratio between the density of the material and the density of water.
13. 0.15 lb/in.^{3}
15. SG = 6.64; metal is antimony.

Problem Set 2–3

1. 160 ft-lb
3. 40.5 ft-lb
5. 55°
7. Shear pins, cold chisel tips, jackhammer parts, landing gears of aircraft
9. Flexure strength in handles, hardness of jaws, shear strength of pivot pin

Problem Set 2–4

1. 89,600 lb/in.^{2}
3. 38,200 lb/in.^{2}
5. Rockwell G scale
7. a. Reduce the stress to below the critical point by making the part larger.
   b. Change the part every 250,000 to 300,000 cycles.
   c. Reduce the frequency of the cycles to prolong the part life.
   d. Change to a different material with longer critical stress life.
9. a. Any part undergoing vibration such as automotive springs, crankshafts, cam shafts.
   b. Aircraft wings that flex in flight.
   c. Parts undergoing cyclic stresses in torsion such as drive shafts and axles.
11. 686, hard

Chapter 3

Problem Set 3–1

1. The gauge could be accurate, reliable, and valid but not very precise and have some subjectivity allowed in reading the gauge and in running the test.
3. 72 mph.
5. 4800 ft/min
7. 150 ft/min
9. 712 oz.
11. 17,600 ft-lb/sec
13. 18,650 W
15. \$24.17

Problem Set 3–2

1. $A = \pi r^2, A = 254$ in.^{2}
3. $C = 2\pi r$, $C = 50.3$ in.
5. $A = \pi D^2/4, A = 50.3$ ft^2
7. $A = 2[(LW)+(LH)+(WH)], A = 146$ in.^{2}
9. $A = \pi Dh + 2\pi r^2, A = 207$ ft^2
11. $A = 4\pi r^2, A = 804$ in.^{2}
13. a. $V = \pi r^2 h$, $V = 754$ in.^{3}
    b. $V = 3.25$ gallons
    c. $A = 2\pi r^2 + \pi Dh$, $A = 477$ in.^{2}
15. 5 gallons = 1160 in.^{3}

$$V = \frac{4}{3}\pi r^3, r = \sqrt[3]{\frac{3V}{4\pi}}, D = 2r = 2\sqrt[3]{\frac{3V}{4\pi}}$$

$D = 13$ in.

Problem Set 3–3

1. 70.4 kg
3. 1100 lb
5. 38 L
7. 89 km/hr
9. 40,468 m^2
11. 2.33 gm/cm^3
13. 73 ft/s
15. 779 N
17. 110,000 MPa

Problem Set 3–4

1. 52,987 psi
3. Both the 52,000 to 52,999 and the 53,000 to 53,999 have three data each. Therefore, the mode is 53,000 psi.
5. 2,250,000
7. 54,488 and 51,486
9. Mean = 157.6 lb, median = 154 lb, range = 110 to 225 lb, standard deviation = 34.3 lb

Section II

Chapter 4

Problem Set 4–1

1. Equation: $N = C_s/\pi D$. From Table 4–2, $C_s = 550$ SFPM, so $N = 525$ rpm.
3. Yes, the recommended rpm calculates to be 1528 rpm. From Table 4–2, $C_s = 1200$ SFPM
5. 2.51 HP
7. High-speed steel. It's inexpensive, easily resharpened, hard enough to cut aluminum, and can be given the large rake angle needed for cutting aluminum, and the aluminum doesn't generate enough heat to damage the HSS tool.
   Carbide tools could be used, but they are more expensive than high-speed steel. The gain in productivity (because of higher cutting speeds) would be only 43%. This may not be enough to justify the higher cost of the carbide tools. The final decision would depend on the prices of the tools, the expected tool life, and the labor costs involved.
9. Copper is soft enough for HSS and the HSS is inexpensive. However, the recommended cutting speed for carbide tool for cutting copper stock is 250% that of a high-speed steel cutting tool. The difference in production rate could justify the cost of the carbide tool. Either a yes or no answer could be defended.

Chapter 5

Problem Set 5–1

1. $(1 \times 10^{-6}$ m)(100 cm/m)/2.54 cm/in. = 0.00003937 in. = 39.37×10^{-6} in. = 39.37 microinches.
3. EDG is a gentle process so it is excellent for thin materials. Since the grinding wheel can have any desired shape, it could be used to form a series of identically shaped, closely spaced, parallel "fins" by grinding crosswise to the fins with a shaped wheel, then forming the slots by a wheel formed of thin parallel disks.
5. *RAM:* Any shaped hole with a uniform cross section can be made by RAM. An example would be cutting irregular shaped sockets in wheel lug nuts, which only a special tool would fit. These are used for security on tires.
   Wire: Any surface that can be generated by moving a straight line can be made by wire EDM. It can be used to cut matching parts that require a very close fit.
7. If the dielectric fluid didn't move through the gap, the particles would not be removed and some of them would reattach to the workpiece. This could reattach the two parts or "short" out the spark and stop the EDM.
9. *Dielectric:* A material that doesn't conduct electricity very well.
   Erode: The removal and carrying away of small particles of a material by a moving fluid.
   Overcut: In RAM EDM, overcut is the gap between the electrode and the workpiece.
   Undercut: The eroding of the material beneath the mask. Any hole that increases in cross-sectional size as it goes deeper is said to be "undercut."
   Wear ratio: The ratio of the material removed from the electrode divided by the material removed from the stock.
11. The erosion of a RAM EDM, electrode is essentially at the end. In a through-hole, the hole will be "sized" by the sides of the electrode, which have not been worn significantly.
13. 1.27 hr = 1 hr 16 min
15. *Broaching:* This would require a broach made especially for this application. The corners of the broach (which make the points of the crescent) would be difficult to make strong enough to keep from breaking, but it would make a hole every second or two.
    Ram EDM: Die wear would make it difficult to keep a sharp corner for the points of the crescent. Graphite dies would be easy to make and keep sharp, but erosion process is a slow one.

Chapter 6

Problem Set 6–1

1. a. Any situation where only very small amounts of material need be removed.
   b. For materials not easily cut by mechanical or electrical means.
   c. When extremely fine tolerances are needed.
3. Not easily. Ceramics are stable around most acids, bases, and solvents. Hydrofluoric acid could be used on a ceramic if only etching or small amounts of material is to be removed.
5. In electroplating, the workpiece is the cathode of the electric cell. Small voltages and currents are applied. In electrochemical machining, the workpiece is the anode; larger voltages and currents are applied.
7. ECM and CM work on materials not easily cut by conventional machining. Close tolerances are obtained and it can be used to cut blind or hidden holes where conventional tools won't reach.
9. a. Mask the entire key blank with wax or plastic.
   b. Cut away the wax where the grooves are to appear.
   c. Place the blank in an acid bath.
   d. Periodically remove the blank and check for depth of cut.
   e. When the depth of cut is sufficient, remove the blank, wash and dry it, recoat the key, and cut away the wax to expose the edges of the metal to be cut.
   f. Place the blank back in the bath until the edge metal is eroded away.
   g. Remove the blank, remove the wax, and polish the key.

Chapter 7

Problem Set 7–1

1. a. Cutting Styrofoam patterns for investment casting
   b. Cutting block wax
   c. Cutting thermoplastic, noninsulating materials that do not stick to the wire
   d. Cloth or paper could conceivably be cut by hot wire
3. a. Cutting wood to fine tolerances
   b. Cutting sheets of metals
   c. Cutting very small holes in hard materials where drilling is difficult
5. Very fast; cheap; not extremely fine tolerances; melting point limited to about 3500°F
7. No, rubber absorbs the energy of the abrasive and would not be cut very well.
9. a. Where very fine tolerances must be maintained
   b. For cutting very small diameter holes deeply into a metal
   c. Where surface oxidation and distortion must be prevented

Section III

Chapter 8

Problem Set 8–1

1. Electroplating uses an electric current to cause the metallic ions to attach to the surface of a part. The plated metal starts out in a solution. In dipping, the coating metal is melted and the part is immersed the molten metal.
3. Cathode. Metal ions are positive. The negative charge on the cathode attracts the positive metal ions to the surface of the part.
5. a. Plating small parts
   b. Plating one-of-a-kind items
   c. Plating delicate parts to ensure even coatings
   d. Replating parts to cover areas where plating has worn off
   e. Jobs done at home where little equipment is at hand
7. Disposal of chemicals; relatively slow process; expensive process; requires considerable skill and training of operator.
9. No, zinc has a higher chemical activity than copper.
11. Items might include the following: bumpers, ignition points, hub caps, decorative parts such as door handles, window levers, turn signal levers, radio and other control knobs, window trim, interior trim, and grills.

Problem Set 8–2

1. a. To apply a hard cutting surface to a cutting tool.
   b. To build up a shaft with metal. The shaft could then be reground to fit a bearing.
   c. To provide a hard protective surface to any part (such as a plow blade) that must resist abrasion.
3. Factors that must be considered are the melting points of the workpiece and add material; the

oxy-fuel mixture ratio; the flow rate of the metal or powder; the thermal expansion coefficients of the stock and the add material; the velocity of the droplets; and the compatibility of the stock and add material. This compatibility often depends on the crystal structures of the two materials.

5. Hardfacing is applied by melting a material and fusing it to a surface. Metallizing sprays a molten material onto a hot surface.
7. The possibilities would include the following:
   a. Coating small parts.
   b. Coating parts that are nonmetallic and could not be electroplated.
   c. Coating very intricate parts that have a lot of small crevices.
   d. Applying a thin coating of silver or other conductive material to a nonconductive surface so that it can be electroplated or electroformed.
9. The purpose of "grooving" is to provide a surface that will mechanically "grip" the material being sprayed on it.
11. This could be done either by vacuum coating or by dipping. The procedure for vacuum coating would include these steps:
   a. Clean the surface.
   b. Place the material in the "bell" along with the coating material.
   c. Melt the coating material, and pull the vacuum.
   d. Remove the part and clean it.
13. Vacuum coating would be difficult to use since it would be difficult to pull a vacuum only on the inside of an oven. Hardfacing with a weld rod would also be difficult to apply to the inside of an oven.

Section IV

Chapter 9

Problem Set 9–1

1. a. A pattern is a shape in the form of the final product from which the mold is made.
   b. In green sand casting, water is used as the bonding material for the sand. Dry sand casting uses oil, resin, molasses, or other organic bonding material for the sand.
   c. The cope is the top half of the casting flask. The drag is the bottom part of the flask.
   d. Mulling is the mixing of the sand with the bonding agent. Riddling is the sifting of the sand into the flask.
   e. Sprues are vertical shafts placed in the mold through which the molten casting material is poured. Runners are the horizontal holes in the mold through which the molten casting material passes.
   f. In centrifugal casting, the flask is spun about its own longitudinal axis. In centrifuged casting, the entire flask is spun about an axis perpendicular to the axis of the flask or sprue.
3. a. Risers let out trapped air in the mold.
   b. Risers provide a hydrostatic head of molten material, which prevents surface shrinkage in the part.
   c. Molten material coming up the riser indicates the mold is full.
5. Investment casting is a one-shot, one-mold-per-part method in which the temporary mold is poured around the pattern. The investment mold is destroyed after casting.

   Die casting uses permanent, reusable molds.
7. Dry sand casting would be a good choice since only one part is to be made, the casting size is large, and the casting is made of brass.
9. The size, shape, material, production rate, and precision required would probably dictate a die cast for the handle. The part could be cast around the steel insert.

Chapter 10

Problem Set 10–1

1. The list might include alternator brushes, ceramic magnets, fuel filters, bushings for small motors, small gears (especially mechanical fuses).
3. Sintering is the "fusing" of the surfaces of particles, without melting the entire particle.
5. Wood screw; no; sintered materials lack sufficient torsion strength

   Bearing; yes; could be permanently impregnated with oil.

   Screwdriver; no; not enough torsion strength, also too brittle.

   Coil spring; no; not enough ductility in sintered parts.

Fuel filter; yes; if made of large-size particles, the sintered part would allow fuel to pass through trapping the large solid particles in the fuel. Hammer; no; insufficient impact strength, too brittle.
Electrical contact; yes; a powdered material with good electrical conductivity and high corrosion resistance could be made into electrical contacts. Knife blade; no; too brittle.
Housing for small electric motor; yes; bearings could be built into the housing.
Gear housing for jewelers' lathe; possibly; this housing may be too large for powder metallurgy techniques. (Depends on the size of the lathe.) Gear for hand drill; possibly; for small drills the gears would probably have sufficient shear strength. For larger drills, the sintered gears would be easily stripped.

7. Sintering is the fusing of the particles. Briquetting is the forming of the particles into a shape. The particles could be sintered after briquetting.

Chapter 11

Problem Set 11-1

1. No, to use magneforming, the material must be able to carry an electric current and thus generate an opposing magnetic field about the material. Lucite and Plexiglas do not conduct electricity and therefore could not be magneformed.
3. The wrench could be close-forged from a blank of steel. The forging process would provide a cold-worked tool that had high strength with close tolerances. The part could even be deburred by a second drop forge with cutting edges on the mold.
5. Products that can be magneformed might include the following:
   a. Metal cooking bowls
   b. Metal toys
   c. Hubcaps for automobiles (if metal)
   d. Metal lamp reflectors
7. The nozzle could be die cast, made by spin forming, or made by flaring a ductile metal in a press.
9. A finger brake is a hinged brake that can make bends of very small radius. It operates by holding one part of the sheet stock steady and bending the free end of the stock about the "fingers." A finger brake makes one bend at a time.
   A press brake can be used to make bends of large radius by forcing the sheet stock into a mold. A press brake can be set up to make several bends at a time.
11. Examples of flanging, seaming, and hemming are shown in the figure.

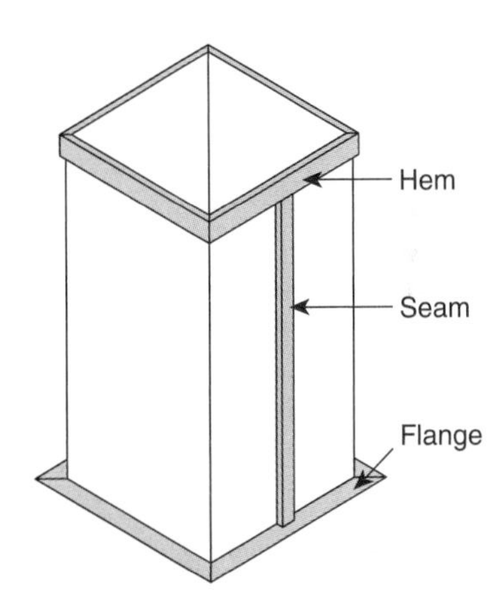

13. The advantages of explosion forming:
    a. The forming is fast.
    b. It can work on very large pieces of stock.
    c. It requires no electric power source.
    d. It does not significantly distort the grain structure.
    The disadvantages are as follows:
    a. It requires care in handling explosives.
    b. It is a noisy procedure.
    c. The cycle time may be slow.

Section V

Chapter 12

Problem Set 12-1

1. The reason steel can be hardened significantly is that it changes its crystal structure when heated. Other metals do not change crystal structure on heating and, therefore, cannot be significantly hardened by quenching.

3. The purpose of the Jominy test is to determine the hardenability or depth to which a piece of steel can be hardened.
5. A file could be made into a hunting knife by the following steps:

   Step 1. Place the file in an oven, heat it to about 1600°F, leave it in the oven at the high temperature for about 15 to 20 minutes, then cool it very slowly. (Either box or full anneal the file.)

   Step 2. Grind the softened file into the blade and handle shape desired. Drill any holes necessary for attachment of the handle in the file. Grind the cutting edge to a reasonably sharp edge.

   Step 3. Reheat the file to 1600°F, leave it at this temperature 15 to 20 minutes, then quench the blade section in oil.

   Step 4. Reheat the blade to about 1100° for 15 minutes to temper it. Let it cool slowly in air.

   Step 5. Sand, polish, and sharpen the blade to its final condition.

   Step 6. Form and attach a handle using bolts through the holes drilled in step 2.
7. Several precautions could be taken:
   a. Order only a month's supply of aluminum rivets at a time. Make sure the rivets are not overly aged by the vendor.
   b. Keep the rivets in the refrigerator to slow down the age-hardening process.
   c. Melt down old rivets and recast them (not recommended).
9. The list of parts requiring hardening might include the following:
   a. Ball and roller bearings
   b. Cam shafts
   c. Gears
   d. Engine valves
   e. Knives
   f. Firing pins for rifles, shotguns, and pistols
   g. Bearing parts of axles
   h. Springs
11. A high-carbon steel is more brittle than a low-carbon steel. Using a flame-hardened low-carbon steel would make the table tougher and would not break as easily if accidentally hit.
13. Advantages:

    Work hardening of steel beams by cold rolling

    Work hardening of cold-formed rivets (strengthens the rivet)

    Work-hardened parts that must resist abrasion, such as plow blades

    Disadvantages:

    Parts that are undergoing flexure (aircraft wings, levers)

    Parts that are undergoing torsion (drive shafts, drills)

    Parts in which ductility is more important than hardness (shock absorbers, landing gears)
15. a. The eutectic composition is at 4.3% carbon, 95.7% iron.
    b. The eutectoid composition is 0.8% carbon.
    c. Cementite has a composition of 6.67% carbon, 93.33% iron.
    d. The melting temperature of a steel with 0.5% C is about 2800°F.

Section VI

Chapter 13

Problem Set 13–1

1. *Epoxy glues:* Good for almost everything but some plastics.

   Urea formaldehyde glues: Good for wood.

   Aliphatic resin glues: Good for wood.

   Casein glues: Good for wood, leather, paper.

   Contact cements: Good for wood or other material. Has high tensile strength but weak in shear.

   Cyanoacrylic glues (super glue): Good for porous materials (including skin).
3. Both soldering and brazing are adhesive bonds. Soldering is generally referred to as a low-melting-temperature metal adhesive, whereas brazing refers to high-temperature copper alloyed bonds.
5. Exterior plywood is glued together with a waterproof glue. Interior plywood does not always have a waterproof glue.
7. 800 lb/120 lb/in.^{2} = 6.7 in.^{2}
9. The surface of the parts must be cleaned, roughened, and the glue worked into all parts of the surface. The parts must be held together until the glue has finished setting. The temperatures must be within the workable range of the specified glue. Proper mixing of the glue is essential.

Chapter 14

Problem Set 14-1

1. The cheapest equipment of all would be the forge weld. All that is needed is a heat source, hammer, and anvil. The cheapest commonly used weld would be a choice of oxy-acetylene welding or arc welding.
3. Electron-beam welding must be done in a vacuum.
5. Reducing flames are those in which there is not enough air or oxygen to burn the fuel completely. Reducing flames are used in melting metals to prevent corrosion of the molten metal.
7. Advantages of GMAW:
   a. It is a fast method of welding.
   b. It provides welds relatively free of oxidation.
   c. It is relatively inexpensive.
   d. It is a relatively easy weld to do.
   e. It is applicable to robotics and computer control.

   Disadvantages of GMAW:
   a. Many variables must be controlled for a good weld.
   b. It requires special equipment and an inert gas.
   c. It is more expensive than SMAW or oxy-fuel welding.
9. Advantages of SMAW:
   a. It is an easy weld to make.
   b. It is a quick weld.
   c. It is an inexpensive method of welding.
   d. The equipment is portable but a power source is required.

   Disadvantages of SMAW:
   a. It is a relatively "dirty" weld but cleaner than oxy-acetylene.
   b. Leaves a slag on the surface.
   c. Strength of weld not as high as a GMAW weld, but better than gas welds.
11. a. Size and type of filler wire
    b. Type of shielding gas
    c. Joint preparation
    d. Gas flow rate
    e. Voltage
    f. Arc length
    g. Polarity
    h. Wire stick out
    i. Nozzle angle
    j. Wire-feed rate
13. Visual checks should include the following:
    a. To see if the welds have an even thickness.
    b. To make sure the welds have no pits in them.
    c. To determine if fusion was complete.
    d. To make sure there are no cracks left.
    e. To check the general appearance of the weld.
15. Plasma arc welding could be used for ceramics or cermets.
17. A 7025 welding rod is rated at 70,000 psi, can be used in flat or horizontal positions, and uses dc reverse polarity.
19. Using a solvent to join polymers results in a weld because the surfaces of the plastics are dissolved and allowed to fuse into a single piece. A cohesive bond is produced.

Chapter 15

Problem Set 15-1

1. Needs of the application of attaching a wheel to an automobile axle:
   a. The attachment must be temporary.
   b. Relative motion not permitted.
   c. Material is steel.
   d. The fastener must survive vibration and be noncorrosive.
   e. Tools must be simple for attaching and disassembly.
   f. Minimum skill needed for installation.
   g. Cost of fastener should be kept low.
   h. Parts should have over-the-counter availability.
   i. Part cannot project beyond the wheel.
   j. Fastener need not provide secondary function.
   k. The fastener must resist fatigue, be reliable and tough, and withstand shear and tension.

   A steel bolt could be used to keep the wheel on the axle. A tapered end to the end of the axle with a square or Pratt and Whitney key to prevent rotational freedom about the axle could be used. (Many other choices would also work.)
3. Glue, dovetailing, mortise and tenon, and dowel pins could all be defended.
5. 10-24 UNC-2B means that the thread is an internal thread (B) for a size 10 bolt having 24 Unified National coarse threads per inch with a class 2 or free fit.

7. The UNC, UNF, and UNEF are threads that differ in their strength and closeness of fit. For high-strength applications the UNC or coarse thread should be selected. For applications where fine adjustments need to be made, the finer threads would be better. If both fine adjustment and strength are needed, a UNF or UNEF thread with a wider contact number of threads between the shaft and the hole would be needed. (A wider nut would be needed.)
9. Machine screws have threads on a cylindrical shaft. Wood screws are designed to provide their own internal thread and are on a tapered shaft. Sheet metal screws can be either tapered or cylindrical but have a flat area between the threads.
11. Since the "pence system" was the price in "pence" for 100 nails, the amount of metal required for 100 nails went up as the *square* of the diameter times the length of the nail. In larger nails this meant that the cost of 100 nails went up 5 pence for each additional 1/4 inch in the larger nails.
13. Whereas rivets are for permanent joints, many types of pins are removable. Like pins, rivets are good in shear, but differ from pins in that they can also take some tension. The final formation of rivets comes *after* they are placed in the joint, but the final formation of most pins (cotter pins excepted) occurs *prior* to placement in the joint.
15. 3/8-16UNC-2A-HEX HD.CAP SCREW X 3.
17. A typical example would be as follows: The five lug bolts holding a wheel on an automobile could be replaced using a tapered shaft, key, and a single large nut on the shaft. (This is what is done on racing cars.)

Section VII

Chapter 16

Problem Set 16–1

1. The goals are (a) to achieve closer tolerances, (b) to provide a protective coating, and (c) to improve the appearance of the part.
3. Sandpapers and belts are available from very coarse "60-grit" to very fine "600-grit" in both wet or dry forms. Even finer papers and cloths such as "crocus" cloths are also available.
5.

	Thickness of Material Removed	**Abrasive Size and How Applied**
Honing	A few thousandths of an inch	Small, in stones
Lapping	Several thousandths of an inch	Fine, put between surfaces
Belt sanding	Up to 0.5 inch or more in soft materials	Fine to coarse; abrasive is on the belt.
Polishing	None	Extremely fine; applied to cloth or wheel.
Buffing	Very little	Fine; abrasive applied between surface and buffing tool

7. Buffing is an abrasive process using very fine compounds to produce a luster or to remove oxides. Buffing does remove a very small amount of material from the surface. Polishing does not remove material from the surface but "smears" the high spots into the low ones to achieve a glossy surface.
9. Honing uses an abrasive stone to remove material from a surface of a part, whereas lapping puts the abrasive material between two surfaces.

Chapter 17

Problem Set 17–1

1. Conversion processes chemically convert the surface atoms into an oxide, chromate, phosphate, cyanide, or other stable and inert chemical compound. Other finishing processes simply coat, polish, or buff the surface.
3. A refractory coating is designed to protect the surface from heat. Porcelain coatings are meant to make the surface hard and/or attractive.
5. Porcelain must be heated to high temperatures to fuse the frit. Plastic underlayment would melt, char, or burn at the temperatures required to fuse the particles of porcelain in the manufacturing process.
7. *Brushing:* Any surface that can be reached with the bristles of a brush. Flat, external surfaces with no sharp crevices are the best candidates. Walls, boat hulls, smooth masonry, and ductwork can all be brushed.

Rolling: Large flat surfaces. Walls, ship hulls, aircraft bodies and wings, and concrete and brick surfaces.
Dipping: Anything that can be picked up and dipped into a liquid coating material can be dipped. Automobile parts, kitchen utensils such as electric mixer bodies, pots and pans, and tools are examples.
Electrocoating: Surface must conduct electricity. Metal automobile bodies, electric motor housings, door knobs, oven and refrigerator cabinets are but a few.
Spray painting: Almost anything can be sprayed. It is especially good for surfaces with sharp crevices and small radius corners. Metal screening, toys, automobile bodies, concrete walls, furniture, light fixtures, and many many more.
Printing: Any bossed or raised surface. License plates for automobiles. Road signs, bossed identification tags, picture frames, and raised surfaces on jewelry.
Curtain coating: Automobile parts, bath tubs and sinks, tools, camping equipment.
Powder coating: Any material that can withstand a moderate amount of heating. Instrument panels, metal toys, chainsaw housings, metal deck furniture.

9. Most automobile body shops use various types of enamels, acrylics, and epoxies. Spray painting is the most common body shop technique, although a few now use electrocoating. Some brushing is done.
11. a. Containment of cleaning solvents, to prevent them from getting into the atmosphere
    b. Containment of painting, especially spray paint overflow to prevent escape into the atmosphere
    c. Disposal of solvents and waste paint
    d. Protective clothing, respirators and other health-related items for the workers
    e. Fire hazards of the solvents and paints
    f. Cleanup places for the workers
    g. Disposal of paint rags, cleanup rags, paint brushes, and other related waste

Section VIII

Chapter 18

Problem Set 18–1

1. *Thermoplastic:* Trashbags (polyethylene), milk cartons (polyethylene), Saran wrap (polyvinylidine chloride), PVC pipe (polyvinyl chloride), plastic drinking cups (Styrofoam).
   Thermosetting: Handle on an iron (Bakelite), plastic dishes (Melmac), skillet handles (Bakelite), Formica top on counters.
3. Milk cartons are thermoplastic. Hold a match or cigarette lighter under it and see if it melts. (*Be careful:* Polyethylene drips a burning drop and can burn you or set things afire.)
5. In resin transfer molding, the resin and the catalyst (hardener) are mixed and the mixture poured into the mold. In reaction injection molding, the catalyst and resin are forced into a reaction chamber and on into the mold in one operation.
7. The epoxy and hardener could be mixed in a beaker or glass then poured into a preformed mold.
9. A syntactic foam is one that has small "microspheres" of glass mixed in the polymer.
11. Oxygen, chlorine or sulphur will react with the opened double bond and not bond to another molecule. It puts an "end" on the molecule preventing further polymerization.
13. Chloro propane or more properly, 1-chloro propane.

Chapter 19

Problem Set 19–1

1. An advanced composite has long directional fibers in the matrix. In ordinary composites, the fibers can be short and randomly placed as to direction.
3. Any items requiring thermal or electrical conductivity. Radiators, electromagnets, trumpets and bells (composites deaden sound), skillets.
5. Pultrusion is the drawing of continuous fibers through the resin and then through a heated die to form shaped beams.
7. In vacuum bagging, the laid-up product is placed in a closed bag, porous stand-offs installed in the bag, and a vacuum created in the bag. Atmospheric pressure can apply an even pressure of up to one atmosphere on the part. The part is cured in the evacuated bag.
9. a. Fiberglass has the highest density.
   b. Fiberglass has the highest tensile strength.
   c. Kevlar has the highest strength-to-weight ratio.
   d. Kevlar is the best in abrasion resistance.
   e. Graphite has the highest rigidity.

Section IX

Chapter 20

Problem Set 20–1

1. Even in a small shop a safety program should be implemented.

   A meeting (or two or three) of all production personnel should be held periodically to (a) outline the dangers of the machinery, (b) outline safety precautions that will be followed, and (c) get feedback from the workers concerning safety and health problems on the job.

   Safety posters including posters on electric shock hazards, procedures, and treatments, the Heimlich maneuver, artificial respiration, lifting properly, and hazardous materials should be posted.

   Everyone should know where first-aid kits are located.

   Telephones should be accessible for emergency calls from anywhere in the shop.

   Safety training sessions should be held for all new personnel and periodically for older personnel.

   Training sessions for fire safety should be held periodically.

   Even the office staff (secretary) should know safety, first-aid, and emergency procedures. In a small shop, the secretary may be the only one available for help.

   Periodic checks for equipment and safety should be conducted. Check that electric cords are not frayed and that safety shields are in place and operating properly.

3. Eyewash stations; fire blankets; fire extinguishers; telephone; artificial respiration or breathing apparatus; emergency lighting.

5. a. The table rest must be within 1/8th inch of the grinding wheel.
   b. Never grind on the side of the wheel.
   c. Keep the wheel balanced and dressed.
   d. Use the correct type of wheel for the material being ground.

7. Besides the normal safety procedures for electrical equipment and eye shields, paint shop personnel must also be provided with respirators or breathing equipment. Places must be provided for the personnel to clean up and remove any paint and solvents from their person. Gloves and protective garments should also be provided to protect them from the spray.

9. a. Periodic checks for grounded electric plugs and frayed electric cords
   b. Proper storage of office supplies
   c. Periodic checks of fire extinguishers and other safety equipment
   d. Periodic checks to determine if chairs and other office furniture are in good repair and properly fit the personnel
   e. Checks for proper lighting and air circulation
   f. Checks for glare from computer screens and proper hand rests for the computer operators
   g. Checks to determine that electric circuits are not overloaded by the use of too many machines, computers, fans, heaters, coffee pots, etc.
   h. Checks to determine if extension cords are being improperly used

11. Human factors engineering = Ergonomics
    = Human engineering
    = Bioengineering
    = Biotechnology
    = Engineering psychology

13. Human factors engineering can help prevent accidents in the following ways: (a) By making sure that controls fit the operator and that the machine can be operated by a human being; (b) by making dials and controls that are not confusing or contrary to commonly accepted usage; (c) by designing machines that do not have controls in such a place that they might accidently be activated.

15. a. Cleaning solvents
    b. Lead-based paints
    c. Used motor oil
    d. Swimming pool acid (muriatic and others)

Section X

Chapter 21

Problem Set 21–1

1. A *manufacturing facility* is a building (or collection of buildings) in which incoming raw materials or incomplete products are brought to a higher stage of completion by one or more manufacturing processes. Examples would be a facility that manufactures automobile engines from raw materials or a facility that assembles the

complete car by processing subassemblies made elsewhere.

A *processing plant* is a building (or a collection of buildings) in which products are sorted, graded, counted, or subjected to other passive processes such as packaging, labeling, etc. The essential concept is that the incoming product is not changed. An example of a processing plant would be one that receives dried fruit and grades, sorts, and packages the dried fruit for market.

3. ($0.40/piece) × (500 pieces/hr) × (8 hr./day) = $1600 per day
($1600/day) × (5 days/week)= $8000 per week
($8,000/week) × (50 weeks/year) = $400,000 per year
5. A fixture is a device that is individually designed to hold a specific workpiece in place. The part may be held in place for a specific machine or machining operation, or held in place relative to other parts for assembly. Its specific job is to facilitate setup or make holding easier.

   A jig is a fixture that additionally establishes certain critical location dimensions on the workpiece. In fact, dimensioning reference planes for irregularly shaped workpieces are usually the planes on the jig against which the workpiece mounts.

   Workholder is the general term for either a jig or a fixture.
7. If the part to be tested is mounted in a lathe, the dial indicator would be mounted on the ways with the quill resting against the diameter to be measured. Then the workpiece would be rotated (by hand) for maximum indicator reading, and the reading recorded. The workpiece then would be rotated for minimum indicator reading, and the reading recorded. The total indicated runout would be the difference between the two readings.
9. Since the sine of 30° (the complement of 60°) is 0.5 (exactly), then use a stack of gauge blocks equal to exactly half of the distance between the centers of the two cylinders. The sine bar would then be at an angle of 60° to the vertical. If an angle of 60° to the *horizontal* is desired: The sine of 60° is exactly equal to the square root of 3, divided by 2 (approximately 0.866). Use a stack of gauge blocks that represents this number multiplied by the distance between the centers of the cylinders.
11. Computerized numerical control (CNC) is an improvement over NC because a computer can be programmed to control several production machines and can store programs and data that would allow it to keep track of the wear of the cutting tools and compensate for that wear in the instructions to the machine. The computer can detect trends in the final dimensions of the product pieces and anticipate service needs of the machine. The computer can also be programmed to respond to tool breakage and other malfunctions, to keep track of the progress of the order, to call for a new supply of raw materials, and to keep track of maintenance needs. One computer can control a plotter, which can be drawing the plans for a part at the same time that it is controlling the machine that is making some other part.
13. CAD stands for computer-aided drafting. CAD programs/equipment are merely an electronic drafting board with no analysis capability.

    CAE stands for computer-aided engineering. CAE programs/equipment take the description of the workpiece that has been created by CAD and subject it to various analyses of value to engineering decision-making. For example, the equipment can locate the center of mass of the workpiece, or, with loading specified by the designer, determine the force and deflections in the parts of a loaded workpiece.

    CAM stands for computer-aided manufacturing or computer-automated manufacturing. CAM programs/equipment takes the description of the part that has been created by CAD and decides how it will be made, writes the instructions for a machine to make the part, then supervises the fabrication of the part, including inspection and statistical quality control.
15. Robots are often cheaper to operate than hiring humans. They can operate in hazardous environments and at temperatures, pressures, and atmospheric conditions that would be fatal to humans. They tend to perform their function with consistent precision beyond the capability of human beings. Their performance capabilities can be greater than that of a human in strength, size, speed, and accuracy. It is difficult to program a robot to handle an unexpected emergency or solve

an unanticipated problem. That's where a human worker has the advantage. Vision, especially "pattern recognition," is still at a relatively primitive level in robots. Further, it is not practical to try to anticipate every task that a robot might be asked to perform. Therefore, it might be an advantage to use intelligent humans for these tasks. Robots do not earn a salary, buy the goods they produce, or pay taxes.

17. See Figure 21–23.
19. Sources of robot "muscle power" would be air, hydraulic fluid, electricity, magnetism, or a vacuum. The advantages of each follow.
    Air and vacuum: Air-operated and vacuum-operated (pneumatic) mechanisms are usually small; they are inexpensive to construct and operate.
    Hydraulic: Hydraulic mechanisms are capable of exerting large forces, are high powered, and are amazingly quick for their size. They can be made very large and can be safely and reliably used in hazardous environments.
    Electrical or magnetic: Electrically or magnetically driven mechanisms are appropriate for light duty. They are very accurate, precise, and clean.
    Vacuum: Vacuum-driven mechanisms are appropriate for handling small or fragile parts.
21. (See Figure 21–31.) Assume that an instruction ("command") is sent to the control unit from an antenna to "look" 5° to the right of its current position. The control unit decides which actuator should operate and how much; those instructions are sent to the energy source, which moves the antenna (the "device"). A sensor (the "monitor") on the antenna mount reports back the instantaneous position of the antenna, which is constantly compared to the initial instruction. As the antenna approaches the new position, the difference between the instruction and the response gets smaller and smaller, so the control unit keeps reducing the signal to the energy source. Consequently, the antenna settles down smoothly into the new position.
23. Materials, utilities, people, and information.

Materials include all items of:

- Raw materials
- Parts between production operations
- Subassemblies
- Completed assemblies
- Fasteners
- Finishing materials: thinners, other expendable supplies
- Tools to be sharpened
- Tools that have been sharpened
- Tools for temporary use
- Air
- Fumes
- Trash
- Recovered scrap
- Food
- Safety equipment: protective equipment, medical supplies, emergency equipment

The flow of *utilities* includes:

- Electricity
- Fuel gases
- Oil
- Steam
- Water: chilled, cold, hot, sterile
- Pressurized air
- Other gases

Items that concern the flow of *people* include:

- Workers: equipment operators, supply movers, foremen, inspectors
- Personnel for the next shift (or prior shift)
- Safety personnel
- Custodians
- Shop stewards
- Maintenance/repair personnel
- Engineers
- Technicians
- Visitors
- Couriers
- Administrative personnel

Flow of *information* includes:

- Instructions for fabrication
- Change orders
- State and local codes
- Machine operations instructions
- Repair orders
- Time cards
- Production schedules
- Workpiece status reports
- Machinery status updates

Appendix

E

Conversion Factors

English System Units

Length

12 inches = 1 foot
3 feet = 1 yard
5280 feet = 1 statute mile

Weight

16 ounces = 1 pound
2000 pounds = 1 ton

Volume

16 ounces = 1 pint
2 pints = 1 quart
4 quarts = 1 (U.S.) gallon
1 cubic foot = 1728 cubic inches
1 gallon = 232 cubic inches

Metric-to-English System Conversions

2.54 centimetres = 1 inch (exactly)
30.48 centimetres = 1 foot (exactly)
91.44 centimetres = 1 yard (exactly)
39.37 inches = 1 metre
1 mile = 1.609 kilometres
1 kilometre = 0.621 mile
1 kilogram ≈ 2.2 pounds
1 pound ≈ 454 grams
1 ton ≈ 909 kilograms
0.946 liter = 1 quart
1.057 quarts = 1 liter
3.784 liters = 1 (U.S.) gallon
1 cubic inch = 16.39 cubic centimetres
1 cubic foot = 28.32 liters
1 pound (force) = 4.45 newtons
1 pound/inch^2 = 6896 pascals
1 horsepower = 550 foot-pounds/second
1 horsepower = 746 watts
1 horsepower = 746 joules/second
1 foot-pound = 1.3547 joules
1 mile/hour = 1.467 feet/second
1 mile/hour = 1.609 kilometres/hour
1 mile/hour = 44.71 centimeters/second

To convert Fahrenheit temperature to Celsius temperature, the equation is

$$°C = 5(°F - 32)/9$$

To convert Celsius temperature to Fahrenheit temperature the equation is

$$°F = 1.8\,(°C) + 32$$

Index

MILLIMETRE/INCH/DECIMAL-INCH EQUIVALENTS

mm	inches	n/64	n/32	n/16	n/8	n/4
0.5	.019685					
0.7938	.03125	(2)	1			
1	.039370					
1.1906	.046875	3				
1.5	.059055					
1.5875	.0625	(4)	(2)	1		
1.9844	.078125	5				
2	.078740					
2.3812	.09375	(6)	3			
2.5	.098425					
2.7781	.109375	7				
3	.118110					
3.175	.125	(8)	(4)	(2)	1	
3.5	.137795					
3.5719	.140625	9				
3.9688	.15625	(10)	5			
4	.157480					
4.3656	.171875	11				
4.5	.177165					
4.7625	.1875	(12)	(6)	3		
5	.196850					
5.1594	.203125	13				
5.5	.216535					
5.5562	.21875	(14)	7			
5.9531	.234375	15				
6	.236220					
6.35	.25	(16)	(8)	(4)	(2)	1
6.5	.255906					

mm	inches	n/64	n/32	n/16	n/8	n/4
6.7469	.265625	17				
7	.275591					
7.1438	.28125	(18)	9			
7.5	.295276					
7.5406	.296875	19				
7.9375	.3125	(20)	(10)	5		
8	.314961					
8.3344	.328125	21				
8.5	.334646					
8.7312	.34375	(22)	11			
9	.354331					
9.1281	.359375	23				
9.5	.374016					
9.525	.375	(24)	(12)	(6)	3	
9.9219	.390625	25				
10	.393701					
10.3188	.40625	(26)	13			
10.5	.413386					
10.7156	.421875	27				
11	.433071					
11.1125	.4375	(28)	(14)	7		
11.5	.452756					
11.5094	.453125	29				
11.9062	.46875	(30)	15			
12	.472441					
12.3031	.484375	31				
12.5	.492126					
12.7	.5	(32)	(16)	(8)	(4)	(2)